PRINCIPLES
AND METHODS
OF TEMPERATURE
MEASUREMENT

PRINCIPLES AND METHODS OF TEMPERATURE MEASUREMENT

Thomas D. McGee
Materials Science and Engineering Department
Iowa State University
Ames, Iowa

A WILEY-INTERSCIENCE PUBLICATION
JOHN WILEY & SONS
New York • Chichester • Brisbane
Toronto • Singapore

Library of Congress Cataloging in Publication Data:

McGee, Thomas D. (Thomas Donald), 1925–
   Principles and methods of temperature measurement / Thomas D.
McGee
      p.   cm.
   "A Wiley-Interscience publication."
   Includes bibliographies and index.
   ISBN 0-471-62767-4
   1. Thermometers and thermometry.   I. Title.
QC271.M384   1988
536′.5′028—dc19

ISBN 0-471-62767-4

Printed in the United States of America

10 9 8 7 6 5 4 3 2

To Clara McGee
and to my parents,
Nacy W. McGee, Sr. and Maude R. McGee

PREFACE

During a scientific career as an engineering professor I have observed that the most common mistake a scientist makes is to accept a temperature reading without question. The scientific literature is full of papers about otherwise fine research that is based on temperature measurements of uncertain quality. I am convinced that every scientist needs access to a comprehensive text as a reference and that courses on temperature measurements should be offered in the universities, instead of being treated as a small part of a physics or measurements course. This book is offered, then, as both a textbook for upper-level undergraduate courses and as a reference text for practicing scientists and engineers.

I thank Ms. Barbara Dubberke for her care and skill in typing the manuscript. I thank Dr. Clayton Swensen for constructive criticism of the first draft of the manuscript.

THOMAS D. McGEE

Ames, Iowa
March 1988

CONTENTS

18   INSTALLATION EFFECTS   493

19   DYNAMIC RESPONSE OF SENSORS   504

20   TEMPERATURE INSTRUMENTATION AND CONTROL   518

PRINCIPLES
AND METHODS
OF TEMPERATURE
MEASUREMENT

1

THE CONCEPT OF
TEMPERATURE

Modern temperature measurement is truly sophisticated. Standards laboratories around the world routinely measure temperature to one thousands of a degree. This can even be done automatically with computer-controlled equipment. The methods of temperature measurement have changed rapidly in recent years and are continuing to do so. Inexpensive commercial equipment for measurement and control often will indicate temperatures to one-tenth of a degree. Because of adverse environments, sensor or instrument instability, vibration, electrical noise, or design compromises, many such instruments really have errors a hundred or more times greater. Therefore, anyone who measures temperature needs to understand the principles of temperature measurement and the methods available for any particular application. The purpose of this book is to provide that information.

The concept of temperature must have originated during the earliest stages of human development. Perhaps we can appreciate that early development through watching the changes taking place during the growth of a child. One of the first things we must teach a toddler is that some things are *too hot* and others are *too cold*. It must have always been this way. Yet the sensual detection of temperature is deceiving. The concept of temperature has been the most difficult of the common properties of matter to define clearly. It is helpful to review the historical development of temperature measurement because only after crude methods of temperature measurement were developed could the concept of temperature really be defined.

1.1 HISTORICAL PERSPECTIVE

The concept of temperature, the scientific principles on which temperature measurement is based, and the practical methods used for measurement evolved as a part of the development of science. The history of temperature measurement is a part of the history of development of science—so that current methods are based on definitions of terms and on sensing devices that can best be understood in a historical perspective. For example, the Kelvin degree is defined in a completely arbitrary way that depends on scientific progress in the eighteenth, nineteenth, and twentieth centuries. It is necessary, then, to start with an explanation of the fundamental concepts from a historical perspective.

1.1.1 The Thermoscope

The first instrument actually used to estimate temperature was the thermoscope. It was developed in the first few years of the seventeenth century and is usually credited to the astronomer Galileo.[1] Sanitorio, Drebbel, and Fludd are also sometimes given credit for "inventing the thermometer," although a thermoscope is not a thermometer but is its predecessor. The

Figure 1.1. The first published figure of a thermoscope, from "Sphaera mundi" by Giuseppe Biancani, by 1617. (From Ref. 2)

logical distinction between the two is that the thermometer possesses a scale, while a thermoscope does not. A thermoscope is a device which traps air in a bulb so that as the air expands or contracts in response to a temperature increase or decrease, it moves a liquid column in a long tube (Fig. 1.1).[2] Since the device was not sealed, it was sensitive to changes in barometric pressure and to losses by evaporation. Water, "spirits of wine," and oils were used as the liquid. Of course, the liquid expanded and contracted too, but not nearly as much as the air.

Galileo's and Fludd's inventions were based in part of the experiments conducted by the ancient Greeks. In 1594, Galileo had read Hero's "Pneumatics," a manuscript probably written in the first century before Christ. Philo of Byzantium made similar experiments about the end of the second century B.C. and Fludd had access to Philo's manuscript.

In that manuscript Philo described an experiment in which a tube from a hollow sphere was extended over a jug of water. If the sphere was placed in the sun, bubbles were released as air expanded out of the sphere. When moved to the shade, water rose in the tube as air in the sphere contracted (see the left side of Fig. 1.2). Fludd's contribution was to place the sphere

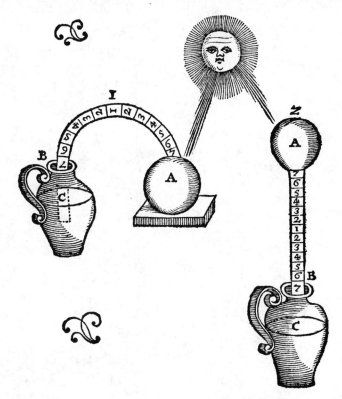

Figure 1.2. Fludd's figure, 1638, showing the development of the thermometer from Philo's experiment. Note that this is a true air thermometer, as it possesses a measuring scale, but is unsealed. (From Ref. 1; courtesy of the Wellcome Institute Library, London.)

vertically above the jug, converting it to a thermoscope, as shown on the right side of Fig. 1.2. Fludd's invention was apparently independent from, but later than Galileo's.

Thermoscopes were constructed and used by Galileo's contemporaries. During the year 1613 Sagredo used the thermoscope to compare the temperature of lakes of different sizes as they cooled in the winter. He found that the small ones were cooler than the large ones. He made a daily record of temperatures during the hot summer of 1613. Sagredo also compared the response of the instrument to winter snow and summer heat, and described his readings as "degrees of heat". Thus he was using his thermoscope as a true thermometer.

Every thermoscope was different. Almost as soon as they became available, users realized that each needed to be calibrated in some way. The first thermoscopes probably were completely uncalibrated and contained colored water. At first, temperatures were marked with colored string wrapped around the tube (Fig. 1.3). However, according to a manuscript by Telioux, a scale was added as early as 1611.

Figure 1.3. One of Santorio's thermoscopes. (Courtesy of the Wellcome Institute Library, London.)

Because of their awareness of the importance of fever, physicians were keenly interested in the measurement of body temperature. In fact, the early Greek physician Galen introduced the idea of degrees of heat and cold long before temperature could actually be measured.[4] His scale had 4 "degrees" above a neutral point and 4 below. Scales were added to many early thermoscopes even though they were not calibrated.

Air thermometers are thermoscopes modified by adding a scale. They were often made by people versed in astronomy and navigation. So it was natural for many of them to have scales divided into 60 degrees, similar to angular measurements in degrees.

1.1.2 The Liquid-in-Glass Thermometer

The air thermometer was subject to variations in atmospheric pressure. It was a major improvement to seal the instrument and to use the thermal dilation of a liquid instead of air to sense temperature changes. (The liquid

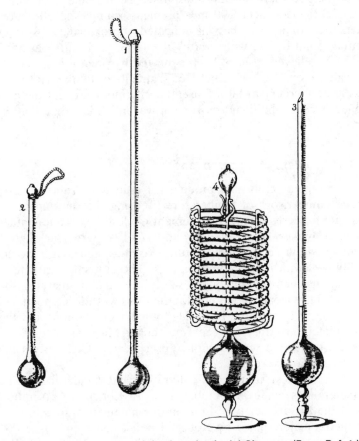

Figure 1.4. Thermometers of the Accademia del Cimento. (From Ref. 1.)

dilation was much less than that of air, so a smaller-diameter tube was needed if the instrument was to have the same sensitivity.) This invention has been attributed to Ferdinand II, Grand Duke of Tuscany, in about 1654.

Three models were produced using spirits of wine as the fluid: the 50-degree, the 100-degree, and the 300-degree models. They were marvelously crafted by a skilled Florentian glassblower, Mariani, who drew out the fine tube, divided it into major parts, marked them with white enamel buttons, added black buttons to divide each major division into 10 parts, filled the bulb with spirits of wine to the proper level, and sealed the end of the tube with a flame (Fig. 1.4). The 300-degree models were made as a demonstration of skill rather than as working thermometers because the tube could not be drawn out uniformly enough for accurate measurements. Some of these remarkable instruments are still in existence. Florentian thermometers were exported to Rome and Paris, and reached England in 1661.

The 50-degree thermometers must have been remarkably uniform. In 1830, G. Libri made over 200 measurements comparing 50-degree thermometers and found that they all indicated approximately $13\frac{1}{2}$ degrees at the melting point of ice, with "zero" at $-18\frac{3}{4}°C$ and "fifty" at 55°C. This uniformity is believed to have been achieved by precision in manufacture, not by calibration. In November 1657, members of the Accademia del Cimento used mercury to fill a Florentine thermometer, but the effort was abandoned because the liquid movement was much less than that of spirits of wine.

1.1.3 The Problem of Comparisons

Every thermometer craftsman produced a unique instrument. One could only make comparisons of temperature by using the same instrument, or perhaps with a nearly identical instrument made by the same craftsman. A thermometer maker could only promise that a particular thermometer would move within certain temperature extremes, and that it would agree reasonably well with identical instruments. The craftsman tried to achieve this by making each in exactly the same way, and by marking the scale using one point as a calibration point. Often this was the ice point, achieved by packing snow around the bulb. The number of divisions and their size were completely arbitrary for each craftsman. Many used scales inverted with respect to current scales—with zero at the top. Obviously, calibration and standardization were needed.

One of the earliest attempts at calibration and standardization between thermometers was made in October 1663 in London. According to the records of the Royal Society of London, the members agreed to use one of several thermometers made by Robert Hooke, and which agreed "at the figure 8," as a standard so that readings of the others could be adjusted to it. Thus a member in one laboratory could compare a temperature to that of a person in another laboratory through the standard corrections.[5]

Thermometers were made by craftsmen in many countries. They used many different scales and calibration points, and this resulted in much confusion. Often, a maker would mark more than one scale on a thermometer to increase versatility. After about two centuries, one thermometer, dated 1841, had the following 18 scales:[1]

1. Old Florentine	7. Deslisle	13. Amontons
2. New Florentine	8. Fahrenheit	14. Newton
3. Hales	9. Reaumur	15. Societé Royal
4. Fowler	10. Bellani	16. De le Hire
5. Paris	11. Christin	17. Edenburg
6. H. M. Poleni	12. Michaelly	18. Cruquiu

1.1.4   Linear Calibration with Two Fixed Points

Early craftsmen had marked one calibration point, often the ice point, the blood point, or the boiling point. Ole Romer, a Danish astronomer, was the first to divide the space between the ice point and steam point into a linear scale. He chose glass tubing of uniform bore to make a linear division reasonable. (He used the length of a drop of mercury to verify that the tubing had a uniform bore all along its length.) Daniel Gabriel Fahrenheit, a Danzig instrument maker, visited Romer in 1708 and saw him calibrate thermometers. He begun producing thermometers, first with Romer's scale, then with a modified Romer scale, and finally with the Fahrenheit scale as used today. Fahrenheit substituted mercury for spirits because it had a more nearly linear thermal expansion with temperature. He made his bulb as a cylinder rather than a globe. These improvements, and his high quality of workmanship, made his thermometers famous as scientific instruments. It is interesting that Fahrenheit used two fixed points for calibration. He used a mixture of sea salt, ice, and water to produce the zero point. He used the ice point for calibration, but did not use the boiling point. He probably used armpit temperature as his other calibration point. When this scale was adopted by Great Britain, the temperature of 212° was established as the boiling point; this temperature and the ice point were used as the two fixed calibration points.

1.1.5   The Centigrade Scale (Now Celsius)

The word "centigrade" means a scale that is divided into 100 parts. There is some uncertainty as to where and by whom the centigrade scale was invented. In about 1740, Anders Celsius in Uppsala, Sweden, invented a scale with 0 at the steam point and 100 at the ice point. A few years later this scale was inverted to the scale now so well known. Either Daniel

Ekström or Marten Strömer probably produced the first such thermometer in Sweden. Some believe that Carl von Linne probably suggested to Celsius the possibility of dividing the space between the ice point and the steam point into 100 parts. Celsius had his "upside-down" thermometer in 1742. Jeane Pierre Christin independently produced a "right-side-up" thermometer in Lyons, France, in 1743. The scale originally proposed by Celsius, but inverted, became the standard for scientific measurements. It was given the name "centigrade" to indicate a scale divided into 100 parts. Since 1948 the official name of that scale has been the "Celsius" scale.

As we shall see later, the adoption of the name "Celsius" instead of "centigrade" came about at the time the triple point of water was substituted for the ice point as a primary calibration point. When that was done there were no longer exactly 100 degrees between the lower (ice point) and the upper (steam point) calibration points.

1.1.6 The Need for a Standard of Temperature

As more thermometers became available from different makers, it was natural to compare them for various applications. When this was done, they often did not agree even though they were calibrated at two fixed points with a linear scale between. There were three types of errors: (1) errors in

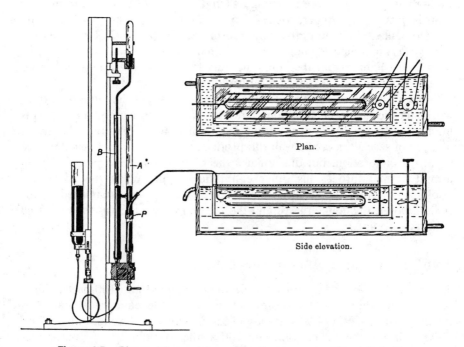

Figure 1.5. Chappuis's precision calibration apparatus. (From Ref. 6.)

primary calibration, where a particular primary reference point (such as a boiling point) was improperly established; (2) errors in workmanship, such as variation in the diameter of the capillary, which gave nonlinear response; and (3) errors in the temperature-indicating response function, such as the difference in thermal expansion coefficients when alcohol is substituted for mercury, or one glass composition for a thermometer is replaced by another. Much of the science of temperature measurement has been concerned with controlling such errors.

To determine the causes of disagreement between thermometers, it was necessary to have a standard for establishing temperatures. For reasons explained later, gas thermometers have become the standard of comparison. In 1884, Chappuis of the International Bureau of Weights and Measures used a constant-volume hydrogen thermometer to investigate eight different mercury thermometers (Fig. 1.5).[6] His hydrogen gas thermometer was taken as linear, and temperatures were determined between 0 and 100°C with an uncertainty of 0.002°C. He also calibrated nitrogen and carbon dioxide gases against hydrogen. Based on this work, the maximum difference between the mercury thermometers and the hydrogen thermometer was 0.107°C, and occurred at 40°C. This was the beginning of the use of gas thermometry as a precision standard for temperature.

1.2 EARLY DEFINITIONS OF TEMPERATURE

The invention of the thermometer made it possible to measure temperature, although the definition of temperature was very difficult. Precise definitions of temperature and a scientific basis for the thermodynamic temperature scale will be developed in later chapters.

Many distinguished physicists have proposed simple definitions of temperature. For example, Max Planck defined temperature as the degree of hotness or coldness of a body. Another often cited definition, attributed to R. H. Fowler, is that if three bodies A, B, and C have A and C in thermal equilibrium with B, then A is in thermal equilibrium with C. That is, all have the same temperature. This has been designated "the zeroth law of thermodynamics". Obviously, such "definitions" are difficult to use.

Part of the difficulty in developing a simple definition of temperature is the result of heat transfer characteristics. We often think of something that feels cold as having a low temperature. However, if we touch a wooden door and a metal doorknob, both of which are at the same low temperature, our senses tell us that the doorknob is much colder than the door only because the metal conducts heat away faster, and that makes our fingers colder. The fingers in contact with the metal are colder, but not because the doorknob was colder than the door. As we will learn later, heat transfer considerations are very important in every temperature measuring situation. After the thermometer was developed, it became possible to make measurements and distinguish between heat transfer characteristics and temperature effects.

1.3 A SIMPLE QUALITATIVE DEFINITION OF TEMPERATURE

Temperature can be considered to be the *level of thermal energy*. As such, it is analogous to voltage as the level of electrical energy, or to elevation as the level of mechanical potential energy. Temperature, then, is the driving force for heat flow, just as voltage is the driving force for electrical flow and hydrostatic head is the driving force for fluid flow. The differential equations describing heat flux, electrical current, and fluid flow have basically the same form. Although not satisfactory to fundamental scientists, defining temperature as the level of thermal energy provides a useful qualitative definition because it aids in understanding practical situations.

1.4 UNITS OF TEMPERATURE FOR VARIOUS TEMPERATURE SCALES

As will be explained later, the size of the degree, the unit for measuring temperature, is completely arbitrary. The two primary degree units in use today were based on the original centigrade and Fahrenheit scales. There are 180 degrees between the ice point and the boiling point on the Fahrenheit scale, and 100 degrees for the same interval on the centigrade scale. So the Fahrenheit degree is smaller and 1 degree centigrade (now Celsius) equals 1.8 degrees Fahrenheit.

The temperature of an object must be represented on some sort of a scale. Not only the size of the temperature unit, but also the zero point of the scale must be designated in order to have a usable scale. Early scientists used the Fahrenheit and centigrade scales. The centigrade scale had absolute zero at 273.1 centigrade degrees below the ice point. The original Kelvin scale used centigrade degrees with its origin at absolute zero, the lowest temperature it is possible to obtain. The Fahrenheit degree has absolute zero at 460° below the arbitrary zero of the Fahrenheit scale or 492 Fahrenheit degrees below the ice point. The absolute scale based on Fahrenheit degrees is called the Rankine scale. Thus in this country we have four temperature scales in use: Celsius, Kelvin, Fahrenheit and Rankine. The latter two are slowly becoming obsolete. Some would argue that centigrade makes a fifth scale because the ice point–poiling point interval is no longer precisely 100 Celsius degrees. But for all practical applications, there is no difference.

In 1975 the U.S. Congress passed the Metric Conversion Act, which committed the United States to voluntary conversion to the metric system using Système International (SI) units. The SI system has seven dimen-

TABLE 1.1 Système International Base Units

Unit	Symbol	Quantity Measured
Meter	m	Length
Kilogram	kg	Mass
Second	s	Time
Ampere	A	Electric current
Kelvin	K	Temperature
Mole	mol	Amount of substance
Candela	cd	Luminous intensity
Radian	rad	Plane angle (dimensionless)
Steradian	sr	Solid angle (dimensionless)

sioned base units, two dimensionless base units (the radian for plane angle and the steradian for solid angle), and 17 derived units. The Kelvin is one of the base units (Table 1.1).

Unfortunately, the rate of conversion to SI units is very slow. It appears that we will need to know all four scales of temperature for a long time to come.

1.5 CONVERSION BETWEEN TEMPERATURE SCALES

Conversion from one scale to another requires only simple algebra. The relationships illustrated are shown in Fig. 1.6. Obviously, for conversion the

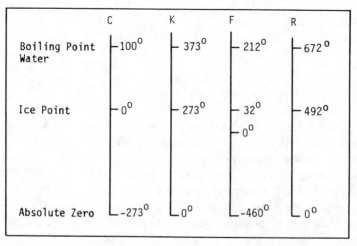

Figure 1.6. Four common temperature scales: Celsius, Kelvin, Fahrenheit, and Rankine.

following simple formulas apply:

$$°C = (°F - 32)\tfrac{5}{9} \tag{1.1}$$

$$°F = \tfrac{9}{5}(°C) + 32 \tag{1.2}$$

$$K = 273 + °C \tag{1.3}$$

$$°R = 460 + °F . \tag{1.4}$$

The $\tfrac{5}{9}$ term comes from $100 \div 180$, the range between the ice point in °C and °F. Note that $\tfrac{9}{5} = 1.8 = (2.0 - 0.2) = 2(1 - 0.1)$, so that a convenient way to get °F from °C is to double °C, subtract 10%, and add 32. Thus

$$1500°C = 2(1500) - 0.1(2)(1500) + 32$$

$$= 3000 - 300 + 32$$

$$= 2732°F .$$

The point $-40°$ is common to both Fahrenheit and Celsius. So alternative conversion formulas are

$$°C = (°F + 40°)\tfrac{5}{9} - 40° \tag{1.5}$$

$$°F = (°C + 40°)\tfrac{9}{5} - 40° . \tag{1.6}$$

For example,

$$173°C = ?°F$$

$$°F = (173 + 40)\tfrac{9}{5} - 40$$

$$= 213(\tfrac{9}{5}) - 40$$

$$= 383.4 - 40 = 343.4°F .$$

1.6   THE THERMAL EXPANSION OF GASES

In 1660, Robert Boyle reported on his studies of air trapped in a U tube. One end was open and the other end was closed. His purpose was to determine what supported the mercury column of a barometer. He trapped various amounts of air in the closed arm with mercury, and varied the pressure on the air by varying the amount of mercury in the open end of the tube. He found, over the limited range of pressure that he could reach, that $PV = \text{constant}$ (P is pressure and V is volume). His experiments were performed at almost constant temperature. They confirmed Torricelli's contention that atmospheric pressure supports the mercury column of a barometer. Because the volume changed at constant pressure when the temperature was not quite constant, it was apparent that the volume at constant pressure was a function of temperature.

In 1787, Jacques-Alexandre Charles studied the volume of gases as a function of temperature with pressure held nearly constant. Jospeh L. Gay-Lussac extended the work and reported, in 1802, that all his gases—air, oxygen, nitrogen, hydrogen, and carbon dioxide—expanded the same amount when heated from the ice point to the boiling point. Taking that interval as 100°C, he reported that at constant pressure,

$$V_{100°C} = V_{0°C}(1 + 100\bar{\alpha}) . \tag{1.7}$$

He solved for $\bar{\alpha}$:

$$\bar{\alpha} = \frac{V_{100} - V_0}{100 \, V_0} = \frac{1}{100} \frac{\Delta V}{V_0}$$

$$= \frac{\Delta V}{\Delta t} \frac{1}{V_0} . \tag{1.8}$$

Note that α has units of $(\text{degree})^{-1}$. He calculated $\bar{\alpha}$, the mean volumetric coefficient of thermal expansion, to be 1/267. In 1847, Victor Regnault obtained a better value of 1/273.

Later experiments conducted with better precision revealed that all gases had very slightly different thermal expansion coefficients. However, all were found to approach a common value as the pressure in the experiment approached zero (Fig. 1.7).[7]

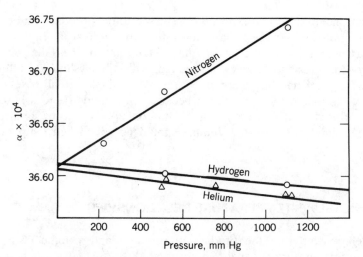

Figure 1.7. Extrapolation of thermal expansion coefficient to zero pressure. (From Walter J. Moore, "Physical Chemistry," © 1950, renewed 1978, p. 11. Reprinted by permission of Prentice-Hall, Inc., Englewood Cliffs, N.J.)

Refinements have continued to the present day. At the present time the "best" value for $\bar{\alpha}$, when extrapolated to zero pressure, is

$$\bar{\alpha} = 36.609 \times 10^{-4} = \frac{1}{273.15°C} .$$

Now it is apparent that the experiments described here permit us, at low pressure, to measure temperature with a volumetric gas thermometer.

1.7   ABSOLUTE ZERO BY THE GAS THERMOMETER

From the general relationship in Eq. (1.7) it is apparent that the volume will approach zero on cooling:

$$\left[V_{0°C}\left(1 + \frac{t}{273.15}\right)\right]_P = 0 \tag{1.9}$$

when $t = -273.15°C$. Therefore, we can measure temperature down from the ice point by -273.15, to zero volume, establish a new zero of temperature, and we have the concept of absolute zero from perfect-gas-law reasoning. Using T as the symbol for absolute temperature, we have the ice point on the absolute temperature scale as

$$T_{\text{ice point}} = 273.15 \text{ K} .$$

This concept of absolute zero is considered to be a fundamental concept. As one cools any system toward absolute zero, thermal energy diminishes and disappears. The energy of the system approaches the zero-point energy—that residual energy associated with the system structure that remains even when no thermal energy is available. Real gases cannot actually be cooled to zero volume because they condense to a liquid or a solid before absolute zero is reached. But conceptually, using a perfect gas at infinitesimal pressure, the volume should be zero when temperature is zero.

1.8   PRESSURE–TEMPERATURE RELATIONSHIPS

The relationship that the product of pressure and volume of a gas at constant temperature is equal to a constant,

$$(PV)_t = k_t , \tag{1.10}$$

can also be used as the starting point for the concept of absolute zero based on pressure–temperature concepts. At constant volume, the pressure of a gas has been found to be a function of temperature:

$$\bar{\alpha}_p = \frac{1}{P_0} \cdot \frac{P - P_0}{t - 0°C}. \tag{1.11}$$

Here $\bar{\alpha}_p$ is the average thermal expansion coefficient in terms of pressure. Then

$$P_{100°C} = P_0[1 + \bar{\alpha}_p(100)] \tag{1.12}$$

at constant volume, or

$$P_v = [P_0(1 + \bar{\alpha}_p t)]_v . \tag{1.13}$$

By reasoning similar to that above, the pressure of a perfect gas will seem to be zero at absolute zero.

It was apparent to the early investigators that if, as experimentally indicated, the Boyle's law constant in Eq. (1.11) changed in the same way with temperature, regardless of which was held constant, then

$$\bar{\alpha}_p = \bar{\alpha}_v .$$

Absolute zero was the same regardless of whether the pressure or the volume was changed.

1.9 THE PERFECT GAS LAW

The phenomena described above led to the concept of a perfect gas, one that obeys the temperature–volume relationship exactly at constant pressure (and the pressure–volume relationship exactly at constant volume). We know this, for 1 mole, as

$$PV = RT , \tag{1.14}$$

where T is absolute temperature and R is the universal gas constant. All gases at extremely low pressure approach perfect gas behavior. Light monatomic gases such as helium approach it at moderate pressures. Gases with large complex molecules, such as CO_2, deviate most from perfect gas behavior.

Emile Clapeyron, in 1834, was the first to combine Boyle's law and Gay-Lussac's law to obtain the equation of state for a perfect gas. Using the new absolute temperature scale, a gas at constant pressure has a volume

$$V_T = V_0 \bar{\alpha}_p T = \frac{V_0}{T_0} T \,, \tag{1.15}$$

where the zero subscript indicates the ice point. Boyle's law states that

$$PV = P_0 V_{T, P_0} \,, \tag{1.16}$$

where the right side represents the PV product referred to a standard pressure P_0 at which the constant $\bar{\alpha}$ and the volume V_0 were measured in Eq. (1.15). Substituting in Eq. (1.16) for V_T gives

$$PV = \frac{P_0 V_0}{T_0} T = (\text{constant})(T) \,. \tag{1.17}$$

The value of the constant, $P_0 V_0 / T_0$, depends on the size of the gas specimen. If 1 mole is used,

$$\frac{P_0 V_0}{T_0} = R \,, \tag{1.18}$$

where R is the universal constant used in the equation of state for gas, and Eq. (1.17) is the "perfect gas law." Clapeyron realized that real gases did not conform exactly to this concept. But he proposed the concept of a perfect gas as a simple relationship useful for theoretical analysis.

1.10   GAS THERMOMETRY

With the realization that temperatures could be determined on the absolute scale with a gas thermometer at the limit $PV \rightarrow 0$, research on gas thermometry was begun. For example, Kew Observatory at Richmond, England, was set up near London in 1851 following a gift from Regnault of a calibrated gas thermometer and associated equipment. This research is still continuing. Standard laboratories in major countries are still attempting to improve the precision of temperature measurement. For example, the U.S. National Bureau of Standards recently reported eliminating some previously unrecognized errors in gas thermometry which will change the steam point by 30 millikelvin (mK), where the error previously was thought to be only about 5 mK.[8]

The method of measuring absolute temperature with the gas thermometer is simple in principle. Measurements are made by successively reducing the amount of gas in either a constant-volume or constant-pressure gas thermometer and extrapolating to $PV = 0$. Thus if we start at a finite PV and use T_2 as the ice point, we can find T_1 by measuring P at constant V, reducing the gas charge, remeasuring P at T_2 and T_1, repeating with less gas, and so on. Then $T_2/T_1 = (PV)_{T_2}/(PV)_{T_1}$ as PV in the numerator and denominator approach zero.

Obviously, T_2 must not change during the experiment. So accurate gas thermometry turns out, practically, to be very difficult. The temperature can only be measured at highest precision under conditions where T_2 can be held constant. Therefore, gas thermometry can most precisely establish absolute temperatures at special temperatures which are fixed by the experimental conditions (i.e., at thermodynamic temperature-invariant points). Measurements at temperatures not at invariant points are necessarily more difficult. To measure a temperature such as 20°C using the calibration temperatures of the ice point and the steam point, in principle, requires only that we determine PV at 20° and calculate the temperature. For a constant-volume thermometer,

$$t = 100 \times \frac{(PV_0)_t - (PV_0)_0}{(PV_0)_{100} - (PV_0)_0},$$

where V_0 is the constant volume of the bulb containing the gas, and the product PV_0 is determined using successively smaller amounts of gas until the limiting value of PV is obtained. Note that three measurements are required at each change in the quantity of gas present. Actually, there are several difficulties in using a gas thermometer: The volume of the bulb changes because of thermal expansion; the pressure is transmitted with a capillary to the pressure measuring apparatus, so there are "dead" volume and capillary volume corrections that must be applied; the response is slow; the bulb volume is large and must be in a large zone of uniform temperature; and the correcting equations depend on adequate representation of real gas properties to determine the volume of an ideal gas. These many complicating factors have made gas thermometry a highly sophisticated and expensive procedure suitable only for government standards laboratories or other large research organizations. As a result, gas thermometry is used mainly to establish temperature scales and to calibrate other devices.

1.11 HIGH-TEMPERATURE GAS THERMOMETRY

Extension of the gas thermometer to high temperatures was begun by Prinsep in 1828 when he measured the melting points of precious metal alloys using a thermometer with a gold bulb filled with air at constant pressure, connected to a sensitive manometer containing olive oil. In 1837, Pouillet improved the instrument, using a platinum bulb. Regnault made further improvements in 1847. Then procelain was substituted for platinum, and I_2 gas for air, by Deville and Troost, and although I_2 was soon replaced

TABLE 1.2.   Melting Points (°C)[a]

Investigator	Year	Silver	Gold	Copper	Nickel	Palladium	Platinum	Iridium
Prinsep	1828	999						
Pouillet	1836	1000	1200					
Ed. Becquerel	1863	960	1092					
Violle	1879	954	1045	1054		(1500)	(1775)	(1950)
Erhardt and Schertel	1879	954	1075					
Barus	1892	985	1093	1097	(1517)	(1643)	(1855)	
Holborn and Wien	1892	971	1072	1082	(1484)	(1587)	(1780)	
Callendar	1892	(961)	(1061)					
D. Berthelot	1898	962	1064					
Holborn and Day	1900	961.5	1064	1084				

[a]Values in parentheses are extrapolated.

Source: Day and Sosman.[9]

by air or nitrogen, the leaky porcelain bulb was retained for the rest of the century. It was not until Holburn and Day, in 1900, returned to a platinum bulb, filled with nitrogen, that the melting point of gold was accurately known. The history of the high-temperature gas thermometry was reviewed in 1900 by Barus, who presented a table of melting-point determination. It is reproduced in Table 1.2 as modified by Day and Sosman.[9] Improvements in high-temperature gas thermometry have resulted in more accurate determinations, which are discussed in Chapter 4.

In recent years it has been recognized that precision gas thermometry above 500°C is very difficult, and other means are used to establish temperature above the gold point, 1337 K. Those temperatures could be established more accurately if precision gas thermometry could be extended to the gold point. Progress toward this goal was reported in 1982 by Guildner and Edsinger of the U.S. National Bureau of Standards.[10]

1.12   SUMMARY

In this chapter we have traced the history of temperature measurement from its early beginnings to the gas thermometer. We have learned that either a constant-volume or a constant-pressure gas thermometer can be used, in principle, to measure temperature, but that in practice gas thermometry is very difficult.

Temperature has been defined in a qualitative way as the level of thermal energy, and it has been identified as a state function, one that must be specified in order to characterize the state of a material or system. Because it is so important to the properties of materials and because all properties of materials depend on temperature, it may be helpful to include a table of

TABLE 1.3   Properties Sometimes Used to Measure Temperature

	Example
Thermal expansion	
Liquid	Mercury thermometer
Solid	Piezoelectric crystal resonant frequency
Gas	Gas thermometer
Electrical resistance	
Metal	Platinum resistance detector
Semiconductor	Thermistor, transistor
Thermoelectric	Thermocouple
Vapor pressure	Liquid hydrogen
Magnetic susceptibility	Cerium-magnesium nitrate
Thermal noise	Josephson junction
Flow rate	Pyrometric cone
Light intensity	Optical pyrometer
Heat emission	Total radiation or infrared pyrometer
Superconductors	Aluminum
Sound velocity	Acoustic thermometry
Nuclear vibrations	Nuclear quadrupole resonance
Color	Crayons for marking
Heat content	Calorimetry plus specific heat

some of the properties that can be used to measure temperature (Table 1.3).

The remainder of this book is devoted to explaining the concept of temperature in greater detail, to presenting the international conventions associated with temperature measurement, and to describing the principles utilized in measuring temperature by a particular method. It will become apparent that measurement of temperature at one regime of temperature may require an entirely different device than that in another regime of temperature. The device selected for a particular application should be the one giving the best compromise among the various factors affecting that particular measurement. Some of the factors to be considered are size, accuracy, reliability of calibration, cost, stability, type of output signal, versatility, damage resistance, response time, and simplicity of operation. There is no ideal temperature measuring instrument. As will be developed in later chapters, several different temperature measuring devices can be used at most temperatures. (The choices become more limited at very high or very low temperatures.) Selection of the best instrument for a particular application requires technical judgment that can be acquired from this book.

REFERENCES

1. W. E. Knowles Middleton, "A History of the Thermometer and Its Use in Meteorology. Johns Hopkins Press, Baltimore, Maryland, 1966.

2. According to Reference 1, this figure came from: "Biancani [Blancamus], Spaera mundi, seu cosmographia demonstrativa [etc.] (Bologna, 1620), p. 111. The dedication to P. F. Malaspina ends 'Vale Parmae Idibus Febr. MDCXVII' (February 13, 1617)."

3. Sanctorii Sanctorii Iustinopolitani . . . . commentaria in primam fen primi libri canonis Avencennae . . . Venus, 1625. Cited in reference 1.

4. F. S. Taylor, *Ann. Sci.* **5**, 129 (1942) gives an interesting analysis of Galen's ideas of degrees of hot and cold.

5. T. Birch, "The History of the Royal Society of London," Vol. 1, p. 320. Royal Society, London, 1756.

6. E. Griffiths, "Methods of Measuring Temperature." Griffin, London, 1918.

7. W. J. Moore, "Physical Chemistry," 2nd ed. Prentice-Hall, Englewood Cliffs, New Jersey, 1956.

8. T. J. Quinn, Temperature standards. *Conf. Ser.—Inst. Phys.* **26**, 1–16 (1975).

9. A. L. Day and R. B. Sosman, "High Temperature Gas Thermometry." Carnegie Institution, Washington, D.C., 1911.

10. L. A. Guildner and R. E. Edsinger, Progress in NBS gas thermometry above 500°C. *Temp.: Its Meas. Control Sci. Ind.* **5** (Part 1), 43–48 (1982).

2

THE THERMODYNAMIC
TEMPERATURE SCALE

The thermodynamic scale of temperature is a theoretical concept based on thermodynamic principles. It cannot be realized in practice, although it can be approximated quite closely. It is the basis for all temperature measurements, so it is necessary for us to understand the concepts that lead to the thermodynamic, or Kelvin scale. To do so, we must have some basic definitions.

2.1 THERMODYNAMIC TERMS DEFINED

System. A system is defined as any region of the universe the boundaries of which can be clearly established. A system may be imaginary in that we imagine a clearly defined portion of the universe.

Thermodynamic System. A thermodynamic system is any system that is selected for thermodynamic analysis. Thermodynamic systems are often imaginary model systems that can be used for thermodynamic analysis.

Isolated System. An isolated system is one which can be separated from the rest of the universe so that no interaction exists across its boundaries. Obviously, this is an intellectual concept. We can imagine an isolated system for the purpose of analysis, but we cannot achieve a perfectly isolated system. We can, however, achieve systems in the laboratory that are sufficiently isolated for the laws of thermodynamics to be useful.

External Effect. An external effect is something delivered across a boundary to a system to alter the system. This is often energy in the form of heat or work. It can be an external pressure, a magnetic field, or some other influence. Note that these external changes alter the properties of the system. They are not properties of the system and, mathematically, are not exact differentials. (See Chapter 3 for a discussion of exact differentials.)

Internal Properties. Properties of a substance within a system are internal properties, that is, properties of the materials within the system. Many of the most valuable of the internal properties are state properties, and, mathematically, are exact differentials. (See Chapter 3 for exact differentials.)

State Properties. State properties are those properties which can be used to describe the state of a system in such a way that no ambiguity or uncertainty is possible. State properties are internal properties, that is, properties of the material, not of the external effects. They are always exact differentials and are often used in an equation of state. For example, $PV = nRT$ is the equation of state for a perfect gas. The state of the gas is described when P, V, n, and T are given. Note that P is an external stress which is in equilibrium with the gas within the system. It is the pressure within the gas that is an internal state property.

Perfect Gas. A perfect gas is an imaginary gas that obeys the perfect gas law, $PV = nRT$. It can be approximated by light monatomic gases at very low pressure.

Heat, Q. Heat is thermal energy that is in the process of transfer across the boundaries of a system or from one portion of a system to another. Heat is usually measured in joules (J) or in British thermal units (Btu). Heat, dQ, is not an exact differential.

Work, W. Work is mechanical energy transmitted across the boundaries of a system, or from one portion of a system to another. It is usually measured in joules or in foot-pounds. Work, dW is not an exact differential.

Internal Energy, U. Internal energy includes all forms of energy within a body and is often considered to be the thermal energy of the body, arbitrarily taking the internal energy at absolute zero to be zero. The total energy of a body includes the sum of the internal energy, the potential energy, and the bulk kinetic energy. The latter two can be related to special coordinates and macroscopic parameters. Internal energy must be obtained under special conditions eliminating potential energy and bulk kinetic energy. Actually, we can never know internal energy exactly. It is the difference in internal energy that can be obtained. Thermodynamic tables of internal energy are based on zero internal energy at absolute zero, so that changes from one state to another can readily be calculated. Internal energy is an extensive variable, can be specified at a point in a body, is a state property, and is an exact differential. For a perfect gas at constant volume, the internal energy per mole depends only on the absolute temperature.

Enthalpy, H. Enthalpy is an important extensive property of a system given by

$$H = U + PV. \tag{2.1}$$

It is important because many gaseous systems having no changes in kinetic or potential energy are subjected to constant pressure. Changes in such a system include both changes in internal energy and work against the external pressure. (Many practical systems occur under these conditions. Often the external pressure is the atmospheric pressure of the earth.) Even though changes in enthalpy are restricted to the conditions of no change in kinetic or potential energy of the bulk system, it is a state property because U, P, and V are state properties. Enthalpy can be defined at a point in a system and is an exact differential.

Specific Heat. Specific heat is the energy required to raise the temperature of one unit mass of a substance (usually, 1 kilogram or 1 gram mole, or 1 pound or 1 pound mole) by 1 degree. Specific heat can be measured at constant volume or constant pressure. The specific heat at constant pressure is greater than that at constant volume because work must be performed against the external pressure if thermal expansion is positive (the usual case is that materials expand when temperature is increased).

Specific heat at constant volume is defined mathematically as

$$C_v \equiv \left(\frac{\partial U}{\partial T}\right)_V, \tag{2.2}$$

where the V indicates constant volume.

Specific heat at constant pressure is defined as

$$C_p \equiv \left(\frac{\partial H}{\partial T}\right)_P, \tag{2.3}$$

where the P indicates constant pressure.

The Mechanical Equivalent of Heat. In SI units, all units of energy are expressed in joules, J. In the English system, energy in the form of work is measured in foot-pounds and energy in the form of heat is measured in British thermal units. The conversion from one set of English units to the other is accomplished using 778 foot-pounds per Btu as the mechanical equivalent of heat.

Isothermal. Isothermal means that the temperature is homogeneous throughout a selected system. The system has a single temperature at a particular moment in time. If the temperature is constant during a period of time, that must be specified additionally.

Adiabatic. Adiabatic means "without exchange of heat." An adiabatic process of a system is one in which no heat is added to or subtracted from the system.

The First Law of Thermodynamics. This law states that all energy must be conserved. It can be neither created nor destroyed, although it can be converted from one form to another. The first law is often written in differential form as

$$dU = dQ - dW, \tag{2.4}$$

where dU is an infinitesimal change in internal energy which occurs when an infinitesimal amount of heat, dQ, is added to or subtracted from a system, and dW is the infinitesimal amount of work performed on or subtracted from the system. The work can be mechanical, electrical, or any other form and the minus sign is used because work out of a system is considered positive.

The Second Law of Thermodynamics. The original deriviation that led to the concept of the second law of thermodynamics is explained in this chapter. In its original form, the second law is stated: "No heat engine is more efficient than a Carnot engine operated reversibly between the same conditions of temperature and pressure."

Since that derivation, the concepts of entropy and statistical mechanics (Chapter 3) have resulted in other statements of the second law. One of these is: "The entropy of a closed system tends to remain constant or to increase but never decreases." A more general statement appropos to temperature measurement is: "Heat always flows spontaneously from a high temperature to a low temperature, never from a low temperature to a high temperature unless external energy is provided."

2.2　THE CARNOT CYCLE

Prior to the invention of the steam engine in 1769, power had been generated by animal exertion, wind, or water. After Watts invented the steam engine, people soon realized that other working gases might be substituted for steam. They also realized that the steam engine was converting heat energy into mechanical energy. They began to wonder what fundamental principles governed the efficiency of the conversion.

To compare engines it was necessary to determine their efficiencies. This was done by comparing the work generated to the heat supplied:

$$\eta = \frac{W}{Q}. \tag{2.5}$$

Engineers soon realized that the work delivered by the existing engines was not the ultimate that could be generated. Many different types of engines were possible (Fig. 2.1).

Because of the many different ways to produce work from heat engines, there was a need to determine thermodynamic efficiency, that is, to determine which was most efficient thermodynamically. For the same heat input, we need to know what, in principle, is the maximum work that can be produced from it. This problem was ingeniously solved by a French engineer, Sadi Carnot, in 1824.[1] His solution makes it possible to compare engines, but it also introduces a fundamental basic principle of thermodynamics which leads to the concept of a thermodynamic temperature scale.

Carnot recognized that all heat engines depend on the high temperature of a working fluid, that the fluid was cooled in extracting work from it, that the heat was delivered to the fluid at a high temperature, and that some heat was exhausted to a reservoir, or receiver, at a low temperature. He proposed a simple ideal cycle to represent this process and studied it in

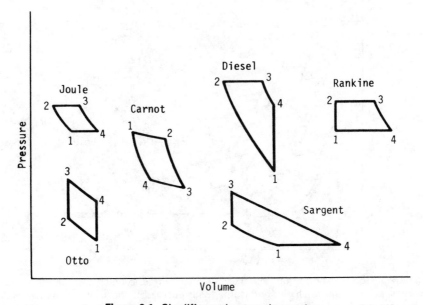

Figure 2.1. Six different heat engine cycles.

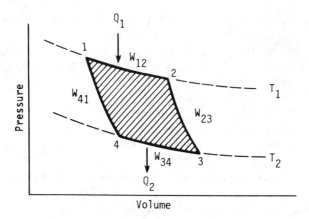

Figure 2.2. The Carnot cycle.

detail. This resulted in the fundamental law of thermodynamic efficiency, which is, in itself, a statement of the second law of thermodynamics. The Carnot cycle, as it is now called, proved useful in understanding all possible heat engine cycles. It is a simple statement of the maximum energy that it is possible to extract for given conditions of operation.

The Carnot cycle consists of isothermal delivery of heat at a high temperature, T_1, adiabatic expansion, isothermal rejection of heat at a low temperature, T_2, and adiabatic compression back to the high temperature (Fig. 2.2). Carnot realized that the most efficient possible heat engine must

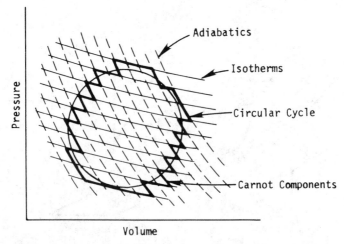

Figure 2.3. Completely arbitrary circular cycle converted to equivalent Carnot cycles. As the spacing of the adiabatics and isotherms is decreased, the arbitrary circular cycle is approached. Then summing all cycles inside the circle gives the Carnot equivalent.

be thermodynamically reversible, that is, that no energy would be lost to friction, turbulence, or any other nonproductive energy-consuming process. He proposed this cycle, operating reversibly, as the standard to which all other engines should be compared. Since any other cycle can be converted, in principle, to a series of Carnot cycles, all engines and all cycles can be compared to the most efficient Carnot equivalent (Fig. 2.3).

For the Carnot cycle any working fluid can be used. It is the heat delivered and extracted that determines the Carnot efficiency. The analysis is easier using perfect gas laws, so in the following analysis, 1 mole of a perfect gas will be assumed to be the working fluid. (Problem 2.3 gives an example of the use of an imperfect gas.)

2.3 ANALYSIS OF THE CARNOT CYCLE

Referring to Fig. 2.4, we recognize that as the cycle moves through its complete range $1 \to 2 \to 3 \to 4 \to 1$, the net work is

$$W_{\text{net}} = \oint P \, dV , \qquad (2.6)$$

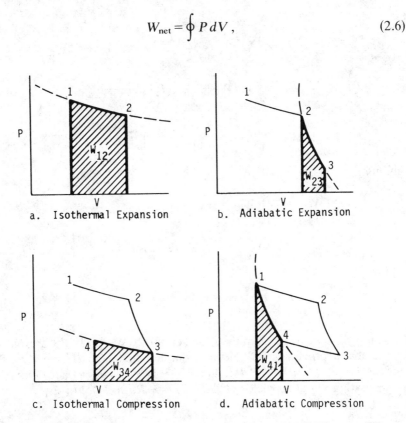

a. Isothermal Expansion   b. Adiabatic Expansion

c. Isothermal Compression   d. Adiabatic Compression

Figure 2.4. Work of each portion of the Carnot cycle shown on a *PV* diagram. The sum of a, b, c, and d is the crosshatched area of Fig. 2.2.

as shown in Fig. 2.2. We need to determine the net work in order to determine efficiency. We can determine the net work (the PV area enclosed by the cycle) as $Q_1 - Q_2$, since all energy must be accounted for and we have lost none to friction, or by summing,

$W_{12} + W_{23} + W_{34} + W_{41}$:

$$1 \rightarrow 2: \quad W_{12} = \int P \, dV = \int_{V_1}^{V_2} \frac{RT}{V} \, dV = RT_1 \ln \frac{V_2}{V_1} = Q_1 \qquad (2.7)$$

$2 \rightarrow 3$: W_{23} can be found:

$$dU = dQ - dW \qquad \text{(first law)}$$

$$dQ = 0 \qquad \text{(adiabatic)}$$

Therefore, $dU = -dW = C_v \, dT$, so

$$W_{23} = -\Delta U_{23} = -\int_{T_1}^{T_2} C_v \, dT \qquad (2.8)$$

$$3 \rightarrow 4: \quad W_{34} = \int P \, dV = -Q_2 = RT_2 \ln \frac{V_4}{V_3} \qquad (2.9)$$

$$4 \rightarrow 1: \quad W_{41} = -\Delta U_{41} = -\int_{T_2}^{T_1} C_v \, dT = +\int_{T_1}^{T_2} C_v \, dT \qquad (2.10)$$

Summing, we have

$$\sum W_i = W_{12} + W_{23} + W_{34} + W_{41}$$

$$= RT_1 \ln \frac{V_2}{V_1} - \int_{T_1}^{T_2} C_v \, dT + RT_2 \ln \frac{V_4}{V_3} + \int_{T_1}^{T_2} C_v \, dT$$

$$= RT_1 \ln \frac{V_2}{V_1} + RT_2 \ln \frac{V_4}{V_3}. \qquad (2.11)$$

Equation (2.11) can be simplified further. Both W_{23} and W_{41} are obtained for adiabatic processes for which, from the first law $dU = -dW$. Then, since $dU = C_v \, dT$, $dW = P \, dV$, and $P = RT/V$, $C_v \, dT = -(RT/V) \, dV$.
Separating variables yields $(C_v/T) \, dT = -R d(\ln V)$. For $2 \rightarrow 3$,

$$\int_{T_1}^{T_2} \frac{C_v}{T} \, dT = -R \int_{V_2}^{V_3} d(\ln V).$$

For $4 \rightarrow 1$,

$$\int_{T_2}^{T_1} \frac{C_v}{T} \, dT = -R \int_{V_4}^{V_1} d(\ln V).$$

By changing signs and interchanging the limits of the left side of the preceding equation, the left sides of the two preceding equations become identical. Therefore,

$$-R \int_{V_2}^{V_3} d(\ln V) = +R \int_{V_4}^{V_1} d(\ln V),$$

so

$$\ln \frac{V_2}{V_3} = \ln \frac{V_1}{V_4} \quad \text{and} \quad \frac{V_2}{V_3} = \frac{V_1}{V_4}.$$

Then

$$\frac{V_2}{V_1} = \frac{V_3}{V_4}. \tag{2.12}$$

Substituting in Eq. (2.11) gives us

$$W = RT_1 \ln \frac{V_2}{V_1} - RT_2 \ln \frac{V_2}{V_1}$$

$$= R \ln \frac{V_2}{V_1} (T_1 - T_2). \tag{2.13}$$

Now we can calculate the Carnot efficiency.

$$\eta = \frac{Q_1 - Q_2}{Q_1} = \frac{W}{Q_1} = \frac{R \ln (V_2/V_1)(T_1 - T_2)}{RT_1 \ln (V_2/V_1)} \tag{2.14}$$

$$= \frac{Q_1 - Q_2}{Q_1} = \frac{R \ln (V_2/V_1)}{R \ln (V_2/V_1)} \left[\frac{T_1 - T_2}{T_1} \right] = \frac{T_1 - T_2}{T_1} \tag{2.15}$$

$$= 1 - \frac{Q_2}{Q_1} = 1 - \frac{T_2}{T_1}. \tag{2.16}$$

Therefore, the efficiency of a perfect reversible heat engine using a perfect gas working fluid is dependent only on the absolute temperature of the source and the receiver. This led to the second law of thermodynamics: "No engine is more efficient than a Carnot engine operating between the same conditions of pressure and temperature." This is important to the engineering analysis of heat engines. It is also important to the understanding of temperature and was used by Lord Kelvin to propose the thermodynamic temperature scale.

2.4   THE KELVIN THERMODYNAMIC TEMPERATURE SCALE

Since, for a Carnot engine, the efficiency

$$\eta = 1 - \frac{Q_2}{Q_1} = 1 - \frac{T_2}{T_1}$$

can equal 1 only when $T_2 = 0$, and because we cannot actually achieve $T_2 =$ absolute zero, no practical engine will have a perfect efficiency and convert all its thermal input into work. Note that this efficiency equation proves that T can never be less than zero. If it were, the efficiency would be greater than 1 and we would get more work out than we put energy in. We would be creating energy, an impossibility.

Based on this consideration, in 1848 Lord Kelvin proposed that the Carnot principles be the basis for a new absolute thermodynamic temperature scale.[2] This new scale would be based on the universal properties of energy, not on any material or measuring system. He proposed that for any temperature T, a series of Carnot cycles could be set up between it and absolute zero (Fig. 2.5). Each cycle would deliver an equal amount of heat, W. Then for n cycles, $Q_1 = nW$. Note that Q_2 is zero for the lowest-temperature cycle, where T_2 is zero. Thus the increments of equal work set

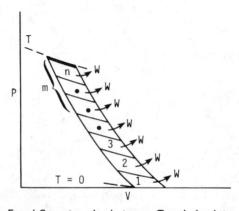

Figure 2.5. Equal Carnot cycles between T and absolute zero.

up a scale where it is only necessary to decide on the size of the increments, W. If we number the Carnot isotherms from the bottom up, from zero to τ, we have a scale of thermodynamic temperatures from zero to τ. If we select a group of m cycles near the top (from the top down), the heat rejected at the lowest of the m cycles will be $Q - mW$, where Q is the energy taken in by the top cycle. Since all Q is converted to work if all n cycles are used and no heat is rejected at the last cycle, $Q = nW$. Then the efficiency for the m cycles at the top is

$$\eta = \frac{Q - (Q - mW)}{Q} = \frac{mW}{nW} = \frac{m}{n}, \tag{2.17}$$

and since we defined our new thermodynamic temperature as equal to the number of cycles, $\tau = n$.

$$\eta = \frac{m}{\tau} = \frac{\Delta\tau}{\tau},$$

where τ represents temperature on the Kelvin scale and $\Delta\tau$ represents m. However, we also know that $\eta = \Delta T / T$ for a Carnot cycle, so

$$\eta = \frac{\Delta\tau}{\tau} = \frac{\Delta T}{T} = \frac{\tau_1 - \tau_2}{\tau_1} = \frac{T_1 - T_2}{T_1},$$

or

$$\eta = 1 - \frac{\tau_2}{\tau_1} = 1 - \frac{T_2}{T_1} \tag{2.18}$$

for any arbitrary states 1 and 2. Therefore,

$$\frac{\tau_2}{\tau_1} = \frac{T_2}{T_1} = 1 - \eta. \tag{2.19}$$

But this is true for any temperature pair T_1 and T_2. This is possible only if $\tau = CT$, where C is a constant. If we choose W to be the right size, $\tau = T$ and $C = 1$. So by choosing the size of the degree for τ the same as that for T, we do not need τ, only T. But we now have a scale that is thermodynamically sound. We have an energy basis for temperature. The absolute perfect gas scale is the same as the thermodynamic Kelvin scale. The size of the degree must be chosen, then, so that $C = 1$.

In making the absolute degree fit the Celsius degree, Kelvin saw two alternatives. One was to establish one fixed point accurately and choose the degree size for the thermodynamic scale so that the thermodynamic and the practical scale had the same number. In effect, this uses absolute zero as one point so that the two points establish a straight line with a zero intercept. The alternative was to choose two points on the practical scale and pass the thermodynamic scale through it, choosing the size of the degree so that the temperatures on each scale would be numerically equal. This he chose as the most practical method at the time (1854), but he preferred the first alternative in principle.

Kelvin knew that his theoretical thermodynamic scale would not be realized in practice because perfect gases were not available. He knew that various gas thermometers would give different temperatures for the same isotherms. However, the gas thermometer appeared to be the closest approximation to the desired relationship, and Kelvin proposed that it be used as the closest practical approximation to the absolute thermodynamic scale. The gas thermometer is still used for this purpose.

Kelvin's second preference, to base the absolute thermodynamic scale on two fixed points to establish the degree size, was used until 1954. The two fixed points were the ice point and the steam point. That scale was then extrapolated to establish absolute zero. The hydrogen thermometer at low pressures was used to approximate the thermodynamic scale. In 1954, the triple point of water was chosen as the single fixed point above absolute zero on which the thermodynamic scale is based—in agreement with the first preference of Lord Kelvin.

A single defining point has two principal advantages. First, the value of any revised Celsius temperature standard can readily be converted to the correct absolute temperature just by adding the temperature of triple point of water. This makes it easier to revise those tables of property relationships which depend upon absolute temperature, such as thermodynamic property tables. Second, adopting a fixed absolute value for the triple point of water prevents the necessity for changing all temperatures based on it (such as Celsius temperatures) every time an improvement in measuring the triple point is achieved.

The practical temperature scale that we actually use to measure temperature has been revised many times as better practical devices have become available, and as better approximations to the true thermodynamic temperature scale have been obtained. In recent years the agreement between the practical scale and the thermodynamic scale is believed to be excellent, within fractions of a degree, over most common temperatures. In Chapter 4 we discuss the International Temperature Scale and its agreement with the thermodynamic temperature scale.

2.5   FUNDAMENTAL TEMPERATURE SCALES

The thermodynamic temperature scale is based on the perfect gas law. A perfect gas always has the same PV product, per mole, as a function of temperature. For n moles,

$$\frac{PV}{n} = RT = N_A kT .$$
(2.20)

Here R is a physical constant based on two other fundamental physical constants, Avogadro's number N_A, and Boltzmann's constant k. Because N_A and k are fundamental constants, whose values can be determined by other physical methods, the PV product per mole has a fundamental relationship to the absolute temperature. In experiments in which real gas behavior can be extrapolated to represent perfect gas behavior, there is a fundamental relationship between the PV product per mole and temperature. This establishes the concept of a *fundamental temperature scale*. It is a scale in which the relationship of the measured variable to temperature depends only on fundamental physical constants, not on arbitrary calibrating constants. (Some authors identify fundamental scales as primary scales, but this causes confusion between primary and secondary calibration methods.) The thermodynamic temperature scale is a fundamental temperature scale.

A fundamental temperature scale is ideal because it eliminates calibration factors. However, for the perfect gas law, real gases must be used and there still exists the need to verify that the measured variable really does have a fundamental relationship to temperature. Obviously, it would be desirable to be able to use fundamental scales for the entire range of temperatures available. Also, if one fundamental method cannot be used over the entire range of temperatures, it would be desirable for various fundamental relationships to overlap, so that they could be compared. At the present time this is not possible. Much research is being conducted on possible fundamental relationships, to extend their range or improve their accuracy, to achieve a system of overlapping fundamental temperature scales.

As the concepts of temperature measurement are developed in later chapters we will discuss other fundamental scales. Some of the possible fundamental relationships include:

1. *Thermal radiance as a function of temperature*:
   a. Planck's law

$$I_\lambda = \frac{2hc^2}{n^2 \lambda^5} \left[\exp\left(\frac{hc}{nk\lambda T}\right) - 1 \right]^{-1}$$
(2.21)

Here h is Planck's constant, c is the velocity of light, n is index of refraction, λ is wavelength, and k is Boltzmann's constant.

b. The Stefan–Boltzmann law

$$W = \frac{2\pi^5 k^4}{15 c^2 h^3} T^4 \tag{2.22}$$

2. *Velocity of sound in a gas*:

$$v = (C_p RT / C_v M)^{1/2} \tag{2.23}$$

Here C_p and C_v are the specific heats at constant pressure and constant volume, respectively, and M is the molecular weight of the gas.

3. *Noise in an electrical resistor.* The root-mean-square voltage variation in an electrical resistor of resistance R is given by

$$\mathscr{V} = [4kR(\Delta u)T]^{1/2}. \tag{2.24}$$

where Δu is the bandwidth and k is Boltzmann's constant.

Other methods of interest include refractive index or dielectric constant measurements in gases and magnetic susceptibility of paramagnetic salts. Of the methods listed above, only Planck's law is now established as a practical method of general interest.

REFERENCES

1. S. Carnot, "Réflexions sur la puissance motrice der feu." Paris, 1824.
2. W. Thompson (Lord Kelvin), On an absolute thermodynamic scale founded on Carnot's theory of the motive power of heat. *Philos. Mag.* [3] **33**, 313–17 (1848).

RECOMMENDED READING

M. W. Zemansky, M. M. Abbott, and H. C. Van Ness, "Basic Engineering Thermodynamics," 2nd ed. McGraw-Hill, New York, 1975.

W. C. Reynolds and H. C. Perkins, "Engineering Thermodynamics," McGraw-Hill, New York, 1970. (Chapter 7 is especially pertinent.)

S. Carnot, "Reflections on the Motive Power of Fire," (R. H. Thurston, transl.), Dover, New York, 1960.

R. Fox, "The Caloric Theory of Gases." Oxford Univ. Press, London and New York, 1971. (Pages 177–191 are especially pertinent.)

CHAPTER 2 PROBLEMS

2.1 Consider the arbitrary circular pressure–volume diagram in Fig. 2.3. Draw a Carnot cycle to approximate the area of the PV diagram. Then construct a diagram using four Carnot cycles, then eight. Note the improved fit as the number of cycles is increased.

2.2 A gas expands, doing 350 J of work while receiving 1.5×10^{10} ergs of heat from the surroundings. Use Eq. (2.1) in its integral form to calculate the change in internal energy.

$$U = Q - W$$

2.3 The perfect gas law was used in demonstrating the Carnot cycle. Suppose that $P\,(5V + 10{,}500\,\text{Pa}) = RT$. Analyze the Carnot cycle using this gas law instead of the perfect gas law.

2.4 Calculate the maximum work possible in going from T_1 to T_2 if Q_1 is given:

(a) $T_1 = 1710\,\text{K}$, $T_2 = 300\,\text{K}$, $Q_1 = 10^6\,\text{ergs}$

(b) $T_1 = 3000°\text{R}$, $T_2 = 600°\text{R}$, $Q_1 = 1500\,\text{Btu}$

2.5 If material considerations limit the highest temperature of a metallic gas turbine to 1100 K, what is the maximum improvement in efficiency that could occur if ceramic components capable of 1300 K are substituted, if the exhaust temperature remains at 450 K?

2.6 If a heat engine operating at 900°C has an exhaust temperature on the earth of 90°C, and −250°C on the cold side of the moon, what is the maximum efficiency possible on the earth and on the moon?

CHAPTER 2 ANSWERS

2.1

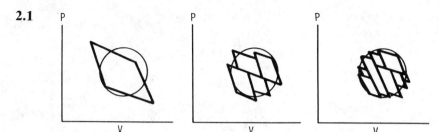

Prob. 2.1 Answer

2.2 $U = Q - W = 1.5 \times 10^{10}\,\text{ergs}\left(\dfrac{\text{joule}}{10^7\,\text{ergs}}\right) - 350\,\text{J}$

$$= 1500 - 350 = 1150\,\text{J}$$

2.3 We start by analyzing the components shown in Fig. 2.4:

(a) Isothermal expansion

$T_1 = $ constant

$$dW = dQ \qquad W_{12} = Q_1 = \int_{V_1}^{V_2} P \, dV = \int_{V_1}^{V_2} \frac{RT_1}{5V + 10{,}500} \, dV$$

$$= RT_1 \int_{V_1}^{V_2} \frac{dV}{5V + 10{,}500} = RT_1 \ln \frac{5V_2 + 10{,}500}{5V_1 + 10{,}500}$$

(b) Adiabatic expansion

$Q = $ constant

$$dU = -dW \qquad W_{23} = -\int dU = -\int_{T_1}^{T_2} C_v \, dT$$

(c) Isothermal compression

$T_2 = $ constant

$$dW = dQ \qquad W_{34} = Q_2 = \int_{V_3}^{V_4} P \, dV = \int_{V_3}^{V_4} \frac{RT_2}{5V + 10{,}500} \, dV$$

$$= RT_2 \int_{V_3}^{V_4} \frac{dV}{5V + 10{,}500} = RT_2 \ln \frac{5V_4 + 10{,}500}{5V_3 + 10{,}500}$$

(d) Adiabatic compression

$Q = $ constant

$$dU = -dW \qquad W_{41} = -\int dU = -\int_{T_2}^{T_1} C_v \, dT$$

We can simplify further. For an adiabatic process,

$$dW = -dU$$

$$P \, dV = -C_v \, dT$$

$$\frac{RT \, dV}{5V + 10{,}500} = -C_v \, dT$$

$$R \frac{dV}{5V + 10{,}500} = \frac{-C_v \, dT}{T}$$

$$R \int \frac{dV}{5V + 10{,}500} = -C_v \int \frac{dT}{T}$$

For the adiabatic processes, 2 to 3 and 4 to 1,

$$R \int_{V_2}^{V_3} \frac{dV}{5V + 10{,}500} = -\int_{T_1}^{T_2} C_v \, dT$$

and

$$R \int_{V_4}^{V_1} \frac{dV}{5V + 10,500} = - \int_{T_2}^{T_1} C_v \, dT$$

The right side of the two equations above can be made identical by changing the sign and inverting the limits in the second equation. Then

$$R \int_{V_2}^{V_3} \frac{dV}{5V + 10,500} = -R \int_{V_4}^{V_1} \frac{dV}{5V + 10,500}$$

or

$$\ln \frac{5V_3 + 10,500}{5V_2 + 10,500} = -\ln \frac{5V_1 + 10,500}{5V_4 + 10,500}$$

and

$$\frac{5V_3 + 10,500}{5V_2 + 10,500} = \frac{5V_4 + 10,500}{5V_1 + 10,500}$$

Inverting the two sides gives

$$\frac{5V_2 + 10,500}{5V_3 + 10,500} = \frac{5V_1 + 10,500}{5V_4 + 10,500},$$

so

$$\frac{5V_2 + 10,500}{5V_1 + 10,500} = \frac{5V_3 + 10,500}{5V_4 + 10,500}.$$

We know that, by definition, the efficiency is

$$\eta = \frac{\sum W_i}{Q_1},$$

and by the first law,

$$\eta = \frac{Q_1 - Q_2}{Q_1}.$$

Substituting for $\sum W_i$, and noting that $W_{23} = -W_{41}$, we have

$$\eta = \frac{RT_1 \ln \dfrac{5V_2 + 10,500}{5V_1 + 10,500} - RT_2 \ln \dfrac{5V_1 + 10,500}{5V_4 + 10,500}}{RT_1 \ln \dfrac{5V_2 + 10,500}{5V_1 + 10,500}}.$$

Substituting from above for the second term in the numerator gives

$$\eta = \frac{RT_1 \ln \dfrac{5\,V_2 + 10{,}500}{5\,V_1 + 10{,}500} - RT_2 \ln \dfrac{5\,V_2 + 10{,}500}{5\,V_1 + 10{,}500}}{RT_1 \ln \dfrac{5\,V_2 + 10{,}500}{5\,V_1 + 10{,}500}}$$

$$\eta = \frac{RT_1 - RT_2}{RT_1} = \frac{T_1 - T_2}{T_1}.$$

This is the same result as for the perfect gas law. Instead of $P(5\,V + 10{,}500) = RT$, we might have used any other suitable relationship.

2.4 (a) $\dfrac{W}{Q_1} = \dfrac{T_1 - T_2}{T_1}$

$$W = Q \frac{T_1 - T_2}{T_1}$$

$$= 10^6 \left(\frac{1710 - 300}{1710} \right) = \frac{1410}{1710} (10^6) = 825{,}000 \text{ ergs}$$

$$= 0.0825 \text{ J}$$

(b) $W = 1500 \left(\dfrac{3000 - 600}{3000} \right) = \dfrac{2400}{3000} (1500)$

$$= 1200 \text{ Btu}$$

$$= (1200)778 \frac{\text{ft} - \text{lb}}{\text{Btu}} = 934{,}000 \text{ ft} - \text{lb}$$

2.5 $\eta_1 = \dfrac{T_1 - T_2}{T_1} = \dfrac{1100 - 450}{1100} = 0.59 = 59\%$

$$\eta_2 = \frac{1300 - 450}{1300} = 0.65 = 65\%$$

2.6 Earth: $\eta = \dfrac{(900 + 273) - (90 + 273)}{900 + 273}$

$$= \frac{810}{1173} = 0.690 = 69\%$$

Moon: $\eta = \dfrac{(900 + 273) - (273 - 250)}{900 + 273}$

$$= \frac{1150}{1173} = 0.98 = 98\%$$

The cold side of the moon allows 29% more energy to be extracted.

3

ENTROPY, TEMPERATURE, AND STATISTICAL MECHANICS

In this chapter we explore a more sophisticated understanding of temperature through both classical and statistical thermodynamics. In Chapter 2 we learned that temperature is measured on the thermodynamic temperature scale by equal increments of thermal energy, beginning at absolute zero and working upward in kelvin, with the size of the degree arbitrarily chosen to be the same as the degree Celsius. We learned that $(Q_1 - Q_2)/Q_1$ is equal to $(T_1 - T_2)/T_1$ if a perfectly efficient, reversible cycle is used. We identified temperature with thermal energy on a macroscopic scale through the use of the imaginary Carnot cycle. We will now develop the concept of entropy in the macroscopic sense from extensions of Carnot reasoning, then develop the microscopic relationship of thermal energy to temperature using mathematical and statistical mechanical arguments.

3.1 THE CONCEPT OF ENTROPY

Rudolf Clausius gave a new interpretation to the thermodynamic equations of the Carnot cycle.[1] He was intrigued by the implications of the Carnot equations when trying to achieve maximum efficiency. He rewrote the Carnot equation to determine the minimum amount of heat to be rejected to the cold receiver as

$$Q_2 = \frac{Q_1}{T_1} T_2 . \tag{3.1}$$

He noted that when T_2 is zero the heat rejected Q_2 is also zero and all of

Q_2 is converted into work. However, he recognized that for practical reasons, T_2 could never be zero. It must be a temperature in the easily accessible range. For many heat engines it might be a little above room temperature. If he assumed that T_2 was fixed in that way, then the amount of heat rejected depended on Q_1/T_1. For reasons discussed in the next paragraph, he considered this to be a property of the system. He called that property entropy, S. Thus

$$S_1 = \frac{Q_1}{T_1}.$$ (3.2)

For the Carnot cycle he could write

$$\frac{Q_1}{T_1} + \frac{Q_2}{T_2} = 0.$$ (3.3)

or

$$S_1 + S_2 = 0.$$ (3.4)

Clausius reasoned that for the Carnot cycle, $S_1 = S_2$, and the entropy change of the working fluid for the complete cycle was zero. He deduced that for a nonreversible cycle, where energy would be wasted, the entropy change must be less than zero. Some of the energy supplied could not be converted to work, so some entropy would be lost. Therefore, for any cycle

$$\oint \frac{dQ}{T} \le 0,$$ (3.5)

where only a perfectly efficient, reversible cycle would have the integral equal to zero.*

The Carnot cycle applies regardless of the material used as the working fluid. So Clausius's concept of entropy is applicable to any material used in a heat engine. We shall see later that entropy is an intensive property which can be used to describe the thermodynamic state of a material (Section 3.3).

Clausius plotted the Carnot cycle to show the change in the thermodynamic state properties S and T, as shown in Fig. 3.1. Note that the adiabatic expansion, $2 \to 3$, and the adiabatic compression, $4 \to 1$, are constant entropy (isentropic) processes because $Q = 0$. There is no change in entropy, although P and V change, because no heat exchange occurs.

Clausius performed an imaginary experiment in which he extracted heat from a main reservoir at a constant high temperature and delivered it through a series of heat engines and intermediate reservoirs to a lower

---

* The symbol $\oint$ indicates that the integral is taken for the complete cycle, returning to the starting state.

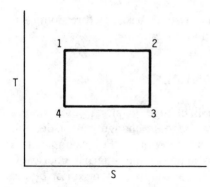

Figure 3.1. Carnot cycle of Fig. 2.2 plotted on T, S coordinates

temperature, extracting work at each engine in the process, then returning heat through a series of refrigerator cycles and intermediate reservoirs to the main reservoir, performing work at each refrigerator. Then if net work were delivered from the system, it could only come from the heat extracted from the main reservoir. But for a cycle returning the heat back to the original reservoir, the work could not be positive because that would violate the first law of thermodynamics. Each individual component would deliver a quantity of heat in the cycle in accordance with Eq. (3.1). For reversible cycles he reasoned that the summation of the Q/T terms for all the cycles must be zero, so that all heat was returned to the reservoir. By imagining each reversible cycle to be delivering a smaller increment of heat, the summation could be replaced by a differential. This permitted him to define entropy in the differential form,

$$dS = \frac{dQ_{\text{rev}}}{T}.$$ (3.6)

For a simple change in state, for any material going from state 1 to state 2, if we can write an expression for dQ_{rev}, we can calculate the entropy change from the integral,

$$\Delta S = \int_1^2 \frac{dQ_{\text{rev}}}{T}.$$ (3.7)

This allows us to calculate ΔS relative to any initial state. Often, absolute zero is used as an arbitrary zero of entropy, with entropy increasing with temperature. Then a scale of entropy for any material can be calculated if dQ_{rev} can be approximated. This is often done using the specific heat of the material as a measure of dQ_{rev}. Thus Eq. (3.7) is used in a powerful and versatile way to produce tables of entropy for a wide variety of materials used in chemical thermodynamic applications.

Having now considered the macroscopic basis for entropy, we need to understand it further by looking at its microscopic basis. Before doing so, let us consider the mathematical background for entropy.

3.2 ENTROPY AS AN EXACT DIFFERENTIAL

When a given material is subjected to a change in state, some of the thermodynamic quantities depend on how the state change was made, and others depend only on the initial and final conditions. For example, if a perfect gas is changed from state 1 to state 2, the internal energy U depends only on the initial and final states, but the work done on or by the gas depends on how the change in state was accomplished. In other words, some properties depend only on the endpoints in a change of state, whereas others depend on the path between. Those properties that depend only on the endpoints are particularly useful for describing the state of a material, and are called state properties.

Clausius was the first to realize that entropy is a state function, and he discovered this from extensions of the thermodynamic analysis made possible by the Carnot cycle using purely macroscopic thermodynamic reasoning. The heat exchanged in a heat engine is not a state quantity. It depends on path and on whether it is exchanged under equilibrium (reversible) conditions. Since the integral of dQ from state 1 to state 2 depends on path, and the integral of dQ_{rev}/T from state 1 to state 2 does not, dividing by T converts a nonstate function to a state function. We need to examine this further.

Mathematically, the characteristic that a change in state (in the value of a function) depends only on the endpoints is that of an exact differential.[2] An exact differential has these properties:

1. A change in state from state 1 to state 2 depends only on states 1 and 2, not on the path between.
2. The line integral over a closed path is zero, that is,

$$\oint dL(x, y) = 0 .$$
(3.8)

We already know that this is true for the Carnot cycle, where

$$\oint dS(T, P) = 0 .$$
(3.9)

There is a simple mathematical test for exactness. Because the order of taking partial derivatives with respect to two variables is not important, for an exact differential, considering $Z = f(x, y)$,

$$\frac{\partial}{\partial y}\left(\frac{\partial Z}{\partial x}\right) = \frac{\partial}{\partial x}\left(\frac{\partial Z}{\partial y}\right)$$

or

$$\frac{\partial^2 Z}{\partial x\, \partial y} = \frac{\partial^2 Z}{\partial y\, \partial x}. \tag{3.10}$$

The total differential dZ is

$$dZ = \frac{\partial Z}{\partial x}\, dx + \frac{\partial Z}{\partial y}\, dy.$$

Let $M = \partial Z/\partial x$ and $N = \partial Z/\partial y$. Then the total differential can be written as

$$dZ = M\, dx + N\, dy. \tag{3.11}$$

The test for exactness is to determine if

$$\frac{\partial M}{\partial y} = \frac{\partial N}{\partial x}. \tag{3.12}$$

This test is subject only to the condition that Z be a continuous function of x and y. Equation (3.12) is a test for exactness because if it is true, it is another form of Eq. (3.10).

The test for exactness can be applied to various thermodynamic functions to determine if they are thermodynamic state functions. For example, consider a gaseous system that depends on the variables P and T. The internal energy will be a function of P and T,

$$dU = \frac{\partial U}{\partial T}\, dT + \frac{\partial U}{\partial P}\, dP. \tag{3.13}$$

The volume will be a function of P and T also,

$$dV = \frac{\partial V}{\partial T}\, dT + \frac{\partial V}{\partial P}\, dP. \tag{3.14}$$

If our gaseous system undergoes a change in state as the result of reversibly supplied heat, from the first law, $dQ = dU + dW$. For dW we can write $P\, dV$ and express dQ as a function of P and T:

$$dQ_{\text{rev}} = \left(\frac{\partial U}{\partial T} + P\frac{\partial V}{\partial T}\right) dT + \left(\frac{\partial U}{\partial P} + P\frac{\partial V}{\partial P}\right) dP. \tag{3.15}$$

Dividing by T yields

$$dS = \frac{dQ_{\text{rev}}}{T} = \frac{1}{T}\left(\frac{\partial U}{\partial T} + P\frac{\partial V}{\partial T}\right) dT + \frac{1}{T}\left(\frac{\partial U}{\partial P} + P\frac{\partial V}{\partial P}\right) dP. \tag{3.16}$$

Each of the foregoing equations is in the form

$$dZ = M(x, y)\, dx + N(x, y)\, dy .$$ (3.17)

Obviously, Eq. (3.13) and (3.14) are exact because

$$\frac{\partial^2 U}{\partial T\, \partial P} = \frac{\partial^2 U}{\partial P\, \partial T}$$

and

$$\frac{\partial^2 V}{\partial T\, \partial P} = \frac{\partial^2 V}{\partial P\, \partial T} .$$

Equation (3.15) is not exact because

$$\frac{\partial^2 U}{\partial T\, \partial P} + \frac{\partial V}{\partial T} + P \frac{\partial^2 V}{\partial T\, \partial P} \neq \frac{\partial^2 U}{\partial P\, \partial T} + P \frac{\partial^2 V}{\partial P\, \partial T}$$ (3.18)

and the second term on the left is always finite.

Equation (3.16) is an exact differential, although considerable manipulation is required before this is apparent. First we test to see if $\partial M/\partial y = \partial N/\partial x$:

$$\frac{1}{T}\left(\frac{\partial^2 U}{\partial T\, \partial P}\right) + \frac{1}{T}\left(\frac{\partial V}{\partial T}\right) + \frac{P}{T}\frac{\partial^2 V}{\partial T\, \partial P}$$

$$= \frac{1}{T}\frac{\partial^2 U}{\partial P\, \partial T} - \frac{1}{T^2}\left(\frac{\partial U}{\partial P}\right) - \frac{P}{T^2}\left(\frac{\partial V}{\partial P}\right) + \frac{P}{T}\left(\frac{\partial^2 V}{\partial P\, \partial T}\right) .$$ (3.19)

Canceling and multiplying by $-T^2$, we get

$$-T\left(\frac{\partial V}{\partial T}\right)_P = \left(\frac{\partial U}{\partial P}\right)_T + P\left(\frac{\partial V}{\partial P}\right)_T .$$ (3.20)

It is not obvious that these are equal. Rather than manipulate further, we merely note that both the left side and the right side can be shown to be equal to the latent heat at constant temperature, $L_p \equiv (dQ/dp)_T$. With this result we have

$$L_p = L_p$$ (3.21)

and entropy is an exact differential, as was true based on the reasoning of Clausius from macrothermodynamic considerations.

3.3 RECIPROCAL TEMPERATURE AS AN INTEGRATING FACTOR IN TWO VARIABLES

In previous sections we learned that we could multiply dQ_{rev} by a factor, $1/T$, and convert it from an inexact differential to an exact differential, dS. The factor $1/T$ in mathematical terms is called an integrating factor because it converts an inexact differential, which may be difficult to integrate, into an exact differential, which can readily be integrated depending only on the endpoints. In principle, every first-order linear equation in two independent variables can be integrated by use of an integrating factor. Sometimes it can be integrated immediately without multiplying through by an integrating factor, although we can consider the integrating factor to be equal to 1 in that case.

When any suitably differentiable function of two variables is found, $Z(x, y)$, there must be an equation[3]

$$dZ = \frac{\partial Z}{\partial x}\, dx + \frac{\partial Z}{\partial y}\, dy\,. \tag{3.22}$$

If the equation

$$M(x, y)\, dx + N(x, y)\, dy = 0$$

is exact, it can be rewritten as

$$\frac{\partial Z}{\partial x}\, dx + \frac{\partial Z}{\partial y}\, dy = dZ = 0\,. \tag{3.23}$$

Then $Z(x, y) = k$, a constant, is a solution for all values of k, and

$$\int_a^x M(x, y)\, dx + \int_b^y N(x, y)\, dy = k \tag{3.24}$$

is a solution to the equation. Since this is true for all values of k, we have a family of solution curves for Eq. (3.23). (When we take the integral, the values for $x = a$ and $y = b$ at the lower limit serve to modify k so that we have a solution in x, y and the modified constant k.)

The integral curves of Eq. (3.24) can be expressed as a function of the k's by writing them in the form $y = Z(x, k)$. Then we can solve for k in terms of a parameter u through

$$u(x, y) = k\,. \tag{3.25}$$

This is an implicit form of the family of integral curves. Therefore, along any integral curve,

$$du = \frac{\partial u}{\partial x}\, dx + \frac{\partial u}{\partial y}\, dy = 0 \tag{3.26}$$

and

$$M\, dx + N\, dy = 0. \tag{3.27}$$

Then

$$\frac{dy}{dx} = \frac{-M}{N} = \frac{-\partial u/\partial x}{\partial u/\partial y}, \tag{3.28}$$

so that

$$\frac{\partial u/\partial x}{M} = \frac{\partial u/\partial y}{N}. \tag{3.29}$$

If we set the ratios in Eq. (3.29) equal to u, then substituting $\partial u/\partial x = uM$ and $\partial u/\partial y = uN$ in Eq. (3.26) gives

$$du = uM\, dx + uN\, dy = 0 \tag{3.30}$$

or

$$u(M\, dx + N\, dy) = 0. \tag{3.31}$$

Therefore, u is an integrating factor. This means that for any function like Eq. (3.22) written in the general form

$$dZ = M\, dx + N\, dy,$$

there is a family of curves for which $uM\, dx + uN\, dy = 0$. One of the curves passes through any point P in the x, y plane and has a line element dx and dy perpendicular to a vector of components uM and uN. Because Eq. (3.31) is integrable, only one such solution curve passes through any such point $P = Z(x, y)$. Thus the integration factor for a function of two variables produces a unique solution. As we have previously learned that $1/T$ serves that purpose for dQ_{rev} to produce a unique solution in dS, thus S is a state function.

3.4 RECIPROCAL TEMPERATURE AS AN INTEGRATING FACTOR WITH MORE THAN TWO VARIABLES

When a heat differential dQ is a function of more than two variables, the methods of Section 3.3 are not sufficient because in three-dimensional space, there is not a unique solution to the linear equation[4]

$$df = M(x, y, z)\, dx + N(x, y, z)\, dy + L(x, y, z)\, dz \ . \qquad (3.32)$$

No integrating factor exists such that

$$df = uM\, dx + uN\, dy + uL\, dz = 0 \ . \qquad (3.33)$$

Therefore, although a vector $uM + uN + uL$ will define a point P in x, y, z space, an infinite number of curves can be drawn perpendicular to that vector. These are not unique solution curves. The mathematical consequences of this always apply to differential forms of more than two variables.

Yet we know that regardless of the number of independent variables involved, entropy is a unique state function and dQ_{rev} is rendered exact when divided by T. This was studied by Carathéodory to determine, in a mathematical sense, what is unique about reciprocal temperature as an integrating factor. He was able to show that division by T rendered dQ_{rev} an exact differential, regardless of the independent variables involved. Since this is not generally necessary for a differential of three or more independent variables, he concluded that (for three variables) points in x, y, z space in the neighborhood of P_0 are not accessible from solution curves for the equation,

$$M\, dx + N\, dy + L\, dz = 0 \ , \qquad (3.34)$$

if, and only if, the equation is integrable. Thus the requirement that Eq. (3.34) be integrable is what establishes entropy as a unique state function with only one solution curve through a given point $P_0 = S$. Then for thermodynamic applications we can write the first law as

$$dQ_{rev} = M'\, dx + N'\, dy + L'\, dz = T\, dS \ . \qquad (3.35)$$

The fact that entropy is exact, that every state has only one value of entropy, and that $1/T$ converts the inexact differential dQ to the exact differential dS is especially important. We find that reciprocal temperature as an integrating factor for differentials of more than two variables is not required in a mathematical sense, but it is a special result of natural law. Note that this mathematical relationship can be shifted by an arbitrary constant, so that entropy is really defined within the limits of that constant. We arbitrarily take $S = 0$ at absolute zero for thermodynamic data tabulations.

3.5 QUALITATIVE RELATIONS BETWEEN MACRO CONCEPTS AND MICRO CONCEPTS FOR ENTROPY

We have just learned that entropy is a thermodynamic state function. Therefore, any change in state may have associated with it an entropy change for which energy is required. For a perfect reversible process the heat required for that energy change $Q_{rev} = \int_1^2 T\, dS$. However, if the change is not performed in a reversible way, more heat than the theoretical minimum will be required; $Q > \int_1^2 T\, dS$

If we examine the effect of heat delivered to a real material, say a working fluid like steam, the heat will bring about an enthalpy change, $\Delta H = Q$. Some of this heat will be used to change the entropy of the fluid. J. Willard Gibbs proposed that the enthalpy could be partitioned, part of it to change the entropy and part of it to be available for useful work.[5] For an isothermal process he wrote, in present-day notation,

$$\Delta H = \Delta G + T\, \Delta S . \tag{3.36}$$

Then the Gibbs free energy G is that portion of the enthalpy which can be used for useful work. Since H, T, and S are exact, so is G. The $T\, \Delta S$ term in Eq. (3.36) represents that portion of the enthalpy change which is unavailable for use. This is the reason that entropy is sometimes called the index of the unavailability of energy.

Writing eq. (3.36) as

$$\Delta G = \Delta H - T\, \Delta S \tag{3.37}$$

emphasizes the importance of the $T\, \Delta S$ term. When ΔG is negative, a reaction can be spontaneous. When $\Delta G = 0$ an equilibrium exists, and all ΔH goes to change ΔS. And when ΔG is positive, the reverse reaction should occur.

For an isothermal reaction, such as the melting of ice,

$$\Delta G = 0 = \Delta H - T\, \Delta S .$$

Therefore,

$$\Delta S = \frac{\Delta H_{mp}}{T_{mp}} , \tag{3.38}$$

where ΔH_{mp} is the heat of fusion and T is the absolute temperature of the melting point. The melting of ice required delivery of energy to the system. The highly ordered crystalline ice broke down, on melting, to a disordered liquid state. When examining many such examples the change from order to disorder associated with entropy increase appears to be generally true.

Therefore, the change $\Delta S = \Delta H / T_{mp}$ which causes a large increase in entropy (because ΔH is large) is associated with an increase in the degree of randomness of the system. This concept of an increase of the microscopic state of disorder with an increase in entropy is helpful in understanding why the heat absorbed in a state change (such as expanding a gas at constant pressure by supplying heat) can be considered to be the sum of free energy and entropy energy. The latter, $\int_1^2 T\,dS$, is that portion of the heat that is consumed in changing the randomness of the system. Now we see that the macroscopic analysis of energy can be related to microscopic changes in the system. We now have entropy qualitatively related to the degree of disorder or randomness of the system, defined again within the limits of an arbitrary constant.

3.6 BACKGROUND FOR THE QUANTITATIVE RELATIONSHIP OF MICROSCOPIC TO MACROSCOPIC THERMODYNAMICS

The kinetic theory of heat was the basis for the development of the quantitative mathematical relationships between microscopic and macroscopic properties. That theory had its origins in the sixteenth century with Francis Bacon, who wrote:

> When I say of motion that it is the genus of which heat is a species I would be understood to mean, not that heat generates motion or that motion generates heat (although both are true in certain cases) but that itself, its essence and quiddity, is motion and nothing else.... Heat is a motion of expansion, not uniformly of the whole body together, but in the smaller parts of it... the body acquires a motion alternative, perpetually quivering, striving and struggling, and initiated by reprercussion, whence springs the fury of fire and heat.

The development in the late nineteenth century of Boltzmann, Maxwell, Clausius, and others put the kinetic theory, which holds that temperature and pressure are manifestations of molecular motion, on a strong physical base.[6]

The final development of thermal physics was made possible by the appearance, at the beginning of this century, of Planck's theory of quantum energy in thermal radiation (that thermal radiant energy has certain allowed, quantitized energy states) which appeared in 1901; and by elegant theoretical derivations based on statistical mechanics by J. Willard Gibbs in his famous treatise, "Elementary Principles in Statistical Mechanics, Developed with Especial Reference to the Rational Foundation of Thermodynamics," published in 1902 by the Yale University Press.[7]

3.7   BASIC STATISTICAL MECHANIC PREMISES

The statistical mechanics analysis proposed by Gibbs gave better under-standing for the macrothermodynamics of Carnot and Kelvin, and gave additional insight and new meaning to the concept of entropy. From a statistical mechanic point of view it is entropy that is the fundamental independent variable on which many thermodynamic properties depend.[8]

The theory of statistical mechanics is far too complex to be included here. Rather, we give a qualitative explanation of the methods used, followed by presentation of the quantitative results, which pertain to the subject of temperature. We hope that the reader will be motivated to study further and have included recommended readings at the end of the chapter.

Statistical mechanics can be explained more readily from the viewpoint of quantum mechanics than from the viewpoint of classical mechanics. A model system is often useful in the explanation. Usually, this is an isolated system (a closed system in the terminology of thermal physics—where the system has constant energy, constant volume, and contains a constant number of particles). The particles in the system are assumed to be able to reach certain accessible states. Each state has a certain energy level, and the levels are quantitized, that is, they are separated one from another by discrete increments of energy.* Then statistical mechanics is essentially a statistical analysis of the consequences of distributing the particles in the energy states available to them.

The statistical analysis is based on two important concepts. One is that the probability of finding a system in a given state is given by observation of the state at many times, divided by the number of observations, that is,

$$P(l) = \frac{n(l)}{q} \tag{3.39}$$

where $P(l)$ is the probability the system will be in state l, $n(l)$ is the number of times it actually was in state l when observed a large number of times, and q is the total number of observations. The other concept is that if we imagine many systems, selected in a random way over all the accessible states, we will get the same result (probability) as we would if we observed the system microscopically over many intervals of time. When many similar systems are imagined in that way, they are described as an ensemble of systems. This concept of statistical mechanics was introduced by G. W. Gibbs and is an application of the ideas of the French mathematician Liouville. This contribution of Gibbs makes it possible to imagine many systems, constructed alike, rather than to attempt to deal with a time average. This method makes it possible to study the statistics of the ensemble of systems—something that is theoretically attainable—instead of

---

* This is a convenient but not essential feature of the derivation.

a time average, which is not actually attainable on either a theoretical or an experimental basis. When we measure a property it is the result of the time average of the distribution of molecules among their accessible states, but we cannot actually study the distribution of molecules among states because we cannot "stop the action" to determine what the distribution is. All we can do is measure the time-average property. By visualizing an ensemble of identical systems randomly, Gibbs made it possible for us to study theoretically the distribution of molecules among the various states—and thus to predict the time average, which can then be compared with experiment.

The simplest model system that can be visualized is one in which only two accessible states are available to the molecules. The distribution of molecules among accessible states is first approached as a purely random phenomena.

An example of this, lucidly presented by Kittel,[8] is the distribution of molecules having a magnetic orientation that can be either up or down. (For simplicity, only the Z direction is considered.) If no interaction between molecules occurs, each molecule has the possibility of spin-up or spin-down orientation. Only a limited number of spin-up or spin-down arrangements are possible for a given number of molecules. For example, for only two molecules, the arrangements could be both with spin up, both with spin down, or one spin up and the other spin down. Each of these arrangements is an accessible state for the system. For the two molecules there is only one way to produce both spins up; there is only one way to produce both spins down; but there are two ways to produce one spin up and one spin down, depending on whether the first molecule has spin up or spin down. Thus the number of molecules and the orientations possible provide a generating function that describes the number of accessible states in the system. For N molecules, considering only two possible orientations, the number of possible arrangements, that is, the number of accessible states, is 2^N. When there is more than one way to produce a given state, as for one spin up and one spin down, the energy level associated with that state is said to be degenerate. Our example has a degeneracy of two. Real systems with millions of molecules have enormously large values of degeneracy.

If we associate a magnetic moment u with spin up and $-u$ with spin down, summing the moments in each possible arrangement will give a total moment for each such arrangement. There are only $N+1$ different total moments possible. These moments are identified with each possible accessible state. Because we assumed no interaction between molecules, each of the accessible states has the same energy level. Nevertheless, because of the number of different ways the states can be arranged, the probability of a particular state, of a particular total moment, is not the same for each state. In our example there is one way in four to make both spins up, one way in four to make both spins down, but two ways in four to make one spin up and one spin down. The probability of a total moment of $2u$ is 0.25, of $-2u$ is 0.25, and of zero is 0.5. If the number of molecules is increased, the probability of the 50:50 arrangement increases enormously, to the point where finding that state approaches a certainty.

Calculation of the number of states possible depends on the model. For Kittel's model with magnetic spins up or down, as the number of molecules in the model changes, the number of states increases as 2^N. For two molecules the total is four, for three it is eight, etc. Each of these is considered a microstate of equal probability. However, the total moment associated with the states has only $N+1$ possible values, because of degeneracy. The two molecules had only three total moment values, $m = -2u$, 0, and $2u$. With three molecules the eight states could be distributed between four total moment values, $-3u$, $-2u$, $2u$, and $3u$. With 10 molecules, there would be $2^{10} = 1024$ states distributed over 11 values. As the number of molecules becomes very large, the number of states and the number of values become enormous, but the number of states per value increases so much more rapidly that with no interaction between molecules, the 50:50 probability is so high that it amounts to a certainty. Only very small deviations are possible. For that reason it is advantageous to count the moments deviating from the expected value, $N/2$.

The total moments, m, of each state in the two-molecule model each had the same energy level. The number of molecules in each state depended only on the number of molecules and their spin. The number of molecules in each state depended on probability alone and was given by a degeneracy function, $g(N, m)$. If we add all the molecules in each state, we get the total. This is the summation of all the molecules in each state. The total number for Kittel's model is

$$(1 + 1)^N = 2^N = \sum_{m=-N/2}^{m=N/2} g(N, m) . \tag{3.40}$$

The value of $g(N, m)$ is a binomial coefficient from the familiar binomial theorem expansion. It can be evaluated for m equal to any integer when N is even, or any half integer when N is odd. Tables of binomial coefficients are readily available.

Kittel goes on with the argument that it is easier to analyze the deviations from the most probable state and shows that for a large number of molecules, the distribution is very sharp around the most probable state where $m = 0$ because half have spin up and half have spin down.

When a magnetic field is applied, the total moment of each state reacts with the fields and the energy level is not the same for each state. The energy of interaction of each state is given by the scalar product of the magnetic field and its total moment. The molecules orient in response to the field and their potential energy is changed. Therefore, models such as Kittel's can be compared with actual properties, and the statistics verified.

The occupancy of the energy states depends on the degeneracy function. The degeneracy function depends on the number of different ways of arranging N items among the N_0 sites,

$$g(N_0, N) = \frac{N_0!}{(N - N_0)! \, N!}.$$ (3.41)

The probability of finding a particular state, then, if the states are determined at random, is given by the binomial coefficient from Eq. (3.41) for that particular state, divided by the total number of states. The state that is the most probable will thus be the one with the most possible number of ways of achieving it, that is, the state with the highest value of degeneracy. When N is large this turns out to be an extremely high probability. For example, the possibility of having all gas molecules on the left side of a container at the same time is effectively zero. In fact, when N is very large, as it is in real systems, the probability of even a very small departure from the most probable state becomes negligible. Because of this, any property of a particular closed system, such as the specific volume of a gas, always has a constant value (the time-average value, which is the same as the probability average value). It is the most probable value for the conditions imposed (temperature, pressure, volume).

The result of this discussion is that, if N is large, a system can always be represented by statistical treatment, for a given group of accessible states, as being in its most probable energy state.

3.8 ENTROPY FROM A STATISTICAL-MECHANICAL VIEWPOINT

As explained above, the total number of accessible states for a system can be computed from the degeneracy function. For the two-state systems given previously, using ε instead of m to represent the energy states, it is

$$g(N, \epsilon) = \sum_{r=0}^{N} \frac{N!}{(N - r)! \, r!}.$$ (3.42)

The logarithm of this function is defined as the physical entropy, σ:

$$\sigma \equiv \log g(N, \varepsilon).$$ (3.43)

The physical entropy is a very simple function. It is related to the conventional macrothermodynamic entropy by

$$k\sigma = S,$$ (3.44)

where k is Boltzmann's constant, so that

$$S = k \log g(N, \varepsilon) . \tag{3.45}$$

Thus we see that the entropy of a system is merely a function of the total number of accessible states. So the greater the number of states (i.e., the greater the disorder), the greater will be the entropy.

3.9   TEMPERATURE FROM A STATISTICAL-MECHANICAL VIEWPOINT

Even from a statistical-mechanical point of view, temperature is difficult to define. Two approaches are generally used. One is to make a mathematical analysis of the distribution of energy among oscillators. The resulting equations can contain a Lagrange multiplier which, on the basis of the fact that $1/T$ is an integrating factor for dQ, is set equal to $1/kT$. The other approach is to make a theoretical analysis of the exchange of energy between two parts of a closed system. When this is done, the exchange of energy will depend on the statistical properties of the two systems, and will stop if

$$\left(\frac{\partial \sigma}{\partial U}\right)_{N_1} = \left(\frac{\partial \sigma}{\partial U}\right)_{N_2} . \tag{3.46}$$

But from experiment, the second law of thermodynamics tells us that exchange of energy stops when the two parts of the system have the same temperature. So a physical temperature τ^* is defined as

$$\frac{1}{\tau^*} = \left(\frac{\partial \sigma}{\partial U}\right)_{N} . \tag{3.47}$$

Then

$$\tau^* = kT . \tag{3.48}$$

Thus we see that temperature is not derived as a fundamental quantity. Rather, it is developed as an indication of the behavior observed in experiment. The statistical-mechanical analysis of thermal equilibrium does give additional insight into the relationships of heat, entropy, and temperature. A brief derivation is presented here to provide that insight, even though it does not add much in defining temperature.

3.10   ENERGY EXCHANGE BETWEEN TWO BODIES

When two bodies are held in a closed system, with body 1 much larger than body 2, we can represent the available states in each by a degeneracy function for each:

$$g_1(N_1,\ U_1) \quad \text{and} \quad g_2(N_2,\ U_2)\,, \tag{3.49}$$

where the N's are the number of molecules in bodies 1 and 2, and the U's are their internal energies. In a closed system $N_1 + N_2 = N$, so $N_2 = N - N_1$. Then the total number of accessible states of the combined system is given by the product of g_1 and g_2 because for any state in body 1 we could have all of the states in body 2. For the total system,

$$g(N,\ U) = g_1(N_1\,U_1)g_2(N_2,\ U_2)\,. \tag{3.50}$$

If we now let our two bodies come into thermal contact and exchange energy but not particles, energy can flow from one to the other. If this happens, the degeneracy function for the total system will become

$$g(N,\ U) = \sum_1 g_1(N_1\,U_1)g_2(N_2,\ U - U_1)\,, \tag{3.51}$$

where the sum is over all values of U_1 up to a maximum of U. The maximum value of $g(N,\ U)$ will be the most probably configuration and we wish to find out what it is. This we can do by setting the differential of g to zero:

$$dg = g_2\left(\frac{\partial g_1}{\partial U_1}\right)_{N_1} dU_1 + g_1\left(\frac{\partial g_2}{\partial U_2}\right)_{N_2} dU_2 = 0\,. \tag{3.52}$$

We know that $dU_1 = -dU_2$ because total energy is constant, so

$$\frac{1}{g_1}\left(\frac{\partial g_1}{\partial U_1}\right)_{N_1} = \frac{1}{g_2}\left(\frac{\partial g_2}{\partial U_2}\right)_{N_2}. \tag{3.53}$$

This we recognize as the partial derivatives of the logarithms of the respective degeneracy functions:

$$\left(\frac{\partial \ln g_1}{\partial U_1}\right)_{N_1} = \left(\frac{\partial \ln g_2}{\partial U_2}\right)_{N_2}. \tag{3.54}$$

Since we defined physical entropy σ as $\ln g$, we have

$$\left(\frac{\partial \sigma_1}{\partial U_1}\right)_{N_1} = \left(\frac{\partial \sigma_2}{\partial U_2}\right)_{N_2}. \tag{3.55}$$

Before the two bodies were allowed to exchange heat, the physical entropy was given by

$$\sigma = \ln{(g_1 g_2)} = \ln{g_1} + \ln{g_2} = \sigma_1 + \sigma_2. \tag{3.56}$$

In other words, the entropies were additive. When the two bodies are not at the same temperature, energy will flow from the higher-temperature body to the lower-temperature body until they come to the same temperature. Then the degeneracy function of each body will reach new values of $(g_1 g_2)$ as the energy is exchanged. One of the important conclusions of thermal physics is that the exchange will result in a larger value for the product $(g_1 g_2)$ so that $\ln{(g_1 g_2)}$ is greater than before. The entropies σ_1 and σ_2 are additive, but the sum is greater than that before thermal contact. Before contact, U_1 and U_2 had restrictions that were removed when contact was established, being replaced with a single restriction $U = U_1 + U_2$ for the system in contact. This makes a larger number of configurations possible. Only if the two bodies were at the same temperature would $(g_1 g_2) = (g_1 g_2)$ and the flow of heat stop. This occurs when the rate of change of entropy with energy is the same for both bodies in accordance with Eq. (3.55) (when the product of the degeneracy functions has reached its maximum and most probable value).

When no more heat will flow from one body to the next, experience tells us that they are at the same temperature. Therefore, we can define temperature in terms of Eq. (3.55) as

$$\frac{1}{\tau^*} \equiv \frac{1}{kT} \equiv \left(\frac{\partial \sigma}{\partial U}\right)_N, \tag{3.57}$$

where the subscript N indicates an unchanged total number of atoms or molecules. Here τ^* is physical temperature, which equals the product of Boltzmann's constant and absolute temperature. This equation defines temperature in terms of the entropy–energy relationship. From a statistical-mechanical point of view it is the entropy–energy relationship that is fundamental, and the thermal physics temperature τ^* has units of energy. Note that the use of $1/\tau^*$ instead of τ^* ensures that heat will flow from the high temperature to the low temperature.

There are systems in which negative absolute temperatures appear to occur.[9] Such systems are not normal systems; that is, they are not true thermal equilibrium systems. When such a system is found, if sufficient time is allowed for thermal equilibrium to be established, normal thermodynamic behavior then occurs. An example of a system with apparently negative thermodynamic temperature is found in a laser. Without excitation the population of electrons in their energy levels is in accordance with semi-conductor thermal equilibria. In some materials, the electronic states are not coupled directly with normal lattice vibrations. Then, if electrons in a lower-energy state are excited to a higher allowed energy state (e.g., by excitation with intense white light), it may be possible to transfer some or all the electrons to the higher-energy state. This *lowers* the entropy and requires a negative temperature in Eq. (3.57). Such a laser has an inverted population of energy states. The lasing action results from this population inversion. When properly stimulated, the electrons fall back to a lower allowed state and emit photons characteristic of the energy differences between the upper and lower states. Negative temperatures are then very useful in describing the system and then appear to be hotter than positive temperatures. Energy flows spontaneously from the higher-energy state (more negative temperature) to the lower-energy state.

In closing, we note that temperature could not be defined by statistical-mechanical considerations without reverting to experimental observation. Thus our original definition of temperature as the level of thermal energy is qualitatively satisfactory.

REFERENCES

1. R. Clausius, Ober die begende Krofte der Warme und die Gesetze, avelche sich darous fur die Warmelehre selbst ableiten lassen. Ann. *Phys.* (*Leipzig*) [3] **79**, 388, 500 (1850).
2. F. Crawford, "*Heat, Thermodynamics, and Statistical Physics*." Harcourt, Brace World, New York, 1963. (See Chapter 9 for an excellent discussion of entropy and the second law)
3. M. Golomb and M. E. Shanks, "*Elements of Ordinary Differential Equations*," pp. 52–53. McGraw-Hill, New York, 1965.
4. C. Carathéodory, *Math. Ann.* **67**, 335 (1909), see Crawford,[2] pp. 257–259.
5. E. A. Guggenheim, "Modern Thermodynamics by the Methods of Williard Gibbs." Methuen, London, 1933.
6. E. H. Kennard, "Kinetic Theory of Gases." McGraw-Hill, New York, 1938.
7. J. Willard Gibbs, "*Elementary Principles in Statistical Mechanics, Developed with Especial Reference to the Rational Foundation of Thermodynamics*." Yale Univ. Press, New Haven, Connecticut, 1902.
8. C. Kittel, "Thermal Physics." Wiley, New York, 1969.
9. G. N. Hatsopoulos and J. H. Keenan, "Principles of General Thermodynamics," Wiley, New York, 1965, pp. 397–409.

4

THE INTERNATIONAL PRACTICAL TEMPERATURE SCALE

4.1 HISTORICAL BACKGROUND

Prior to World War I it was very difficult to compare scientific results in one laboratory with those from another laboratory because there was no standard temperature scale. Therefore, a property measurement in this country might not agree with the same property measurement in another country, because the two laboratories could not produce the same desired temperature of measurement. Just before World War I the National Physical Laboratory in England, the National Bureau of Standards in the United States, and the Physikalisch-Technische Reichsanstalt in Germany began cooperation to attempt to establish a common temperature scale. However, the onset of hostilities interrupted the effort.[1]

After World War I, efforts at cooperation were more successful. The International Committee of Weights and Measures coordinated the effort and at a meeting in Paris in 1927, the International Conference of Weights and Measures approved the establishment of an International Temperature Scale (ITS).[2] In 1960 the name of the scale was changed to the International Practical Temperature Scale (IPTS). The International Committee on Weights and Measures has a subcommittee, the Consultative Committee on Thermometry, which is responsible for continuing improvements and revisions in the temperature scale. Revisions of the scale were made in 1948, 1954, 1960, and 1968. The revisions in 1948 and 1968 were major revisions. The text of the 1968 conference was revised in 1975 to clarify meaning and to remove or replace instructions that no longer appear suitable. A new ITS is to take place in 1990. See Section 4.7.

Six international conferences have resulted from worldwide interest in improved temperature measurement and control. The proceedings of the last five were published shortly after the year of the conference as: "Temperature, Its Measurement and Control in Science and Industry" and are dated 1949, 1955, 1962, 1972, and 1982.[3-7] In April 1975, another international conference, the European Conference on Temperature Measurement, was held at the National Physical Laboratory of England. Its proceedings were published as: "Temperature Measurement, 1975" by the Institute of Physics, London, 1975.[8] These conference proceedings allow one to trace the development of improved methods and the evolution of the current practices.[9]

The most recent, the Sixth International Conference, was held March 15–18 1982 in Washington, D.C.[7] It was cosponsored by the American Institute of Physics, the Instrument Society of America, and the National Bureau of Standards. Meetings of the CCT during and after the conference provided useful discussions which were expected by some to provide changes in IPTS by 1987.[10] These will be discussed after the current IPTS is presented.

At the Sixth International Conference, the keynote address, given by R. P. Hudson, presented a brief review of the history of IPTS, gave an analysis of the attributes and deficiencies of the existing IPTS, and reviewed research developments that might lead to modifications in the near future.[11] In that paper he gives four necessary features needed by a temperature standard scale: (1) definition, (2) realization, (3) transfer, and (4) utilization.

The modes of defining and realizing the IPTS are described in Section 4.2. Standard sensing devices are used to define the IPTS. One of the important requirements of such a scale is that the measurement be unique. Anyone using a method specified by the scale should get exactly the same temperature as anyone else would if both are measuring the same temperature with the same specified method. In practice, some difference will always exist, so such a scale is nonunique, even though the differences will be very small with the present-day IPTS. It is necessary to specify methods to define and realize the scale to minimize the "nonuniqueness" of the scale.

In a given application it is often not feasible to use the IPTS instruments for the actual measurement. This means that it is necessary to transfer the measurement from the IPTS method to another, practical instrument for the measurement. Such transfers always reduce accuracy in determining the thermodynamic temperature as represented by IPTS. Often, then, transfer is required to allow utilization of the scale. IPTS must be more precisely defined as a standard than most of the applications require, in part to allow for transfer before utilization. Fortunately, the present IPTS is suitable for all but the most unusual, exacting applications.

4.2 BASIS FOR THE INTERNATIONAL PRACTICAL TEMPERATURE SCALE

The IPTS is a standard scale that is intended to be the best approximation of the thermodynamic temperature scale to which the various countries of the world can agree. Since there is a political element in worldwide agreements the IPTS may not be the scale that actually deviates the least from the thermodynamic scale. But even so, temperatures on the IPTS can be established and utilized anywhere in the world.

The IPTS has three components:

1. Fixed points based on thermodynamic invariant points. These are highly reproducible phase transitions (e.g., freezing points, triple points, etc.) of specific systems. The temperatures of the fixed points are established as accurately as possible on the thermodynamic temperature scale by international cooperation and agreement. They are classified as defining and secondary, depending on the precision to which they are known.
2. The instruments to be used in interpolating temperatures between the defining fixed points.
3. The equations to be used to calculate the intermediate temperatures (i.e., the interpolating equations). These equations pass through the defining fixed points.

4.3 EVOLUTION OF THE INTERNATIONAL PRACTICAL TEMPERATURE SCALE

The first IPTS scale had six fixed points, required three instruments for measuring temperature in four different temperature ranges, and specified an interpolating equation for each of the four ranges. The ranges were:

Range I: the oxygen point to the ice point (−182.97 to 0°C)
Range II: the ice point to the aluminum point (0 to 660°C)
Range III: the aluminum point to the gold point (660 to 1063.0°C)
Range IV: above the gold point (1063.0°C and higher)

The platinum resistance thermometer was used for ranges I and II, the platinum–platinum 90 rhodium 10 thermocouple was used for range III, and the optical pyrometer was used for range IV. These instruments are still used, although the interpolating equations and the limits of some of the ranges have been modified.

The first IPTS-27 was based on the ice point and the steam point as true temperatures on the thermodynamic scale, and the size of the degree was taken as 1/100 of the difference between the steam point and the ice point. In 1954 a major change in philosophy was made: to use the triple point of water and absolute zero as the two points defining the thermodynamic temperature scale. This is the method originally preferred by Lord Kelvin when he proposed the thermodynamic scale. The method made cryogenic temperature measurement easier, although it reduced the accuracy to which the size of the degree was known.

Since major revisions were made in 1948 and 1968, and a new scale takes effect in 1990, one must be careful in using instruments calibrated years ago. Instruments still exist that were calibrated before 1968. The 1968 IPTS is still in use, although changes in the interpolating equations were adopted since 1968.[12] All such changes have been made to improve agreement with the thermodynamic temperature scale or to improve the calculating method. An important amendment took place in 1975. As a part of that amendment the kelvin was adopted as the standard temperature unit, replacing Celsius. The amended introduction to the IPTS-68 reads:

The unit of fundamental physical quantity known as the thermodynamic temperature, symbol T, is the kelvin, symbol K, defined as 1/273.16 of the thermodynamic temperature of the triple point of water.

For historic reasons, connected with the way temperature scales were originally defined, it is common practice to express temperature in terms of a thermal state 0.01 kelvin lower than the triple point of water. A thermodynamic temperature expressed in this way is known as Celsius temperature, symbol t, defined by

$$t = T - 273.15 \text{ K}.$$

The unit of Celsius temperature is the degree Celsius, symbol °C, which is, by definition, equal in magnitude to the kelvin. A difference in temperature may be expressed in kelvin or in degrees Celsius.

The International Practical Temperature Scale of 1968 (IPTS-68) has been constructed in such a way that any temperature measured on it is a close approximation to the numerically corresponding thermodynamic temperature. Moreover such measurements are easily made and highly reproducible; in contrast, direct measurements of thermodynamic temperatures are both difficult to make and imprecise.

The IPTS uses both International Practical Kelvin Temperatures, symbol T_{68}, and International Practical Celsius Temperatures, symbol t_{68}. The relationship between T_{68} and t_{68} is the same as between T and t; in other words,

$$t_{68} = T_{68} - 273.15 \text{ K}.$$

The foregoing statement makes it clear that the fundamental scale is the absolute thermodynamic scale, and that the Celsius scale is just a shift of the zero point.

4.4 THE IPTS-68 AS AMENDED

The IPTS-68 was amended in 1975. The defining points and secondary points are given in Tables 4.1 and 4.2. Table 4.2 was prepared by working group 2 of the CCT and was approved for publication in June 1984 as an improvement over the previous list.[13] The group used the same basic

(text continued on page 77)

TABLE 4.1 Defining Fixed Points of the IPTS-68[a]

Equilibrium state	Assigned Value of International Practical Temperature	
	T_{68} (K)	t_{68} (°C)
Equilibrium between the solid, liquid, and vapor phases of equilibrium hydrogen (triple point of equilibrium hydrogen)	13.81	−259.34
Equilibrium between the liquid and vapor phases of equilibrium hydrogen at a pressure of 33330.6 Pa (25/76 standard atmosphere)	17.042	−256.108
Equilibrium between the liquid and vapor phases of equilibrium hydrogen (boiling point of equilibrium hydrogen)	20.28	−252.87
Equilibrium between the liquid and vapor phases of neon (boiling point of neon)	27.102	−246.048
Equilibrium between the solid, liquid, and vapor phases of oxygen (triple point of oxygen)	54.361	−218.789
Equilibrium between the solid, liquid, and vapor phases of argon (triple point of argon)	83.798	−189.352
Equilibrium between the liquid and vapor phases of oxygen (dew point of oxygen)	90.188	−182.962
Equilibrium between the solid, liquid, and vapor phases of water (triple point of water)	273.16	0.01
Equilibrium between the liquid and vapor phases of water (boiling point of water)	373.15	100
Equilibrium between the solid and liquid phases of tin (freezing point of tin)	505.1181	231.9681
Equilibrium between the solid and liquid phases of zinc (freezing point of zinc)	692.73	419.58
Equilibrium between the solid and liquid phases of silver (freezing point of silver)	1235.08	961.93
Equilibrium between the solid and liquid phases of gold (freezing point of gold)	1337.58	1064.43

[a]Except for the triple points and one equilibrium hydrogen point (17.042 K) the assigned values of temperature are for equilibrium states at a pressure p_0 = 1 standard atmosphere (101.325 Pa).

TABLE 4.2 Recommended Values of Temperature for a Selected Set of Secondary Reference Points (SPP). Part A: First-Quality Points

Equilibrium state[a]	T_{68} (K)	t_{68} (°C)	Uncertainty[b] (K)	Purity of material[c] (wt % or vol %)	References[d]
Equilibrium between the liquid and vapour phases of equilibrium hydrogen[e]					
$p = p_0 \exp[A + \dfrac{B}{T_{68}} + CT_{68}] + \sum_{i=0}^{5} b_i T_{68}^i$	13.81	−259.34	0.001	99.99	2
$A = 4.030\ 610\ 6$	to	to			
$B = -101.803\ 28\ \text{K}$					
$C = 4.877\ 861\ 1 \times 10^{-2}\ \text{K}^{-1}$					
$b_0 = 984.374\ 89$					
$b_1 = -293.525\ 87\ \text{K}^{-1}$					
$b_2 = 34.778\ 619\ \text{K}^{-2}$					
$b_3 = -2.046\ 488\ 3\ \text{K}^{-3}$					
$b_4 = 5.979\ 981\ 1 \times 10^{-2}\ \text{K}^{-4}$	20.28	−252.87			
$b_5 = 6.941\ 649\ 2 \times 10^{-4}\ \text{K}^{-5}$					
Triple point of neon isotope ^{20}Ne	24.546	−248.604	0.001	99.95[f]	9, 10
Triple point of natural neon[g,ee]	24.5622	−248.5878	0.0004	99.99[f]	7, 8 (93)[h] 10 (11)[b], 12–18
Equilibrium between the liquid and vapour phases of neon[g,i]					
$\ln \dfrac{p}{p_0} = A + \dfrac{B}{T_{68}} + CT_{68} + DT_{68}^2$	24.562	−248.588	0.002	99.99[f]	7, (8, 10)
$A = 11.534\ 31$	to	to			
$B = -252.593\ 8\ \text{K}$					
$C = -1.216\ 916 \times 10^{-1}\ \text{K}^{-1}$	27.102	−246.048			
$D = 1.475\ 653 \times 10^{-3}\ \text{K}^{-2}$					

TABLE 4.2A (Continued)

Equilibrium state[a]	T_{68} (K)	t_{68} (°C)	Uncertainty[b] (K)	Purity of material[c] (wt % or vol %)	References[d]
Triple point of nitrogen[ee]	63.1462	−210.0038	0.0007	99.999	2, 15, 21, 22
Boiling point of nitrogen	77.344	−195.806	0.002	99.999	21
Equilibrium between the liquid and vapour phases of nitrogen					
$\ln\dfrac{p}{p_0} = A + \dfrac{B}{T_{68}} + CT_{68} + DT_{68}^2$	63.146	−210.004	0.002	99.999	21
	to	to			
	77.344	−195.806			
$A = 14.974\ 170$					
$B = -874.467\ 650$ K					
$C = 6.655\ 806 \times 10^{-2}$ K^{-1}					
$D = 2.473\ 924 \times 10^{-4}$ K^{-2}					
Triple point of argon[j,ee]	83.7976	−189.3524	0.0003	99.999	12, 17, 23–30
Boiling point of argon[k]	87.2946	−185.8554	0.0003	99.999	23, 27, 134
Equilibrium between the liquid and vapour phases of argon					
$\ln\dfrac{p}{p_0} = A + \dfrac{B}{T_{68}} + CT_{68}$	83.798	−189.352	0.001	99.999	23
	to	to			
	87.294	−185.856			
$A = 10.593\ 120$					
$B = -864.700\ 28$ K					
$C = -7.876\ 11 \times 10^{-3}$ K^{-1}					
Equilibrium between the liquid and vapour phases of oxygen[i]					
$\ln\dfrac{p}{p_0} = A + \dfrac{B}{T_{68}} + \dfrac{C}{T_{68}^2} + DT_{68} + ET_{68}^2 + FT_{68}^2$ 65[l]		−208.15[l]	0.001	99.999	33 (8)

$$A = 33.935\ 826$$
$$B = -1724.524\ 0\ K$$
$$C = 9846.889\ 7\ K^2$$
$$D = -3.131\ 257\ 6 \times 10^{-1}\ K^{-1}$$
$$E = 1.927\ 509\ 1 \times 10^{-3}\ K^{-2}$$
$$F = -4.720\ 598\ 4 \times 10^{-6}\ K^{-3}$$

	T_{68}		t_{68}		Uncertainty	Purity %	References
Triple point of methane (CH$_4$)^k	90.188	to 90.6854	−182.962	to −182.4646	0.0003	99.995	26, 29, 34–37
Triple point of mercury	234.3083		− 38.8417		0.0005	99.9999	43 (32)
Freezing point of mercury^m	234.3137		− 38.8363		0.0005	99.9999	43 (32)
ICE POINT^n	273.15		0				
Melting point of gallium^{dd}	302.9220		29.7720		0.0002	99.99999	47–50, 136–139
Triple point of gallium^{dd}	302.9240		29.7740		0.0002	99.99999	47, 49–51, 136, 137
Triple point of indium^m	429.7795		156.6295		0.0004	99.9999	56, 140
Freezing point of indium^m	429.7848		156.6348		0.0003	99.9999	55, 56, 141
Freezing point of bismuth^m	544.592		271.442		0.001	99.9999	57
Freezing point of cadmium^m	594.2582		321.1082		0.0003	99.9999	58 (59)^h, 60, 61
Freezing point of lead^m	600.652		327.502		0.001	99.9999	62
Freezing point of antimony^m	903.905		630.755		0.003	99.9999	59, 67
Freezing point of aluminium^{m,o}	933.607		660.457		0.005	99.9999	68, 69
Melting point of the copper/71.9% silver eutectic^p	1,052.70		779.55		0.07	99.999	70, 71, 142
Melting point of the copper/71.9% silver eutectic^o	1,053.03		779.88		0.05	99.999	72. 73
Freezing point of copper	1,358.03		1,084.88		0.010	99.999	74–78
Freezing point of palladium^q	1,828.5		1,555.3		0.2	99.999	94, 95
Freezing point of platinum	2,041.9		1,768.7		0.4	99.99	100–104
Freezing point of rhodium	2,236		1,963		5	99.99	105
Melting point of iridium	2,720		2,447		6	99.99	113
Melting point of molybdenum	2,897		2,624		5	99.9	116–119
Melting point of tungsten	3,693		3,420		10	99.9	121–124

TABLE 4.2 Part B: Second-Quality Points

Equilibrium state[a]	T_{68} (K)	t_{68} (°C)	Uncertainty[b] (K)	Purity of material[c] (wt % or vol %)	References[d]
Triple point of normal hydrogen	13.958	−259.192	0.002	99.99[r]	2
Boiling point of normal hydrogen	20.397	−252.753	0.002	99.99[r]	2
$\alpha-\beta$ transition point of solid oxygen	23.873	−249.277	0.005	99.999	3–6, 15
Equilibrium between the solid and vapour phases of neon[g,i]					
$\ln\dfrac{p}{p_0} = A + \dfrac{B}{T_{68}} + CT_{68} + DT_{68}^2$	20	−253.15	0.002	99.99	7 (8)
$A = 11.038\ 09$	to	to			
$B = -266.917\ 2\ \mathrm{K}$					
$C = -7.875\ 422 \times 10^{-2}\ \mathrm{K}^{-1}$					
$D = 1.516\ 646 \times 10^{-3}\ \mathrm{K}^{-2}$	24.562	−248.588			
$\alpha-\beta$ transition point of solid nitrogen	35.621	−237.529	0.006	99.999	19
$\beta-\gamma$ transition point of solid oxygen	43.8021	−229.3479	0.0006	99.999	3–6, 15, 20
Equilibrium between the solid and vapour phases of nitrogen					
$\ln\dfrac{p}{p_0} = A + \dfrac{B}{T_{68}} + CT_{68}$	56	−217.15	0.002	99.999	21
$A = 12.189\ 891$	to	to			
$B = -861.621\ 597\ \mathrm{K}$					
$C = -1.006\ 522 \times 10^{-2}\ \mathrm{K}^{-1}$	63.146	−210.004			
Equilibrium between the solid and vapour phases of argon					
$\ln\dfrac{p}{p_0} = A + \dfrac{B}{T_{68}}$	81	−192.15	0.001	99.999	23
$A = 11.022\ 51$	to	to			
$B = -955.992\ \mathrm{K}$	83.788[s]	−189.362[s]			

	K	°C		%	Ref.
Triple point of krypton	115.764	−157.386	0.001	99.995	12, 26, 38
Triple point of xenon[t]	161.3897	−111.7603	0.0006	99.995	26, 38–40, 143
Sublimation point of carbon dioxide	194.673	−78.477	0.003	99.99	26, 41
Triple point of carbon dioxide	216.580	−56.570	0.001	99.99	15, 26, 42, 144
Triple point of bromobenzene[u]	242.417	−30.733	0.010	99.998	44
Triple point of phenoxybenzene (diphenyl ether)	300.021	26.871[v]	0.002	99.9999	45, 46
Triple point of succinonitrile	331.230	58.080	0.005	99.9995	52
Freezing point of sodium	370.969	97.819	0.005	99.99	53
Triple point of benzoic acid	395.520	122.370	0.004	99.998	54
Freezing point of benzoic acid[w]	395.532	122.382	0.004	99.998	54
Boiling point of mercury	629.811	356.661	0.004	99.9999	63, 31

Equilibrium between the liquid and vapour phases of mercury[j]

$$\frac{t_{68}}{K} = A + B\left(\frac{p}{p_0} - 1\right) + C\left(\frac{p}{p_0} - 1\right)^2 + D\left(\frac{p}{p_0} - 1\right)^3$$

	K	°C		%	Ref.
	622.15	349	0.010	99.9999	63 (31)
	to	to			
	636.15	363			

$A = 356.657$
$B = 55.552$
$C = -23.03$
$D = 14.0$

Boiling point of sulphur | 717.824 | 444.674 | 0.002 | 99.995 | 64

Equilibrium between the liquid and vapour phases of sulphur

$$\frac{t_{68}}{K} = A + B\left(\frac{p}{p_0} - 1\right) + C\left(\frac{p}{p_0} - 1\right)^2 + D\left(\frac{p}{p_0} - 1\right)^3$$

	K	°C		%	Ref.
	708.15	435	0.005	99.99	65
	to	to			
	726.15	453			

$A = 444.674$
$B = 69.010$
$C = -27.48$
$D = 19.14$

TABLE 4.2B (Continued)

	821.406	548.256	0.010	99.9999	66
Melting point of the copper/66.9% aluminium eutectic (Al-CuAl₂)					
Radiance temperature[x] (0.653 μm) of melting iron	1,670	1,397	6	99.95	79
Radiance temperature[x] (0.653 μm) of melting palladium[y]	1,688	1,415	5	99.99	80
Radiance temperature[x] (0.997 μm) of melting titanium	1,711	1,438	6	99.9	81
Freezing point of nickel[z]	1,728	1,455	3	99.9	82, 83, 84
Freezing point of cobalt[z]	1,768	1,495	3	99.9	82
Radiance temperature[x] (0.653 μm) of melting titanium	1,800	1,527	3	99.9	81, 85, 86
Freezing point of iron	1,811	1,538	3	99.99	87 (88–92)
Radiance temperature[x] (0.993 μm) of melting vanadium	1,875	1,602	7	99.9	96
Radiance temperature[x] (0.650 μm) of melting zirconium	1,940	1,667	8	99.98	99
Melting point of titanium	1,943	1,670	3	99.9	85, 86
Radiance temperature[x] (0.653 μm) of melting vanadium[aa]	1,990	1,717	5	99.9	96, 97 (98)
Melting point of zirconium	2,128	1,855	8	99.98	99
Radiance temperature[x] (0.65 μm) of melting ruthenium	2,294	2,021	8	99.98	98
Melting point of aluminium oxide (Al₂O₃)	2,327[bb]	2,054[bb]	6	99.9	106
Radiance temperature[x] (0.995 μm) of melting molybdenum	2,331	2,058	8	99.95	107
Radiance temperature[x] (0.650 μm) of melting niobium[aa]	2,429	2,156	5	99.9	98, 108, 109
Radiance temperature[x] (0.653 μm) of melting molybdenum[aa]	2,529	2,256	6	99.95	98, 107

Equilibrium state	T_{68} (K)	t_{68} (°C)		Purity (%)	References
Melting point of ruthenium	2,607	2,334	10	99.98	110
Radiance temperature[x] (0.995 μm) of melting tantalum	2,620	2,347	8	99.99	111
Melting point of yttria (Y_2O_3)	2,712	2,439	12	99.999	112
Melting point of niobium	2,746	2,473	7	99.9	114, 115
Radiance temperature[x] (0.653 μm) of melting tantalum	2,846	2,573	8	99.99	111
Radiance temperature[x] (0.653 μm) of melting tungsten	3,208	2,935	10	99.95	120

TABLE 4.2 Part C: Possible Useful Points Deserving Further Study[cc]

Equilibrium state	Approximate temperature		References
	T_{68} (K)	t_{68} (°C)	
Triple point of equilibrium deuterium	18.68	−254.47	125 (135), 126, 127
Triple point of propane	85.52	−187.63	128
Triple point of ethane	90.35	−182.80	129, 130
Boiling point of methane	111.66	−161.49	27, 34
Boiling point of krypton	119.80	−153.35	27
Boiling point of xenon	165.05	−108.10	27, 40
Equilibrium between the liquid and vapour phases of xenon	161−165	−112 to −108	40
Equilibrium between the solid and vapour phases of carbon dioxide	170−195	−103 to −78	41, 131
Various solid-solid transitions			
Triple point of rubidium	312.42	39.27	132
Triple and freezing points of various organic substances, mostly in the range 100 K to 200 °C; typical examples recently studied are:			133
Triple point of 1,3-dioxolan-2-one ($C_3H_4O_3$)	309.47	36.32	46
Triple point of n-icosane ($C_{20}H_{42}$)	309.64	36.49	46

Notes of the Tables 4.2A, B, and C

a   The equilibrium states in this table are at a pressure $p_0 = 101\,325$ Pa (one standard atmosphere), except for the triple points, the radiance temperatures and those cases where a range of pressure. is explicitly allowed

b   The uncertainties listed are (where possible) the standard deviation of the consensus values. Individual experimental uncertainties are best estimates based upon the information presented in the references and upon the agreement between results of comparable experiments. For some of the points below 273 K that are relatively distant from a defining fixed point the largest contributor to the uncertainty is non-uniqueness of the IPTS-68

c   The minimum purity of the material to which the listed values of temperature and uncertainty apply is given in percent by volume when the material is liquid or gaseous at 0 °C and one standard atmosphere and in percent by weight when the material is solid at 0 °C and one standard atmosphere. (There are a few exceptions: for example, mercury)

d   Only those references that contributed directly to the choice of the value of temperature are listed. Other determinations (usually either of lesser accuracy, with insufficient information to be useful, or quoted on a temperature scale for which the relation to the IPTS-68 is insufficiently well established) are cited in [1] or in previous reports of WG 2 to the CCT

e   The term $\sum_{i=0}^{5} b_i T_{68}^i$ in the equation adds to the value of p a pressure amounting to the equivalent of up to 1 mK

f   If nitrogen is an impurity in neon, 100 ppm of nitrogen will lower the triple point ~ 1.5 mK

g   These values are for neon with an isotopic composition close to that specified in Section III.6 of the IPTS-68, Amended Edition of 1975. Triple point temperatures and vapour pressure equations for ^{20}Ne and ^{22}Ne are given by Furukawa [10]

h   Parenthetic reference provides more information but appears to relate to the same experiment as described in the non-parenthetic reference

i   There is no convenient way to obtain a "mean value" for vapour-pressure equations from different references. We have found it preferable to list, for each substance. an equation from a particular reference. In every case the differences between it and the equivalent equations in the references in parentheses are within the uncertainty listed

j   The triple point of argon is an alternative defining fixed point of the IPTS-68 (Amended Edition of 1975). The current recommended value for this fixed point differs by 0.4 mK from the value used in the IPTS-68 (Amended Edition of 1975). There will therefore be systematic differences of about this magnitude between IPTS-68 realizations based upon the triple point of argon and those based upon the condensation point of oxygen in the range near and between these two fixed points

k   Some of the measured values used to calculate the recommended temperature derive from IPTS-68 realizations based upon the triple point of argon and some from realizations based upon the condensation point of oxygen. There are small systematic differences between these two sets (se footnote j)

l   Below about 65 K the vapour pressure of oxygen is so low as to be of little use as a secondary reference for accurate thermometry

m   See Table 5 of the IPTS-68 (Amended Edition of 1975) for the effect of pressure on these freezing points

n   The ice point is a very close approximation to the temperature defined as being 10 mK below the triple point of water o   The listed temperature is the value obtained by extrapolation of the IPTS-68 defining equations for the platinum resistance thermometer

p   The listed value of temperature is that measured with a platinum 10% rhodium/platinum thermocouple according to the definition of the IPTS-68

q   Freezing point when oxygen is not present

r   One must be cautious if neon is a substantial impurity (which normally it is not); 100 ppm of neon will lower the triple point of normal hydrogen ~ 0.2 mK

s   This equation does not apply within the melting range of argon [23], so the maximum temperature of application is (arbitrarily) set 10 mK below the triple point

t   This point lies approximately mid-way between two IPTS-68 defining fixed points, so could be useful for reducing non-uniqueness in any scale revision. At present it is placed in category B because its reproducibility is not good enough for inclusion in category A. The irreproducibility may be associated with isotope effects. Further study is highly desirable

u   The published value [44] has been converted to the IPTS-68 and lowered 2 mK to obtain the temperature corresponding to $F^{-1} = 1$

v   The listed temperature is estimated to be the triple-point temperature of the ideally pure substance. In practice it is difficult to achieve a purity higher than 99.999%, for which the triple-point temperature is $26.868 \pm 0.002\,°C$

w   Freezing point given is the value under one atmosphere of dry air. Different values are obtained under an atmosphere of nitrogen or oxygen [54]

x   Radiance temperature values are given for the particular wavelengths indicated. The radiance temperatures are determined with the specimens in vacuum or inert-gas environment at atmospheric pressure. Pressure corrections are negligible, if compared with measurement uncertainties

y   Temperature measured with the specimen in one atmosphere of pure argon

z   Temperature measured with the specimen in vacuum [83] or in purified helium [82]

aa Measured temperatures adjusted to a common wavelength before averaging

bb Melting point in vacuum

cc The reference points listed in this Part are suggested as examples only. There is no implication that these are the only such points nor that the accompanying references constitute a complete set

dd Measured temperatures were based upon samples ranging from 0.999 995 to 0.999 999 9 in purity and were adjusted to the equivalent of purity 0.999 999 9

ee The inherent accuracy of several triple points (and probably of some other points as well) is better than indicated. One of the chief contributors to the tabular uncertainties is the non-uniqueness of the IPTS-68 (see also note b). In a recent intercomparison of cryogenic triple-point cells where it was possible to exclude the non-uniqueness component, it was found that a group of each of neon, nitrogen, and argon triple-point realizations of various manufactures agreed to within ± 0.15 mK. A full report of this work will be published in *Metrologia*, and the complete set of data on which the report is based will be published by the Bureau International des Poids et Mesures

Source: **Refs. 13 and 13a. Reprinted with permission.**

References to Tables 4.2A, B, and C.

2. J. Anscin: *Thermometric Fixed Points of Hydrogen*: Metrologia **13**, 79–86 (1977).

3. J. Ancsin: *Crystalline transformations of oxygen*: In: *Temperature Measurement 1975*, B. F. Billing, T. J. Quinn (eds.). *Institute of Physics Conference Series* **26**, 57–64 (1975).

4. W. R. G. Kemp, C. P. Pickup: *The Transition Temperatures of Solid Oxygen*: *TMCSI* **4**, 217–223 (1972).

5. M. P. Orlova: *Temperatures of phase transitions in solid oxygen*: *TMCSI* **3**, Part 1, 179–183 (1962).

6. R. Muijlwijk, M. Durieux, H. Van Dijk: *The Temperatures at the Transition Points in Solid Oxygen*: Physica **43**, 475–480 (1969). (see also R. Muijlwijk: *Vapour Pressures of Oxygen and Platinum Thermometry below 100 K*: Ph. D. Thesis, Leiden (1968)).

7. J. Ancsin: *Vapour Pressures and Triple Point of Neon and the Influence of Impurities on These Properties*: Metrologia **14**, 1–7 (1978).

8. J. L. Tiggelman: *Low Temperature Platinum Thermometry and Vapour Pressures of Neon and Oxygen*: Ph. D. Thesis, Leiden (1973).

9. R. C. Kemp, W. R. G. Kemp: *The Triple Point of ²⁰Ne*: Metrologia **17**, 67–68 (1981).

10. G. T. Furukawa: *Vapor Pressures of Natural Neon and of the Isotopes ²⁰Ne and ²²Ne from the Triple Point to the Normal Boiling Point*: Metrologia **8**, 11–27 (1972).

11. G. T. Furukawa, W. G., Saba, D. M. Sweger, H. H. Plumb: *Normal Boiling Point and Triple Point Temperatures of Neon*: Metrologia **6**, 35–37 (1970).

12. R. C. Kemp, W. R. G. Kemp: *The Triple Points of Krypton, Argon, Nitrogen and Neon*: Metrologia **14**, 83–88 (1978).

13. J. P. Compton: *The Realisation of the Normal Boiling Point of Neon. II Vapour Pressure Measurements*: Metrologia **6**, 103–109 (1970).

14. V. M. Khnykov, L. I. Rabukh, L. B. Belyansky, T. S. Pankiev, M. P. Orlova, D. N. Astrov: *Realization of the IPTS in the 18.7 to 273 K range using the triple point temperatures*: *Comité Consultatif de Thermométrie*, 11e Session, Document CCT/76–39 (1976).

15. F. Pavese, D. Ferri: *Ten years of research on sealed cells for phase transition studies of gases at IMGC*: *TMCSI* **5**, 217–227 (1982).

16. S. D. Ward: *Realizations of the triple points of argon and neon at NPL*: *Comité Consultatif de Thermométrie*, 13e Session, Document CCT/80–51 (1980).

17. P. Seifert: *Tripelpunkte von Gasen als Fixpunkte im Tieftemperaturbereich*; *Metrologische Abhandlungen* **3**, 133–146 (1983). (see also P. Seifert: *Sealed triple point cells for low temperature thermometer calibration*: *Comité Consultatif de Thermométrie*, 14e Session, Document CCT/82–6 (1982)).

18. E. R. Grilly: *The Vapour Pressure of Solid and Liquid Neon*: Cryogenics **2**, 226–229 (1962).

19. R. C. Kemp, W. R. G. Kemp: *The α–β Transition in Nitrogen*: Metrologia **15**, 87–88 (1979).

20. J. A. Cowan, R. C. Kemp. W. R. G. Kemp: *An Investigation of the β-γ Transition in Oxygen*: Metrologia **12**, 87–91 (1976).

21. J. Ancsin: *Some Thermodynamic Properties of Pure and Impure Nitrogen*: Can. J. Phys. **52**, 1521–1531 (1974).

22. J. L. Tiggelman, M. Durieux: *Vapour Pressures of Liquid Oxygen and Nitrogen*: *TMCSI* **4**, 149–157 (1972). (see also M. R. M. Moussa: *On Thermometry between 63 and 273.15 K*: Ph.D. Thesis, Leiden (1966)).

23. J. Ancsin: *Studies of Phase Changes in Argon*: Metrologia **9**, 147–154 (1973).

24. G. T. Furukawa, W. R. Bigge, J. L. Riddle: *Triple Point of Argon*: *TMCSI* **4**, 231–243 (1972).

25. G. T. Furukawa: *Reproducibility of the triple point of argon in sealed transportable cells*: *TMCSI* **5**, 239–248 (1982).

26. D. R. Lovejoy: *Some Boiling and Triple Points Below 0°C*: Nature **197**, 353–354 (1963).

27. R. C. Kemp, W. R. G. Kemp, J. A. Cowan: *The Boiling Points and Triple Points of Oxygen and Argon*: Metrologia **12**, 93–100 (1976).

28. V. M. Khnykov, M. P., Orlova, L. B. Belyanskii, L. N. Rabukh: *The Argon Triple Point: a New Reference Point for the Temperature Scale*: Zhurnal Fizicheskoi Khimii **52**, 1483–1484 (1978). [Trans.: Russian Journal of Physical Chemistry **52**, 849–850 (1978)].

29. W. Blanke: *The temperatures of the triple points of methane and argon on the IPTS-68*: Comité Consultatif de Thermométrie, 15e Session, Document CCT/84-7 (1984).

30. F. Pavese: *The Triple Point of Argon and Oxygen*: Metrologia **14**, 93-103 (1978).

31. D. Ambrose, C. H. S. Sprake: *The Vapour Pressure of Mercury*: J. Chem. Thermodynamics **4**, 603-620 (1972).

33. G. T. Furukawa, J. L. Riddle, W. R. Bigge, E. R. Pfeiffer: *Application of Some Metal SRM's as Thermometric Fixed Points*: NBS Special Publication 260-77 (1982).

33. J. Ancsin: *Vapor Pressure Scale of Oxygen*: Can. J. Phys. **52**, 2305-2312 (1974).

34. F. Pavese, G. Cagna, D. Ferri: *The Triple Point of Pure Methane*: Proceedings of the Sixth International Cryogenic Engineering Conference (IPC Science and Technology, London) 281-285 (1976).

35. F. Pavese: *On the IPTS-68 Temperature Value of the Triple Point of Methane*: Metrologia **15**, 47-49 (1979).

36. J. Bonhoure, R. Pello: *Temperature of the Triple Point of Methane*: Metrologia **14**, 175-177 (1978).

37. J. Bonhoure, R. Pello: *Points Triples de l'Argon et du Methane: Utilisation de Cellules Scellées*: Metrologia **16**, 95-99 (1980).

38. A. Inaba, K. Mitsui: *Réalisation des points triples du krypton et du xénon au moyen de cellules scellées*: Comité Consultatif de Thermométrie, 12e Session, Annexe T17, T111-T113 (1978).

39. R. C. Kemp, W. R. G. Kemp, P. W. Smart: The triple point of natural xenon: TMCSI **5**, 229-230 (1982).

40. J. Ancsin: *Note Concerning the Suitability of Xenon as Temperature Fixed Point*: Metrologia **14**, 45-46 (1978).

41. C. R. Barber: *The Sublimation Temperature of Carbon Dioxide*: Br. J. Appl. Phys. **17**, 391-397 (1966).

42. D. Ambrose: *The Triple Point of Carbon Dioxide as a Thermometric Fixed Point*: Br. J. Appl. Phys. **8**, 32-34 (1957).

43. G. T. Furukawa, W. R. Bigge: *Le point triple du mercure comme étalon thermométrique*: Comité Consultatif de Thermométrie, 11e Session, Annexe T14, T138-T144 (1976).

44. J. F. Masi, R. B. Scott: *Some Thermodynamic Properties of Bromobenzene from 0 to 1500 K*: J. Res. Natl. Bur. Stand. **79A**, 619-628 (1975).

45. J. D. Cox, M. F. Vaughan: *The Triple Point of Phenoxybenzene*: Metrologia **16**, 105-109 (1980).

46. J. D. Cox, M. F. Vaughan: *Temperature fixed points: Evaluation of four types of triple-point cell*: TMCSI **5**, 267-280 (1982).

47. B. W. Mangum, D. D. Thornton: *Determination of the Triple-Point Temperature of Gallium*: Metrologia **15**, 201-215 (1979).

48. D. D. Thornton: *The Gallium Melting-Point Standard: A Determination of the Liquid-Solid Equilibrium Temperature of Pure Gallium on the International Practical Temperature Scale of 1968*: Clin Chem **23**, 719-724 (1977).

49. M. V. Chattle, R. L. Rusby, G. Bonnier, A. Moser, E. Renaot, P. Marcarino, G. Bongiovanni, G. Frassineti: *An intercomparison of gallium fixed point cells*: TMCSI **5**, 311-316 (1982).

50. M. Borovicka, J. Demian: *Realization of the triple point and melting point of gallium*: Comité Consultatif de Thermométrie, 14e Session, Document CCT/82-24 (1982).

51. B. W. Mangum: *Triple point of gallium as a temperature fixed point*: TMCSI **5**, 299-309 (1982).

52. M. E. Glicksman, P. W. Voorkees: *The triple-point equilibria of succinonitrile: Its assessment as a temperature standard*: TMCSI **5**, 321-326 (1982).

53. J. Butkiewicz, W. Gizmajer: *Etude sur le point thermométrique du sodium*: Comité Consultatif de Thermométrie, 13e Session, Document CCT/80-1 (1980).

54. F. W. Schwab, E. Wickers: *Freezing Temperature of Benzoic Acid as a Fixed Point in Thermometry*: J. Res. Natl. Bur. Stand. **34**, 333-372 (1945).

55. E. H. McLaren: *The Freezing Points of High-Purity Metals as Precision Temperature Standards. IV. Indium: Thermal Analyses on Three Grades of Cadmium*: Can. J. Phys. **36**, 1131-1147 (1958).

56. S. Sawada: *Realization of the triple point of indium in a sealed glass cell*: TMCSI **5**, 343-346 (1982).

57. E. H. McLaren, E. G. Murdock: *The Freezing Points of High-Purity Metals as Precision Temperature Standards. VII. Thermal Analyses on Seven Samples of Bismuth*

References to Tables 4.2A, B, and C (Continued)

with Purities Greater than 99.999 + %: Can. J. Phys. **41**, 95–112 (1963).

58. E. H. McLaren: Intercomparison of 11 Resistance Thermometers at the Ice, Steam, Iron, Cadmium, and Zinc Points: Can. J. Phys. **37**, 422–432 (1959).

59. E. H. McLaren, E. G. Murdock: The Freezing Points of High-Purity Metals as Precision Temperature Standards. VIIIb. Sb: Liquidus Points and Alloy Melting Ranges of Seven Samples of High-purity Antimony; temperature-scale realization and reliability in the range 0–631°C: Can. J. Phys. **46**, 401–444 (1968).

60. J. V. McAllan, J. J. Connolly: The use of the cadmium point to check calibrations on the IPTS: TMCSI **5**, 351–353 (1982).

61. G. T. Furukawa, E. R. Pfeiffer: Investigation of the Freezing temperature of cadmium: TMCSI **5**, 355–360 (1982).

62. E. H. McLaren, E. G. Murdock: The Freezing Points of High-Purity Metals as Precision Temperature Standards. VI. Thermal Analyses on Five Samples of Lead with Purities Greater than 99.999 + %: Can. J. Phys. **38**, 577–587 (1960).

63. J. A. Beattie, B. E. Blaisdell, J. Kaminsky: An Experimental Study of the Absolute Temperature Scale. IV. The Reproducibility of the Mercury Boiling Point. The Effect of Pressure on the Mercury Boiling Point: Proc. Am. Acad. Arts Sci. **71**, 375–385 (1937).

64. R. J. Berry: The Reproducibility of the Sulphur Point. Can. J. Phys. **38**, 1027–1047 (1960).

65. J. A. Beattie, M. Benedict, B. E. Blaisdell: An Experimental Study of the Absolute Temperature Scale, II. The Reproducibility of the Sulphur Point. The Effect of Pressure on the Sulphur Point: Proc. Am. Acad. Arts. Sci. **71**, 327–360 (1937).

66. J. V. McAllan: Metal Binary Eutectics as Fixed Temperature Points: TMCSI **4**, 265–274 (1972).

67. G. Bongiovanni, L. Crovini, P. Marcarino: Comparaison de thermomètres à résistance de platine au point de congélation de l'antimoine: Comité Consultatif de Thermométrie, 9e Session, Annexe T16, T66–T74 (1971).

68. J. V. McAllen, M. M. Ammar: Comparison of the Freezing Points of Aluminium and Antimony: TMCSI **4**, 275–285 (1972).

69. G. T. Furukawa, W. R. Bigge, J. L. Riddle, M. L. Reilly: The freezing point of aluminium as a temperature standard: In: Temperature Measurement 1975, B. F. Billing, T. J. Quinn (eds.). Institute of Physics Conference Series **26**, 389–397 (1975).

70. L. Crovini, P. Marcarino: Point de fusion de l'eutectique cuivre-argent: Comité Consultatif de Thermométrie, 10e Session, Annexe T7, T79–T82 (1974).

71. R. E. Bedford, C. K. Ma: Measurement of the melting temperature of the copper 71.9% silver eutectic alloy with a monochromatic optical pyrometer: TMCSI **5**, 361–369 (1982).

72. P. Marcarino, L. Crovini: Characteristics of platinum resistance thermometers up to the silver freezing point: In: Temperature Measurement 1975, B. F. Billing, T. J. Quinn (eds.). Institute of Physics Series **26**, 107–116 (1975).

73. J. V. McAllen: Reference temperatures near 800°C: TMCSI **5**, 371–376 (1982).

74. F. Righini, A. Rosso, G. Ruffino: New determination of the freezing point of copper: High Temp.-High press. **4**, 471–475 (1972).

75. T. Ricolfi, F. Lanza: The silver and copper freezing points as accurate reference standards for radiation pyrometry: High Temp.-High Press. **9**, 483–487 (1977).

76. P. B. Coates, J. W. Andrews: A precise determination of the freezing point of copper: J. Phys. F. **8**, 277–285 (1978).

77. M. Ohtsuka, R. E. Bedford: Measurement of the thermodynamic temperature interval between the freezing points of silver and copper: TMCSI **5**, 175–181 (1982).

78. T. P. Jones, J. Tapping: A photoelectric pyrometer temperature scale below 1064.43°C and its use to measure the silver point: TMCSI **5**, 169–174 (1982).

79. A. Cezairliyan, J. L. McClure: Radiance Temperature (at 653 nm) of Iron at Its Melting Point: J. Res. Natl. Bur. Stand. **79A**, 541–544 (1975).

80. A. P. Miiller, A. Cezairliyan: Radiance Temperature (at 653 nm) of Palladium at its Melting Point: High Temperature Science **11**, 41–47 (1979).

81. F. Righini, A. Rosso, L. Coslovi, A. Cezairliyan, J. L. McLure: *Radiance Temperature of Titanium at its Melting Point: Proc. of Seventh Symposium on Thermophysical Properties*, A. Cezairliyan (ed.) (Am. Soc. Mech. Engrs., New York) 312–318 (1977).

82. M. S. Van Dusen, A. I. Dahl: *Freezing Points of Cobalt and Nickel*: J. Res. Natl. Bur. Stand. **39**, 291–295 (1947).

83. H. T. Wensel, W. F. Roeser: *Freezing Point of Nickel as a Fixed Point on the International Temperature Scale*: J. Res. Natl. Bur. Stand. **5**, 1309–1318 (1930).

84. P. Ratanapupech, R. G. Bautista: *Normal Spectral Emissivities of Liquid Iron, Liquid Nickel, and Liquid Iron-Nickel Alloys*: High Temp. Sci. **14**, 269–283 (1981).

85. A. Cezairliyan, A. P. Miiller: *Melting Point, Normal Spectral Emittance (at the Melting Point), and Electrical Resistivity (above 1900 K) of Titanium by a Pulse Heating Method*: J. Res. Natl. Bur. Stand. **82**, 119–122 (1977).

86. B. Y. Berezin, S. A. Kats, M. M. Kenisarin, V. Y. Chekhovskoi: *Heat and Melting Temperature of Titanium*: Teplofizika Vysokikh Temperatur **12**, 524–529 (1974) [Trans: High Temperature **12**, 450–455 (1974)].

87. W. F. Roeser, H. T. Wensel: *Freezing Temperatures of High-Purity Iron and of Some Steels*: J. Res. Natl. Bur. Stand. **26**, 273–287 (1941).

88. T. H. Schofield, A. E. Bacon: *The Melting Point of Titanium*: J. Inst. Metals **82**, 167–169 (1953–54).

89. C. A. Bristow: *The Constitutional Diagram of the Alloys of Iron and Nickel. Part I. – The Delta Region*: Iron and Steel Institute Special Report **24**, 1–8 (1939).

90. F. Adcock: *An Investigation of the Iron-Carbon Constitutional Diagram. Part I. – Preliminary Survey of the Delta Region*: J. Iron and Steel Institute **135**, 281–292 (1937).

91. R. A. Oriani, T. S. Jones: *An Apparatus for the Determination of the Solidus Temperatures of High-Melting Alloys*: Rev. Sci. Instr. **25**, 248–250 (1954).

92. A. Cezairliyan, J. L. McClure: *Thermophysical Measurements on Iron Above 1500 K Using a Transient (Subsecond) Technique*: J. Res. Natl. Bur. Stand. **78A**, 1–4 (1974). [see remarks re sample purity in L. J. Swartzendruber: *The Fe (Iron) System*: Bulletin of Alloy Phase Diagrams **3**, 161–165 (1982)]

93. J. L. Tiggelman, C. Van Rijn, M. Durieux: *Vapour Pressures of Liquid and Solid Neon between 19 K and 30 K*: TMCSI **4**, 137–147 (1972).

94. T. P. Jones, K. G. Hall: *The Melting Point of Palladium and Its Dependence on Oxygen*: Metrologia **15**, 161–163 (1979).

95. P. B. Coates, T. R. D. Chandler, J. W. Andrews: *A new determination of the freezing point of palladium*: High Temp – High Press **15**, 573–582 (1983).

96. A. Cezairliyan, A. P. Miiller, F. Righini, A. Rosso: *Radiance Temperature of Vanadium at its Melting Point*: High Temperature Science **11**, 223–232 (1979).

97. B. Y. Berezin, V. Y. Chekhovskoi, A. E. Sheindlin: *Heat of Fusion of Vanadium*: Dokl. Akad. Nauk SSSR **201**, 583–585 (1971). [Trans.: Sov. Phys. – Dokl. **16**, 1007–1009 (1972)]

98. B. Y. Berezin, S. A. Kats, V. Y. Chekhovskoi: *Spectral Emissivities of Molten Refractory Metals*: Teplofizika Vysokikh Temperatur **14**, 497–502 (1976). [Trans.: High Temperature **14**, 448–452 (1976)]

99. A. Cezairliyan, F. Righini: *Measurement of Melting Point, Radiance Temperature (at Melting Point), and Electrical Resistivity (above 2,100 K) of Zirconium by a Pulse Heating Method*: Revue Internationale des Hautes Tempér. et Réfractaires **12**, 201–207 (1975).

100. T. J. Quinn, T. R. D. Chandler: *The Freezing Point of Platinum Determined by the NPL Photoelectric Pyrometer*: TMCSI **4**, 295–309 (1972).

101. T. J. Quinn: *Corrections in Optical Pyrometry for the Refractive Index of Air*: Metrologia **10**, 115 (1975).

102. T. P. Jones, J. Tapping: *The Freezing Point of Platinum*: Metrologia **12**, 19–26 (1976).

103. H. Kunz, J. Lohrengel: *Nouvelle détermination du point de congélation du platine*: Comité Consultatif de Thermométrie, 11e Session, Annexe T25, T197–199 (1976). (see also Jahresbericht der PTB, p. 64 and p. 190 (1975).

104. F. Lanza, T. Ricolfi: *The IMGC determination of the freezing point of platinum*: High Temp.–High Press. **8**, 217–224 (1976).

105. W. F. Roeser, H. T. Wensel: *Freezing Point*

References to Tables 4.2A, B, and C   (Continued)

of Rhodium: J. Res. Natl. Bur. Stand. **12**, 519–526 (1934).

106. J. Hlavac: IUPAC Report: *Melting Temperatures of Refractory Oxides: Part I*: Pure Appl. Chem. **54**, 682–688 (1982). [See also S. J. Schneider: *IUPAC Report PAC 21/1- N.* London: Butterworth 1970, based upon S. J. Schneider, C. L. McDaniel: *Effect of Environment upon the Melting Point of* Al_2O_3: J. Res. Natl. Bur. Stand. **71A**, 317–333 (1967)]

107. A. Cezairliyan, L. Coslovi, F. Righini, A. Rosso: *Radiance temperature of molybdenum at its melting point*: In: *Temperature Measurement 1975*, B. F. Billing, T. J. Quinn (eds.). *Institute of Physics Conference Series* **26**, 287–296 (1975).

108. A. Cezairliyan: *Radiance Temperature of Niobium at its Melting Point*: J. Res. Natl. Bur. Stand. **77A**, 333–339 (1973).

109. F. Righini, A. Rosso: *Ten years of high speed pyrometry at IMGC*: TMCSI **5**, 433–438 (1982).

110. M. M. Kenisarin, B. Y. Berezin, N. B. Gorina, S. A. Kats, V. P. Polyakova, E. M. Savitskii, V. Y. Chekhovskoi: *The Melting Point of Ruthenium*: Teplofizika Vysokikh Temperatur **12**, 1309–1310 (1974). [Trans.: High Temperature **12**, 1159–1160 (1974)].

111. A. Cezairliyan, J. L. McClure, L. Coslovi, F. Righini, A. Rosso: *Radiance temperature of tantalum at its melting point*: High Temp.-High Press. **8**, 103–111 (1976).

112. M. Foex: *Recherche sur le point de fusion de l'oxyde d'yttrium*: High Temp.-High Press. **9**, 269–282 (1977).

113. F. Henning, H. T. Wensel: *The Freezing Point of Iridium*: J. Res. Natl. Bur. Stand. **10**, 809–821 (1933).

114. A. Cezairliyan: *Measurement of melting point, normal spectral emittance (at melting point), and electrical resistivity (above 2650 K) of niobium by a pulse heating method*. High Temp.-High Press. **4**, 453–458 (1972).

115. B. Y. Berezin, M. M. Kenisarin, V. Y. Chekhovskoi: *Melting Point of Niobium*: Teplofizika Vysokikh Temperatur **10**, 1214–1217 (1972). [Trans.: High Temperature **10**, 1092–1094 (1972)].

116. A. Cezairliyan, M. S. Morse, C. W. Beckett: *Measurement of Melting point and Electrical Resistivity (above 2,840°K) of Molybdenum by a Pulse Heating Method*: Rev. Int. Hautes Tempér. et Réfract. **7**, 382–388 (1970).

117. R. E. Latta, R. E. Fryxell: *Determination of Solidus-Liquidus Temperatures in the* UO_{2+x} *system (−0.50 < x < 0.20)*: J. Nucl. Mat. **35**, 195–210 (1970).

118. M. M. Kenisarin, B. Y. Berezin, V. Y. Chekhovskoi: *The Melting Point of Molybdenum as a Secondary Fixed Point on the International Practical Temperature Scale*: High Temp.-High Press. **4**, 707–713 (1972).

119. E. Rudy, J. Progulski: *A Pirani Furnace for the Precision Determination of the Melting Temperatures of Refractory Metallic Substances*: Planseeber. Pulvermet. **15**, 13–45 (1967).

120. A. Cezairliyan, A. P. Miiler: *Radiance Temperature (at 653 nm) of Tungsten at Its Melting Point*: Int. J. Thermophys. **3**, 89–99 (1982).

121. A. Cezairliyan: *Measurement of Melting Point and Electrical Resistivity (above 3600 K) of Tungsten by a Pulse Heating Method*: High Temp. Sci. **4**, 248–252 (1972).

122. R. D. Allen: *Techniques for Melting-Point Determination on an Electrically Heated Refractory Metal*: Nature **193**, 769–770 (1962).

123. E. Rudy, St. Windisch, J. R. Hoffman: *Ternary Phase Equilibria in Transition Metal-Boron-Carbon-Silicon System Part I. Related Binary Systems. Volume VI. W-C System: Supplemental Information on the MO-C System: Report AFML-TR-65-2*, Air Force Materials Laboratory Research and Technology Division, Dayton, Ohio (1966).

124. C. F. Zalabak: *The Melting Points of Tantalum Carbide and of Tungsten: NASA Technical Note D-761* (1961).

125. F. Pavese: *The Triple Point of Equilibrium-Hydrogen Isotopes*: Physica **107B**, 333–334 (1981).

126. J. Ancsin: *About the usefulness of deuterium in thermometry*: Comité Consultatif de Thermométrie, 14e Session, Document CCT/82-8 (1982).

127. R. C. Kemp: *The triple points of equilibrium and normal deuterium*: TMCSI **5**, 249–250 (1982).

128. F. Pavese, L. M. Besley: *Triple-point tem-*

perature of propane: measurements on two solid-to-liquid transitions and one solid-to-solid transition: J. Chem. Thermodynamics **13**, 1095–1104 (1981).

129. F. Pavese: *Some thermodynamic properties of ethane between its double solid-to-solid transition and its triple-point temperature*: J. Chem. Thermodynamics **10**, 369–379 (1978).

130. G. C. Straty, R. Tsumura: *Phase transition and melting pressures of solid ethane*: J. Chem. Phys. **64**, 859–861 (1976).

131. J. Ancsin: *Sur quelques propriétés du dioxyde de carbone*: Comité Consultatif de Thermométrie, 12e Session, Annexe T18, T114–T115 (1978).

132. J. M. Figueroa, B. W. Mangum: *The triple point of rubidium: A temperature fixed point for biomedical applications*: TMCSI **5**, 327–337 (1982).

133. L. A. K. Stavely, L. Q. Lobo, J. C. G. Calado: *Triple-points of low melting substances and their use in cryogenic work*: Cryogenics **21**, 131–144 (1981).

134. F. Pavese: *The Use of Triple Point of Gases in Sealed Cells as Pressure Transfer Standards: Oxygen (146.25 Pa), Methane (11,696 Pa), and Argon (68,890 Pa)*: Metrologia **17**, 35–42 (1981).

135. F. Pavese, C. Barbero: *The Triple Point of Pure Normal Deuterium*: Cryogenics **19**, 255–260 (1979).

136. J. Bonhoure, R. Pello: *Température du Point Triple du Gallium*: Metrologia **19**, 15–20 (1983).

137. S. Lau: *Entwicklung und Anwendung des Gallium – Fixpunktes in ASMW*: Metrologische Abhandlungen des ASMW **4**, 1–5 (1984).

138. H. E. Sostman: *Melting Point of Gallium as a Temperature Calibration Standard*: Rev. Sci. Instrum. **48**, 127–130 (1977).

139. B.N. Oleinik, A. G. Ivanova, V. A. Zamhovetz, N. N. Ergardt: *Point fixe de référence de fusion du gallium*: Comité Consultatif de Thermométrie, 14e Session, Document CCT/82–32 (1982).

140. J. Ancsin: *Melting Curves and Heat of Fusion of Indium*: Metrologia **21**, 7–9 (1985).

141. B. N. Oleinik, A. G. Ivanova, M. M. Dvinianinov, V. A. Zamkovets: *Realization of the indium freezing point*: Comité Consultatif de Thermométrie, 15e Session, Document CCT/84–1 (1984).

142. H. Itoh: *The Ag-Cu Eutectic Point as a Reference Temperature*: Comité Consultatif de Thermométrie, 15e Session, Document CCT/84–14 (1984).

143. R. C. Kemp, W. R. G. Kemp, P. W. Smart: *Realization of the triple point of natxenon and experiments with 136xenon in a sealed cell of novel design*: Comité Consultatif de Thermométrie, 15e Session, Document CCT/84–10 (1984).

144. G. Bonnier, Y. Hermier, Wu Bi Qin: *Triple Point of Carbon Dioxide in a Multicompartment Sealed Cell*: In: *Proc. 2nd IMEKO Symposium on Temperature Measurement in Industry and Science*, F. Bernhard (ed.), Suhl, GDR, October 1984; p. 39–54.

criteria as for the previous list which are:

Part (a) The SRP must be highly reproducible.

Part (b) There must be more than one experimental measurement of the experimental temperature.

Part (c) An estimate of the accuracy of the assigned temperature should be given.

Part (d) The purity of the material corresponding to the assigned temperature should be stated.

Part (e) References to the original experiments should be given.

Part (f) Information regarding the effects of other physical quantities on the value of the temperature should given, where appropriate.

Part A of Table 4.2 lists those points that fulfill requirements (a) through (e). Part B includes points that do not fit all the criteria or are inconvenient to use. Part C includes as examples, only, of points that could be of value if established. Tables following Tables 4.2A, B, and C explain footnotes and list references.

Secondary points are not part of IPTS but are presented for the user's convenience. The IPTS-68 has four ranges, uses three interpolating instruments, and is described as follows.

Range I: 13.8 to 273.15 K (−259.34 to 0°C). Use the platinum resistance thermometer and interpolating equations:

$$T_{68} = \sum_{i=0}^{20} B_i \left(\frac{\ln W + 3.28}{3.28} \right)^i \tag{4.1}$$

$$W = W_m + \Delta W, \tag{4.2}$$

where W_m is the measured resistance ratio, R/R_0, and the B_i are coefficients tabulated for the purpose (Table 4.3). The subscript on R_0 indicates the resistance at 0°C. The ΔW term is a correction term which depends on the temperature in range I. Range I is divided into four parts and a different ΔW equation is used for each part. The four parts and equations are:

Part (a), 13.81 to 17.042 K: $\Delta W_1 = A_1 + B_1 T + C_1 T^2 + D_1 T^3$
$$\tag{4.3}$$
Part (b), 17.042 to 54.361 K: $\Delta W_2 = A_2 + B_2 T + C_2 T^2 + D_2 T^3$
$$\tag{4.4}$$
Part (c), 54.361 to 90.188 K: $\Delta W_3 = A_3 + B_3 T + C_3 T^2$ $\tag{4.5}$
Part (d), 90.188 to 273.15 K: $\Delta W_4 = B_4 t + C_4 t^3 (t_{68} - 100)$ $\tag{4.6}$

In addition to the coefficients B_i in Table 4.3, two other sets of coefficients A_i and C_i are shown. They can be used as alternative methods for calculating and are discussed in Chapter 8. The coefficients A_i were replaced by B_i in 1975 to reduce rounding-off errors. The coefficients C_i are for Chebyshev polynomial interpolation.

In part (a) the constants for ΔW in Eq. (4.3) are determined by the measured resistance ratio deviations at the triple point of equilibrium hydrogen, the temperature of 17.042 K, the boiling point of equilibrium hydrogen, and by the derivative at the boiling point of equilibrium hydrogen as derived from Eq. (4.4).

In part (b) the constants in Eq. (4.4) are determined by measuring the deviations at the boiling point of equilibrium hydrogen, the boiling point of

TABLE 4.3 Interpolation Coefficients[a]

i	A_i	B_i	C_i
0	0.273 150 000 000 000 0 E +3	+38.59276	+161.82457
1	0.250 846 209 678 803 3 E +3	+43.44837	+101.20451
2	0.135 099 869 964 999 7 E +3	+39.10887	+51.52727
3	0.527 856 759 008 517 2 E +2	+38.69352	+24.77384
4	0.276 768 548 854 105 2 E +2	+32.56883	+10.11559
5	0.391 053 205 376 683 7 E +2	+24.70158	+3.53282
6	0.655 613 230 578 069 3 E +2	+53.03828	+0.91392
7	0.808 035 868 559 866 7 E +2	+77.35767	+0.14666
8	0.705 242 118 234 052 0 E +2	−95.75103	−0.00327
9	0.447 847 589 638 965 7 E +2	−223.52892	+0.00265
10	0.212 525 653 556 057 8 E +2	+239.50285	+0.01172
11	0.767 976 358 170 845 8 E +1	+524.64944	+0.01013
12	0.213 689 459 382 850 0 E +1	−319.79981	+0.00366
13	0.459 843 348 928 069 3 E +0	−787.60686	+0.00007
14	0.763 614 629 231 648 0 E −1	+179.54782	−0.00071
15	0.969 328 620 373 121 3 E −2	+700.42832	−0.00066
16	0.923 069 154 007 007 5 E −3	+29.48666	−0.00085
17	0.638 116 590 952 653 8 E −4	−335.24378	−0.00028
18	0.302 293 237 874 619 2 E −5	−77.25660	+0.00034
19	0.877 551 391 303 760 2 E −7	+66.76292	+0.00026
20	0.117 702 613 125 477 4 E −8	+24.44911	+0.00005

[a]The values for A_i are written in the form: decimal number E signed integer exponent, where E denotes the base 10.

Source: Ref. 8. Reprinted with permission.

neon, the triple point of oxygen, and by the derivative at the triple point of oxygen as determined with Eq. (4.5).

In part (c) the constants in Eq. (4.5) are determined by measuring the deviations at the triple point of oxygen and the boiling point of oxygen, and by the derivative of Eq. (4.6) at the boiling point of oxygen. In part (d) the constants are determined by the deviations measured at the boiling point of oxygen and the boiling point of water.

Note that, in practice, a platinum resistance thermometer in part (a) of range I must be calibrated at the three higher temperature parts (b), (c), and (d) to ensure that the derivatives will be constant at the transition from one part to another. This requires the user to determine 13 coefficients using eight fixed points. Any error in the higher-temperature measurements will produce error in the lower-temperature measurements. Kirby et al. have proposed a new deviation function.[14] If it or a similar version is adopted, the calibration process will be simpler. They recommend

$$\Delta W = a(w - 1) + \sum_{i=1}^{5} b_i (\ln W)^i , \qquad (4.7)$$

where the a, b_i coefficients are determined from the seven defining points plus one near the xenon triple point. Electrical resistance temperature measurement is discussed in more detail in Chapter 8.

Range II: 273.15 to 903.89 K (0.00 to 630.74°C). Use the platinum resistance thermometer with equations

$$t_{68} = t' + \Delta t$$

$$t' = \frac{1}{\alpha}[W(t') - 1] + \delta\left(\frac{t}{100}\right)\left(\frac{t}{100} - 1\right) \tag{4.8}$$

$$\Delta t = 0.045\left(\frac{t'}{100}\right)\left(\frac{t'}{100} - 1\right)\left(\frac{t'}{419.58} - 1\right)\left(\frac{t'}{630.74} - 1\right), \tag{4.9}$$

where $W(t') = R(t')/R(0°C)$, $R(t')$ is the resistance at temperature t, and the constants $R(0°C)$, α, and δ are obtained by measuring the resistance at the triple point of water, the boiling point of water (or the freezing point of tin), and the freezing point of zinc. Note that 419.58°C is the zinc point and 630.74°C is the antimony point.

Range III: 903.89 to 1337.58 K (630.74 to 1064.43°C). Use the platinum–platinum 90 rhodium 10 thermocouple with the equation

$$E = a + bt + ct^2, \tag{4.10}$$

where E is the electromotive force generated by the thermocouple when its cold junction is at 0°C and its hot junction is at temperature t. The constants are obtained for each thermocouple by calibrating it at 630.74 ± 0.2°C as determined with a platinum resistance thermometer, and at 961.93°C and 1064.43°C, the freezing points of silver and gold, respectively.

For such thermocouples, it is required that the thermocouple have the following voltage characteristics:

$$E_{t_{Au}} = 10{,}334 \ \mu V \pm 30 \ \mu V \tag{4.11}$$

$$E_{t_{Au}} - E_{t_{Ag}} = 1186 \ \mu V + 0.17(E_{t_{Au}} - 10{,}334 \ \mu V) \pm 3 \ \mu V \tag{4.12}$$

and

$$E_{t_{Au}} - E_{(630.74)} = 4782 \ \mu V + 0.63(E_{t_{Au}} - 10{,}334 \ \mu V) \pm 5 \ \mu V. \tag{4.13}$$

Thermoelectric pyrometry is discussed in Chapter 10.

Range IV: All Temperatures above 1337.58 K (1064.43°C). No particular thermal radiation instrument is specified. The scale is defined by the ratio of the radiant flux at temperature to the flux at the gold point, using Planck's law,

$$J_\lambda = \frac{2hc^2/n^2\lambda^5}{\exp\left(hc/kn\lambda T\right) - 1}.$$

(4.14)

Here h is Planck's constant, c is the velocity of light in a vacuum, k is Boltzmann's constant, λ is the wavelength in a vacuum, and n is the index of refraction of the medium through which the radiant flux occurs. The temperature is determined by comparing the ratio of the flux from a blackbody at that temperature to that from a blackbody at the gold point. Thus the defining equation for air is

$$\frac{J_{\lambda,T}}{J_{\lambda,Au}} = \frac{e^{1.4388/1.0028\lambda T_{Au}} - 1}{e^{1.4388/1.0028\lambda T} - 1},$$

(4.15)

where 1.0028 is the index of refraction of air. Measurement of the radiant flux requires some type of an optical pyrometer. Historically, the disappearing filament optical pyrometer was used. Now photoelectric optical pyrometers are usually preferred. Radiation methods are discussed in Chapters 11 to 14.

Note that temperatures are represented in range IV by a fundamental relationship. Temperature is obtained from Planck's law and the ratio of the radiance from the unknown body to the radiance of a blackbody at the gold point. If both the unknown and the gold point standard behave as blackbodies, there are no arbitrary constants in the defining equation. All the constants are either fundamental physical constants or measurable parameters, such as wavelength and index of refraction. Temperatures in the other three ranges are obtained by comparison with the gas thermometer and the fundamental perfect gas law. The other three IPTS defining equations are not fundamental.

4.5 EPT-76: A PROVISIONAL SCALE OF IPTS

Over most of its range the IPTS is believed to be much more accurate than is required for most applications as a representation of true thermodynamic temperature. However, at the time this book is being written, the IPTS is not defined below the triple point of hydrogen, 13.81 K. The platinum resistance thermometer is not as accurate as desired below the triple point of argon, 54.361 K. It is known to have significant error at the lower end of the scale. This is very important because many exciting physical phenomena exist below the triple point of argon. Practical devices such as many superconductors operate at extremely low temperatures, so there is a great need for improved temperature measurement at temperatures approaching absolute zero.

Part of the impetus for accuracy in temperature measurement in this temperature range is the fact that the temperature is so low. An error of 0.1 K at 1000 K is 0.01% An error of 0.1 K at 10 K is 1%. As absolute zero is approached, temperature measurements are often needed within about 0.1 mK (0.0001 K).

There had been great difficulty in establishing a continuous and accurate scale. Constant-volume gas thermometry had been used as a thermo-dynamic scale representation to calibrate fixed points down to about 4 K. Helium vapor pressure had often been used as the best practical device to represent true thermodynamic temperature. However, difficulties exist because most of the auxiliary methods cannot be used over a wide range of temperatures, temperatures of fixed points may be less accurate than required, and the scales or instruments for various methods may not overlap or be readily available to others. Early attempts at improving the low end of IPTS revealed problems with nonuniqueness (different platinum resistance thermometers gave different temperatures between the same calibrating points), with conflicting results and with conflicting opinions within the CCT.

The IPTS is not defined below 13.81, but several helium vapor pressure scales have been recommended. The most recent were based on research at the National Bureau of Standards for ^{4}He vapor pressure published in 1958[15] and for ^{3}He published in 1962.[16] In 1965, measurements at NBS by acoustic thermometry in the range 4–20 K indicated that the ^{4}He temperatures of 1958 were lower than the thermodynamic temperatures.[17] Research at several metrology laboratories confirmed the acoustic thermometry results.[18] An especially significant experiment was conducted by T. C. Cetas and C. A. Swenson at Iowa State University in which magnetic susceptibility measurements of nearly ideal paramagnetic salts were used to obtain 10 times better precision than previous methods for the range 0.9–18 K.[19] The scale was thermodynamically smooth. Improvements by Swenson provided a corrected scale, known as $T_{XAC'}$, that was smooth and consistent with thermodynamic fixed points.[20] T. C. Cetas made additional measurements and proposed a smooth scale from 1 to 83 K.[21] Because of the inconsistencies between helium temperatures, IPTS-68, and thermo-dynamic temperature, in 1974 the CCT decided to continue an inter-national comparison of platinum resistance thermometers (begun in 1971), to compare scales from 1 to 30 K, to publish a best estimate for 1–30 K as soon as possible, and to revise the 1958 ^{4}He and 1962 ^{3}He scales as soon as possible.[22] The results of the comparisons,[23,24] and the availability of new pressure–volume thermometry and constant-volume gas thermometry measurements on helium (NPL-75)[25] showed that a scale could be established that would represent thermodynamic temperatures more ac-curately and that would smoothly join IPTS-68 at 27.1 K.[26,27] The CCT then adopted the "Echelle Provisoire de Temperature de 1976 entre 0.5 K et 30 K.[28] This provisional temperature scale (EPT-76) is to be used until a new IPTS is adopted or until further improvements are made. In 1974, a new IPTS was anticipated about 1983, but the new scale, ITS-90 is scheduled to appear in 1990.

The EPT-76, published in English in 1979, is defined in terms of reference points (Table 4.4). Approved methods of interpolating between reference points include:

1. A thermodynamic interpolating instrument such as a gas thermometer or a magnetic thermometer.

2. Conversion of IPTS-68 temperatures above 13.81 by corrections from "Table 2" (Table 4.5).

3. Correction of ^{4}He or ^{3}He vapor pressures results using "Table 3" (Table 4.6) at temperatures up to 5 K.

4. Use of other metrology scales for which corrections are provided in "Tables 5 to 7" (Tables 4.7 to 4.9).

In 1983 a method of calculating EPT-76 from saturation vapor pressures of ^{3}He or ^{4}He was published by Durieux and Rusby.[29] The equations are

TABLE 4.4 Reference Points of the 1976 Provisional 0.5–30 K Temperature Scale (EPT-76)[a]

Reference Point	Assigned Temperature T_{76} (K)
Superconducting transition point of cadmium	0.519
Superconducting transition point of zinc	0.851
Superconducting transition point of aluminum	1.1796
Superconducting transition point of indium	3.4145
Boiling point of ^{4}He[b]	4.2221
Superconducting transition point of lead	7.1999
Triple point of equilibrium hydrogen[c]	13.8044
Boiling point of equilibrium hydrogen at a pressure of 33 330.6 Pa (25/76 standard atmosphere)[c]	17.0373
Boiling point of equilibrium hydrogen[b,c]	20.2734
Triple point of neon[d]	24.5591
Boiling point of neon[b,c,d]	27.102

[a]Superconducting transition point: the transition temperature between the superconducting and the normal state in zero magnetic field as given by NBS-SRM 767.
[b]Boiling point under a pressure $p_0 = 101\,325$ Pa (1 standard atmosphere).
[c]These are the four lower defining points of the IPTS-68. (Note: The values of temperature assigned to these points in EPT-76 are not the same as those assigned in IPTS-68.) The term equilibrium hydrogen means here that the hydrogen should have its equilibrium ortho-para composition at the relevant temperature.
[d]The two neon points are for neon with the natural isotopic composition of 2.7 mmol of ^{21}Ne and 92 mmol of ^{22}Ne per 0.905 mol of ^{20}Ne.

TABLE 4.5 Differences between EPT-76 (T_{76}) and the IPTS-68 (T_{68})

T_{68} (K)	$T_{68}-T_{76}$ (mK)	T_{68} (K)	$T_{68}-T_{76}$ (mK)	T_{68} (K)	$T_{68}-T_{76}$ (mK)
13.81	5.6	19.0	7.4	24.5	2.1
14.0	4.6	19.5	7.3	25.0	1.6
14.5	3.0	20.0	6.9	25.5	1.1
15.0	2.0	20.5	6.4	26.0	0.7
15.5	2.2	21.0	5.8	26.5	0.3
16.0	2.6	21.5	5.3	27.0	0.0
16.5	3.6	22.0	4.8	27.1	0.0
17.0	4.6	22.5	4.2	28.0	0.0
17.5	5.6	23.0	3.7	29.0	0.0
18.0	6.5	23.5	3.2	30.0	0.0
18.5	7.2	24.0	2.7		

TABLE 4.6 Differences between the EPT-76 (T_{76}) and the Helium Vapor Pressure Scales [the 1958 ^{4}He Scale (T_{58}) and the 1962 ^{3}He Scale (T_{62})][a]

T_{vp} (K)	$T_{vp}-T_{76}$ (mK)	T_{vp} (K)	$T_{vp}-T_{76}$ (mK)	T_{58} (K)	$T_{58}-T_{76}$ (mK)
0.5	−1.9	1.8	−3.9	3.2	−6.6
0.6	−2.1	2.0	−4.1	3.4	−6.8
0.8	−2.5	2.2	−4.4	3.6	−7.0
1.0	−2.9	2.4	−4.9	3.8	−7.0
1.2	−3.2	2.6	−5.4	4.0	−7.1
1.4	−3.5	2.8	−5.9	4.2	−7.1
1.6	−3.7	3.0	−6.3	4.5	−7.1
		3.2	−6.6	5.0	−7.1

[a] T_{vp} means an average of T_{62} and T_{58} up to 3.2 K.

TABLE 4.7 Differences between the NBS Provisional Temperature Scale 2–20 K (1965) ($T_{NBS2-20}$) and EPT-76 (T_{76})

$T_{NBS\,2-20}$[a] (K)	$T_{NBS\,2-20}-T_{76}$ (mK)	$T_{NBS\,2-20}$[a] (K)	$T_{NBS\,2-20}-T_{76}$ (mK)
2.3	2.1	11.0	−1.0
2.8	−1.1	12.0	0.2
3.2	1.0	13.0	−1.8
4.2	2.5	14.0	−2.2
5.0	3.0	15.0	−0.6
6.0	1.7	16.0	0.9
7.0	4.8	17.0	1.7
8.0	2.2	18.0	−0.9
9.0	−1.5	19.0	−0.2
10.0	−2.1	20.0	−0.8(+7[b])

[a] These temperatures are very close to the actual temperatures (acoustic points) at which the NBS 2–20 scale is defined.
[b] With $T_{NBS\,2-20}$ as given in early NBS calibrations.

TABLE 4.8   Differences between the Magnetic Scales of ISU, KOL, NML, PRMI, NPL, the NPL-75 Gas Thermometer Scale, and the EPT-76[a]

Laboratory	Scale	a $(10^{-3}\,K)$	b (10^{-3})	c $(10^{-3}/K)$	d $(10^{-3}\,K^2)$	Range (K)
ISU	$T_{XAc'}$	0	0	0.0025	0	1.1–30
KOL	$T_m(III)$	−8.0	1.5	−0.0413	8.3	2–27
NML	T_{XNML}	−1.5	0.41	−0.0109	0	1.1–30
NML	T_{MAS}	−1.5	0.49	−0.0125	0	1.1–30
PRMI	T_{XPRMI}	0	0.51	−0.0125	0	4.2–27
NPL	T_{X1}	0	0	0	0	0.5–3.1
NPL	T_{NPL-75}	0	0	−0.0056	0	2.6–27.1

[a]Coefficients are given for the relationship

$$T_i - T_{76} = a + bT_i + cT_i^2 + d/T_i, \tag{4.16}$$

where T_i represents temperatures on the various laboratory scales.

TABLE 4.9   Differences between the NBS Version of the IPTS-68 (T_{NBS-68}) and the EPT-76 (T_{76})

T_{NBS-68} (K)	$T_{NBS-68}-T_{76}$ (mK)	T_{NBS-68} (K)	$T_{NBS-68}-T_{76}$ (mK)	T_{NBS-68} (K)	$T_{NBS-68}-T_{76}$ (mK)
13.8	1.7	19.0	4.9	24.5	2.9
14.0	1.8	19.5	4.9	25.0	2.5
14.5	2.1	20.0	4.9	25.5	2.1
15.0	2.0	20.5	4.8	26.0	1.7
15.5	2.2	21.0	4.7	26.5	1.3
16.0	2.4	21.5	4.6	27.0	1.1
16.5	3.0	22.0	4.5	27.5	1.1
17.0	3.5	22.5	4.2	28.0	1.1
17.5	3.9	23.0	4.0	29.0	1.1
18.0	4.4	23.5	3.7	30.0	1.0
18.5	4.8	24.0	3.4		

more convenient than difference Table 3 (Table 4.6) from EPT-76. These equations were adopted by CCT at its meeting in 1982 to replace Table 3.[30] The equations, for pressure P in pascal, are:

^{4}He

1.4 K to the λ point:

$$\ln P = \sum_{k=-1}^{n} a_k T^k \tag{4.17}$$

Above the λ point:

$$\ln P = \sum_{k=-1}^{n} a_k t^k + b(1-t)^{1.9}, \tag{4.18}$$

where $t = T/T_c$ and T_c is the critical temperature.

^{3}He:

$$\ln P = \sum_{k=-1}^{n} a_k T^k + b \ln T \tag{4.19}$$

The coefficients for the tables are given in Table 4.10. Vapor pressures and their derivatives are given in Table 4.11.

TABLE 4.10   Coefficients of the Vapor Pressure Equations

^{4}He 0.5 to 2.1768 K	^{4}He 2.1768 K to T_c	^{3}He 0.2 K to T_c
$a_{-1} = -7.41816$ K	$a_{-1} = \quad -30.93285$	$a_{-1} = -2.50943$ K
$a_0 = 45.42128$	$a_0 = \quad 392.47361$	$a_0 = \quad 9.70876$
$a_1 = \quad 9.903203$ K^{-1}	$a_1 = -2{,}328.04587$	$a_1 = -0.304433$ K^{-1}
$a_2 = -9.617095$ K^{-2}	$a_2 = \quad 8{,}111.30347$	$a_2 = \quad 0.210429$ K^{-2}
$a_3 = \quad 6.804602$ K^{-3}	$a_3 = -17{,}809.80901$	$a_3 = -0.0545145$ K^{-3}
$a_4 = -3.0154606$ K^{-4}	$a_4 = -25{,}766.52747$	$a_4 = \quad 0.0056067$ K^{-4}
$a_5 = \quad 0.7461357$ K^{-5}	$a_5 = -24{,}601.4$	$b = \quad 2.25484$
$a_6 = -0.0791791$ K^{-6}	$a_6 = \quad 14{,}944.65142$	
	$a_7 = -5{,}240.36518$	
	$a_8 = \quad 807.93168$	
	$b = \quad 14.53333$	
	$T_c = \quad 5.1953$ K	

4.6   THE RELATIONSHIP OF IPTS-68 TO THERMODYNAMIC TEMPERATURES

In the evolution of the International Practical Temperatures Scale, the goal has been to produce a working scale as close as possible to the theoretical thermodynamic temperature scale by making it conform as close as possible to the current best estimate of that theoretical scale. Except at temperatures near absolute zero, gas thermometry has been considered the method most nearly approaching the theoretical scale. Over the past 60 years, gas thermometry, particularly at cryogenic temperatures, has improved. Some systematic errors have been recognized and corrected. The IPTS, then, has been attempting to conform to a moving target. Decisions in the early years, such as the exact value of absolute zero as 273.16°C below the triple point of water are, at present, locked in by international agreement. Future

TABLE 4.11 Vapor Pressure in Pascal and the Derivative dP/dT in Pascal/Kelvin

T (K)	0.00	0.01	0.02	0.03	0.04	0.05	0.06	0.07	0.08	0.09
					^{4}He (EPT-76)					
0.50	0.002063	0.002874	0.003958	0.005389	0.007261	0.009685	0.012798	0.016760	0.021762	0.028028
	0.069678	0.093589	0.124326	0.163446	0.212771	0.274413	0.350807	0.444734	0.559354	0.698236
0.60	0.035820	0.045444	0.057251	0.071644	0.089086	0.110099	0.135277	0.165287	0.200879	0.242887
	0.865381	1.065255	1.302815	1.583534	1.913427	2.299075	2.747648	3.266925	3.865312	4.551863
0.70	0.292242	0.349974	0.417220	0.495232	0.585382	0.689173	0.808240	0.944364	1.099473	1.275653
	5.336292	6.228987	7.241026	8.384181	9.670929	11.11446	12.72867	14.52819	16.52834	18.74518
0.80	1.475154	1.700399	1.953985	2.238698	2.557515	2.913611	3.310367	3.751377	4.240451	4.781625
	21.19548	23.89669	26.86700	30.12526	33.69103	37.58451	41.82659	46.43878	51.44325	56.86276
0.90	5.379167	6.037580	6.761609	7.556249	8.426746	9.378604	10.41759	11.54975	12.78138	14.11907
	62.72069	69.04097	75.84813	83.16722	91.82383	99.44402	108.4544	118.0819	128.3540	139.2987
1.00	15.56969	17.14038	18.83859	20.67205	22.64878	24.77709	27.06562	29.52328	32.15930	34.98322
	150.9441	163.3189	176.4520	190.3728	205.1109	220.6959	237.1582	254.5279	272.8356	292.1118
1.10	38.00487	41.23441	44.68228	48.35927	52.27645	56.44521	60.07725	65.58459	70.57956	75.87478
	312.3876	333.6936	356.0611	379.5210	404.1044	429.8426	456.7665	484.9074	514.2961	544.9637
1.20	81.48320	87.41806	93.69293	100.3217	107.3184	114.6977	122.4742	130.6630	139.2795	148.3392
	576.9409	610.2585	644.9469	681.0365	718.5574	757.5396	798.0126	840.0058	883.5484	928.6689
1.30	157.8582	167.8526	178.3389	189.3338	200.8545	212.9181	225.5422	238.7447	252.5436	266.9573
	975.3958	1023.757	1073.781	1125.493	1178.922	1234.094	1291.033	1349.766	1410.318	1472.713
1.40	282.0041	297.7031	314.0730	331.1333	348.9034	367.4028	386.6516	406.6697	427.4774	449.0952
	1536.975	1603.127	1671.191	1741.190	1813.145	1887.077	1963.007	2040.953	2120.934	2202.969
1.50	471.5437	494.8436	519.0160	544.0820	570.0627	596.9796	624.8542	653.7082	683.5633	714.4413
	2287.074	2373.267	2461.563	2551.978	2644.525	2739.217	2836.068	2635.089	3036.291	3139.683
1.60	746.3643	779.3542	813.4332	848.6233	884.9469	922.4262	961.0836	1000.941	1042.021	1084.346
	3245.275	3353.074	3463.086	3575.318	3689.775	3806.459	3925.373	4046.519	4169.894	4295.499

TABLE 4.11 (Continued)

T (K)	0.00	0.01	0.02	0.03	0.04	0.05	0.06	0.07	0.08	0.09
1.70	1127.939	1172.820	1219.014	1266.541	1315.424	1365.684	1417.344	1470.425	1524.949	1580.937
	4423.330	4553.383	4685.650	4820.126	4956.800	6095.661	5236.696	5379.890	5525.227	5672.686
1.80	1638.410	1697.389	1757.894	1819.947	1883.567	1948.773	2015.586	2084.024	2154.105	2225.848
	5822.247	5973.885	6127.573	6283.284	6440.984	6600.638	6762.209	6925.654	7090.929	7257.985
1.90	2299.270	2374.389	2451.220	2529.780	2610.083	2692.144	2775.977	2861.593	2949.006	3038.226
	7426.769	7597.224	7769.288	7942.897	8117.979	8294.458	8472.253	8651.277	8831.437	9012.634
2.00	3129.262	3222.123	3316.818	3413.352	3511.731	3611.958	3714.034	3817.962	3923.739	4031.363
	9194.762	9377.707	9561.351	9745.564	9930.211	10115.15	10300.12	10485.27	10670.12	10854.60
2.10	4140.829	4252.131	4365.259	4480.202	4596.948	4715.480	4835.780	4957.827	5081.617	5207.358
	11038.50	11221.63	11403.77	11584.69	11764.17	11941.94	12117.74	12291.30	12472.54	12676.07
2.20	5335.146	5465.001	5596.946	5731.003	5867.192	6005.537	6146.059	6288.780	6433.721	6580.904
	12881.78	13089.66	13299.72	13511.95	13726.37	13942.97	14161.76	14382.72	14605.86	14831.17
2.30	6730.351	6882.084	7036.125	7192.494	7351.214	7512.306	7675.792	7841.692	8010.029	8180.024
	15058.66	15288.31	15520.13	15754.11	15990.24	16228.52	16468.95	16711.51	16956.20	17203.02
2.40	8354.097	8529.870	8708.164	8888.999	9072.398	9258.380	9446.967	9638.179	9832.036	10028.56
	17451.95	17702.99	17956.14	18211.38	18468.70	18728.10	18989.57	19253.11	19518.70	19786.33
2.50	10227.77	10429.69	10634.33	10841.72	11051.88	11264.83	11480.58	11699.16	11920.59	12144.89
	20056.01	20327.72	20601.44	20877.19	21154.94	21434.69	21716.44	22000.17	22285.89	22573.57
2.60	12372.07	12602.16	12835.17	13071.13	13310.06	13551.97	13796.88	14044.81	14295.79	14549.83
	22863.22	23154.83	23448.39	23743.90	24041.36	24340.74	24642.07	24945.31	25250.49	25557.57
2.70	14806.95	15067.17	15330.51	15596.98	15866.61	16139.42	16415.42	16694.64	16977.09	17262.79
	25866.58	26177.49	26490.31	26805.04	27121.66	27440.19	27760.61	28082.93	28407.14	28733.24
2.80	17551.76	17844.02	18139.59	18438.48	18740.72	19046.33	19355.32	19667.71	19983.53	20302.78
	29061.23	29391.11	29722.88	30056.53	30392.08	30729.51	31068.83	31410.04	31753.14	32098.13
2.90	20625.50	20951.69	21281.38	21614.58	21951.32	22291.62	22635.49	22982.95	23334.02	23688.72
	32445.01	32793.78	33144.45	33497.01	33851.47	34207.82	34566.08	34926.24	35288.30	35652.27

x										
3.00	24047.07	24409.09	24774.80	25144.21	25517.35	25894.23	26274.88	26659.31	27047.54	27439.60
	36018.14	36385.93	36755.63	37127.24	37500.77	37876.22	38253.59	38632.88	39014.10	39397.25
3.10	27835.49	28235.25	28638.89	29046.42	29457.88	29873.27	30292.62	30715.94	31143.26	31574.60
	39782.32	40169.33	40558.27	40949.15	41341.96	41736.71	42133.40	42532.03	42932.61	43335.12
3.20	32009.97	32449.40	32892.90	33340.49	33792.20	34248.04	34708.03	35172.19	35640.54	36113.10
	43739.59	44145.99	44554.35	44964.65	45376.89	45791.09	46207.23	46625.31	47045.35	47467.33
3.30	36589.89	37070.93	37556.24	38045.84	38539.74	39037.97	39540.55	40047.49	40558.82	41074.55
	47891.25	48317.13	48744.94	49174.70	49606.40	50040.04	50475.62	50913.14	51352.59	51793.98
3.40	41594.70	42119.30	42648.36	43181.90	43719.94	44262.51	44809.61	45361.27	45917.50	46478.33
	52237.30	52682.55	53129.72	53578.83	54029.85	54482.80	54937.66	55394.44	55853.14	56313.74
3.50	47043.78	47613.87	48188.60	48768.01	49352.11	49940.93	50534.47	51132.76	51735.82	52343.66
	56776.26	57240.68	57707.00	58175.23	58645.35	59117.37	59591.29	60067.09	60544.79	61024.37
3.60	52956.31	53573.78	54196.10	54823.28	55455.34	56092.30	56734.17	57380.99	58032.75	58689.50
	61505.84	61989.20	62474.44	62961.56	63450.56	63941.44	64434.20	64928.84	65425.36	65923.76
3.70	59351.24	60017.99	60689.77	61366.60	62048.50	62735.48	63427.58	64124.79	64827.16	65534.68
	66424.04	66926.20	67430.24	67936.17	68443.98	68953.68	69465.27	69978.75	70494.13	71011.41
3.80	66247.39	66965.30	67688.43	68416.80	69150.43	69889.34	70633.54	71383.06	72137.92	72898.14
	71530.59	72051.69	72574.70	73099.63	73626.48	74155.27	74686.00	75218.67	75753.30	76289.89
3.90	73663.73	74434.71	75211.11	75992.95	76780.24	77573.01	78371.27	79175.05	79984.37	80799.24
	76828.45	77369.00	77911.53	78456.07	79002.62	79551.19	80101.80	80654.45	81209.16	81765.94
4.00	81619.69	82445.75	83277.42	84114.73	84957.71	85806.36	86660.73	87520.82	88386.67	89258.28
	82324.81	82885.78	83448.85	84014.06	84581.40	85150.91	85722.58	86296.44	86872.50	87450.78
4.10	90135.69	91018.91	91907.98	92802.90	93703.71	94610.43	95523.08	96441.69	97366.27	98296.85
	88031.30	88614.07	89199.11	89786.43	90376.05	90967.99	91562.27	92158.89	92757.89	93359.27
4.20	99233.46	100176.1	101124.9	102079.7	103040.6	104007.7	104981.0	105960.5	106946.2	107938.1
	93963.06	94569.26	95177.90	95788.98	96402.54	97018.57	97637.10	98258.15	98881.72	99507.82
4.30	108936.4	109940.9	110951.7	111968.9	112992.5	114022.5	115058.9	116101.7	117151.1	118206.9
	100136.5	100767.7	101401.5	102037.9	102676.9	103318.5	103962.7	104609.6	105259.1	105911.2
4.40	119269.3	120338.2	121413.8	122495.9	123584.7	124680.2	125782.3	126891.2	128006.8	129129.2
	106566.0	107223.5	107883.6	108546.4	109211.9	109880.0	110550.8	111224.3	111900.5	112579.3

TABLE 4.11 (Continued)

T (K)	0.00	0.01	0.02	0.03	0.04	0.05	0.06	0.07	0.08	0.09
4.50	130258.4	131394.4	132537.3	133687.1	134843.8	136007.4	137177.9	138355.5	139540.1	140731.7
	113260.8	113945.0	114631.9	115321.5	116013.7	116708.6	117406.1	118106.3	118809.2	119514.7
4.60	141930.4	143136.2	144349.1	145569.1	146796.3	148030.8	149272.4	150521.3	151777.5	153041.0
	120222.8	120933.6	121647.1	122363.2	123081.9	123803.3	124527.3	125254.0	125983.3	126715.3
4.70	154311.8	155590.0	156875.6	158168.6	159469.0	160776.9	162092.3	163415.2	164745.7	166083.8
	127450.0	128187.4	128927.6	129670.4	130416.1	131164.6	131915.9	132670.2	133427.4	134187.7
4.80	167429.5	168782.8	170143.8	171512.6	172889.1	174273.3	175665.4	177065.4	178473.3	179889.1
	134951.1	135717.7	136487.6	137261.0	138037.9	138818.6	139603.1	140391.7	141184.6	141982.0
4.90	181313.0	182744.8	184184.8	185632.9	187089.3	188553.9	190026.9	191508.3	192998.2	194496.7
	142784.3	143591.6	144404.3	145222.8	146047.6	146879.0	147717.6	148564.0	149418.8	150282.7
5.00	196003.9	197519.9	199044.7	200578.7	202121.8	203674.2	205236.1	206807.6	208389.1	209980.7
	151156.6	152041.4	152938.1	153847.9	154772.2	155712.4	156670.3	157647.9	158647.7	159672.4
5.10	211582.6	213195.3	214819.0	216454.1	218101.1	219760.7	221433.4	223120.3	224822.5	226542.3
	160725.4	161810.8	162933.5	164099.9	165318.4	166600.4	167962.4	169430.7	171054.5	172963.1
$^3He(EPT-76)$										
0.20	0.001473	0.002982	0.005689	0.010304	0.017831	0.029640	0.047540	0.073854	0.111500	0.164058
	0.108698	0.201059	0.352024	0.587583	0.940590	1.451226	2.167267	3.144113	4.444630	6.138790
0.30	0.235843	0.331964	0.458384	0.621966	0.830515	1.092810	1.418624	1.818748	2.304991	2.890186
	8.303170	11.02031	14.37801	18.46849	23.38763	29.23412	36.10860	44.11296	53.34952	63.92037
0.40	3.588183	4.413839	5.382996	6.512467	7.820004	9.324274	11.04482	13.00204	15.21713	17.71206
	75.92674	89.46842	104.6433	121.5468	140.2718	160.9079	183.5415	208.2553	235.1287	264.2370
0.50	20.50955	23.63301	27.10649	30.95467	35.20282	39.87673	45.00272	50.60755	56.71843	63.36296
	295.6517	329.4404	365.6668	404.3910	445.6687	489.5525	536.0907	585.3286	637.3076	692.0659
0.60	70.56913	78.36522	86.77986	95.84193	105.5806	116.0251	127.2051	139.1503	151.8905	165.4559
	749.6386	810.0576	873.3517	939.5471	1008.667	1080.733	1155.763	1233.773	1314.778	1398.789

0.70	179.8764	195.1823	211.4039	228.5715	246.7155	265.8663	286.0544	307.3101	329.6640	353.1465
	1485.817	1575.869	1668.953	1765.073	1864.234	1966.438	2071.685	2179.976	2291.309	2405.683
0.80	377.7878	403.6185	430.6687	458.9688	488.5490	519.4394	551.6701	585.2711	620.2723	656.7036
	2523.095	2643.540	2767.014	2893.511	3023.027	3155.554	3291.085	3429.613	3571.130	3715.628
0.90	694.5948	733.9755	774.8753	817.3237	861.3501	906.9839	954.2542	1003.190	1053.821	1106.175
	3863.098	4013.532	4166.921	4323.254	4482.523	4644.719	4809.831	4977.850	5148.765	5322.567
1.00	1160.282	1216.170	1273.867	1333.403	1394.806	1458.103	1523.324	1590.496	1659.648	1730.806
	5499.247	5678.793	5861.196	6046.446	6234.533	6425.447	6619.179	6815.718	7015.056	7217.181
1.10	1804.000	1879.257	1956.605	2036.070	2117.682	2201.466	2287.450	2375.662	2466.128	2558.875
	7422.085	7629.759	7840.192	8053.377	8269.303	8487.961	8709.344	8933.441	9160.245	9389.746
1.20	2653.932	2751.323	2851.077	2953.219	3057.777	3164.776	3274.244	3386.207	3500.691	3617.722
	9621.936	9856.808	10094.35	10334.56	10577.43	10822.94	11071.10	11321.89	11575.30	11831.33
1.30	3737.326	3859.530	3984.359	4111.840	4241.998	4374.859	4510.449	4648.794	4789.919	4933.849
	12089.98	12351.23	12615.08	12881.51	13150.53	13422.13	13696.30	13973.03	14252.31	14534.15
1.40	5080.610	5230.228	5382.727	5538.134	5696.473	5857.770	6022.050	6189.337	6359.657	6533.034
	14818.53	15105.45	15394.90	15686.88	15981.38	16278.39	16577.91	16879.93	17184.45	17491.46
1.50	6709.494	6889.062	7071.761	7257.618	7446.656	7638.900	7834.375	8033.105	8235.115	8440.430
	17800.96	18112.94	18427.39	18744.32	19063.71	19385.55	19709.86	20036.61	20365.81	20697.44
1.60	8649.072	8861.068	9076.441	9295.215	9517.415	9743.064	9972.187	10204.81	10440.95	10680.64
	21031.51	21368.02	21706.94	22048.29	22392.06	22738.23	23086.82	23437.80	23791.19	24146.97
1.70	10923.90	11170.75	11421.22	11675.33	11933.11	12194.57	12459.75	12728.66	13001.33	13277.79
	24505.14	24865.69	25228.63	25593.94	25961.63	26331.69	26704.11	27078.90	27456.05	27835.55
1.80	13558.05	13842.15	14130.09	14421.92	14717.64	15017.29	15320.89	15628.45	15940.01	16255.59
	28217.40	28601.60	28988.14	29377.03	29768.26	30161.82	30557.71	30955.93	31356.48	31759.35
1.90	16575.21	16898.89	17226.66	17558.53	17894.54	18234.71	18579.06	18927.60	19280.38	19637.40
	32164.54	32572.06	32981.88	33394.03	33808.48	34225.24	34644.31	35065.69	35489.37	35915.36
2.00	19998.69	20364.28	20734.18	21108.43	21487.04	21870.03	22257.44	22649.27	23045.57	23446.33
	36343.64	36774.23	37207.11	37642.30	38079.78	38519.56	38961.63	39406.00	39852.67	40301.63

TABLE 4.11 (Continued)

T (K)	0.00	0.01	0.02	0.03	0.04	0.05	0.06	0.07	0.08	0.09
2.10	23851.61	24261.40	24675.74	25094.65	25518.16	25946.28	26379.04	26816.46	27258.57	27705.39
	40752.89	41206.45	41662.31	42120.46	42580.92	43043.67	43508.73	43976.10	44445.77	44917.75
2.20	28156.93	28613.23	29074.31	29540.19	30010.90	30486.45	30966.87	31452.19	31942.42	32437.60
	45392.05	45868.66	46347.59	46828.84	47312.42	47798.33	48286.57	48777.15	49270.08	49765.35
2.30	32937.74	33442.87	33953.01	34468.18	34988.41	35513.73	36044.16	36579.71	37120.42	37666.31
	50262.98	50762.97	51265.33	51770.07	52277.18	52786.69	53298.59	53812.90	54329.62	54848.77
2.40	38217.41	38773.73	39335.30	39902.15	40474.31	41051.78	41634.62	42222.82	42816.43	43415.47
	55370.34	55894.37	56420.84	56949.78	57481.20	58015.10	58551.51	59090.42	59631.87	60175.85
2.50	44019.96	44629.93	45245.40	45866.40	46492.96	47125.09	47762.84	48406.23	49055.27	49710.01
	60722.39	61271.50	61823.20	62377.49	62934.41	63493.96	64056.16	64621.04	65188.60	65758.88
2.60	50370.46	51036.66	51708.62	52386.39	53069.98	53759.43	54454.76	55156.01	55863.20	56576.36
	66331.89	66907.65	67486.19	68067.52	68651.67	69238.67	69828.53	70421.28	71016.96	71615.58
2.70	57295.52	58020.71	58751.97	59489.31	60232.78	60982.40	61738.21	62500.23	63268.50	64043.05
	72217.18	72821.78	73429.40	74040.09	74653.88	75270.78	75890.85	76514.10	77140.58	77770.32
2.80	64823.92	65611.13	66404.72	67204.73	68011.18	68824.12	69643.58	70469.59	71302.20	72141.42
	78403.35	79039.71	79679.45	80322.60	80969.19	81619.28	82272.90	82930.10	83590.92	84255.40
2.90	72987.32	73839.91	74699.24	75565.35	76438.27	77318.05	78204.72	79098.33	79998.92	80906.52
	84923.59	85595.55	86271.31	86950.93	87634.46	88321.95	89013.45	89709.02	90408.72	91112.60
3.00	81821.18	82742.95	83671.86	84607.96	85551.30	86501.91	87459.85	88425.17	89397.90	90387.10
	91820.71	92533.13	93249.91	93971.10	94696.79	95427.02	96161.87	96901.41	97645.70	98394.82
3.10	91365.81	92361.09	93363.98	94374.54	95392.82	96418.87	97452.74	98494.49	99544.17	100601.8
	99148.83	99907.82	100671.8	101441.0	102215.4	102995.0	103780.0	104570.5	105366.4	106168.0
3.20	101667.5	102741.4	103823.3	194913.5	106012.0	107118.8	108234.0	109357.7	110489.9	111630.8
	106975.4	107788.5	108607.5	109432.5	110263.5	111100.8	111944.3	112794.2	113650.6	114513.6
3.30	112780.2	113938.5	115105.5							
	115383.3	116259.8	117143.2							

changes must, of necessity, be made in conformance with the understandings produced by the previous decisions. If a new IPTS were to be developed, it probably would not conform to the ranges, instruments, or interpolating equations now in use. Unfortunately, it is not possible to start over now. Fortunately, the existing scale continues to be improved, and the precision and accuracy now available are more than sufficient for most practical applications. It is only in the ultrahigh-precision applications that the foibles of IPTS have been serious concerns, such as in establishing the values of fundamental physical constants or in achieving the highest possible accuracy in calibrations. All parts of IPTS-68 are believed to have been reproducible to better than ±30 mK from 13.81 K to 630°C, and to about ±0.2°C from 630 to 1064.43°C. Range IV reproducibility is best near the gold point and may be ±50 mK, becoming progressively poorer as temperature increases.

In establishing the IPTS using defining points and using different instruments in different ranges, one of the problems is the continuity of the scale at the range limits. Within a given range, the defining equation is chosen to be smooth and the first and second derivatives of the functions with respect to temperature have appropriate values. Where one range meets another, no changes in those derivatives should occur. If changes do occur, physical measurements for a temperature interval extending from one range to another will be altered by the temperature scale inconsistency. This is particularly undesirable for measurements such as specific heat, where an apparent discontinuity may be the result of IPTS equations rather than the result of changes within the material. Unfortunately, discontinuities do occur. The most serious in IPTS-68 has been at the junction of platinum resistance measurements with thermocouple measurements, about 630°C. The first derivatives differ by about 0.2%

The IPTS equations pass through the defining points, so the greatest error in representing thermodynamic temperatures occurs about midway between the points. The accuracy of temperature measurement is discussed in detail in Chapter 17.

4.7   THE INTERNATIONAL TEMPERATURE SCALE OF 1990 (ITS-90)

During its meeting June 9–11, 1987, the Comité Consultatif de Thermométrie (CCT) adopted a new International Temperature Scale,[31] which is expected to be approved by the 1990 General Conference on Weights and Measures to take effect in 1990. The word "practical" was dropped from the title because no further changes appear to be needed in the foreseeable future. The scale is believed to represent true thermodynamic temperatures within about ±2 mK from 1 to 273 K, increasing to about ±3.5 mK standard deviation at 730 K, then linearly to about ±7 mK at 900 K.

Adoption of the new scale will eliminate the type S thermocouple and greatly reduce the error between 630°C and 1064°C, where IPTS-68 is believed to be about 0.25 K low at 750°C, and 0.3 K high at 1050°C. It will also eliminate the discontinuity in slope at 630°C where IPTS-68 has range II joining range III.

ITS-90 will retain the triple point of water, taken exactly as 273.16 K, as its defining point. It will retain T_{90} as the scale equivalent to thermo-dynamic temperature and t_{90} as a scale shift to Celsius;

$$t_{90} = T_{90} - 273.15 \text{ K} . \tag{4.20}$$

The scale is constructed to be thermodynamically smooth and very close to existing data for true thermodynamic temperature.

4.7.1 General Principles—ITS-90

The ITS-90 extends upwards from 0.5 K. In certain regions there are alternative definitions, but they give virtually equivalent values of T_{90}.

Between 0.5 and 4.2 K, the vapor pressure temperature relations of ^{3}He and ^{4}He are used. Between 3.2 K and 962°C defining points and standard instruments calibrated at the fixed points are used. Above 962°C Planck's law is used.

The standard instruments are the gas thermometer from 3.2 to 24.6 K, and the platinum resistance thermometer from 13.8 K to 962°C. The instrument above 962°C is not specified, nor is the wavelength to be used in Planck's law specified because choice of wavelength is not expected to make a significant difference.

4.7.2 Definition of ITS-90 in Various Temperature Ranges

1. 0.5 to 4.221 K: Helium vapor pressure/temperature relations. The following analytical equations relate the vapor pressure P of ^{3}He or ^{4}He to T_{90}:

    ^{3}He between 0.5 K and 3.2 K:

    $$\ln(P) = \sum_{k=-1}^{n} a_k T_{90}^k + d \ln(T_{90}) \tag{4.21}$$

    where the coefficients a_k and d are given in Table 4.12.
    ^{4}He between the λ point (2.1768 K) and the normal boiling point (4.2221 K):

    $$\ln(P) = \sum_{k=1}^{n} C_k (T_{90}/T_c)^k + e(1 - T_{90}/T_c)^{1.9} \tag{4.22}$$

TABLE 4.12   Coefficients of the Helium Vapor Pressure Equations

^{3}He 0.2 K to 3.2 K Coefficients in Eq. (4.31)	^{4}He 2.1768 K to 4.2 K Coefficients in Eq. (4.32)
$a_{-1} = -2.50943$ K	$c_{-1} =$ -30.93285
$a_0 =$ 9.70876	$c_0 =$ 392.47361
$a_1 = -0.304433$ K^{-1}	$c_1 =$ $-2,328.04587$
$a_2 =$ 0.210429 K^{-2}	$c_2 =$ 8,111.30347
$a_3 = -0.0545145$ K^{-3}	$c_3 = -17,809.80901$
$a_4 =$ 0.0056067 K^{-4}	$c_4 =$ 25,766.52747
$d =$ 2.25484	$c_5 = -24,601.4$
$T_c =$ 3.3162 K	$c_6 =$ 14,944.65142
	$c_7 =$ $-5,240.36518$
	$c_8 =$ 807.93168
	$e =$ 14.53333
	$T_c =$ 5.1953 K

where the coefficients C_k and e are given in Table 4.12. (Note that these compare with Eqs. 4.17, 4.18, and 4.19, and Table 4.10.)

2. **4.2 K to the Triple Point of Neon (24.6 K).** Interpolating Instrument: Gas Thermometer. In this range, T_{90} is defined in terms of an interpolating ^{4}He or ^{3}He gas thermometer, calibrated at three defining fixed points: the triple point of neon (24.6 K), the triple point of equilibrium hydrogen (13.8 K), and the normal boiling point of ^{4}He (4.2 K).

3. **13.81 K to 962°C (1235 K).** Interpolating Instrument: Platinum Resistance Thermometer.

   13.8 to 303 K. The thermometer is calibrated at the triple point of equilibrium hydrogen (13.81 K), neon (24.6 K), oxygen (54.4 K), argon (83.8 K), water (273.16 K) and gallium (303 K), and in addition within 0.1 K of 17 K and 20 K. The temperatures at 17 and 20 K are determined by either gas thermometer (see number 2) or from the vapor pressure of hydrogen at approximately 33 and 101 kPa, using:

$$\ln(P/P_0) = A + B/T_{90} + CT_{90} + DT_{90}^2 \qquad (4.23)$$

where $P_0 = 101.325$ kPa, $A = 3.94080$, $B = -101.3378$ K, $C = 0.0543200$ K^{-1} and $D = -0.00011056$ K^{-2}.

The reference function and the deviation functions are to be prepared before the CCT meeting in 1989.

0°C to 660°C or 962°C.   The resistance thermometer is calibrated at the freezing points of gallium, tin, zinc, aluminum, and silver. The reference and difference equations are to be prepared before the CCT meeting in 1989. If calibration is terminated at 660°C, the silver point is not used.

83.8 to 693 K.   The resistance thermometer is calibrated at the triple point of water and two other points selected from the triple points of argon, carbon dioxide, mercury or gallium, and the freezing points of indium, tin, and zinc. It applies only for the range selected. If the freezing points of tin or zinc are selected, then either both must be selected or the freezing point of indium must be included. The reference and difference equations are to be prepared before the CCT meeting in 1989.

4.   Above 961.93.   The temperature is defined by

$$\frac{L_\lambda(T_{90})}{L_\lambda(T_{90,\,x})} = \frac{\exp(C_2/\lambda T_{90,\,x}) - 1}{\exp(C_2/\lambda T_{90}) - 1} \tag{4.24}$$

where $T_{90,\,x}$ refers to the silver point or the gold point, L is the spectral radiance at the wavelength λ and $C_2 = 0.014388$ mK.

4.7.3   Supplemental Information

The apparatus, methods and procedures to be used will be described in a document, "Supplementary Information for the ITS-90," that is to be available when the new scale goes into effect.

The standards community would have preferred to have the resistance thermometer extend all the way to the gold point. But fragility and instability prevented this. Because of the fragility of the standard primary platinum thermometers at high temperatures it is expected that the Au/Pt and the Pd/Pt thermocouples will be used as secondary standards. They will be the basis for the calibration services provided by standards laboratories such as the National Bureau of Standards.

4.8   INTERNATIONAL MEASUREMENT CONFEDERATION

The International Measurement Confederation (IMEKO) has held two international meetings on Temperature Measurement as an activity of Technical Committee 12. The first "Symposium on Temperature Measurement in Industry and Science" was held at Karlovy Vory, Czechoslovakia, in 1981. The second was held in Suhl, German Democratic Republic, on October 16–18, 1984.[32] Although there were no U.S. participants, the European, Chinese, Canadian, Soviet, and Australian participants gave a wide range of papers of interest to the temperature measurement community.

REFERENCES

1. J. A. Hall, The early history of the international practical temperature scale. *Metrologia* **3**, 25 (1967).

2. International Committee of Weights and Measures, *Conf. Gen. Poids Mes.* **7**, 94 (1927).

3. Q. C. Wolfe, ed., "Temperature: Its Measurement and Control in Science and Industry," Vol. 1. Reinhold, New York, 1941. (The conference was held in New York City in November 1939.)

4. C. M. Herzfeld, ed., "Temperature: Its Measurement and Control in Science and Industry," Vol. 2, Reinhold, New York, 1955. (The conference was held in Washington, D.C. in October 1954.)

5. C. M. Herzfeld, ed., "Temperature: Its Measurement and Control in Science and Industry," Vol. 3, Parts 1, 2, and 3. Reinhold, New York, 1962. (The conference was held in Columbus, Ohio in March 1961.)

6. H. H. Plumb, ed., "Temperature: Its Measurement and Control in Science and Industry," Vol. 4, Parts 1, 2, and 3. Instrum. Soc. Am., Pittsburgh, Pennsylvania, 1972. (The conference was held in Washington, D.C. in June 1971.)

7. J. F. Schooley, "Temperature: Its Measurement and Control in Science and Industry," Vol. 5, Part 1 and 2. Am. Inst. Phys. New York, 1982.

8. B. F. Billing and T. J. Quinn, eds., "Temperature Measurement, 1975," Conf. Ser. N. 26. Inst. Phys. London.

9. See also C. R. Barber, ed., "Temperature Measurement at the National Physical laboratory," Collect. Pap. 1934–1970. Nat. Phys. Lab., London, 1973.

10. P. Giacomo, News from BIPM. *Metrologia* **19**, 77–81 (1983).

11. R. P. Hudson, Temperature scales, the IPTS, and its future development. *Temp.: Its Meas. Control Sci. Ind.* **5** (Part 1), 1–8 (1982).

12. The text of the IPTS 1968 is available in English in a National Physical Laboratory monograph, "The International Practical Temperature Scale of 1968." H. M. Stationery Office, London, 1969. The official French text is published in *Comptes rendus des séances de la Treizième Conférence Générales des Poids et Measures, Annexe 2* (1967). The 1975 amended edition is published in English: H. Preston-Thomas, The international practical temperature scale of 1968 amended edition of 1975. *Metrologia* **12**, 7–17 (1976).

13. L. Crovini, R. E. Bedford, and A. Moser, Extended list of secondary reference points. *Metrologia* **13**, 197–206 (1977).

13a. R. E. Bedford, G. Bonnier, H. Maas, and F. Pavese, Recommended values of temperature for a selected set of secondary reference points. *Metrologia* **20**, 145–155 (1984).

14. C. G. Kirby, R. E. Bedford, and J. Kathnelson, A proposal for a new deviation function in the IPTS-68 below 273 K. *Metrologia* **11**, 117–124 (1975).

15. F. G. Brickwedde, H. Van Dijk, M. Durieux, J. D. Clement, and J. K. Logan, The "1958 He^4 Scale of Temperatures. *J. Res. Natl. Bur. Stand., Sect. A* **64A**, 1–17 (1960).

16. R. H. Sherman, S. G. Sydoriak, and T. R. Roberts, The 1962 He^3 scale of temperatures. *J. Res. Natl. Bur. Stand., Sect. A* **68A**, 579–588 (1964).

17. H. H. Plumb and G. Cataland, An absolute temperature scale from 4°K to 20°K

determined from measurements with an acoustical thermometer. *J. Res. Natl. Bur. Stand.*, *Sect. A* **69A**, 375–377 (1969).

18. J. S. Rogers, R. J. Tainsh, M. S. Anderson, and C. A. Swenson, Comparison between gas thermometer, acoustic and platinum resistance temperature scales between 2 and 20 K. *Metrologia* **4**, 7–59 (1968).

19. T. C. Cetas and C. A. Swenson, A paramagnetic salt temperature scale, 0.9 to 18 K. *Metrologia* **8**, 46–64 (1972).

20. C. A. Swenson, Relationship from 1 to 34 K between a paramagnetic salt temperature scale and other scales; an addendum. *Metrologia* **9**, 99–101 (1973).

21. T. C. Cetas, A magnetic temperature scale from 1 to 83 K. *Metrologia* **12**, 27–40 (1976).

22. *Sess. Com. Int. Poids Mes.*, *Com. Consult. Thermom.* **10**, T36 (1974).

23. L. M. Besley and W. R. G. Kemp, An intercomparison of temperature scales in the range 1 to 30 K using germanium resistance thermometry. *Metrologia* **13**, 35–51 (1977).

24. S. D. Ward and J. P. Compton, Intercomparison of platinum resistance thermometers and T_{68} calibrations. *Metrologia* **15**, 31–46 (1979).

25. K. H. Berry, NPL-75: A low temperature gas thermometry scale from 2.6 K to 27.1 K. *Metrologia* **15**, 89–115 (1979).

26. R. L. Rusby and C. A. Swenson, A new determination of the helium vapor pressure scales using a CMN magnetic thermometer and the NPL-75 thermometer scale. *Metrologia* **16**, 73–87 (1980).

27. M. Durieux, D. N. Astrov, W. R. G. Kemp, and C. A. Swenson, The derivation and development of the 1976 provisional 0.5 K to 30 K temperature scale. *Metrologia* **15**, 57–63 (1979).

28. Bureau International des Poids et Mésures, The 1976 provisional 0.5 K to 30 K temperature scale. *Metrologia* **15**, 65–68 (1979).

29. M. Durieux and R. L. Rusby, Helium vapor pressure equations on EPT-76. *Metrologia* **19**, 67–72 (1983).

30. P. Giacomo, News from BIPM. *Metrologia* **19**, 77–81 (1983).

31. C. Swenson, private communication.

32. International Measurement Confederation (IMKO), "Second Symposium on Temperature Measurement in Industry and Science, Parts 1 and 2. Készült az OMIKK házinyomdájában, Budapest, (1984).

5

GENERAL CHARACTERISTICS OF TEMPERATURE MEASURING DEVICES AND TREATMENT OF DATA

For practical use there are certain desirable characteristics for any temperature measuring device. In this chapter we see what those characteristics are and compare some of the common devices in terms of those characteristics. In laboratory analysis it is also necessary to analyze data. This chapter includes a section on data analysis.

5.1 DESIRABLE CHARACTERISTICS OF TEMPERATURE MEASURING DEVICES

Every temperature measuring device consists of a temperature sensing element, an interpretation and display device, and a method of connecting, or relating, one to the other. These may have certain characteristics which we consider desirable.

5.1.1 The Sensing Element

Desirable characteristics for the sensing element include:

1. *Unambiguous response with temperature*. If X is the property being measured at the sensing element and related to temperature, we could have X versus T curves like those shown in Fig. 5.1. Obviously, the linear response is easiest to interpret and is desirable for instrumentation reasons.

2. *High sensitivity at all temperatures*. It is necessary for X to vary with temperature enough to be measured with sufficient accuracy.

99

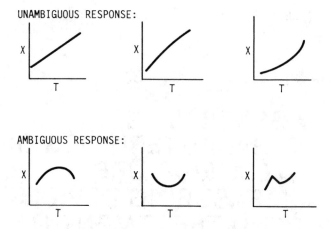

Figure 5.1. Possible temperature response curves.

Obviously, higher sensitivity may lead to high accuracy of measurement if reliability is also obtained. Sensitivity can be defined as the slope of the temperature measuring function,

$$S \equiv \frac{\partial X}{\partial T}.$$ (5.1)

Usually, it varies with temperature (Fig. 5.2).

3. *Stability.* Stability is very important if reproducible and reliable temperatures are to be obtained. It is necessary for the sensing element to remain stable for the life of the device, that is, unchanging with time. Because sensing elements can be altered by high or low temperatures, and by the gases, liquids, and solids with which they come in contact, a great deal of care is exercised in protecting them. However, all such devices have a useful lifetime, In making

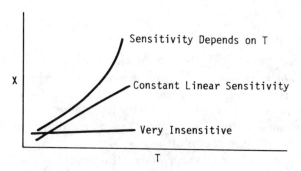

Figure 5.2. Examples of temperature sensitivity.

temperature measurements one must always be aware of the limitations of the sensing element, protect it from abuse, be wary of failure, and be ready to replace it.

4. *Low cost*. Actual instruments vary in cost from a few cents to many thousands of dollars.

5. *Wide range*. A wide range of temperatures, from very high to very low, would make instrumentation easier.

6. *Small size*. Size is important for many applications.

7. *Small heat capacity*. The amount of heat required to change the temperature of a sensor can be very important. If the thermal mass is too large, it may not be possible to make a measurement even if the physical dimensions are suitable.

8. *Rapid response*. Whenever the temperature changes, the sensing element must also change temperature before temperature measurement can be achieved. The output response must be rapid enough for the particular application.

9. *Usable output*. A very common temperature sensor is an ordinary liquid-in-glass thermometer. It is very convenient, self-contained, and versatile. Although anyone can read a thermometer, there is no recordable output that could be used for unattended applications. This is a severe limitation. A usable output is desirable.

5.1.2 Readout Instruments

The interpretation and display of temperature by an instrument is just as important as the sensing device. Obviously, the instrument must be at least as sensitive and stable as the sensing device, but the resolution of temperature cannot be improved beyond that of the sensing device. Desirable properties include:

1. Sufficiently sensitive for the sensing element and the application
2. Stable, unchanging with time
3. Automatic, not requiring manual adjustment
4. Capable of output suitable for control and recording instruments
5. Low cost

5.2 ACTUAL CHARACTERISTICS OF TEMPERATURE MEASURING DEVICES

The problem associated with selection and calibration of actual measuring devices can be illustrated by considering two properties of three metals that depend on temperature: copper, tungsten, and platinum. The temperature of such a metal sensor can be written in a general way as

$$T = f(x, y),\qquad(5.2)$$

where x and y are properties of the metal that could be used to indicate temperature. For our illustration we choose x as the electrical resistivity and y as the linear thermal expansion. For experimental purposes we can simply measure the electrical resistance and the length of a metal-sensing element as we vary the temperature. Suppose that we perform an experiment in which we measure the electrical resistivity of copper, tungsten, and platinum rods at the ice point, the steam point, and at various temperatures in between; and that we measure the length of the copper and the tungsten rods at the same temperatures. At each intermediate temperature we hold the temperature constant until all five measurements have been completed. Let us assume that the temperature–property relationship is

$$T = a + bx, \qquad T = a' + b'y \qquad(5.3)$$

under the conditions that pressure and all other variables affecting the relationship of T and (x, y) are held constant. Our experimental measurements allow us to make five different equations in the form of Eq. (5.3). For example, if the length of the tungsten rod is halfway between its length at the ice point and the steam point, the temperature would appear to be 50°C. When we compare the indicated temperatures for the five measurements, we find that they do not agree exactly. If we assume that platinum has the correct $T = f(x)$ relationship (i.e., if $T = a + bx$, where x is electrical resistivity), we can calculate the error resulting from using one of the other four methods (Fig. 5.3). The actual experimental result is that the five methods do not agree, and that we really do not know whether the platinum temperature is accurate. In principle, interpretation of $T = f(x, y)$ must be based on the thermodynamic temperature scale. Only then can we calibrate our platinum, copper, or tungsten sensing elements. We use the thermodynamic temperature scale in principle and the International Practical Temperature Scale (IPTS) in practice for interpolation between fixed points such as the ice point and the steam point. The gas thermometer at low pressures is usually taken as the closest approximation to the thermodynamic temperature scale. It was the basis for establishing the IPTS. We can use any of the properties and metals above if we calibrate them against IPTS. Each function will be different, and each will not be a linear function because the thermal expansions and electrical resistivities are not actually linear functions. All deviate slightly from linearity.

From this illustration we see that each method of sensing temperature has its own response function. Experience with different temperature measuring methods has confirmed that some methods are suitable in certain temperature ranges, and others are better in different temperature ranges.

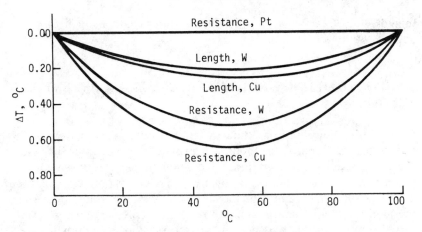

Figure 5.3. Error in the indicated temperature as compared to the platinum temperature, assuming a linear relationship between temperature and the function measured.

Much of the rest of this book is devoted to particular temperature measuring devices and their principles of operation, construction, and use.

5.3 THE REDUCTION OF DATA: CURVE FITTING

In laboratory experiments relating a property to temperature, in calibrating sensing devices, and in interpretation of numerical data, the statistical analysis of data is essential if precision is desired. The most common method of analyzing a temperature–property relationship is that of least squares. In that analysis an equation which relates the temperature to a property is obtained in such a way that the difference between the experimental data points and the equation is minimized. (Actually, it is the square of the differences that is minimized because this can be done more easily.) The function that is statistically fit to the data is not limited to any particular mathematical relationship. Modern computers and programmable calculators usually have programs for curve fitting of relationships such as

1. Straight lines (linear regression), $y = a + bx$
2. Exponential curves, $y = ae^{bx}$ $(a > 0)$
3. Logarithmic curves, $y = a + b \ln x$
4. *Polynominals*, $y = ax + bx^2 + cx^3 + \cdots$

These programs allow one to choose a function and fit the data, but do not indicate which function should be selected. Usually, the simpler functions are tried first, often based on theoretical equations. A good fit to the data does not mean a correct mathematical function has been determined—only that it fits the data. Some other mathematical function may actually be the correct one.

5.4 CURVE FITTING FOR TEMPERATURE MEASUREMENT FUNCTIONS

Curve fitting for a sensor output to temperature is basically a calibration procedure. Calibration methods are explained in Chapter 17. However, there are special considerations involved in the curve-fitting process for temperature measurements, especially if the utmost precision is required. These include uncertainty of the fixed-point temperatures, uncertainty of the interpolating equations, and the requirement for extreme precision.

Uncertainty in the fixed-point temperatures comes from two causes. The international standard is continuously being improved, but the exact temperature of a particular fixed point is unknown. If a fixed-point temperature is modified, such as happened recently for the steam point, previous calibrations based on it will be wrong. This is important only when milli-kelvin accuracy is required, unless one is working at extremely high temperatures. The other uncertainty of fixed points is experimental. Any variations in purity or other experimental requirements will mean that a temperature associated with that fixed point will be wrong. This will cause a systematic error that cannot be reduced by replication if the same calibrating materials are used.

The IPTS specifies the interpolating equation to be used to find the sensor–temperature relationship between fixed points. For example, if the platinum resistance thermometer is used in range II of IPTS, the interpolating equation is obtained from the equation for resistance,

$$R_t = R_0(1 + \alpha t + \beta t^2), \tag{5.4}$$

where R_0 is the resistance at 0°C, α and β are constants, and t is temperature (°C). In principle, R_0, α, and β are the unknown constants which can be found by measuring R_t at three fixed points, one of which is the ice point. Then, with three equations and three unknowns we can solve for R_0, α, and β. If we choose the fixed points to be the ice point, the steam point, and the antimony point, our thermometer should be properly calibrated. However, if we change one or more of the fixed points, our solution of the equations will give a different result. If we compare two different thermometers at 50°C, calibrated at the same three points they will not agree exactly. Yet each is mathematically correct! Replication of each calibration experiment will also contribute variation, but that random source of error can be analyzed statistically and limits placed. But we may have a perfectly reproducible temperature sensor that does not agree exactly with another when they are both calibrated by the same procedure. For that reason instruments believed to be especially good in representing the thermodynamic temperature scale are carefully preserved and used as standards by the various national standards laboratory around the world.

As we have seen in Chapter 4, the greatest uncertainty occurs at points somewhere between the fixed points. If we choose to fit a curve to three calibration points, the curve must pass through those points. If we extrapolate the curve, both the systematic and random errors will contribute to deviations accentuated by the extrapolation. Therefore, in curve fitting we should calibrate at points as close as possible to the temperature of interest.

Certain interpolating equations are specified for IPTS. They were selected as the equations most nearly fitting the thermodynamic temperature scale, as estimated from gas thermometry. Whether we use them or other equations for other instruments, we know that they are usually empirical relationships. (In the case of Planck's law, we assume that it is correct in theory and in practice. We have no way to test that assumption.) Our curve fitting for a particular instrument, then, must be done very carefully, with the correct calibration procedure to minimize systematic uncertainty, and with enough replications to have a sound statistical base for analyzing the random errors.

When a thermocouple calibration curve is plotted, Seebeck voltage as a function of temperature, the curve looks linear. If we determine the coefficients for a type S thermocouple in the polynomial

$$V = a + bt + ct^2 + ct^3 + \cdots, \qquad (5.5)$$

we find that they get progressively smaller. The value of c is often less than 10^{-7}, so that even when $t^2 = 10^6$ that term makes a small contribution to the Seebeck voltage. Yet precise representation of standard thermocouple curves often requires a six- to eight-degree polynomial. The voltages and the temperatures are represented to five or more significant figures. When that type of precision is required, the computer used in curve fitting, and the software used with it, must not contribute rounding-off errors that would limit the precision. It is possible to improve the correlation coefficient in accordance with an algorithm and not actually improve the statistical fit to the data.

Experimenters in metrology often try to partition the variance in their data between the various causes and estimate the root-mean-square precision of their results. In polynomial curve fitting the residual error remaining after determining the equation can be determined as the degree of the polynomial is increased. Usually, the higher degrees introduce more oscillation of the fitted curve. If the curvature introduced by a higher degree reduces the residuals, it indicates a better fit to the data. When the residuals stay the same or increase as the degree of the polynomial is increased, no further benefit is being achieved.

Where standard equations exist, as in our thermocouple example above, it may be easier to determine the differences from the standard equation and represent the differences statistically. This usually requires a lower-degree polynomial for the difference equation. But the actual calculation of temperature will then require two calculations, the standard temperature and the difference.

Many commercial devices using microprocessor systems use segmented linear representations of the equations of temperature as a function of the sensor property. The fit of a curve over a given range is improved as the number of linear segments is increased. Instruments with 20 or more linear segments are available. If only moderate precision is required, this is a very satisfactory method.

REFERENCES

Recommended Reading Any good statistics text such as
G. W. Snedecor and W. G. Cochran, "*Statistical Methods,*" 7th ed. Iowa State Univ. Press, Ames, 1980.

CHAPTER 5 PROBLEMS

5.1 An electrical resistance measuring device can be used to measure temperature. Two electrical resistance temperature sensors are available. Suppose that electrical resistances for sensors A and B can be related to temperature by the following equations:

$$R_A = 1000(1 + 0.0049t - 0.000003t^3)$$
$$R_B = 10,000(1 + 0.35t + 0.004t^2)$$

(*a*) Calculate the resistance of A and B at 10°C; at 100°C.
(*b*) Calculate the sensitivity of A and B at 10°C; at 100°C.
(*c*) Which is most sensitive at 100°C?

5.2 From the following data, calculate the equation, fitting it by the linear least-squares method.

Temperature (°C)	Resistance (Ω)
0	2757
12.7	2556
24.1	2354
46.2	1955
63.8	1691
76.9	1464
98.6	1054

5.3 A thermistor is used to measure temperature. The electrical resistance of the thermistor at several temperatures is as follows:

Temperature (K)	Resistance (Ω)
273.1	19,150
285.7	16,610
297.1	6,016
319.2	2,379
336.9	1,258
349.9	762
371.6	387

You have reason to believe that the resistance temperature function has the form

$$R = R_0 e^{\beta(1/T - 1/T_0)}$$

where β is a constant.

(a) Convert this equation to logarithmic form and find the value of β if R_0 is 19,150Ω at a T_0 of 273.15K, using linear least squares.

(b) Determine the sensitivity of this thermistor at 20°C; at 85°C.

CHAPTER 5 ANSWERS

5.1 (a) At 10°C:

$$R_A = 1000[1 + 0.0049(10) - 0.000003(10)^3] = 1046 \ \Omega$$
$$R_B = 10,000[1 + 0.35(10) + 0.004(10)^2] = 45,400 \ \Omega$$

At 100°C:

$$R_A = 1000[1 + 0.0049(100) - 0.000003(100)^3] = -1510 \ \Omega$$
$$R_B = 10,000[1 + 0.35(100) + 0.004(100)^2] = 760,000 \ \Omega$$

(b) $S \equiv \dfrac{\partial R}{\partial T}$

For R_A, $\dfrac{\partial R_A}{\partial t} = 1000(0.0049 - 0.000009\,t^2)$

At 10°C:

$$S_A = 1000[0.0049 - 0.000009(10)^2] = 4 \ \Omega/°C$$

At 100°C:

$$S_A = 1000[0.0049 - 0.000009(100)^2] = -85.1 \ \Omega/°C$$

$$\text{For } R_B, \frac{\partial R_B}{\partial t} = 10,000(0.35 + 0.008t)$$

At 10°C:

$$S_B = 10,000[0.35 + 0.008(10)] = 4300 \ \Omega/°C$$

At 100°C:

$$S_B = 10,000[0.35 + 0.008(100)] = 11,500 \ \Omega/°C$$

(c) B is the most sensitive.

5.2

x	y	xy	x^2
0	2,757	0	0
12.7	2,556	32,461.2	161.29
24.1	2,354	56,731.4	580.81
46.2	1,955	90,321.0	2,134.44
63.8	1,691	107,885.8	4,070.44
76.9	1,464	112,581.6	5,913.61
98.6	1,054	103,924.4	9,721.96
322.3	13,871	505,753.4	22,582.55

$$n = 7$$

$$b = \frac{n\sum xy - \sum x \sum y}{n \sum x^2 - (\sum x)^2}$$

$$= \frac{7(505,753.4) - (322.3)(13,871)}{7(22,582.55) - (322.3)^2}$$

$$= \frac{930,350}{54,200} = -17.17$$

$$a = \frac{\sum y}{n} - b\frac{\sum x}{n}$$

$$= \frac{13{,}871}{7} - (-17.17)\left(\frac{322.3}{7}\right)$$

$$= 1982 + 790$$

$$= 2772$$

$$y = 2772 - 17.17x$$

5.3

$$R = R_0 e^{\beta(1/T - 1/T_0)}$$

$$R_0 = 19{,}150 \ \Omega \ \text{at} \ T = 273.15 \ \text{K}$$

$$\frac{R}{R_0} = e^{\beta(1/T - 1/T_0)}$$

$$\ln \frac{R}{R_0} = \beta\left(\frac{1}{T} - \frac{1}{T_0}\right) = \beta\left(\frac{1}{T} - \frac{1}{273.15}\right)$$

$$= \beta\left(\frac{1}{T} - 0.003661\right)$$

T	$1/T$	$(1/T - 1/T_0)$	R	R/R_0	$\ln(R/R_0)$
273.1	0.003661	0	19,150	1	0
285.7	0.003500	−0.0001609	16,610	0.8674	−0.142
297.1	0.003366	−0.0002949	6,016	0.3142	−1.158
319.2	0.003133	−0.0005279	2,379	0.1242	−2.086
336.9	0.002968	−0.0006929	1,258	0.0657	−2.723
349.9	0.002858	−0.0008029	762	0.0398	−3.224
371.6	0.002691	−0.0009699	387	0.0202	−3.902

Treating the third column as x and the sixth column as y, after multiplying x by 10,000, gives

x	y	xy	x^2
−0	0	0	0
−1.609	−0.142	+0.2285	2.5889
−2.949	−1.158	+3.4149	8.6966
−5.279	−2.086	+11.0120	27.8678
−6.929	−2.723	+18.8677	48.0110
−8.029	−3.224	+25.8855	64.4648
−9.699	−3.902	+37.8455	94.0706
−34.494	−13.107	+97.2541	245.6997

$$b = \frac{n\sum xy - \sum x \sum y}{n\sum x^2 - (\sum x)^2}$$

$$= \frac{7(97.2541) - (-34.494)(-13.107)}{7(245.6997) - (-34.494)^2}$$

$$= \frac{680.7787 - 452.1128}{1719.8979 - 1189.8361} = \frac{228.6658}{530.0619} = 0.43139$$

$$\beta = 10{,}000b = 4313.9$$

$$\ln \frac{R}{R_0} = 4313.9\left(\frac{1}{T} - \frac{1}{273.15}\right)$$

$$R = 19{,}150\,e^{4313.9(1/T - 1/273.15)}$$

This equation is plotted in the accompanying figure. Note that the resistance determined experimentally at 285.7° is above the line, and probably erroneous.

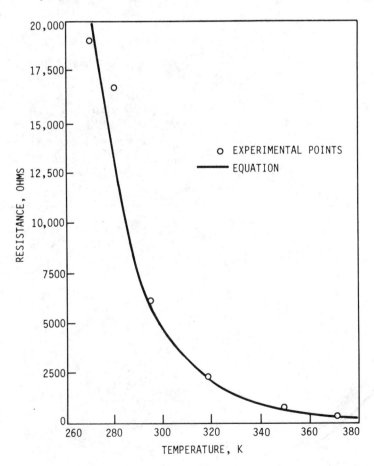

6

LIQUID-IN-GLASS THERMOMETERS

As explained in Chapter 1, the liquid-in-glass thermometer was the first true temperature measuring instrument. That chapter gives a rather complete account of its early history. Some of its more recent technical history is given in this chapter, in which we also describe important features of thermometers and of their calibration and use. These subjects have been included because thermometers are still the most widely used instruments for measuring temperature, and because many who use them are not aware of the errors that result from improper use.

6.1 HISTORICAL PERSPECTIVE

High-quality thermometer manufacture was begun by Fahrenheit in the early eighteenth century.[1] The basic elements of the thermometer included a cylindrical bulb which served as a reservoir for the measuring fluid, a capillary stem for measuring the expansion of the fluid in the bulb, and a calibrated scale to relate the position of the mercury column in the stem to the temperature of the thermometer. Permanently attaching the scale to the capillary stem was one of the important developments in precision thermometry.[2] Techniques improved so that by 1899 a complete book was available describing the instruments and methods for calibration and using precision thermometers.[3] That book was based on research for the International Bureau of Weights and Measures. The techniques described in 1899 for precision thermometry are basically those that are used today, although the calibrating points are now more accurately defined.

6.2 THERMOMETER NOMENCLATURE[4,5]

Thermometers can be described by the following terms (Fig. 6.1).[6]

Bulb. The bulb is the liquid reservoir. The liquid in this reservoir expands or contracts in volume. It is the difference between the volumetric reversible thermal expansion of the liquid and its glass container that makes it possible to measure temperature. The greater the reversible thermal expansion of the liquid as compared to the glass, the more sensitive the thermometer should be. Mercury is most commonly used because its expansion is more nearly linear on the thermodynamic temperature scale. It is less sensitive than many organic liquids, such as ethanol.

To reduce the response time of a thermometer, the bulb should be small and the bulb wall should be as thin as possible. For thermometers to be used at more than 100°C, which have internal gas pressure, the wall thickness should be thick. This is also true for mechanical strength. Selection of bulb wall thickness is a compromise between these conflicting factors.

Stem. The stem of the thermometer is a glass tube containing a tiny capillary connected to the bulb. The change in volume of the liquid in the bulb causes the liquid in the capillary to advance or retreat as the bulb temperature is raised or lowered.

Main Scale. The main scale of the thermometer is a graduated region of the stem calibrated to read degrees of temperature when used for the purpose for which the thermometer was designed.

Auxiliary Scale Some thermometers have an auxiliary scale to use when calibrating the thermometers. This is often a short scale with graduations to check for the ice point and is provided when the main scale does not include the calibration point. The capillary size and the graduations of the auxiliary scale must be identical with the main scale.

Contraction Chamber. Some thermometers have one or more contraction chambers. These serve to shorten the thermometer by providing a small reservoir in the stem of sufficient volume to accommodate the fluid, which

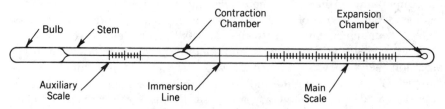

Figure 6.1. Thermometer nomenclature. (After Ref. 6.)

would fill that portion of the stem if the thermometer were not shortened. An auxiliary scale below the reservoir is usually provided for calibration purposes.

Reference Point. Each thermometer should have a reference point for calibration purposes on the main scale or the auxiliary scale. This is most commonly the ice point. Sometimes, it is the steam point.

Immersion Line. Thermometers intended for partial immersion have a line scribed around the stem to indicate the proper immersion depth.

Emergent Stem. The emergent stem is that part of the stem between the top of the liquid column and the uniform temperature region into which the bulb is immersed. For a mercury-in-glass thermometer immersed partway into a water bath, it would be the part of the stem between the waterline and the top of the mercury column.

Expansion Chamber. Most thermometers have an expansion chamber at the top of the stem to serve as a liquid reservoir. This helps prevent breakage when overheated, can be used in uniting a separated column, and provides more volume for gas-filled thermometers.

Differential Thermometers. Some thermometers, called differential thermometers, are intended to indicate changes in temperature rather than the absolute value of temperature. These are especially useful for applications such as calorimetry, where it is the change in temperature of the calorimeter that must be accurately known. Two common types are calorimeter thermometers and Beckmann thermometers.

6.3 TYPES OF IMMERSION

The accuracy of measurement with a thermometer is dependent on the temperature of the entire thermometer, not just the bulb. Because of this, three types of immersion are recognized (Fig. 6.2).

1. *Partial Immersion.* Partial immersion thermometers are inserted to the immersion line in the medium being measured. Usually, this medium is a liquid. The temperature indicated by the thermometer depends not only on the temperature below the immersion line, but also on the temperature of the emergent stem. If the average temperature of the emergent stem is different from its temperature when the thermometer was calibrated, the indicated temperature will be erroneous.

2. *Total Immersion.* The term "total immersion" is used to indicate that

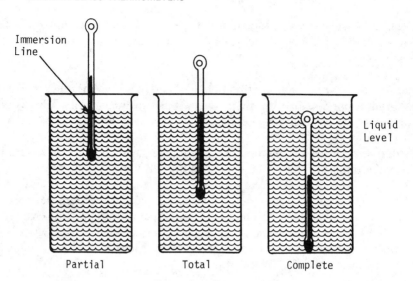

Figure 6.2. Illustration of partial, total, and complete immersion.

the thermometer is immersed sufficient to cover the mercury column. If a total-immersion thermometer is immersed either above or below the top of the mercury column, the indicated temperature will be erroneous. Even when properly immersed, an error may occur if the temperature of the stem above the mercury is not the same as when the thermometer was calibrated. This is especially true for high-temperature thermometers which are designed not to be completely immersed, because the internal gas pressure may rise too high.

3. Complete Immersion. The term "complete immersion" is applied to thermometers completely covered by the medium being measured. If a complete-immersion thermometer is not completely immersed, an error will occur if the temperature of the stem is not the same as that of the bulb.

6.4   THERMOMETER CONSTRUCTION

Thermometers are made in a wide variety of forms and temperature ranges. Mercury is the measuring liquid in most precision thermometers and in high-temperature thermometers. Mercury freezes at −38.9°C, and mercury-in-glass thermometers should not be used below −35°C. Alloying mercury with 8.5% thallium produces a liquid that freezes at −59°C and can be used to −56°C. Organic liquids (usually dyed red for visibility) are used to lower temperatures, although at lesser accuracy. They may be alcohol, toluene, pentane, or other organic liquids or liquid mixtures. Some are used to about 73 K (−200°C).

Early attempts to increase thermometer sensitivity by decreasing the capillary volume per degree Celsius in proportion to bulb volume were successful until the ratio reached about $1:10,000$. Attempts to increase sensitivity further resulted in irreproducible results, and the mercury moved in "jumps" rather than smoothly when temperature was changed. This was caused by the combination of surface tension, thermal stresses, viscous flow, and internal pressure. Modern thermometers of highest sensitivity have a capillary volume per degree Celsius of about $1:6200$. This produces a movement in the stem of about 1 cm per degree. Surface tension may cause the meniscus to flatten with cooling, rather than to move. Tapping the side of these sensitive thermometers may give more consistent performance.

Thermometers intended for high-temperature use are often filled with an inert gas above the mercury to reduce mercury volitalization. Gas pressure is often about 20 atm. These are often designed for partial or total immersion. If completely immersed, errors as high as 20° may be produced by the excessive internal pressure.

6.4.1 Zero Shift

Early thermometers were found to change calibration from time to time. Found by checking the calibration at the ice point, this was designated "zero shift" and depended on the glass used in construction. The bulb and capillary for early thermometers were often made from soft glasses such as "soda-lime" glasses, similar to ordinary window or bottle glass. Certain German glasses made from Thüringen sand were found to be superior. This was attributed to Al_2O_3 in the sand. Studies of zero shift revealed that it was diminished if only one alkali oxide was present (the early glasses often contained K_2O as well as Na_2O). Thermometers of the present day are made from harder, single-alkali glasses and have much less zero shift than the original glasses. The change in zero is time and temperature dependent. It usually affects all temperatures in the same way, so that a correction for zero shift can be applied after calibration at the zero point.

The zero shift is actually the result of changes in the specific volume of silicate glass which accompany changes in temperature. These changes are thermally activated processes which depend not only on the temperature, but also on the previous thermal history of the glass. Such changes take place during annealing and in use, and at a more rapid rate at the higher temperatures. Thus if a thermometer is heated to a high temperature, held for a time (an hour at 200°C, for example) and then is quickly cooled to the ice point, the bulb of the thermometer will be oversize and a zero-point depression will occur. As long as the thermometer is used at relatively low temperatures (to 100°C, for example) the zero shift will affect all values and it will recover only very slowly, taking years to change appreciably. This zero shift can be corrected at low temperatures by adding the ice-point correction to all values.

TABLE 6.1 Maximum Temperature for Various Glasses

Glass	Strain point (°C)	Maximum Recommended Temperature (°C)	
		Continuous	Intermittent
Corning 7500	500	370	430
Kimble R6	490	360	420
Jena 16 III	495	365	425
Corning 8800	529	400	460
Jena 2954	548	420	480
Corning 1720	668	540	600
Jena 2955	665	535	595

Thermometers used at 100°C and below exhibit another type of zero shift, which is more rapid in recovery but of lesser magnitude. For example, heating to 100°C and then returning to the ice-point may produce a depression of 0.01 to 0.10 degree that will recover in about 3 days. These zero-point depressions limit the precision of mercury-in-glass thermometers. Therefore, precision measurements should have a zero-point determination immediately afterward, and a correction should then be applied.

In addition to the zero depression described above, which results in a temporary increase in glass specific volume at high temperatures, a very small permanent decrease in specific volume also occurs, which takes place over many years. This happens more rapidly at higher temperatures and results in a smaller bulb volume, hence a zero increase. This affect depends on the original annealing of the glass and is usually very small. However, improper exposure of gas-filled thermometers to temperatures near the strain point, even for a few minutes, can result in errors of many degrees.

Glasses now used for thermometer applications are intended to minimize zero shift. Special glasses are used for high-temperature thermometers. Both the zero shift and the maximum temperature depend on glass composition. The maximum temperature depends on the glass used in construction (Table 6.1).

Thermometers often are constructed with the stem modified by adding a stripe of colored glass to the stem to enhance visibility. This serves as a background for the liquid column. Different-colored stripes serve to identify different thermometers. Stems of many thermometers are shaped in front of the liquid column to act as an enlarging lens for easier reading. This is often true of small thermometers, such as fever thermometers.

6.4.2 Thermometer Scales and Accuracy

Thermometer scales for precision thermometers are etched on the stem. For low-accuracy thermometers, they are often printed on a separate card. The

TABLE 6.2 Tolerances for Celsius Total-Immersion Mercury Thermometers

Temperature Range (°C)	Graduation Interval (°C)	Tolerance (°C)	Accuracy (°C)	Corrections Stated to:
Thermometer Graduated under 150°C				
0 up to 150	1.0 *or* 0.5	0.5	0.1–0.2	0.1
0 up to 150	0.2	0.4	0.02–0.05	0.02
0 up to 100	0.1	0.3	0.01–0.03	0.01
Thermometers Graduated under 300°C				
0 up to 100	1.0 or 0.5	0.5	0.1–0.2	0.1
Above 100 up to 300		1.0	0.2–0.3	0.1
0 up to 100	0.2	0.4	0.02–0.05	0.02
Above 100 up to 200		0.5	0.05–0.1	0.02
Thermometers Graduated above 300°C				
0 up to 300	2.0	2.0	0.2–0.5	0.2
Above 300 up to 500		4.0	0.5–1.0	0.2
0 up to 300	1.0 or 0.5	2.0	0.1–0.5	0.1
Above 300 up to 500		4.0	0.2–0.5	0.1

TABLE 6.3 Tolerances for Celsius Partial-Immersion Mercury Thermometers

Temperature Range (°C)	Graduation Interval (°C)[a]	Tolerance (°C)	Accuracy[b] (°C)	Corrections Stated to:
Thermometers Not Graduated above 150°C				
0 up to 100	1.0 or 0.5	1.0	0.1–0.3	0.1
0 up to 150	1.0 or 0.5	1.0	0.1–0.5	0.1
Thermometers Not Graduated above 300°C				
0 up to 100	1.0	1.0	0.1–0.3	0.1
Above 100 up to 300	1.0	1.5	0.5–1.0	0.2
Thermometers Graduated above 300°C				
0 up to 300	2.0 or 1.0	2.5	0.5–1.0	0.5
Above 300 up to 500		5.0	1.0–2.0	0.5

[a]Partial-immersion thermometers are sometimes graduated in smaller intervals than shown in this table, but this in no way improves the performance of the thermometers, and the listed tolerances and accuracies still apply.
[b]The accuracies shown are attainable only if emergent stem temperatures are closely known and accounted for.

TABLE 6.4 Tolerances for Low-Temperature Total-Immersion Thermometers

Temperature Range (degrees)	Type of Thermometer	Graduation Interval (degrees)	Tolerance (degrees)	Accuracy (degrees)	Corrections Stated to:
		Celsius Thermometers			
−35 to 0	Mercury	1 or 0.5	0.5	0.1–0.2	0.1
	Mercury	0.2	0.4	0.02–0.05	0.02
−56 to 0	Mercury-thallium	0.5	0.5	0.1–0.2	0.1
	Mercury-thallium	0.2	0.4	0.02–0.05	0.02
−200 to 0	Organic liquid	1.0	2.0	0.2–0.5	0.1
		Fahrenheit Thermometers			
−32 to 32	Mercury	1 or 0.5	1.0	0.1–0.2	0.1
	Mercury	0.2	0.5	0.05	0.02
−69 to 32	Mercury-thallium	1 or 0.5	1.0	0.1–0.2	0.1
	Mercury-thallium	0.2	0.5	0.05	0.02
−328 to 32	Organic liquid	2 or 1.0	3.0	0.3–0.5	0.2

TABLE 6.5 Tolerances for Low-Temperature Partial-Immersion Thermometers

Temperature Range (degrees)	Type of Thermometer	Graduation Interval (degrees)	Tolerance (degrees)	Accuracy (degrees)	Corrections Stated to:
		Celsius Thermometers			
−35 to 0	Mercury	1.0 or 0.5	0.5	0.2–0.3	0.1
−56 to 0	Mercury-thallium	1.0 or 0.5	0.5	0.2–0.3	0.1
−90 to 0	Organic liquid	1.0	3.0	0.4–1.0	0.2
		Fahrenheit Thermometers			
−35 to 32	Mercury	1.0 or 0.5	1	0.3–0.5	0.1
−69 to 32	Mercury thallium	1.0 or 0.5	1	0.3–0.5	0.1
−130 to 32	Organic liquid	2.0 or 1.0	5	0.8–2.0	0.5

line width for the scales of precision thermometers should be more than two-tenths of the distance between lines. Minimum spacing between lines is 0.3 mm for milk glass scales and 0.4 mm for other scales. Wider spacings are recommended. A typical total immersion precision thermometer might have a range from −1.0 to 50.0°C with calibration lines every 0.2 degree spaced at 2 mm so that temperature can be estimated to 0.02°C after suitable corrections are applied. Tables 6.2 to 6.5 of the National Bureau of Standards represent the expected accuracy of thermometers which they calibrate.

Special thermometers are used for a tremendous variety of applications. They may have special size, range, scale, precision, or other requirements. Standards have been established for many of these applications.[7]

6.5 TECHNIQUES FOR USING LIQUID-IN-GLASS THERMOMETERS

As explained earlier, precision thermometer measurements should be followed immediately by a calibration point measurement (such as the ice-point) so that the appropriate calibration shift correction can be applied. It is also possible to calibrate thermometers against a platnium resistance temperature measuring instrument. Calibration of thermometers is described in Refs. 4 and 6. Calibration procedures for all temperature measuring devices are described in Chapter 17.

6.5.1 Immersion Error Corrections

When a partial immersion thermometer is totally or completely immersed, when a total immersion thermometer is not totally immersed, and when a complete immersion thermometer is not completely immersed, an error in temperature will exist. The error is caused by the difference in the capillary column length caused by the mercury being at a different temperature than that which existed when the thermometer was originally calibrated. The sign and magnitude of the correction depends on the deviation of the temperature from the calibration temperature. If that deviation can be estimated, the stem correction can be calculated. Two methods are used to estimate that correction. The most accurate way is to use an auxiliary faden thermometer, one especially constructed to have a long bulb which spans the region of the stem that differs from the calibration conditions. Faden thermometers often are not readily available. The other way is to use a small auxiliary thermometer for each 120 to 150 mm of emergent stem differing from calibration conditions. They should be placed to represent the average temperature of the correction region, midpoint for one thermometer, third point for two, and so on. With either method the tem-

TABLE 6.6   Values of K for Various Glasses

t_s(°C)	Corning 0041	Corning 8800	Corning 8810	Jena 16 III	Jena 59 III
50	0.000157	0.000166	0.000156	0.000158	0.000164
120	0.000159	0.000167	0.000157	0.000158	0.000165
250	0.000163	0.000168	0.000161	0.000161	0.000170
350	0.000168	0.000173	0.000166		0.000177

perature correction is given by

$$\Delta t = KL(t_b - t_s) , \qquad (6.1)$$

where K is usually taken as 0.00016 for mercury in glass (K varies slightly with glass composition and temperature but can be considered constant for most applications; Table 6.6), L is either the number of degrees spanned by the faden thermometer bulb or the deviation in length of the stem to be corrected—measured in degrees and equal to the correct immersion degree minus the actual immersion degree of the mercury column, t_b is the indicated bulb temperature, and t_s is the auxiliary thermometer.

A procedure recommended by Wise is to make the correction by using Eq. (6.1) but using K based on the average of t_b and t_s.[5] She gives tables for normal glass and borosilicate glass, for the mean temperature $(t_b + t_s)/2$, as shown in Table 6.7.

Example 6.1: Faden thermometer. If a total-immersion thermometer indicates 90° when immersed to 80°C, and an attached faden thermometer with a bulb length of 30 degrees, which extends from 60 to 90°, indicates 80

TABLE 6.7   Values of K Based on Mean Temperature

Mean Temperature (°C)	Normal glass	Borosilicate glass
0	0.000158	0.000164
100	0.000158	0.000164
150	0.000158	0.000165
200	0.000159	0.000167
250	0.000161	0.000170
300	0.000164	0.000174
350		0.000178
400		0.000183
450		0.000183

degrees, the correction is calculated as follows:

$$L = 90° - 60° = 30°$$

$$\Delta t = 0.00016(30)(90 - 80) = 0.048° .$$

Then the temperature is $90.0 + 0.05 = 90.05°C$.

Example 6.2: auxilliary thermometer. If the thermometer in Example 6.1 had a small thermometer attached to it with its bulb midpoint at 85° instead of the faden thermometer, and if its temperature reads 60°, the correction would be calculated as follows:

$$L = 90° - 80° = 10°$$

$$\Delta t = 0.00016(10)(90 - 60) = 0.048°$$

$$t = 90.00 + 0.05 = 90.05° .$$

For the calorimeter thermometers the stem correction can be calculated from the equation

$$\Delta t = K(t_2 - t_1)(t_1 + t_2 - S - t_s) , \qquad (6.2)$$

where t_1 is the indicated temperature before the temperature rise, t_2 is the indicated temperature after the temperature rise, S is the scale reading (temperature) to which the thermometer is immersed, and t_s is the average temperature of the emergent stem above the liquid. For example, if a thermometer is immersed to 20°C, its initial reading is 25°C, the final reading is 30°C, and the auxiliary thermometer indicated that the stem temperature was 20°C, then the correction is

$$\Delta t = 0.00016(30 - 25)(25 + 30 - 20 - 20)$$

$$= 0.00016(5)(15) = 0.012°C .$$

Then the temperature change was $(30 - 25) + 0.012$ or $5.012°C$.

Beckmann thermometers are most easily used if they are immersed to zero on the sensitive scale. Then the emergent stem correction can be calculated as

$$\Delta t = K(t_2 - t_1)(t_0 + t_1 + t_2 - t_s) , \qquad (6.3)$$

where t_0 is the temperature corresponding to the zero on the Beckmann scale, and the other terms are as defined previously.

6.5.2 Pressure Corrections

External pressure will slightly increase the indicated temperature of thermometers. In 1899, Guilliaume proposed the correction, β, as the degrees per torr of external pressure applied as

$$\beta = k \frac{R_o^2}{R_o^2 - R_i^2}, \tag{6.4}$$

where k is a constant (now taken as $3.9 \times 10^{-5}\,°C$), R_o is the outside radius of the bulb, and R_i is the inside radius. This correction often amounts to about $0.1°C$ per atmosphere. For internal pressure Eq. (6.4) should be modified by adding 1.5×10^{-5} to β. Where accuracy is essential, corrections should be determined experimentally.

6.5.3 Measuring Techniques

For precise work, thermometers should be measured with a 5 to 10× microscope, mounted so that parallax errors are minimized. A small microscope attached to the thermometer by a spring is helpful when partial immersion thermometers are used, if care is taken not to modify the stem temperature during viewing.

Mercury does not wet glass and a meniscus depression will exist which depends on capillary diameter. However, since thermometers are calibrated against standards of temperature and the capillary is the same diameter all along the stem, no meniscus correction need be applied.

Organic liquids wet glass. This causes difficulties at low temperatures because considerable time may be required for the liquid film to drain down when the thermometer is cooled. It is best to cool the bulb first, so that the liquid film draining down in the stem will have a low viscosity, then cool the stem to the appropriate temperature. The film should be allowed to drain until a constant reading is obtained. Under adverse conditions this could be as long as an hour. Obviously, rapid dynamic temperature changes cannot be measured under such conditions with a liquid-in-glass thermometer.

6.5.4 Dynamic Response Characteristics

Dynamic response characteristics of a thermometer depend on the bulb volume, bulb wall thickness, total weight, type of thermometer, and many other factors. This subject will be explored thoroughly in a subsequent chapter. For many applications where liquid temperature changes are being determined, the response of a conventional scientific thermometer with a thin-walled bulb is rapid enough. Response times (time constants) of such thermometers are often in the range 2 to 10 seconds.

6.5.5 Column Separation

Mercury columns often become separated, especially when jarred in a horizontal position. Then the separated portion may move up and down as the main column moves up and down in response to temperature changes. Rejoining the mercury column is sometimes difficult, especially with gas-filled thermometers. There are several methods that sometimes work.

1. Rapid jarring of the thermometer so that inertia of the separated portion forces it against the main portion of the column. This can sometimes be accomplished with a wrist snap, as with a fever thermometer. Tapping the bulb on a paper pad is sometimes successful.

2. Centrifugal force can sometimes be used if a suitable device is at hand.

3. Cooling the bulb so that all the mercury descends into the bulb, followed by tapping to rejoin the mercury, is often successful. For those calibrated down to a degree or two below the ice point, a salt, ice, and water mixture can be used.

4. When a contraction chamber or expansion chamber exists, heating so that the ends of the separated and the main column are in the expansion chamber, followed by tapping to unite the mercury, will often be successful. Obviously, one must be careful not to break the thermometer by heating enough to fill the expansion chamber completely.

Organic liquids, when separated, can be treated as described above. In addition, cooling the bulb while heating the separated portion can be used to distill the organic film down into the bulb. The glass should not be overheated in this process.

REFERENCES

1. D. G. Fahrenheit, *Philos. Trans. R. Soc. London*, 78–84, (1724).

2. W. E. Knowles Middleton, "A History of the Thermometer." Johns Hopkins Press, Baltimore, Maryland, 1966.

3. C. E. Guilliaume, "Traité practique de la thermométrie de précision." Gauthier-Villars et Fils, Paris, 1889.

4. J. A. Wise, "Liquid-in-Glass Thermometry," Monogr. No. 150. National Bureau of Standards, 26 pages, U.S. Gov. Printing Office, Washington, D.C., 1976.

5. B. W. Mangum and J. A. Wise, "Standard Reference Materials: Description and Use of Precision Thermometers for the Clinical Laboratory, SRM 933 and SRM 934." NBS Special Publication 260–48, 26 pages, May, 1974. Institute of Basic Standards, National Bureau of Standards, Washington, D.C.

6. J. A. Wise and R. J. Soulen, Thermometer calibration: A model for state calibration laboratories. *NBS Monogr.* (U.S.) **174** (1976) (Contains Wise[4]).

7. ASTM designation E1-86, "Standard Specifications for ASTM THERMOMETERS." *1986 Annual Book of ASTM Standards*, **14.01** 56–118 (1986).

CHAPTER 6 PROBLEMS

6.1 If the volume of mercury in the bulb of a thermometer is 150 mm³, and the diameter of the capillary is 0.1 mm, how much will the mercury column move for a 1°C change in the temperature of the bulb?

6.2 If the thermometer in Problem 6.1 is calibrated in degrees Fahrenheit, what will the movement be?

6.3 A thermometer intended for partial immersion, with the immersion line at 33°C on the scale, was actually immersed to 100°C on the scale. Assuming that the emergent stem had a temperature of 30.00°C when it was calibrated, what is the actual temperature of the water in which it is immersed if the indicated temperature is 70.00°C?

CHAPTER 6 ANSWERS

6.1 The change in volume of mercury depends on the difference between the cubical thermal expansion coefficients of the mercury and the glass. Taking this difference as 0.00016 per °C, we can calculate ΔV:

$$\Delta V = 150(0.00016) = 0.024 \text{ mm}^3/°C .$$

The mercury movement in the stem has a cylindrical volume which must equal ΔV. Thus

$$\frac{\pi D^2}{4} L = \Delta V \qquad \text{or} \qquad \frac{\pi (0.1)^2}{4} L = 0.024 .$$

Then the length of mercury column, L, per degree is

$$L = \frac{(0.024)(4)}{\pi (0.1)^2}$$

$$= 3.06 \text{ mm/}°C .$$

6.2 The change in volume ΔV for a degree F will be $\frac{5}{9}$ that of a degree C.

Therefore, the change in length per degree F will be $\frac{5}{9}$ that of the length change per degree C.

$$L = \tfrac{5}{9}(3.06) = 1.70 \text{ mm/°F}.$$

6.3 The thermometer will read too high because the mercury in the emergent stem will expand at 70°C relative to its calibration of 30°C. So the correction is

$$\Delta t = KL(t_b - t_s)$$
$$= 0.00016(70 - 33)(30 - 70)$$
$$= -0.236$$
$$t = t' - \Delta t = 70 + (-0.24) = 69.76\text{°C}.$$

7

SEALED LIQUID OR GAS SENSING INSTRUMENTS AND BIMETALLIC SENSORS

Liquid-in-glass thermometers, as presented in Chapter 6, respond to temperature because the liquid in the bulb has a higher volumetric coefficient of thermal expansion than does the glass shell of the bulb. When the temperature of the bulb is increased, the liquid expands more than the glass, and some must flow up the capillary of the stem to compensate for the difference between the two expansions. This, when calibrated, permits manual observation of temperature but does not easily provide for recording or controlling the temperature. The devices discussed in this chapter are modifications of the thermometer which by pressure or mechanical measuring features make it possible to indicate, record, or control temperature. They are not as accurate as precision thermometers. They are usually accurate to about 1% of full scale. They are widely used by industry and by the general public for temperature measurement and control.[1] The first three devices have been classified by the Scientific Apparatus Makers Association as shown in Table 7.1.[2]

7.1 GAS PRESSURE THERMOMETERS

Gas pressure thermometers are "constant-volume" gas thermometers that use the thermal expansion of a gas at nearly constant volume to produce a pressure which can be used to measure temperature. The principles of the constant-volume thermometer were described in Chapter 1. The typical industrial gas pressure thermometer consists of (1) a rather large bulb as a temperature sensing device, (2) a small-diameter capillary tube to transmit

TABLE 7.1 Scientific Apparatus Makers Association Filled System Guidelines

Filled System Classification (SAMA)	Fill Fluid	Compensation	Scale	Range Limits (°F)		Overrange	Bulb Elevation Errors
				Lower	Upper		
I		Uncompensated	Linear above −100°F	−125	600	100°	Minor below 100 ft
IA	Liquid	Full		−125	600	0–100%[a]	None
IB		Case		−125	600	100%	
IIA		Not required	Scale divisions increase with temp. increase	Amb.	550	Almost always less than 100%	Minor below 100 ft
IIB	Vapor			−430	Amb.		None
IIC				−430 to Amb.	Amb. to 550		Not allowable[b]
IID				−400	500		Minor below 100 ft
IIIA	Gas	Full	Linear above −400°F	−400	1500	100–300%[c]	None
IIIB		Case		−400	1500		None
(SAMA has no class IV)							
VA	Mercury	Full	Linear	−40	1000	100%	Minor below 25 ft
VB		Case		−40	1000		

[a] Depends on bulb length.
[b] Bulb and measuring element must be at the same elevation.
[c] Depends on range.

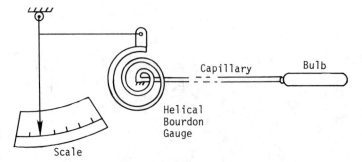

Figure 7.1. Gas bulb thermometer.

pressure changes from the bulb to the pressure sensor, and (3) a pressure sensor calibrated to convert pressure to temperature (Fig. 7.1). The assembly is actually a complex transducer that converts the pressure at nearly constant volume to mechanical movement, which in turn is converted to temperature.

This type of gas pressure temperature measurement requires a fairly large volume of gas to achieve sensitivity, so that the volume of the bulb is usually 75–150 cc (5–10 in.³). It is usually made from thin-drawn tubing and is usually silver-soldered to the capillary tubing. The bulb is made of a metal suitable for the environment in which it is immersed. Invar (a low-thermal-expansion alloy of nickel and iron), nickel-plated copper, Monel, and stainless steel are used.

One of the advantages of a gas thermometer is that it can be constructed with a long, narrow bulb that will physically span a large area, thus averaging the temperature over that large area. Elongated bulbs, up to 61 m (200 ft), are possible. Response is very rapid if the bulb has a large surface area for heat transfer. A coiled capillary is sometimes used as the bulb. Obviously, the bulb is not truly "constant volume." So the device must be empirically calibrated.

The capillary tubing should have a very small diameter to prevent the thermal expansion of the gas in the tubing, at ambient temperature, from being a significant portion of the bulb volume. The tubing is usually cold drawn from ductile metals such as steel or copper. It usually has an inside diameter of 0.127–0.508 mm (0.005–0.020 in.) and an outside diameter of about 4.75 mm ($\frac{3}{16}$ in.). It can be as much as 61 m (200 ft) long and is often covered with a steel, rubber, or copper sheath to make it somewhat less fragile.

The pressure gauge is usually a spiral Bourdon gauge because greater movement is achieved with a spiral gauge than with a simple Bourdon gauge (Fig. 7.2). The internal gas volume within the gauge should also be small, to minimize ambient effects. Larger gauges have greater travel, so a compromise between small gas volume and large size is required. The bulb

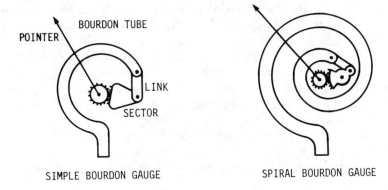

Figure 7.2. Two Bourdon gauges.

volume should be at least eight times the combined volume of the capillary and the Bourdon gauge. Typical values are 1.5 cc/ft of capillary, and 1.0 cc for the gauge. Mechanical or gas expansion compensation for ambient temperature at the gauge is often included in the design.

The simple Bourdon gauge is a flattened tube bent to form about a 270° segment of the circumference of a circle. The tube has an eliptical or flattened cross section. As the pressure is increased, it elastically deforms to become more nearly circular in cross section. That causes the free end to move nearly linearly with pressure. Coupling that movement to electrical or mechanical apparatus gives a temperature indicating, recording, or controlling device.

Almost any nonreactive gas can be used to fill these devices. Dry nitrogen is most common.

7.2 LIQUID-FILLED PRESSURE SENSORS

If the bulb, capillary, and pressure gauge described above are filled with liquid, the same basic system can be used for temperature measurement. Since the volumetric coefficient of thermal expansion of a liquid is less than that of a gas, the change in volume is much less. However, liquids are nearly incompressible, so that higher pressures can easily be produced. Therefore, the bulb can be smaller, often 30–55 cc (2–5 in.^{3}), and a simple Bourdon gauge gives sufficient movement. Pressures in these thermometers may be from 0.5 to 100 MPa (75–1500 psi). Note that it is the net change in volume that produces the pressure indicated by the gauge.

The choice of liquid to fill the thermometer depends on the temperature range and sensitivity desired. For example, mercury has an average volumetric coefficient of thermal expansion between 20 and 100°C of 0.00018 per °C, and ethanol has about six times higher, or 0.0011 per °C. The organic liquids are capable of a wide range in temperature. Mercury can be used from −35 to about 500°C, xylene from −40 to 400°C, and ethanol from −45 to 150°C. When mercury is used, steel, stainless steel, or other nonreactive alloys are used because mercury will react with copper.

When the measuring instrument is at a different level than the sensing element, it may be necessary to correct for the hydrostatic head of the liquid. This must be done for the installation and changed whenever the levels of the instrument or the sensor are changed. This head correction is simply a zero shift, because it must be applied at all temperatures and the density changes of the liquid are small. Provision for adjusting to compensate for head is usually provided with the instrument, usually as a scale movement.

7.3 REMOTE THERMOMETERS

If the bulb and capillary are attached to the stem of a thermometer instead of to a Bourdon gauge, the volumetric changes can be measured in the same way that they are for liquid-in-glass thermometers. Ambient corrections are much more difficult and the accuracy of such systems is far less than that of a conventional precision thermometer. This arrangement is often employed for indicating remote temperatures where low cost is more important than high accuracy, such as in indoor–outdoor thermometers (Fig. 7.3).

In addition to the capillary and bulb size limitations these thermometers must not have vapor trapped in the bulb or capillary. Toluene and alcohol are frequently used as the measuring liquid because they have high thermal expansion coefficients and respond to the range of ambient temperatures usually encountered.

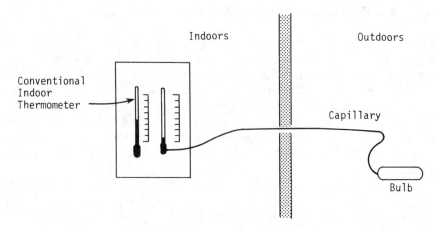

Figure 7.3. Indoor–outdoor thermometer.

7.4 VAPOR PRESSURE THERMOMETERS

The vapor pressure of a volatile liquid is a function of temperature (Fig. 7.4). A metal bulb partially filled with one of these liquids, and connected by a capillary to a Bourdon gauge, will produce a vapor pressure thermometer. Because of the wide variety of liquids that can be used, the range of temperatures which can be measured is extremely wide. Despite the nonlinear nature of the vapor pressure–temperature relationship, these devices are often used because of their versatility. They are accurate to about ±2% of the scale reading despite wide ranges in the capillary and gauge temperature. However, barometric pressure changes can affect the results if the temperature, and hence the vapor pressure, are too low. Each liquid has a minimum temperature below which it should not be used, and it cannot be used above its critical temperature. They are more accurate at the high end of the scale and bulb size can be much smaller than for gas- or liquid-filled pressure devices.

The liquid–vapor interface of a vapor pressure thermometer must be within the bulb, so that the pressure sensed by the instrument corresponds to the equilibrium saturation pressure of the liquid at the desired temperature. If the temperature of the bulb is higher than that of the capillary and gauge, both will be filled with liquid and the cool liquid serves as the pressure-transmitting fluid. Then head corrections may be required because vapor pressure devices must operate at relatively low pressures, and liquid head can be a large fraction of the total pressure sensed by the instrument.

If the temperature of the bulb is less than that of the capillary and gauge, the latter two will be filled with superheated vapor, and the vapor will transmit the pressure from the bulb to the gauge. Then head corrections are not needed.

If the temperature of the bulb varies above and below the capillary or gauge temperature, sluggish response will occur, because each time the difference in temperature changes from positive to negative it will be necessary for the liquid either to vaporize out of the capillary and gauge or to condense into them. For this reason vapor pressure thermometers are not usually recommended for ambient temperatures. However, by filling the capillary, gauge, and part of the bulb with an inert, pressure-transmitting liquid, suitable response can be obtained. It is essential that the pressure-transmitting liquid not be allowed to run out of the capillary and gauge, or an error will result.

Vapor pressure thermometers must absorb or release heat in order to increase or decrease the pressure in response to a change in temperature. Liquids with a small enthalpy of vaporization will respond more quickly because less heat transfer will be required. A small bulb nearly completely filled with liquid is best because heat transfer to liquid is more rapid than to gas, and because less liquid must be vaporized to change the pressure if the gas volume is small. However, if the bulb temperature may be higher than

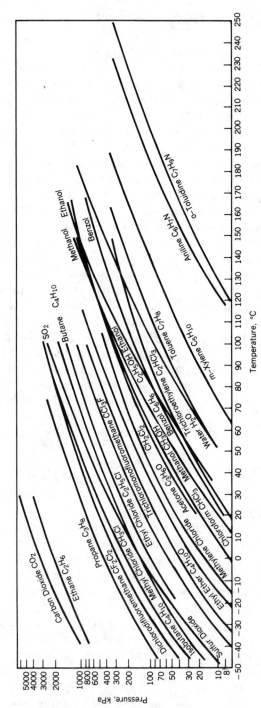

Figure 7.4. Vapor pressure of volatile liquids used in vapor pressure thermometers.

that of the capillary and gauge, sufficient liquid must be available in the bulb to fill them with liquid.

7.5 BIMETALLIC TEMPERATURE SENSING DEVICES

The coefficient of linear reversible thermal expansion is defined as

$$\alpha \equiv \frac{1}{l}\frac{dl}{dt} \qquad (7.1)$$

where dl is the differential change in length accompanying a differential change in temperature dt and division by l normalized α to unit length. Often it is convenient, and approximately correct, to use the initial length l_0 at a standard temperature instead of the actual length l. Then

$$\alpha' = \frac{1}{l_0}\frac{dl}{dt}. \qquad (7.2)$$

The slope of the curve at any temperature is α'. The average slope for any temperature interval is

$$\alpha'_{ave} = \frac{1}{l_0}\frac{\Delta l}{\Delta t}. \qquad (7.3)$$

Thermal expansions must always indicate the temperature, or the temperature range, for which they apply. For the glass shown in Fig. 7.5 the average coefficient of thermal expansion at 400° is considerably different from α' at 400°. For this glass the expansion per degree above the glass transition temperature t_g is greater than that below t_g. Because α' above t_g depends on previous thermal history, it is not truly reversible at the higher temperature.

Metal thermal expansion coefficients depend upon composition and crystallographic structure. They can often be represented by empirical equations such as

$$\alpha' = at + bt^2 + ct^3. \qquad (7.4)$$

When only moderate accuracy is required, the second and third power terms can usually be neglected.

If a metal of known composition expands with temperature then measuring the change in length can be used to calculate temperature. This change in length is very small. However, if a long, low-expansion metal rod is inserted in a high-expansion metal tube the change in length can be used to

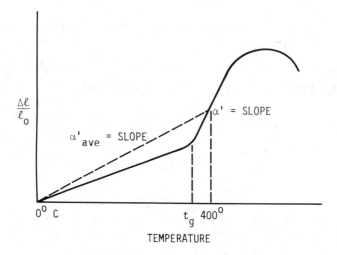

Figure 7.5. Thermal expansion of glass.

actuate a microswitch for controlling temperature. This change in length can be mechanically amplified, however, if thin sheets of two different metals with very different thermal expansion coefficients are bonded together. They bend in response temperature in a useful way.

To explain this bending, we start by considering metals 1 and 2 supported at one end (Fig. 7.6).

If they have an original length l_0 at, say, 0°C and are heated to 10°C each will change in length:

$$l = l_0 + \alpha' l_0 (\Delta t) \tag{7.5}$$

If α' for metal 1 is $2.692 \times 10^{-5}/°C$ and for metal 2 it is $0.09 \times 10^{-5°}/C$ then, at 10°C,

$$l_1 = l_0 + (2.692 \times 10^{-5})(l_0)(10)$$

$$l_2 = l_0 + (0.09 \times 10^{-5})(l_0)(10) \, .$$

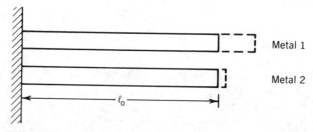

Figure 7.6. Two metal strips of original length l_0. With a change in temperature there will be a change in length for each indicated by the dotted lines.

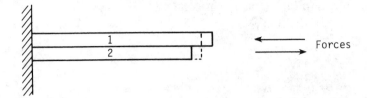

Forces

Figure 7.7. Forces required to cause strips 1 and 2 to be the same length (dotted lines) after heating to produce the unrestrained length (solid lines).

Because l_1 is greater than l_2 we would have to stretch 2 and compress 1 to force them to a common length (Fig. 7.7). The force required will depend upon l_0, α'_A, α'_B and the Young's modulus of elasticity of each. The forces produce a couple which causes a bending moment. Therefore, if we take two such strips 1 and 2 and fasten them tightly together all along their length, they will bend on heating as shown in Fig. 7.8. The amount of deflection will depend upon the thickness a_1 and a_2, the length of the strip, the difference in the thermal expansions $(\alpha'_A - \alpha'_B)$, the temperature interval Δt, and the elastic moduli.[3] The deflection can be precisely calculated from engineering mechanics principles.

The deflection, based upon elastic theory, must satisfy the differential equation

$$EI\frac{d^2y}{dx^2} = 0$$

where E is Young's modulus, I is the moment of inertial of the beam about its bending axis, y is the vertical distance, and x is the longitudinal distance. Solving this equation for a composite beam where the longitudinal strain of each component is calculated from $\Delta x = x\alpha\,\Delta t$ gives the general equation for the radius of curvature ρ as,

$$\frac{1}{\rho} = \frac{(\alpha_1 - \alpha_2)(t_2 - t_1)}{(a_1 + a_2)/2 + [2(E_1 I_1 + E_2 I_2)/(a_1 + a_2)](1/E_1 a_1 + 1/E_2 a_2)} \tag{7.6}$$

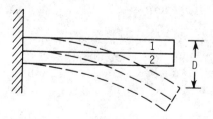

Figure 7.8. The deflection of a bimetallic cantilever strip.

The curvature will be a maximum if

$$\frac{a_1}{a_2} = \left(\frac{E_2}{E_1}\right)^{1/2}. \tag{7.7}$$

Since $E_1 \approx E_2$ for many metals, the thicknesses are usually equal. Then if $(E_2/E_1)^{1/2} = 1$, the curvature is given by

$$\frac{1}{\rho} = \frac{3}{2}\frac{(\alpha_2 - \alpha_1)(t_2 - t_1)}{a_1 + a_2} \tag{7.8}$$

Since, for a cantilever beam,

$$\frac{1}{\rho} = \frac{2D}{l^2},$$

the deflection is given by

$$D = \frac{3}{4}\frac{(\alpha_2 - \alpha_1)(t_2 - t_1)l^2}{a_1 + a_2}.$$

For our purposes we can lump the thermal expansion differences and Young's moduli together into a deflection constant K

$$D = K\frac{(\Delta t)l^2}{d} \tag{7.9}$$

where we have let the thickness of both 1 and 2 be $d/2$. Values of K usually range from 1×10^{-5} to 9×10^{-5}. From this equation we see that D depends on the choice of metals 1 and 2 (through K), is directly proportional to the length squared, and is inversely proportional to the thickness. Thus for a large deflection we want to maximize K and l, and minimize d. The elastic moduli of metals differ only slightly, but the reversible thermal expansion varies a great deal. Most bimetallic strips use a high thermal expansion alloy such as steel or stainless steel for one component, and a low thermal expansion alloy such as Invar (approximately 36% Ni, 64% Fe) as the second component. Some appropriate thermal expansions are given in Table 7.2. Note that Invar has a very low coefficient so it is used as the low expansion element.

By choosing l large or d small we can increase D for a given metal pair to obtain a large deflection. However, usually we want to utilize D to indicate or record temperature. Often a bimetallic strip is used as a control device to cause an electric switch to open or close, or to move a control

TABLE 7.2 Linear Coefficients of Reversible Thermal Expansion

Metal	Temperature, °C	Coefficient, $(\times 10^{-5})$
Iron	40	1.210
Brass	20	1.89
Copper	−191–16	1.409
Zinc	10–100	2.628
Tungsten	20–100	0.336
Tin	18–100	2.692
Nickel	40	1.279
Invar	20	0.09
Steel	40	1.322

component to compensate for ambient temperature variations. For such purposes the strip often must exert a force, and that force also depends upon d and l. Again the force calculation can be solved precisely by engineering mechanics. We will use the simplification, for the force to restore the cantilever to its original straight configuration as,

$$F = \frac{16\,KE(\Delta t)d^2\,W}{l} \qquad (7.10)$$

where $16\,KE$ can be lumped together as a spring constant determined experimentally and W is the width of the strip. The value of $16\,KE$ is from 100 to 900.

From Eqs. 7.6 and 7.10 we see that a bimetallic strip design must compromise between its deflection and the force it can deliver because it has conflicting dependence on l and d. If a large force is required d and W should be large and l small, but D must be large enough for the application. Fortunately in most thermostatic applications this compromise is possible, especially since D is independent of W.

For applications for temperature indication, F need not be large. Then l can be increased by spiral or helical windings. One example is the dial thermometer which is enclosed in stainless steel and is rugged enough to be used in many situations where other thermometers would fail.

The fixed end of the cantilever is fastened to the closed end of a stainless steel tube. The bimetallic strip is wound in a helix so the deflection causes a rotation of the free end. It is attached to the shaft of an indicating pointer. The dial scale is fixed to the tube, so the movement of the pointer indicates temperature. These dial thermometers are inexpensive but have low precision, often 1 or 2°C. (Fig. 7.9.)

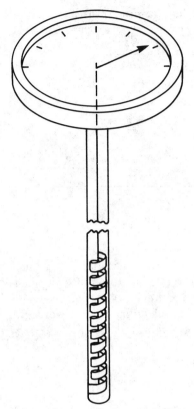

Figure 7.9. A dial thermometer.

REFERENCES

1. T. J. Rhodes, "Industrial Instruments for Temperature and Control." McGraw-Hill, New York, 1941.
2. W. G. Andrew and H. B. Williams, "Applied Instrumentation in the Process Industries," Vol. 1 p. 195. 2nd ed. Gulf Publishing Co., Houston, Texas, 1979.
3. M. Kutz, "Temperature Control," Chapter 8 Wiley, New York, 1968.

CHAPTER 7 PROBLEMS

7.1 Suppose that you want a gas pressure device to measure temperature for process control. If nitrogen gas is used in a 100-cc bulb, the 35-ft-long capillary tube has 1.0 cc/ft, and the Bourdon gauge has a 1.5-cc capacity, what error will occur if the bulb temperature is

230°C, the gauge temperature is 25°C, and the capillary temperature is the average of the gauge and bulb temperatures? Assume that the nitrogen gas has a pressure of 3.45 MPa (500 psi) at 0°C and that its equation of state is $PV = nRT$.

7.2 If a vapor pressure thermometer is used with the bulb above the gauge a distance of 5 m, what error will result if the liquid is ethanol and the uncorrected indicated temperature is 53°C?

CHAPTER 7 ANSWERS

7.1 First we calculate the volume of the capillary and its average temperature:

$$V_c = (1.0 \text{ cc/ft})(35 \text{ ft}) = 35 \text{ cc}$$

$$T_c = 273 + \frac{25 + 230}{2} = 400.5 \text{ K} .$$

We will use the subscripts g for the gauge, c for capillary, b for bulb, 1 for 0°C, and 2 for the pressure at 230°C. Then we can calculate the total number of moles in the various parts of the apparatus as n_t:

$$n_g + n_c + n_b = n_t .$$

From $PV = nRT$ we can substitute, at 0°C;

$$\frac{P_1 V_g}{RT_1} + \frac{P_1 V_c}{RT_1} + \frac{P_1 V_b}{RT_1} = \frac{P_1 V_t}{RT_1} = n_t .$$

At 230°C we have a common pressure, P_2, but the number of moles in each volume has changed. However, the total number of moles has not changed, so we can write

$$\frac{P_2 V_g}{RT_g} + \frac{P_2 V_c}{RT_c} + \frac{P_2 V_b}{RT_b} = n_t = \frac{P_1 V_t}{RT_1} .$$

Solving for P_2 gives

$$P_2 = \frac{P_1 V_1 / RT_1}{V_g / RT_g + V_c / RT_c + V_b / RT_b} = \frac{P_1 V_1 / T_1}{V_g / T_g + V_c / T_c + V_b / T_b} .$$

Substituting yields

$$P_2 = \frac{3.45(1.5 + 35 + 100)/273.15}{1.5/298 + 35/400 + 100/503} = 5.902 \text{ MPa}.$$

If the entire assembly were at 230°C we could write, for n_t,

$$\frac{P_2 V_t}{RT_2} = \frac{P_1 V_t}{RT_1}$$

or

$$P_2 = P_1 \frac{T_2}{T_1} = 3.45 \left(\frac{503}{273.15} \right)$$

$$= 6.353 \text{ MPa}.$$

Thus the pressure was decreased by $[(6.35 - 5.90)/6.35](100) = 7.09\%$.

7.2 The head of 5 m causes a liquid pressure equal to the head times the density of the liquid. Using the density as 0.79 at 25°C, we obtain

$$P_n = h\rho = (500 \text{ cm})(0.79 \text{ g/cc})$$

$$= 395 \text{ g/cm}^2$$

$$= 3950 \text{ Pa}.$$

The vapor pressure P of ethanol at 53°C is found from Fig. 7.4 to be 29.372 Pa. This value is too high by the head pressure, so the true vapor pressure

$$P_t = P_i - P_n$$

$$= 29,372 - 3950$$

$$= 25,422 \text{ Pa}.$$

From Fig. 7.4 this corresponds to a bulb temperature of 49.7°C. The error is 3.3°C.

8

ELECTRICAL RESISTANCE TEMPERATURE MEASUREMENT USING METALLIC SENSORS

8.1 HISTORICAL PERSPECTIVE

In 1821, Sir Humphry Davy studied the electrical resistance of metals as a function of temperature and realized that the resistivity of platinum increases with temperature.[1] The variation of electrical resistance with temperature was first used, about 1871, as a measure of temperature by Carl Wilhelm Siemens, the great German engineer, who became Sir William Siemens in 1911.[2] Siemens preferred platinum for the resistance elements and recommended an interpolation formula. His firm, still in existence, produced the resistance elements and the resistance measuring equipment, but difficulties with stability and consequent inaccuracy caused users to be dissatisfied.

Shortly afterward, the English physicist Hugh Longburne Callendar made elaborate experiments and greatly improved the method. He reported his findings in a magnificent paper published in the *Philosophical Transactions* of the Royal Society in 1887.[3,4] From his comparison of platinum resistance with air thermometer temperature measurements he proposed a parabolic equation for relating platinum resistance to temperature. This became famous in resistance thermometry as the Callendar equation:

$$t = \left(\frac{R_t - R_0}{R_{100} - R_0}\right)100 + \delta\left(\frac{t}{100}\right)\left(\frac{t}{100} - 1\right). \qquad (8.1)$$

Here t is degrees Celsius, R is resistance, and δ is a constant representing the small departure from linearity that is appropriate to the particular thermometer. The subscripts on R indicate the resistance at 0°C, at 100°C, and at the measured temperature.

In use the Callendar equation was solved in two parts. The first term is just a linear interpolation term which gives a nominal platinum temperature t_{pt}. When that temperature is substituted in the second term a small correction term is obtained which is added algebraically to t_{pt}, giving an estimate of the temperature. Callendar prepared tables of the correction term which allowed him to solve for temperature quickly. Substituting the corrected temperature into the correction term allowed him to calculate a better second approximation. A third approximation was seldom necessary.

Later, in 1891, in response to problems of instability associated largely with the Siemens instrument, Callendar gave detailed instructions for construction of stable platinum sensing elements. This included winding the wire on a mica coil form to prevent mechanical strains in use, annealing it, and protecting it in a partially evacuated glass envelope.[5]

Callendar's work put resistance temperature measurement on a firm scientific base. His basic principles of construction, calibration, and interpolation endure to the present time. In laboratory use measurements are often made routinely to an accuracy of ±0.001°C. This accuracy is not achieved below 0°C, however, with the original Callendar equation.

Because the Callendar equation at low temperatures did not give good agreement with gas thermometry, Van Dusen, in 1925, proposed the addition of a third term to the Callendar equation.[6] The additional term improved the accuracy in the range of temperature from the boiling point of oxygen to the ice point and was used from 1927 until 1968 for IPTS measurements. The Callendar–Van Dusen equation,

$$t = \left(\frac{R_t - R_0}{R_{100} - R_0}\right)100 + \delta\left(\frac{t}{100}\right)\left(\frac{t}{100} - 1\right) + \beta\left(\frac{t}{100} - 1\right)\left(\frac{t}{100}\right)^3, \quad (8.2)$$

was well adapted to hand calculations. The constant for the sensing element, β, was evaluated at the oxygen point. In 1968 a new equation was adopted for IPTS range I. Present-day equations will be presented later in this chapter. However, for moderate accuracy, both Eqs. (8.1) and (8.2) are still in use for some applications. Other equations have been proposed to replace Eq. (8.2). It is possible that future modifications of IPTS, range I will give alternatives for moderate-accuracy and high-accuracy purposes.

Callendar also recommended sensor element designs and resistance instrument design, the latter as a modified direct-current Wheatstone bridge. Resistance measuring instruments are devices that either solve Ohm's law directly or compare the resistance of the thermometer to a standard resistance. For precision measurements, comparative instruments have been used almost exclusively in the past, formerly as modified Wheatstone bridges. Standards laboratories in countries around the world now use sophisticated computer-controlled ratio transformer bridges. With these bridges, errors in resistance equivalent to a few microkelvins, one or two orders of magnitude better than the stability of the sensing element, are possible. In the past, most of these bridges used dc circuitry but since about 1970, ac circuitry has become increasingly popular. Bridge circuits will be discussed in detail later.

The use of Ohm's law devices has become practical with the tremendous improvements in high-impedence amplifiers and in precision constant-voltage or constant-current sources. Most of these improvements occurred in the 1960s and 1970s. The sophisticated, more expensive forms are used in standards and in government and industrial laboratories. The less sophisticated forms are available at low cost relative to other, moderate-precision methods of measuring temperature. This, and the small size of solid-state sensors, have made it practical to use resistance temperature measurements in many situations where thermocouples have previously been used, especially in range II of IPTS. There are now many low-cost resistance temperature instruments available, their use is increasing, and the precision available at low cost is improving.

8.2 RESISTANCE OF SENSING ELEMENTS

In common with many other temperature measuring instruments, all electrical resistance pyrometers consist of three components: the temperature sensing element, the connecting wires, and the electrical resistance measuring device. The temperature sensing element is usually a coil of fine wire or a metal film. It is often called a resistance temperature detector (RTD). It is most commonly constructed of platinum, but palladium, nickel, copper, and other alloys are sometimes used. Platinum is used from 13.81 K to about 750°C. Palladium is suitable from 200 to 600°C, nickel from −100 to 300°C, and copper from −100 to 200°C. Copper has a low resistivity which reduces accuracy, and it is subject to oxidation. Nickel has a ferroelectric Curie temperature that prevents it from being used above 300°C.

In general, the resistance of metal is a complex function of temperature. This is especially true near and below the Debye temperature, Θ_D, as shown in Fig. 8.1.[7] For most metals, however, the dependence of resistance on temperature at temperatures near and above the Debye temperature can be written as a power series,

$$R = R_0(1 + At + Bt^2 + C't^3) . \tag{8.3}$$

Here A, B, and C' are constants of the material. At temperatures above the ice point, Callendar found the third term to be unnecessary. At lower temperatures greater accuracy is achieved if a fourth power term is included:

$$R = R_0[1 + At + Bt^2 + C(t - 100)t^3] . \tag{8.4}$$

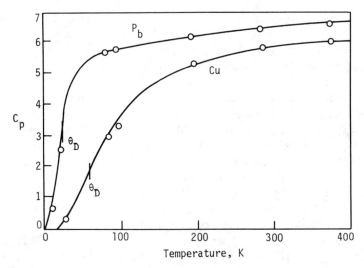

Figure 8.1. Variation of specific heat and (resistance/kelvin) with temperature for lead and copper. The circles are ρ/T; the line is C_p. (From N. F. Mott and N. Jones, "The Theory of the Properties of Metals and Alloys." With permission from Oxford University Press, London, copyright 1958.)

This equation fits the resistivity temperature curve of platinum within ±0.02 K, from the ice point down to the oxygen boiling point of 90.2 K. The Callendar–Van Dusen equation (8.2) is based on it. The coefficients for Johnson, Matthey and Company platinum[8] are

$$A = 3.98 \times 10^{-3}/°C$$
$$B = -5.85 \times 10^{-7}/(°C)^2$$
$$C = -4.35 \times 10^{-5}.$$

Note that B and C are negative so that the curve of resistance versus temperature curves downward slightly (Fig. 8.2). The main temperature dependence is controlled by the value of A, but the slope of the resistance–temperature curve is not constant (Fig. 8.3). In the equation above, R_0 is usually taken at the ice point. For platinum, the last terms in Eq. (8.3) or (8.4) are so small that only the first three terms are used in range II of IPTS.

The resistance of a uniform diameter wire is given by the equation

$$R = \frac{\rho l}{a}, \tag{8.5}$$

where ρ is the resistivity, l is the length, and a is the cross-sectional area. Therefore, dividing Eqs. (8.3) and (8.4) by R_0 produces a resistance ratio,

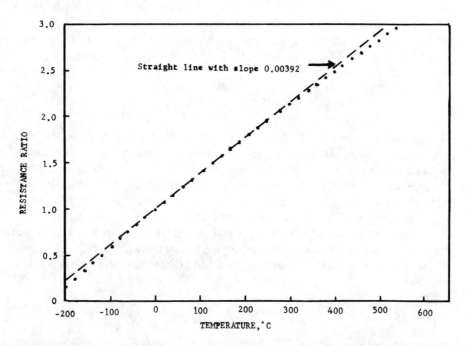

Figure 8.2. Resistance ratio, $W = R_t/R_0$, versus temperature for a platinum resistance thermometer with $\alpha = 0.00392$ and $R_0 = 100\ \Omega$.

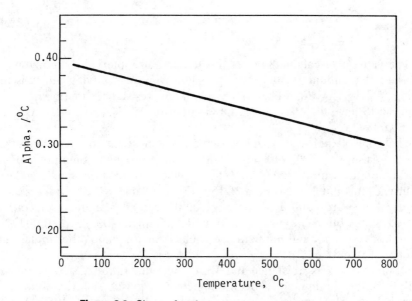

Figure 8.3. Slope of resistance versus temperature.

W, that is almost a resistivity ratio. The value of W is not quite the resistivity ratio, because changes in l/a by thermal expansion change the geometry slightly. The resistance ratio,

$$W = \frac{R}{R_0},$$
(8.6)

is specified in range I of IPTS. Using W instead of R has several advantages. Sensors with quite different values of R_0 can be compared because they should have similar values of W. Flexure, oxidation, and other changes to R_0 will often have about the same effect on R, so that W does not change as much. Changes in resistivity can be more readily related to the theory of metallic conduction than can changes in resistance. And calibration curves for thermometers of slightly different values of R_0 can be reduced to one curve when only moderate accuracy is required.

For lesser accuracy, in many applications the temperature dependence can be written

$$R \approx R_0(1 + At).$$
(8.7)

Then A is called a temperature coefficient of resistance. It is quite sensitive to small amounts of impurities or to mechanical strain. For the range of temperatures from the ice point to the steam point, the mean value of A over that temperature interval, which is designated α, can be calculated by

$$\alpha \approx \frac{R_{100} - R_0}{R_0(100)}.$$
(8.8)

The value of α calculated over this temperature interval is a convenient measure of platinum purity.[9] An equivalent way of specifying purity is the ratio, $W_{100} = R_{100}/R_0$. The IPTS platinum resistance thermometer is required to have $W_{100} \geq 1.3925$. Obviously, the mean α for 0–100°C must be ≥ 0.003925.

The goal of high purity at reasonable cost has caused some manufacturers to use $\alpha = 0.003920$. (Table 8.1) The European community often uses platinum with $\alpha = 0.00385$, as specified in the German standard, Deutsches Institut für Normen (DIN) 43760 (Table 8.2).[10] Older instruments often were calibrated for α of 0.00385; and manufacturers in the United States offer equipment calibrated for either 0.00385 or 0.003920. Values of resistivity and temperature coefficients for some other metals are given in Table 8.3.

Strain also affects resistance and W. Usually, the ratio R_{100}/R_0 decreases slightly if the platinum has been subjected to plastic deformation, even though both R_{100} and R_0 usually increase. (A change of 100 μΩ cor-

TABLE 8.1 $W = R_t/R_0$ as a Function of Temperature for a Typical High Purity Platinum Resistor[a]

$T(°C)$	$W(T_{68})$	$T(°C)$	$W(T_{68})$	$T(°C)$	$W(T_{68})$	$T(°C)$	$W(T_{68})$
−268.95	0.0005487	−212	0.1192758	−163	0.3300113	−114	0.5367356
−260	0.0012617	−211	0.1234086	−162	0.3343007	−113	0.5408911
−259	0.0014988	−210	0.1275638	−161	0.3385869	−112	0.5450446
−258	0.0017936	−209	0.1317395	−160	0.3428700	−111	0.5491959
−257	0.0021531	−208	0.1359339	−159	0.3471498	−110	0.5533452
−256	0.0025848	−207	0.1401457	−158	0.3514265	−109	0.5574925
−255	0.0030964	−206	0.1443731	−157	0.3557000	−108	0.5616377
−254	0.0036953	−205	0.1486148	−156	0.3599703	−107	0.5657809
−253	0.0043879	−204	0.1528696	−155	0.3642375	−106	0.5699222
−252	0.0051804	−203	0.1571361	−154	0.3685014	−105	0.5740615
−251	0.0060783	−202	0.1614132	−153	0.3727623	−104	0.5781989
−250	0.0070865	−201	0.1657000	−152	0.3770200	−103	0.5823344
−249	0.0082089	−200	0.1699953	−151	0.3812746	−102	0.5864680
−248	0.0094486	−199	0.1742983	−150	0.3855260	−101	0.5905996
−247	0.0108073	−198	0.1786080	−149	0.3897744	−100	0.5947295
−246	0.0122865	−197	0.1829238	−148	0.3940198	−99	0.5988574
−245	0.0138868	−196	0.1872447	−147	0.3982621	−98	0.6029835
−244	0.0156086	−195	0.1915702	−146	0.4025014	−97	0.6071079
−243	0.0174519	−194	0.1958996	−145	0.4067377	−96	0.6112304
−242	0.0194159	−193	0.2002322	−144	0.4109711	−95	0.6153511
−241	0.0214996	−192	0.2045676	−143	0.4152015	−94	0.6194700
−240	0.0237011	−191	0.2089052	−142	0.4194290	−93	0.6235872
−239	0.0260182	−190	0.2132445	−141	0.4236536	−92	0.6277026
−238	0.0284480	−189	0.2175850	−140	0.4278754	−91	0.6318163
−237	0.0309873	−188	0.2219265	−139	0.4320944	−90	0.6359283
−236	0.0336325	−187	0.2262684	−138	0.4363106	−89	0.6400385
−235	0.0363801	−186	0.2306104	−137	0.4405240	−88	0.6441471
−234	0.0392260	−185	0.2349523	−136	0.4447346	−87	0.6482540
−233	0.0421666	−184	0.2392936	−135	0.4489426	−86	0.6523592
−232	0.0451979	−183	0.2436341	−134	0.4531479	−85	0.6564627
−231	0.0483160	−182	0.2479737	−133	0.4573506	−84	0.6605646
−230	0.0515171	−181	0.2523120	−132	0.4615506	−83	0.6646649
−229	0.0547973	−180	0.2566487	−131	0.4657480	−82	0.6687635
−228	0.0581529	−179	0.2609838	−130	0.4699429	−81	0.6728605
−227	0.0615801	−178	0.2653171	−129	0.4741353	−80	0.6769559
−226	0.0650753	−177	0.2696483	−128	0.4783251	−79	0.6810497
−225	0.0686348	−176	0.2739773	−127	0.4825125	−78	0.6851420
−224	0.0722551	−175	0.2783041	−126	0.4866975	−77	0.6892326
−223	0.0759327	−174	0.2826284	−125	0.4908800	−76	0.6933217
−222	0.0796642	−173	0.2869501	−124	0.4950601	−75	0.6974092
−221	0.0834464	−172	0.2912693	−123	0.4992379	−74	0.7014951
−220	0.0872759	−171	0.2955857	−122	0.5034133	−73	0.7055795
−219	0.0911497	−170	0.2998993	−121	0.5075864	−72	0.7096624
−218	0.0950650	−169	0.3042100	−120	0.5117572	−71	0.7137438
−217	0.0990188	−168	0.3085178	−119	0.5159258	−70	0.7178236
−216	0.1030084	−167	0.3128226	−118	0.5200922	−69	0.7219019
−215	0.1070313	−166	0.3171244	−117	0.5242563	−68	0.7259787
−214	0.1110850	−165	0.3214232	−116	0.5284182	−67	0.7300541
−213	0.1151672	−164	0.3257188	−115	0.5325780	−66	0.7341279

TABLE 8.1 (cont.)

$T(°C)$	$W(T_{68})$	$T(°C)$	$W(T_{68})$	$T(°C)$	$W(T_{68})$	$T(°C)$	$W(T_{68})$
−65	0.7382003	−15	0.9400645	35	1.1387801	85	1.3344704
−64	0.7422712	−14	0.9440688	36	1.1427233	86	1.3383537
−63	0.7463406	−13	0.9480718	37	1.1466654	87	1.3422358
−62	0.7504085	−12	0.9520737	38	1.1506063	88	1.3461168
−61	0.7544751	−11	0.9560743	39	1.1545459	89	1.3499965
−60	0.7585401	−10	0.9600737	40	1.1584843	90	1.3538750
−59	0.7626038	−9	0.9640718	41	1.1624216	91	1.3577524
−58	0.7666660	−8	0.9680687	42	1.1663576	92	1.3616285
−57	0.7707267	−7	0.9720644	43	1.1702924	93	1.3655035
−56	0.7747861	−6	0.9760589	44	1.1742260	94	1.3693773
−55	0.7788440	−5	0.9800522	45	1.1781584	95	1.3732499
−54	0.7829006	−4	0.9840442	46	1.1820896	96	1.3771213
−53	0.7869557	−3	0.9880350	47	1.1860196	97	1.3809915
−52	0.7910095	−2	0.9920245	48	1.1899484	98	1.3848606
−51	0.7950618	−1	0.9960129	49	1.1938760	99	1.3887284
−50	0.7991128	0	1.0000000	50	1.1978024	100	1.3925951
−49	0.8031624	1	1.0039859	51	1.2017276	105	1.4119107
−48	0.8072106	2	1.0079706	52	1.2056516	110	1.4311966
−47	0.8112575	3	1.0119540	53	1.2095743	115	1.4504530
−46	0.8153030	4	1.0159362	54	1.2134959	120	1.4696799
−45	0.8193471	5	1.0199172	55	1.2174163	125	1.4888772
−44	0.8233899	6	1.0238970	56	1.2213354	130	1.5080451
−43	0.8274314	7	1.0278755	57	1.2252534	135	1.5271836
−42	0.8314715	8	1.0318528	58	1.2291702	140	1.5462926
−41	0.8355103	9	1.0358289	59	1.2330858	145	1.5653723
−40	0.8395477	10	1.0398038	60	1.2370001	150	1.5844226
−39	0.8435839	11	1.0437775	61	1.2409133	155	1.6034436
−38	0.8476187	12	1.0477499	62	1.2448253	160	1.6224354
−37	0.8516522	13	1.0517211	63	1.2487360	165	1.6413978
−36	0.8556844	14	1.0556911	64	1.2526456	170	1.6603310
−35	0.8597152	15	1.0596599	65	1.2565540	175	1.6792350
−34	0.8637448	16	1.0636275	66	1.2604611	180	1.6981098
−33	0.8677731	17	1.0675938	67	1.2643671	185	1.7169554
−32	0.8718001	18	1.0715589	68	1.2682719	190	1.7357719
−31	0.8758258	19	1.0755228	69	1.2721755	195	1.7545592
−30	0.8798502	20	1.0794855	70	1.2760779	200	1.7733174
−29	0.8838733	21	1.0834470	71	1.2799791	205	1.7920464
−28	0.8878952	22	1.0874072	72	1.2838791	210	1.8107464
−27	0.8919157	23	1.0913663	73	1.2877779	215	1.8294173
−26	0.8959350	24	1.0953241	74	1.2916755	220	1.8480591
−25	0.8999531	25	1.0992807	75	1.2955719	225	1.8666719
−24	0.9039699	26	1.1032361	76	1.2994671	230	1.8852557
−23	0.9079854	27	1.1071903	77	1.3033611	235	1.9038104
−22	0.9119996	28	1.1111432	78	1.3072540	240	1.9223360
−21	0.9160127	29	1.1150950	79	1.3111456	245	1.9408327
−20	0.9200244	30	1.1190455	80	1.3150361	250	1.9593003
−19	0.9240349	31	1.1229949	81	1.3189253	255	1.9777390
−18	0.9280442	32	1.1269430	82	1.3228134	260	1.9961486
−17	0.9320522	33	1.1308899	83	1.3267003	265	2.0145293
−16	0.9360589	34	1.1348356	84	1.3305859	270	2.0328809

TABLE 8.1 (cont.)

$T(°C)$	$W(T_{68})$	$T(°C)$	$W(T_{68})$	$T(°C)$	$W(T_{68})$	$T(°C)$	$W(T_{68})$
275	2.0512036	485	2.7944677	695	3.4851492	905	4.1200760
280	2.0694972	490	2.8115330	700	3.5009298	910	4.1344803
285	2.0877619	495	2.8285686	705	3.5166788	915	4.1488505
290	2.1059976	500	2.8455745	710	3.5323961	920	4.1631866
295	2.1242043	505	2.8625505	715	3.5480817	925	4.1774885
300	2.1423820	510	2.8794966	720	3.5637356	930	4.1917563
305	2.1605307	515	2.8964129	725	3.5793576	935	4.2059897
310	2.1786504	520	2.9132993	730	3.5949477	940	4.2201889
315	2.1967411	525	2.9301556	735	3.6105059	945	4.2343536
320	2.2148027	530	2.9469820	740	3.6260321	950	4.2484839
325	2.2328354	535	2.9637784	745	3.6415263	955	4.2625796
330	2.2508390	540	2.9805447	750	3.6569883	960	4.2766408
335	2.2688136	545	2.9972809	755	3.6724182	965	4.2906674
340	2.2867591	550	3.0139869	760	3.6878159	970	4.3046592
345	2.3046755	555	3.0306627	765	3.7031813	975	4.3186163
350	2.3225629	560	3.0473083	770	3.7185144	980	4.3325386
355	2.3404212	565	3.0639236	775	3.7338151	985	4.3464260
360	2.3582504	570	3.0805086	780	3.7490833	990	4.3602785
365	2.3760505	575	3.0970633	785	3.7643191	995	4.3740959
370	2.3938214	580	3.1135875	790	3.7795223	1000	4.3878783
375	2.4115632	585	3.1300813	795	3.7946929	1005	4.4016256
380	2.4292759	590	3.1465446	800	3.8098308	1010	4.4153378
385	2.4469594	595	3.1629774	805	3.8249360	1015	4.4290147
390	2.4646136	600	3.1793796	810	3.8400084	1020	4.4426563
395	2.4822387	605	3.1957512	815	3.8550479	1025	4.4562625
400	2.4998345	610	3.2120921	820	3.8700546	1030	4.4698334
405	2.5174011	615	3.2284023	825	3.8850283	1035	4.4833687
410	2.5349384	620	3.2446817	830	3.8999689	1040	4.4968686
415	2.5524463	625	3.2609304	835	3.9148765	1045	4.5103328
420	2.5699250	630	3.2771481	840	3.9297510	1050	4.5237614
425	2.5873743	635	3.2933350	845	3.9445922	1055	4.5371543
430	2.6047943	640	3.3094909	850	3.9594002	1060	4.5505115
435	2.6221848	645	3.3256159	855	3.9741749	1065	4.5638328
440	2.6395460	650	3.3417097	860	3.9889162	1070	4.5771183
445	2.6568777	655	3.3577725	865	4.0036240	1075	4.5903678
450	2.6741799	660	3.3738042	870	4.0182984	1080	4.6035813
455	2.6914526	665	3.3898046	875	4.0329392	1085	4.6167588
460	2.7086958	670	3.4057738	880	4.0475464	1090	4.6299002
465	2.7259094	675	3.4217117	885	4.0621199	1095	4.6430054
470	2.7430934	680	3.4376182	890	4.0766597	1100	4.6560744
475	2.7602478	685	3.4534933	895	4.0911657		
480	2.7773726	690	3.4693370	900	4.1056378		

[a]Table courtesy of Rosemount, Inc.

TABLE 8.2 Resistance as a function of Temperature for $R_0 = 100\ \Omega$. $\alpha = 0.00385$

°C	−0	−1	−2	−3	−4	−5	−6	−7	−8	−9
−200	18.49	—	—	—	—	—	—	—	—	—
−190	22.80	22.37	21.94	21.51	21.08	20.65	20.22	19.79	19.36	18.93
−180	27.08	26.65	26.23	25.80	25.37	24.94	24.52	24.09	23.66	23.23
−170	31.32	30.90	30.47	30.05	29.63	29.20	28.78	28.35	27.93	27.50
−160	35.53	35.11	34.69	34.27	33.85	33.43	33.01	32.59	32.16	31.74
−150	39.71	39.30	38.88	38.46	38.04	37.63	37.21	36.79	36.37	35.95
−140	43.87	43.45	43.04	42.63	42.21	41.79	41.38	40.96	40.55	40.13
−130	48.00	47.59	47.18	46.76	46.35	45.94	45.52	45.11	44.70	44.28
−120	52.11	51.70	51.29	50.88	50.47	50.06	49.64	49.23	48.82	48.41
−110	56.19	55.78	55.38	54.97	54.56	54.15	53.74	53.33	52.92	52.52
−100	60.25	59.85	59.44	59.04	58.63	58.22	57.82	57.41	57.00	56.60
−90	64.30	63.90	63.49	63.09	62.68	62.28	61.87	61.47	61.06	60.66
−80	68.33	67.92	67.52	67.12	66.72	66.31	65.91	65.51	65.11	64.70
−70	72.33	71.93	71.53	71.13	70.73	70.33	69.93	69.53	69.13	68.73
−60	76.33	75.93	75.53	75.13	74.73	74.33	73.93	73.53	73.13	72.73
−50	80.31	79.91	79.51	79.11	78.72	78.32	77.92	77.52	77.13	76.73
−40	84.27	83.88	83.48	83.08	82.69	82.29	81.89	81.50	81.10	80.70
−30	88.22	87.83	87.43	87.04	86.64	86.25	85.85	85.46	85.06	84.67
−20	92.16	91.77	91.37	90.98	90.59	90.19	89.80	89.40	89.01	88.62
−10	96.09	95.69	95.30	94.91	94.52	94.12	93.73	93.34	92.95	92.55
0	100.00	99.61	99.22	98.83	98.44	98.04	97.65	97.26	96.87	96.48

°C	0	1	2	3	4	5	6	7	8	9
0	100.00	100.39	100.78	101.17	101.56	101.95	102.34	102.73	103.12	103.51
10	103.90	104.29	104.68	105.07	105.46	105.85	106.24	106.63	107.02	107.40
20	107.79	108.18	108.57	108.96	109.35	109.73	110.12	110.51	110.90	111.28
30	111.67	112.06	112.45	112.83	113.22	113.61	113.99	114.38	114.77	115.15
40	115.54	115.93	116.31	116.70	117.08	117.47	117.85	118.24	118.62	119.01
50	119.40	119.78	120.16	120.55	120.93	121.32	121.70	122.09	122.47	122.86
60	123.24	123.62	124.01	124.39	124.77	125.16	125.54	125.92	126.31	126.69
70	127.07	127.45	127.84	128.22	128.60	128.98	129.37	129.75	130.13	130.51
80	130.89	131.27	131.66	132.04	132.42	132.80	133.18	133.56	133.94	134.32
90	134.70	135.08	135.46	135.84	136.22	136.60	136.98	137.36	137.74	138.12
100	138.50	138.88	139.26	139.64	140.02	140.39	140.77	141.15	141.53	141.91
110	142.29	142.66	143.04	143.42	143.80	144.17	144.55	144.93	145.31	145.68
120	146.06	146.44	146.81	147.19	147.57	147.94	148.32	148.70	149.07	149.45
130	149.82	150.20	150.57	150.95	151.33	151.70	152.08	152.45	152.83	153.20
140	153.58	153.95	154.32	154.70	155.07	155.45	155.82	156.19	156.57	156.94
150	157.31	157.69	158.06	158.43	158.81	159.18	159.55	159.93	160.30	160.67
160	161.04	161.42	161.79	162.16	162.53	162.90	163.27	163.65	164.02	164.39
170	164.76	165.13	165.50	165.87	166.24	166.61	166.98	167.35	167.72	168.09
180	168.46	168.83	169.20	169.57	169.94	170.31	170.68	171.05	171.42	171.79
190	172.16	172.53	172.90	173.26	173.63	174.00	174.37	174.74	175.10	175.47
200	175.84	176.21	176.57	176.94	177.31	177.68	178.04	178.41	178.78	179.14
210	179.51	179.88	180.24	180.61	180.97	181.34	181.71	182.07	182.44	182.80
220	183.17	183.53	183.90	184.26	184.63	184.99	185.36	185.72	186.09	186.45
230	186.82	187.18	187.54	187.91	188.27	188.63	189.00	189.36	189.72	190.09
240	190.45	190.81	191.18	191.54	191.90	192.26	192.63	192.99	193.35	193.71
250	194.07	194.44	194.80	195.16	195.52	195.88	196.24	196.60	196.96	197.33
260	197.69	198.05	198.41	198.77	199.13	199.49	199.85	200.21	200.57	200.93
270	201.29	201.65	202.01	202.36	202.72	203.08	203.44	203.80	204.16	204.52

TABLE 8.2   (cont.)

°C	0	1	2	3	4	5	6	7	8	9
280	204.88	205.23	205.59	205.95	206.31	206.67	207.02	207.38	207.74	208.10
290	208.45	208.81	209.17	209.52	209.88	210.24	210.59	210.95	211.31	211.66
300	212.02	212.37	212.73	213.09	213.44	213.80	214.15	214.51	214.86	215.22
310	215.57	215.93	216.28	216.64	216.99	217.35	217.70	218.05	218.41	218.76
320	219.12	219.47	219.82	220.18	220.53	220.88	221.24	221.59	221.94	222.29
330	222.65	223.00	223.35	223.70	224.06	224.41	224.76	225.11	225.46	225.81
340	226.17	226.52	226.87	227.22	227.57	227.92	228.27	228.62	228.97	229.32
350	229.67	230.02	230.37	230.72	231.07	231.42	231.77	232.12	232.47	232.82
360	233.17	233.52	233.87	234.22	234.56	234.91	235.26	235.61	235.96	236.31
370	236.65	237.00	237.35	237.70	238.04	238.39	238.74	239.09	239.43	239.78
380	240.13	240.47	240.82	241.17	241.51	241.86	242.20	242.55	242.90	243.24
390	243.59	243.93	244.28	244.62	244.97	245.31	245.66	246.00	246.35	246.69
400	247.04	247.38	247.73	248.07	248.41	248.76	249.10	249.45	249.79	250.13
410	250.48	250.82	251.16	251.50	251.85	252.19	252.53	252.88	253.22	253.56
420	253.90	254.24	254.59	254.93	255.27	255.61	255.95	256.29	256.64	256.98
430	257.32	257.66	258.00	258.34	258.68	259.02	259.36	259.70	260.04	260.38
440	260.72	261.06	261.40	261.74	262.08	262.42	262.76	263.10	263.43	263.77
450	264.11	264.45	264.79	265.13	265.47	265.80	266.14	266.48	266.82	267.15
460	267.49	267.83	268.17	268.50	268.84	269.18	269.51	269.85	270.19	270.52
470	270.86	271.20	271.53	271.87	272.20	272.54	272.88	273.21	273.55	273.88
480	274.22	274.55	274.89	275.22	275.56	275.89	276.23	276.56	276.89	277.23
490	277.56	277.90	278.23	278.56	278.90	279.23	279.56	279.90	280.23	280.56
500	280.90	281.23	281.56	281.89	282.23	282.56	282.89	283.22	283.55	283.89
510	284.22	284.55	284.88	285.21	285.54	285.87	286.21	286.54	286.87	287.20
520	287.53	287.86	288.19	288.52	288.85	289.18	289.51	289.84	290.17	290.50
530	290.83	291.16	291.49	291.81	292.14	292.47	292.80	293.13	293.46	293.79
540	294.11	294.44	294.77	295.10	295.43	295.75	296.08	296.41	296.74	297.06
550	297.39	297.72	298.04	298.37	298.70	299.02	299.35	299.68	300.00	300.33
560	300.65	300.98	301.31	301.63	301.96	302.28	302.61	302.93	303.26	303.58
570	303.91	304.23	304.56	304.88	305.20	305.53	305.85	306.18	306.50	306.82
580	307.15	307.47	307.79	308.12	308.44	308.76	309.09	309.41	309.73	310.05
590	310.38	310.70	311.02	311.34	311.67	311.99	312.31	312.63	312.95	313.27
600	313.59	313.92	314.24	314.56	314.88	315.20	315.52	315.84	316.16	316.48
610	316.80	317.12	317.44	317.76	318.08	318.40	318.72	319.04	319.36	319.68
620	319.99	320.31	320.63	320.95	321.27	321.59	321.91	322.22	322.54	322.86
630	323.18	323.49	323.81	324.13	324.45	324.76	325.08	325.40	325.72	326.03
640	326.35	326.66	326.98	327.30	327.61	327.93	328.25	328.56	328.88	329.19
650	329.51	329.82	330.14	330.45	330.77	331.08	331.40	331.71	332.03	332.34
660	332.66	332.97	333.28	333.60	333.91	334.23	334.54	334.85	335.17	335.48
670	335.79	336.11	336.42	336.73	337.04	337.36	337.67	337.98	338.29	338.61
680	338.92	339.23	339.54	339.85	340.16	340.48	340.79	341.10	341.41	341.72
690	342.03	342.34	342.65	342.96	343.27	343.58	343.89	344.20	344.51	344.82
700	345.13	345.44	345.75	346.06	346.37	346.68	346.99	347.30	347.60	347.91
710	348.22	348.53	348.84	349.15	349.45	349.76	350.07	350.38	250.69	350.99
720	351.30	351.61	351.91	352.22	352.53	352.83	353.14	353.45	353.75	354.06
730	354.37	354.67	354.98	355.28	355.59	355.90	356.20	356.51	356.81	357.12
740	357.42	357.73	358.03	358.34	358.64	358.95	359.25	359.55	359.86	360.16
750	360.47	360.77	361.07	361.38	361.68	361.98	362.29	362.59	362.89	363.19
760	363.50	363.80	364.10	364.40	364.71	365.01	365.31	365.61	365.91	366.22
770	366.52	366.82	367.12	367.42	367.72	368.02	368.32	368.63	368.93	369.23
780	369.53	369.83	370.13	370.43	370.73	371.03	371.33	371.63	371.93	372.22
790	372.52	372.82	373.12	373.42	373.72	374.02	374.32	374.61	374.91	375.21

TABLE 8.2 (cont.)

°C	0	1	2	3	4	5	6	7	8	9
800	375.51	375.81	376.10	376.40	376.70	377.00	377.29	377.59	377.89	378.19
810	378.48	378.78	379.08	379.37	379.67	379.97	380.26	380.56	380.85	381.15
820	381.45	381.74	382.04	382.33	382.63	382.92	383.22	383.51	383.81	384.10
830	384.40	384.69	384.98	385.28	385.57	385.87	386.16	386.45	386.75	387.04
840	387.34	387.63	387.92	388.21	388.51	388.80	389.09	389.39	389.68	389.97
850	390.26	—	—	—	—	—	—	—	—	—

Note: Resistance can be calculated from the equations

$$0\text{–}850°C: R = 100(1 + 3.90802 \times 10^{-3}t - 0.580195 \times 10^{-6}t^2)$$

$$-200 \text{ to } 0°C: R = 100[1 + 3.90802 \times 10^{-3}t - 0.580195 \times 10^{-6}t^2$$

$$- 4.27350 \times 10^{-12}(t - 100)t^3]$$

Source: Ref. 10, reprinted with permission.

responds to a change of about 0.001°C.) When a platinum resistor has received rough handling or has been plastically deformed by thermal shock or other means, it becomes unstable because annealing can occur at temperatures as low as 100°C. When annealing occurs the resistance changes back toward its original resistance. However, rough handling can make a change in the resistance equivalent to 0.1°C at the triple point of water, and this is one of the main causes of deterioration of precision platinum resistance thermometers.[11] As the temperature approaches absolute zero, the electrical resistivity of metals decreases until, within a few degrees of absolute zero, some of them even become superconductors and

TABLE 8.3 Resistivity and Temperature Coefficient of Resistivity for Several Metals

Metal	Temperature of ρ(°C)	ρ (Ω-cm $\times 10^{-6}$)	α (Ω/K)	Temperature of α(°C)
Al	20	2.6548	0.00429	20
Cu	20	1.6730	0.0068	20
Au	20	2.35	0.004	0–100
Fe	20	9.71	0.00651	20
Pb	20	20.648	0.0036	20–40
Ni	20	6.84	0.0069	0–100
Pd	20	10.8	0.00377	0–100
Pt	20	10.6	0.003927	0–100
Rd	20	4.51	0.0042	0–100
Ag	20	1.59	0.0041	0–100
Sn	0	11.0	0.0047	0–100

have nearly zero electrical resistance. Therefore, it becomes more and more difficult to measure resistance accurately. (A small error becomes a larger percentage of the resistance.) Special alloys have been discovered which do not have such small resistivities. One of them (Fig. 8.4), an alloy of rhodium with 0.5 atomic percent iron, has exceptionally stable resistivity from 0.5 to 20 K.[12]

For precision thermometry, platinum resistance sensors are unequaled in the upper part of range I and in range II. There is a great deal of interest in extending them into range III, but this has not yet occurred because of stability problems. One interesting feature of precision thermometry in range II is that platinum is sensitive to thermal cycling, where the effects amount to about 1 mK at the triple point of water (Fig. 8.5). This has been shown to depend on oxygen partial pressure and to be explained as the result of reducing the metal cross section by about one-quarter monolayer of oxygen.[13] Under more severe oxidizing conditions, PtO_2 can be made to

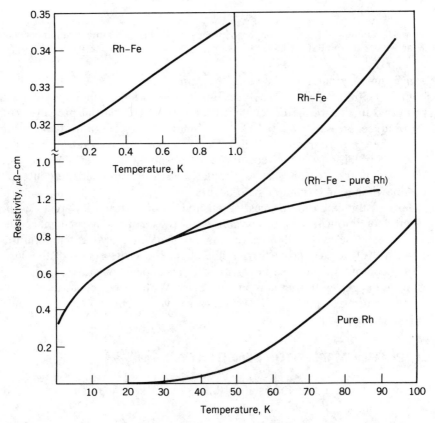

Figure 8.4. Resistance of rhodium–0.5% iron. [From R. L. Rusby, *Platinum Met. Rev.* **25**, 57 (1981).]

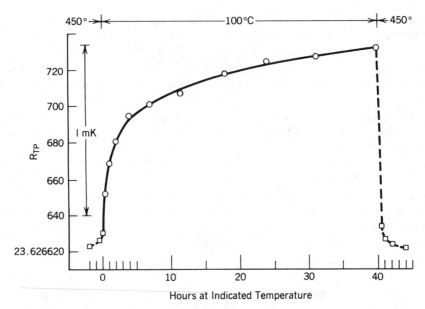

Figure 8.5. Change of R_{TP} in ohms for thermometer K9 due to thermal cycling between 450 and 100°C. (From Ref. 13.)

form in much greater thickness in the critical range in which it is stable, about 250–500°C. This can give serious hysteresis in resistance if elements are cycled in that range of temperatures and oxygen partial pressure, and can produce errors up to 0.2°C.[14] Platinum sensors are not stable in an oxygen-free environment, so the effect cannot be avoided. When this degree of precision is required, the ratio W should be determined for the same oxidation state. That is, R_0 should be measured immediately after R_t before a change in oxidation state occurs.[15]

Because the oxidation process is sensitive to oxygen partial pressure, the presence of impurities or of oxygen-gettering components can make the oxygen response vary with construction details. When used at temperatures in which PtO_2 forms and vaporizes, the change in crosssection of the wire, and concommitant increase in resistance, can depend on previous thermal history if oxygen-gettering components exist. With proper construction, use at temperatures of 650–750°C appears to have excellent stability because PtO_2 does not form.[16]

8.3 PLATINUM RESISTOR SENSITIVITY

The sensitivity of a metal resistance thermometer is defined as

$$\text{sensitivity} = S \equiv \frac{\partial R}{\partial t}, \tag{8.9}$$

so for Eq. (8.9),

$$S = R_0(A + 2Bt + 3C't^2). \qquad (8.10)$$

If we first divide Eq. (8.9) through by R_0 to normalize the relationship,

$$\frac{\partial W}{\partial t} = \frac{\partial(R/R_0)}{\partial t} = A + 2Bt + 3C't^2. \qquad (8.11)$$

Since B and C' are small, $\partial W/\partial t \approx A$. With Eq. (8.11), the sensitivity of thermometers of different resistances can be compared. Obviously, high sensitivity, as well as stability, are desirable.

8.4 ORIGIN OF THE CALLENDAR EQUATION

In his pioneering work, Callendar found that a parabolic equation would fit his data well. If we write the first three terms of Eq. (8.3) as $R_t = R_0(1 + At + Bt^2)$, we have a basis for the Callendar equation (8.1). Thus

$$R_t = R_0(1 + At + Bt^2); \qquad (8.12)$$

when $t = 0°C$, $R_t = R_0$, and when $t = 100°C$,

$$R_{100} = R_0[1 + A(100) + B(10,000)].$$

Then

$$\frac{R_t - R_0}{R_{100} - R_0} = \frac{R_0(1 + At + Bt^2) - R_0}{R_0[1 + A(100) + B(10,000)] - R_0}$$

$$= \frac{R_0}{R_0}\left(\frac{1 + At + Bt^2 - 1}{1 + 100A + 10,000B - 1}\right)$$

$$= \frac{At + Bt^2}{100A + 10,000B} = \frac{t}{100}\left(\frac{A + Bt}{A + 100B}\right).$$

Solving for the first t in the numerator gives

$$t = \left(\frac{R_t - R_0}{R_{100} - R_0}\right)100\left(\frac{A + 100B}{A + Bt}\right). \qquad (8.13)$$

Adding and subtracting Bt in the last term gives

$$t = \left(\frac{R_t - R_0}{R_{100} - R_0}\right) 100 \left(\frac{A + Bt - Bt + 100B}{A + Bt}\right)$$

$$= \left(\frac{R_t - R_0}{R_{100} - R_0}\right) 100 \left[1 + \frac{B(100 - t)}{A + Bt}\right].$$

Let $[(R_t - R_0)/(R_{100} - R_0)]100 = t_{pt}$. Then

$$t \approx t_{pt} + t_{pt}\frac{B(100 - t)}{A + Bt}. \tag{8.14}$$

We know that for platinum $A = 3.92 \times 10^{-3}$ and $B = -5.88 \times 10^{-7}$, so the term Bt in the denominator of the right term is negligible for all practical temperatures.

Then $t = t_{pt} + t_{pt}(B/A)(100 - t)$ and we can let $B/A = -\delta \times 10^{-4} = -\delta/(100)^2$. But $B/A \approx -1.5 \times 10^{-4}$, so $\delta \approx -1.5$. Then

$$t = t_{pt} - (t_{pt})\frac{\delta}{(100)^2}(100 - t)$$

or

$$t = t_{pt} + \frac{t_{pt}}{100}\frac{\delta(t - 100)}{100}$$

$$= t_{pt} + \frac{t_{pt}}{100}\left(\frac{t}{100} - 1\right)\delta$$

$$= \left(\frac{R_t - R_0}{R_{100} - R_0}\right)100 + \delta\left(\frac{t}{100}\right)\left(\frac{t}{100} - 1\right) \tag{8.15}$$

if $t_{pt} \approx t$. (Note that δ is negative. $\delta \approx -1.5$.) This final equation, (8.15), is the Callendar equation.

8.5   SENSING ELEMENT CONSTRUCTION

In his early work Callendar showed that sensing element reliability required high-purity platinum, annealing the wire at a temperature at least as high as the intended use temperature, freedom from mechanical strains in winding, and support in use so that thermal or mechanical dimensional changes would not stress the wire. He showed that iron was an especially serious contaminate at high temperatures. We also know that heating in contact with phosphorus, arsenic, antimony, selenium, and tellurium causes embrittlement; that platinum is attacked by molten metals and their vapors; and that heating in a strongly reducing atmosphere in a system containing silica, silicates, or sulfur will lead to failure. Therefore, the design chosen for a particular sensing element depends on the intended application. It is essential to protect the platinum if reliability is to be achieved.

The original construction recommended by Callendar included a coil of fine annealed platinum wire wound on a mica coil form in such a way that it was free to expand and contract, and a glazed refractory procelain protection tube. The connecting wires were heavy platinum welded to the sensing wires (Fig. 8.6).

Many modifications have been made since then. Some of these are to wind the wire into a tight helix and then wind the helix on the coil form; to wind it on a fused quartz (low-thermal-expansion) frame inside a hermetically sealed quartz envelope (Fig. 8.7); to wind it in a formless coil inside a glass tube and to protect it from movement with fine powdered insulation; and so on.

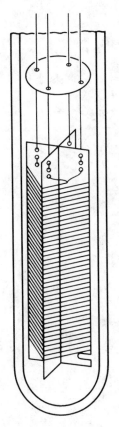

Figure 8.6. Schematic of Callendar's construction.

PLATINUM RESISTANCE SENSING ELEMENT

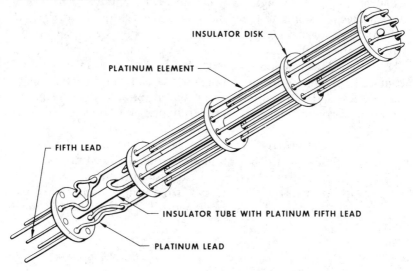

INSULATOR DISK

PLATINUM ELEMENT

FIFTH LEAD

INSULATOR TUBE WITH PLATINUM FIFTH LEAD

PLATINUM LEAD

(a)

SENSING ELEMENT

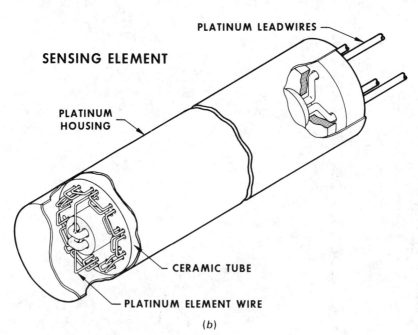

PLATINUM LEADWIRES

PLATINUM HOUSING

CERAMIC TUBE

PLATINUM ELEMENT WIRE

(b)

Figure 8.7 (a) and (b).

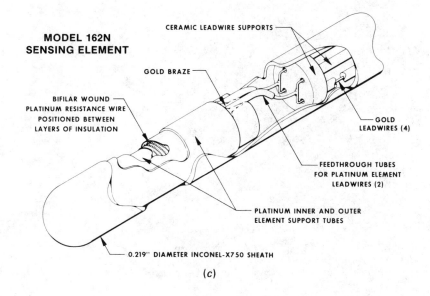

**MODEL 162N
SENSING ELEMENT**

CERAMIC LEADWIRE SUPPORTS

GOLD BRAZE

BIFILAR WOUND
PLATINUM RESISTANCE WIRE
POSITIONED BETWEEN
LAYERS OF INSULATION

GOLD
LEADWIRES (4)

FEEDTHROUGH TUBES
FOR PLATINUM ELEMENT
LEADWIRES (2)

PLATINUM INNER AND OUTER
ELEMENT SUPPORT TUBES

0.219" DIAMETER INCONEL-X750 SHEATH

(c)

DIMENSIONAL DRAWING

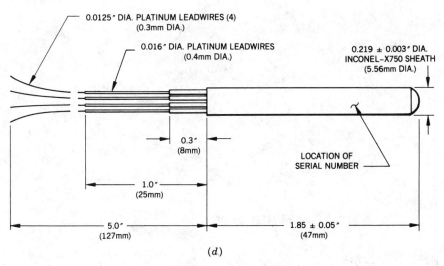

0.0125" DIA. PLATINUM LEADWIRES (4)
(0.3mm DIA.)

0.016" DIA. PLATINUM LEADWIRES
(0.4mm DIA.)

0.219 ± 0.003" DIA.
INCONEL-X750 SHEATH
(5.56mm DIA.)

0.3"
(8mm)

1.0"
(25mm)

LOCATION OF
SERIAL NUMBER

5.0"
(127mm)

1.85 ± 0.05"
(47mm)

(d)

Figure 8.7 (c) and (d).

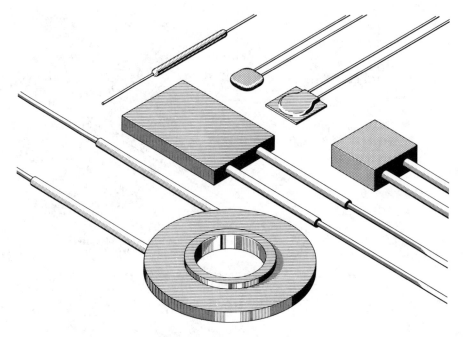

Wide temperature range...−260° C to +1400° C
Miniature size
Rugged, easily attached
Reliable, strain-free
Variety of configurations

(e)

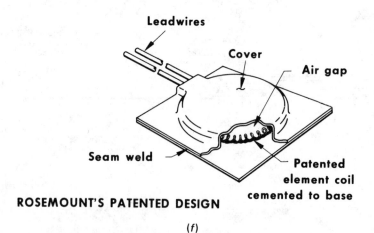

Leadwires

Cover

Air gap

Seam weld

**Patented
element coil
cemented to base**

ROSEMOUNT'S PATENTED DESIGN

(f)

Figure 8.7. Precision platinum resistance thermometers. (Courtesy Rosemount, Inc.)

There is no established combination of wire diameter, length, and resistance that is recognized. Many high-quality research instruments have $R_0 = 25$ or $100\,\Omega$, approximately, and use 0.1-mm-diameter wire. Some recommend 50, 200, and $500\,\Omega$. It takes about 8 m of 0.1-mm-diameter wire to obtain $100\,\Omega$, and about 0.61 m of 0.075-mm wire to obtain $25.5\,\Omega$ at the ice point. Both the size and cost can be reduced if finer wire is used. Some commercial instruments use 0.01- to 0.05-mm-diameter wire and a few have only 10-Ω resistance at 0°C. The finer wire is more susceptible to vibration damage but has faster response time and smaller size.

For aerospace applications a design which is very simple but rugged and free from mechanical strain has been very successful. It is a coil placed loosely inside a ceramic tube, with each turn of the coil anchored at one point inside the tube by a ribbon of ceramic glaze. Another successful design used in aerospace applications at lesser precision is a bifilar winding on the outside of a hollow ceramic tube which is then covered completely with a ceramic glaze. That design has been used to accelerations of $1000\,\text{m/s}^2$ and frequencies of 20 kHz.[17]

In recent years platinum resistors on ceramic substrates have become increasingly popular. Both size and cost have been reduced. Reduction in size has made this versatile device much more useful. Thick-film printing techniques from the semiconductor industry are used to apply thin platinum films on tiny ceramic substrates.[18] After firing on the films and encapsulating in a ceramic glaze, these units can be as small as $1\,\text{mm} \times 2\,\text{mm} \times 2.5\,\text{mm}$, or in cylindrical form, 3 mm diameter $\times 7$ mm.[14,18] This small size and the concommitant rapid response makes it possible to use platinum resistance thermometers in many applications formerly limited to thermocouples or thermistors. Thick-film devices are used for routine temperature measurement where 0.1% accuracy is sufficient.

Platinum sensors are also produced in special shapes for attachment to surfaces, such as a resistive coil near the surface of a ceramic washer, so that the washer can be bolted to a surface. One manufacturer has the sensor in a 16-gauge by $1\frac{1}{2}$-in. hypodermic needle.[19]

There has been a great deal of effort to extend platinum resistance thermometry above 630°C, ideally all the way to the gold point, to eliminate the type S thermocouple in range III of IPTS.[20,21] Design of the sensor to reduce thermal expansion strains is one key element in extending the range upward. Elimination of contamination and using the correct oxygen partial pressure is also important. One design that appears especially promising is a simple bifilar winding with each coil stabilized at 180° but free to expand and contract (Fig. 8.8).[22,23]

Some platinum resistance thermometers have been used successfully for over 50 years. In one application, an element was kept at 540°C for 4 years with less than 1-degree change in calibration. Precision designs are capable of ±0.1 mK. For good precision it is necessary to anneal the platinum after it has been formed. Usually, elements are sheathed in tubes of glass, steel,

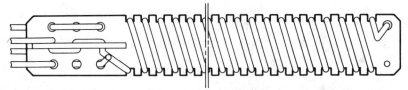

Figure 8.8. High-temperature bifilar platinum winding on a fused quartz or sapphire strip-shaped support. (From Ref. 22.)

copper, brass, stainless steel, or platinum. Sheath design is very important to response time.

8.6   THE CONNECTING WIRES

The resistances of the connecting wires and their junctions are included in the resistance measured. Because these extend from the temperature at the measurement site to the temperature at the resistance measuring instrument, their temperature coefficient of resistance and their size and composition are extremely important. The sensing element itself is constructed with matched massive connecting wires extending from it. These are sometimes as short as 8 mm. Either platinum or gold connecting wires are welded on to extend them to the ambient temperature region. Short lengths of tungsten wire are sometimes used as a part of the hermetic glass-to-metal seal if the resistor is sealed into a soft-glass envelope. Tungsten is used because it has a thermal expansion compatible with soft glass. The external platinum or gold wires are soft soldered to copper to extend them to the measuring instrument. It is very important that the connecting wires have the same thermal gradient and that they be matched in resistance. For precision measurements, if braided wires are used, each strand should be insulated to prevent the noise that could occur if one of the wires in a multiwire strand is broken. The connecting wires should be grouped together in a shielded cable to improve thermal contact and reduce electrical noise.

For precision measurements, the current wires should be coaxial and twisted together in one cable. Standard cables are provided with twisted conductors to reduce inductive coupling, and with foil sheathes and grounding wires to reduce common-mode noise. Voltage wires should be in a separate shielded cable. Especially at cryogenic temperatures, the cold end of the connecting wires should be fastened in good thermal contact with a thermal conductor at the same temperature as the sensor. It is not unusual for a cryogenic sensor to have much lower resistance than the connecting wires themselves. Then it is essential to minimize the errors produced by current flowing in the connecting wires, and by heat conducting inward along the connecting wires. Low-capacitance insulation such as PTFE (polytetrafluoroethylene) reduces leakage and reactance errors.

8.7 PRINCIPLES OF ELECTRICAL RESISTANCE MEASUREMENT

Electrical resistance can be measured with either direct current or alternating current. For many decades direct-current measurements were used because they were more accurate. Recently, standards laboratories have adopted alternating-current methods because they now have greater precision and accuracy. Both ac and dc methods are used in commercial instruments. First we need to consider the differences in these methods.

8.7.1 Direct-Current Methods

Direct-current methods are made with a constant current, not varying in time. The current flows through a resistance in accordance with Ohm's law. If the applied voltage is constant and there is no variation in the resistance caused by external or internal heating (if the temperature of the resistor is constant), the current is also constant and in phase with the voltage. Therefore, in accordance with Ohm's law,

$$V = IR .\qquad(8.16)$$

In principle we have only to apply a small, accurately known, unchanging voltage to the resistor, measure the current, and compute the resistance. Alternatively, we can apply a small, accurately known current, measure the voltage, and compute the resistance. The current in either case must be so small that Joule self-heating, I^2R, is too small to change the resistance by changing the temperature.

In practice, this can be done directly by applying the voltage or the current, and solving Ohm's law. Or it can be done indirectly by comparison methods in which a standard known resistance is compared to the sensor resistance. Instruments for this purpose will be discussed after ac measurement principles are discussed.

8.7.2 Alternating-Current Methods

Alternating current is usually generated with the voltage varying sinusoidally with time. When an ac voltage is applied to a pure resistor, a current flows through the resistor in response to the applied voltage. If the resistor were a perfect resistor, then, at any instant in time, the current flowing in it at that instant in time would follow Ohm's law. Because the voltage varies sinusoidally in time, the current flowing through the resistor would also vary sinusoidally in time and would be in phase with it [Fig. 8.9(b)].

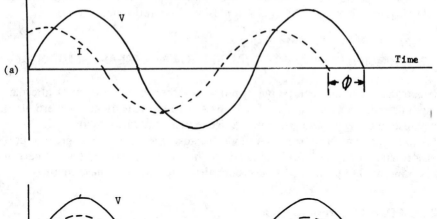

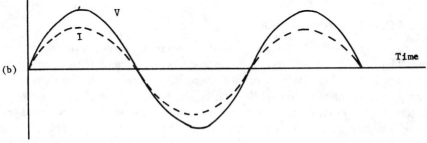

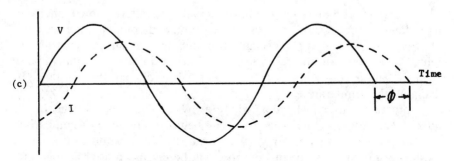

Figure 8.9. Current–voltage relationships for a (*a*) capacitative, (*b*) purely resistive, and (*c*) inductive loads. *V*, Solid line; *I*, dashed line.

For a sinusoidal voltage,

$$V = V_m \sin \omega\theta \tag{8.17}$$

where V_m is the maximum voltage, ω the angular frequency of the alternating current, and θ is time.*

*In this book θ is used for time because lowercase t is reserved for degrees Celsius.

Dividing Eq. (8.17) by R gives the current if the load could be purely a resistance [Fig. 8.9(b)].

It is not possible for a load to be purely resistive. Because alternating current varies with time, both capacitative and inductive effects will occur. If a dc voltage is applied to a pure capacitor, current will flow into the capacitor until it is charged. At the instant of time the electric circuit is closed, the current flows freely into the capacitor as though it were a short circuit. No voltage will appear. Only later will it be possible to find a voltage across the capacitor, after sufficient electrons have flowed in to saturate the capacitor and the voltage increases with time. The current can be calculated for a perfect capacitor by

$$I = C \frac{dV}{d\theta}. \tag{8.18}$$

Substituting Eq. (8.17) and taking the derivative with respect to time gives

$$I = CV_m \frac{d(\sin \omega\theta)}{dt} \tag{8.19}$$

$$= \omega C V_m \cos \omega\theta \tag{8.20}$$

$$= I_m \cos \omega\theta . \tag{8.21}$$

The cosine of an angle leads the sine by 90°, so a perfect capacitor has the current leading the voltage by 90°. An ac voltage applied to a resistor that has some capacitance will also produce a current that leads the voltage by some phase angle ϕ [Fig. 8.9(a)]. Then

$$I = I_m \sin (\omega\theta + \phi) , \tag{8.22}$$

where ϕ is the phase angle between the current and the voltage.

Every ac circuit also has inductive effects. At the instant a voltage is applied to a pure inductive circuit, the current is zero because formation of the magnetic field opposes the flow. For an inductor

$$V = -L \, dI/d\theta . \tag{8.23}$$

Then

$$dI = -\frac{V}{L} \, d\theta . \tag{8.24}$$

Substituting for V from Eq. (8.17) gives

$$dI = -\frac{V_m}{L} \sin(\omega\theta)\, d\theta. \qquad (8.25)$$

Integrating yields

$$I = -\frac{\omega V_m}{L} \cos \omega\theta, \qquad (8.26)$$

which lags the voltage by 90°. Every real circuit has inductive effects associated with it. For a real, not perfect, inductive circuit the current lags the voltage by some phase angle ϕ [Fig. 8.9(c)]. Then

$$I = I_m \sin(\omega\theta - \phi). \qquad (8.27)$$

Both capacitative and inductive affects will occur. One will dominate, so the circuit will be described as inductive or capacitative, depending on which is larger. Both affect the rate of change of the current and are called reactances.

The equivalent to resistance in an ac circuit is called the impedance. For a capacitative circuit, the impedance Z is given by

$$Z = R + jX_c. \qquad (8.28)$$

Z is a complex number with real part the resistance R and imaginary part the capacitance X_c. Similarly, the impedance of an inductive circuit is given by

$$Z = R + jX_L, \qquad (8.29)$$

where X_L is the inductance.

Every real circuit has inductive and capacitative effects that respond to changing current. When a pure resistance is to be measured, as for temperature measurement, it is desirable to minimize capacitative and inductive affects by proper selection of connecting wires, winding to reduce inductance, avoiding materials of high magnetic permeability such as iron, and so on. For comparative methods, it is not possible to obtain balance of a standard and a measuring resistor unless both the resistance and the reactances are balanced. This is sometimes described as resistance and quadrature balance. (Quadrature is that component 90° out of phase with the resistance.)

To calculate the impedance for a real circuit, Eqs. (8.28) and (8.29) can be combined as appropriate. Both X_L and X_c are very small for most

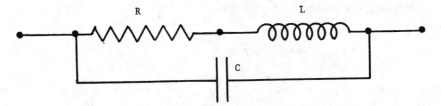

Figure 8.10. Equivalent circuit for an ac resistor.

resistive measurement situations, so that their contribution to the circuit can sometimes be neglected. For example, the equivalent circuit of a resistor is often modeled as shown in Fig. 8.10.

The impedance for this series–parallel network can be calculated easily. The impedance of the RL series portion of the circuit is

$$Z_{RL} = Z_R + Z_L = R + j\omega L .$$
(8.30)

For the capacitance, $X_c = (j\omega C)^{-1}$ and it is in parallel with Z_{RL}. Then

$$\frac{1}{Z} = j\omega C + \frac{1}{R + j\omega L} .$$
(8.31)

Then

$$Z = \frac{1}{j\omega C + 1/(R + j\omega L)}$$
(8.32)

$$= \frac{R + j\omega L}{j\omega C(R + j\omega L) + 1}$$
(8.33)

$$= \frac{R + j\omega L}{j\omega RC + (1 - \omega^2 LC)} .$$
(8.34)

Multiplying both numerator and denominator by the complex conjugate gives

$$Z = \frac{R + j\omega L}{j\omega RC + (1 - \omega^2 LC)} \left[\frac{-j\omega RC + (1 - \omega^2 LC)}{-j\omega RC + (1 - \omega^2 LC)} \right]$$
(8.35)

$$= \frac{R(1 - \omega^2 LC) + \omega^2 RLC + j\omega L(1 - \omega^2 LC) - j\omega R^2 C}{\omega^2 R^2 C^2 + (1 - \omega^2 LC)^2}$$
(8.36)

$$= \frac{R + j\omega[(L - \omega^2 L^2 C) - R^2 C]}{\omega^2 R^2 C^2 + (1 - \omega^2 LC)^2} .$$
(8.37)

The in-phase component of the impedance is

$$R_\omega = \frac{R}{\omega^2 R^2 C^2 + (1 - \omega^2 LC)^2}. \tag{8.38}$$

Because the LC product is so small, this can be approximated by

$$R_\omega \approx \frac{R}{1 + \omega^2 R^2 C^2}. \tag{8.39}$$

From Eq. (8.37) the phase angle is the angle with its tangent the imaginary part over the real part, or

$$\phi = \tan^{-1} \frac{L - \omega^2 L^2 C - R^2 C}{R} \tag{8.40}$$

$$= \tan^{-1}\left(\frac{L}{R} - \frac{\omega^2 L^2 C}{R} - RC\right). \tag{8.41}$$

This can be approximated by

$$\phi \approx \tan^{-1}\left(\frac{L}{R} - CR\right). \tag{8.42}$$

In addition to the self-capacitance of the resistor, there will also be a distributed capacitance C' between the resistor and its surroundings. By an analysis similar to the above, when the surroundings are connected at ground potential, the approximate relationship is[24]

$$R_\omega \approx \frac{R}{(1 + \omega^2 CC'R^2/6)[1 + \omega^2 R^2(C - C'/6)^2]}. \tag{8.43}$$

For most resistances of 500 Ω or less, at low frequencies,

$$\phi \approx \tan^{-1} \frac{\omega L}{R}. \tag{8.44}$$

For higher resistances,

$$\phi \approx \tan^{-1}\left[\omega\left(\frac{C'R}{6} - CR\right)\right]. \tag{8.45}$$

For resistors of 500 Ω or less, at frequencies up to 1.6 Hz, the value of R_ω is constant to 1 part in 10^{-7} and the phase angle is very sensitive to inductance.[25] Therefore, minimizing inductance is very important in resistance thermometer design. Most modern resistance thermometers have small distributed inductance and capacitance. For many resistance thermometers, the dc and ac resistances are indistinguishable at frequencies up to 400 Hz.[25] This makes it possible to use appropriate ac methods, but an unknown sensor should be tested first.

Because of difficulties with capacitive and inductive effects, the preferred methods for temperature measurement using electrical resistors were dc methods up until about 1970. The availability of high-quality operational amplifier components, lock-in amplifiers, and improved ratio transformers has made the ac methods increasingly attractive. Both dc and ac measuring instruments are discussed next.

8.8 RESISTANCE MEASURING INSTRUMENTS REQUIREMENTS

Resistance measuring instruments can operate either directly solving Ohm's law or indirectly by comparison with a standard resistor. Both ac and dc instruments are used. The simplest in concept is the dc Ohm's law device shown in Fig. 8.11. Either a known voltage can be applied and the current measured, or a known current can be applied and the voltage measured. In either case, current must flow through the resistor to obtain a measuring signal. Self-heating effects require that current be minimized to obtain an accurate temperature measurement. The degree of self-heating that can be tolerated, and therefore the current that can be allowed, depends on the accuracy required. Precision resistance measurements are possible with great care and expense to 1 part in 10^7, about two orders of magnitude better than the best temperature measurement now possible, so that electrical resistance measurement is not, fundamentally, a limitation. However, if only 1-degree accuracy is required, the resistance only needs to be measured to about 1 part in 1000. This can be done easily at low cost. Therefore, the type of instrument selected for an application depends on the requirements of the application. In addition to the accuracy required, sensor

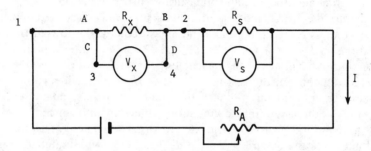

Figure 8.11. Ohm's law direct-current method.

size and geometry, reliability, cost, indication, recording and control functions, telemetry and computer requirements, and other factors must be considered. This field of instrumentation is changing so rapidly that it is not possible to include them all in a single book, nor is it possible for a book to stay up to date for long. We will therefore outline some general principles and give a few examples. But the reader will have a much broader selection of instrumentation available than can possibly be covered here. It will be necessary to study the specifications for the various instruments in terms of the requirements for the application in order to select a suitable instrument.

8.9   OHM'S LAW DEVICES

Ohm's law devices have become increasingly popular as operational amplifiers[26-28] and microprocessor circuits have become available. We will start the discussion using the dc circuit in Fig. 8.11, although ac methods are also used.

The battery supplies a source of direct current that flows through three resistances in series R_A, R_s, and R_X. Adjusting R_A allows the current to be varied in a series of settings, so that if the resistance is determined at each current setting, it can be plotted as a function of current and extrapolated to zero current to eliminate Joule heating effects. The voltmeters have very high internal resistance, so that essentially no current flows through them. The connecting wires A and B are called current wires because they carry the current. Wires C and D are the voltage wires. Voltages are measured at constant current across the standard resistance R_s to compute the current flowing in the circuit. Voltage is measured across the unknown resistance to compute the unknown resistance using the current calculated from the standard resistor. In this system the resistances of both the current leads and the voltage leads are unimportant, because no current is assumed to flow in C and D and the resistances of wires A and B are not being measured.

Either a standard current can be used, and the voltage across R_X measured, or a standard voltage across R_X can be used and the current through it measured. The voltage source, the current measuring instrumentation, and the voltage measuring instrumentation can be incorporated into the instrument, so that it is only necessary to connect the sensor to the instrument and make the measurement. For low accuracy, the connections to R_X for wires A, B, C, and D can be shifted to points 1 and 2 as the output terminals of the instrument. This requires only two terminals and is convenient, but the resistance of the connecting wires is included in the measurement and this reduces the accuracy. It is seldom used. More common is a three-terminal connection where an unbalanced three-wire Wheatstone bridge is used. (Bridges are discussed later.) The three-wire bridge improves the accuracy a great deal because, with matched connecting wires, that error is reduced to the error in matching the connecting

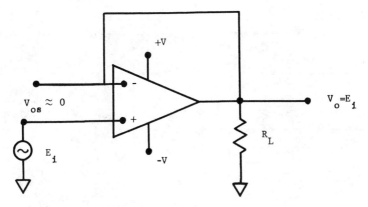

Figure 8.12. Typical noninverting voltage follower circuit. The input can be a dc source instead of the ac source shown here. (After Ref. 26.)

wires. If the wires are long, if contact resistance or broken-strand resistances occur, if the wires do not have identical temperature gradients along their length, both reproducibility and accuracy suffer. If a bridge is unbalanced, the error includes some current in the voltage wires, and errors of 0.2% of full scale are common. (This is 1.5°C for 750° full scale.) Four terminal measurements, 1, 2, 3, and 4 in Fig. 8.11, reduce this error. The limit of error then depends on the current and voltage measurement and on drift and ambient temperature effects on the electronics.

The voltage measurement across both the sensor and the standard resistor (to determine the current) should be done with the highest practical input impedance. Modern voltage-follower operational amplifiers have very high input impedance and are ideally suited to the application. (They are also called isolation amplifiers, or buffer amplifiers, or unity-gain amplifiers; Fig. 8.12.)

8.9.1 Operational Amplifiers

An operational amplifier has a minimum of five terminals, two for the + and − input signals, two for the +V and −V supply voltages, and one for the output. The ideal operational amplifier has infinite input impedance, zero output impedance, zero offset voltage, and infinite open loop (no feedback) gain. When connected without a feedback resistor, in closed loop as in Fig. 8.12, V_0 will follow E_i exactly. The gain is V_0/E_1 and equals 1.

Real operational amplifiers have very high input impedances, a few megohms being common. The output impedance is often 1 to 10 Ω. The voltage between the input terminals V_{os} is not exactly zero, but is often 1–3 mV. Compensation for nonzero offset can produce a unity gain. The voltage follower serves to limit the current in the voltage measuring circuit

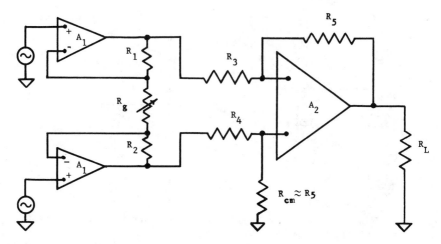

Figure 8.13. Simplified schematic of a buffered instrument amplifier. R_g adjusts gain, R_{cm} balances out common mode. Supply voltages for the voltage followers A_1 and the differential amplifier A_2 not shown. (After Ref. 27.)

to a negligible value and to produce an output signal equal to the input signal with enough current available so that it can operate a meter to indicate voltage. The voltage follower, then, serves as a power amplifier without loading the signal source (without drawing appreciable current from the signal source).

In practice a voltage follower has one serious defect. It reproduces noise signals very well, so noise power is also amplified. Noise can be classified as common mode, where the noise is common to both inputs, or normal mode where it is inherent to the signal. Much of the electrical noise in an industrial environment comes from reactance coupling of the conductors to inductive loads such as motors, fluorescent lights, and spark noise. This noise is picked up by both inputs and is common mode. The 60-Hz noise from lights and machinery and the radio-frequency noise from welding or other spark sources can be filtered. However, a differential operational amplifier will only amplify the difference in voltage between the inputs, and thus rejects common-mode noise very well. Because of the low input impedance, differential amplifiers are usually combined with voltage followers to produce a special amplifier known as an instrumentation amplifier (Fig. 8.13). In it, the differential amplifier is used to reject common-mode noise, and voltage followers are used on each input to prevent loading the signal source.

8.9.2   Instrumentation Amplifiers

Many different types of instrument amplifiers are available. They usually can provide moderate gain without deteriorating input impedance. Gains of

1–1000, externally adjustable with a single feedback resistor, are common. Then errors in gain linearity and from gain temperature dependence must be considered.

Often, the output is divided into three parts: output, sense, and reference. All three can be connected to the load, so that resistance in the wires to the load can be matched by resistance in the wires to the sense terminal. The sense terminal completes the circuit to the feedback resistance network and this helps stabilize the gain. The reference terminal is used to provide an adjustable reference voltage to provide a bias voltage at zero input so that the zero on an indicating instrument can be obtained when it requires an input at apparent zero, for example, a recorder requiring 2–20 mA.

The output of the instrument amplifier can be supplied to a high-input-impedance ohmmeter, calibrated to read temperature. However, a meter can provide only limited resolution. Most modern resistance thermometers have a digital-to-analog converter to provide a signal for light-emitting diodes in a digital display. The display can be easily visible and indicate to 0.1°C or better. The principal disadvantage of a digital display is the difficulty of following rapidly changing temperatures. The meter allows one to see the temperature change and to estimate rates and change of rates.

8.9.3 Digital Display

Digital displays with a digital panel meter require the following:

1. Sufficiently high input impedance to interface with an instrumentation amplifier (or the instrumentation amplifier built into the meter).

2. A measuring circuit to determine the resistance from the amplified voltage signal provided by the sensor output circuit. Usually, this is a dual-slope integration circuit that requires a timing frequency generator and counter.

3. An analog-to-digital converter to provide a digital signal for the display. Usually, this is a binary-coded decimal (BCD) output compatible with the display assembly. This also requires a timing frequency generator and counter.

4. A light-emitting-diode display assembly with a network to convert the BCD signal to the digital display.

5. A power supply capable of providing the voltages necessary for operating the meter.

Most of the requirements are usually met through a large-scale-integrated (LSI) microprocessor chip or a CMOS chip from one of the chip manufacturers. Power for the microprocessor can be the same voltage (e.g.,

TABLE 8.4 Digital Display Terminology and Resolution

Digits	Counts Full Scale	Resolution (%F.S.)
$2\frac{1}{2}$	199	0.5
3	999	0.1
$3\frac{1}{2}$	1,999	0.05
$3\frac{3}{4}$	3,999	0.025
4	9,999	0.01
$4\frac{1}{2}$	19,999	0.005
$4\frac{3}{4}$	39,999	0.0025

±5 V) as that used for the instrumentation amplifier. This can be from a battery or from a rectified 115-V supply as is needed for the application.

Commercial digital panel meters can be very inexpensive, with minimum accuracy and capabilities. Or they may be quite sophisticated, with special circuits for picking and holding maximum or minimum values, for averaging, for linear outputs for analog meters or recorders, for digital interfacing to provide computer indication, control, and response functions, and so on.

The analog-to-digital conversion process converts an analog signal to a digital signal by one of several methods. A binary decimal (0 or 1) signal has a possible range of values from the most significant bit (MSB) to the least significant bit (LSB). For example, "thirteen" as a binary number is 1101, where the 1 on the left is the most significant bit and the 1 on the right is the least significant bit. The most significant bit limits the maximum possible without overrun (1111), and the least significant bit determines the precision (0 or 1) of the conversion. (A number between 0 and 1 must be rounded off to 0 or 1.)

The decimal display is rated in terms of the number of digits in the display. The most significant digit determines the maximum possible without overrun, without regard for decimal point. Displays are rated by their full-scale range, and that also determines their resolution (Table 8.4). The resolution is 1 part per unit of full-scale range. For a $3\frac{1}{2}$-digit display, it is 1/2000 or 0.05%.

Display resolution does not indicate precision or accuracy of the temperature measurement. Displays sometimes indicate much better resolution than the system can produce.

A variety of methods for improving the precision and accuracy of the method is possible. The dual-slope method for scaling the input signal to an analog-to-digital converter is usually considered superior to other methods because of its reliability and its excellent common-mode noise rejection. In this method the input signal is sampled and held by an S/H amplifier (to avoid changes during the conversion), the signal is applied to a integrator circuit for a fixed timing period, producing a linear rise in the voltage on the integrator, the signal is disconnected and a reverse polarity reference voltage is applied to the integrator circuit, and the time required to return the integrator voltage to zero is measured. The slope of the linear charging voltage–time curve depends on the input voltage. The slope of the discharge voltage–time curve is fixed by the reference voltage. The ratio of the charging and discharging time, multiplied by the reference voltage, gives the signal voltage. Variations in capacitance or clock frequency are unimportant, as they do not occur within the time period of charge/discharge. Correction for zero drift can be made each cycle by integrating with the input shorted. A further improvement is the quad-slope method, where a precision reference voltage is also sampled and the unknown voltage is scaled by the reference voltage slopes.

Outputs of analog-to-digital (A/D) converters can also suffer from nonlinearity in the conversion process, zero shifts, and missed digits. Drift in the power supplies with time, drift in the components, the effects of temperature on the electronic components, and the effects of humidity can cause loss of precision and accuracy. Manufacturers specifications must be studied with care and with some skepticism. Different manufacturers write their specifications in different ways, so that specifications such as linearity or precision may not be directly comparable.

Three commercial examples of Ohm's law devices are given here. The Tegam model 869 hand-held, battery-operated digital resistance thermometer has the following specifications:[29]

TEGAM, MODEL 869. (© TEGAM, INC. 1987)

Range	Temperature Span	Resolution	4-Wire 1 Year Accuracy (18°–28°C)
200	−150 to +199.9°C	0.1°C	± .3°C
	−199.9 to −150°C	0.1°C	± 1.5°C
630	−220 to +630°C	1°C	± 1°C

Note that the accuracy is given for the instrument at 18–28°C and does not include sensor errors. The meter is designed for a 100-Ω sensor with $\alpha = 0.00385$ and requires only 500 µA of sensing current. Additional error due to instrument drift outside the range 18–20°C is not more than ±0.015°C/°C in the range −10 to 18°C and 28 to 50°C. The display is $3\frac{1}{2}$-digit LCD with polarity and decimal indications. The conversion rate is 1.5 readings per second using dual-slope A/D conversion with continuous linearization. Common-mode rejection is 0.001°C/V to a maximum of 42 V peak to ground. The meter can be used at 80% relative humidity to 35°C. It must be derated 3% RH/°C from 35 to 60°C. Storage can be up to 90%

RH from −35 to 35°C. From 35 to 60°C, derate storage relative humidity 3% RH/°C.

The instrument is designed for four-wire probes but can be used with three-wire probes at reduced (but unspecified) accuracy. Maximum connecting wire resistance is 50 Ω, four-wire mode, and 10 Ω per wire, three-wire mode.

Keithley 195-T Digital Bench-Type Thermometer. The Keithley 195-T has the following specifications[30]:

KEITHLEY 195-T

Span	Resolution	Four-Wire Accuracy 1 Yr, 18–28°C +(% rdg + counts)	Temperature Coefficient 0–18°C and 28–50°C +(% rdg + counts)/°C
−200.00 to 230.00°C	0.01°C	0.03 + 10	0.003 + 0.4
230.00 to 630.00°C	0.01°C	0.03 + 40	0.003 + 4
−220.00 to −200.00°C	0.01°C	0.03 + 40	0.003 + 4

The meter accepts three- or four-wire 100-Ω platinum probes with up to 25 Ω of connecting wire resistance per wire in four-wire mode. Both α and δ in the Callendar equation are programmable. The meter has nine built-in program modes and a IEEE-488 bus connection for external programming, talk, and listen modes. The sensor current is 1 mA maximum.

Fluke Digital Thermometer. The Fluke model 2189 digital thermometer is an autoranging solid-state instrument designed for 100-Ω platinum or nickel or 10-ohm copper sensors.[31] The $5\frac{1}{2}$-digit LED display operates at $3\frac{1}{2}$ readings per second. It is an example of the types of versatility available in modern instruments. Options include alarms, peak storing, analog and digital outputs, up to 100 sensors from multipoint selector, IEEE-488 translator, printer, battery supply, and a calibrator. Resolution and accuracy using the Fluke 100-Ω platinum sensor are given in Table 8.5.

8.10   BRIDGE MEASURING INSTRUMENTS

Comparative measurements are usually made with some sort of bridge. There are two general types of bridges, out-of-balance bridges and null-balance bridges. Out-of-balance bridges are used to indicate temperature by the deviation from balance. Current flows in the voltage wires of out-of-balance bridges and they perform very much like Ohm's law devices. Since balance is not required, they can respond quickly to temperature for

TABLE 8.5 Data for Fluke Model 2189A Digital Thermometer

Temp.[b] (°C)	Maximum System Error[a] (+°C)		
	At Calibration	90 Days 18–28°C Ambient	1 Year 18–28°C Ambient

Low-Temperature Range: −183 to +204°C;
Resolution 0.01°C or °F

Temp.[b] (°C)	At Calibration	90 Days 18–28°C Ambient	1 Year 18–28°C Ambient
−183	C	C	C
−50	0.04	0.08	0.11
0	0.03	0.07	0.09
50	0.05	0.10	0.13
100	0.07	0.12	0.16
150	0.08	0.15	0.20
200	0.09	0.17	0.23

High-Temperature Range: 204 to +480°C;
Resolution 0.1°C or °F; Periodic Probe Exposure[d]

Temp.[b] (°C)	At Calibration	90 Days 18–28°C Ambient	1 Year 18–28°C Ambient
204	0.14	0.25	0.27
300	0.18	0.32	0.33
400	0.21	0.39	0.40
480	0.29	0.48	0.50

[a]The accuracies in the high-temperature range are based on the user performing an ice-point adjustment in accordance with the following schedule:

Probe Exposure Temperature Range (°C)	Total Exposure Time before Adjustment (hours)
200–350	500
350–480	250

[b]Interpolation should be used for accuracies at intermediate temperatures.

[c]The system operates down to −183°C, but the probe calibration is not verified below −50°C. It is estimated that the accuracy below −50°C is the same as the accuracy at an equal temperature in the positive range. Low-temperature calibrations are available as a special.

[d]Exposure of the Y2039A at high temperature for long periods of time may cause the probe to change its characteristics and require the accuracy specifications to be degraded. For example, there is a 20% probability that exposure at 480°C for 500 hours will require degrading. It is easy for the user to determine if degrading is necessary by measuring the ice-point resistance of the probe. The 2189A Instruction Sheet explains this procedure.

either meter or digital display. Out-of-balance bridges have moderate precision and accuracy, often in the 0.1°C range.

Null-balance bridges are capable of remarkable precision and accuracy. Refined instruments at standards laboratories, such as the National Bureau of Standards, are capable of resistance measurements with an error of a few microkelvin, one or two orders of magnitude better than the stability of the sensing element.

Both null-balance and out-of-balance bridges are used extensively in electronic circuitry. Temperature compensation within an instrument is often accomplished with out-of-balance two-wire bridges, using a thermistor as the temperature-compensating resistor. Inexpensive resistance thermometers often use out-of-balance three-wire bridges with platinum, nickel, or copper as the sensing element. They can be either ac or dc bridges. Modern computer-controlled bridges usually are ac ratio transformer bridges that have evolved from dc Wheatstone bridges over the last 30 years. We start our discussion with the simple dc Wheatstone bridge and proceed to more complex instruments.

8.10.1 Direct-Current Wheatstone Bridges

The simplest Wheatstone bridge has four resistors connected in bridge configuration (Fig. 8.14). The temperature sensing resistor R_x in one branch is parallel to R_s a calibrated precision resistor in the other branch. R_x is external to the bridge and connected by external wires. When a stable voltage V from a battery or another source is applied as shown, and R_s

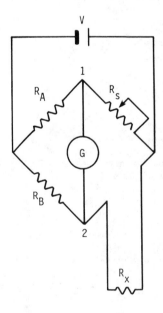

Figure 8.14. Wheatstone bridge.

adjusted until the null detector G does not deflect, the voltage at point 1 equals the voltage at point 2. If $R_A = R_B$, then $R_x = R_s$. If $R_A \neq R_B$, then $R_x = (R_B/R_A)R_s$. By providing several decades of R_B/R_A, the range is increased by several decades. Note that current flows in the wires connected to the resistor (the current wires), but not in those connected to the null detector (the voltage wires). Note also that this is truly a comparison method because the resistance of the sensor is determined in comparison to that of a standard. The result is only as good as the standard.

The two-wire bridge includes the resistance of the wires connected to both the standard and the sensing resistor. One of Callendar's important recommendations was to move the junction on the right side, to which the battery voltage is supplied, into the sensing element capsule (Fig. 8.15). Then, at balance, no current flows through the detector and

$$\frac{R_s + r_b}{R_A} = \frac{R_x + r_a}{R_b}. \tag{8.46}$$

If $R_A = R_b$, solving for R_x gives

$$R_x = R_s - r_a + r_b. \tag{8.47}$$

If the connecting wires are matched in resistance so that $r_a = r_b$, then

$$R_x = R_s. \tag{8.48}$$

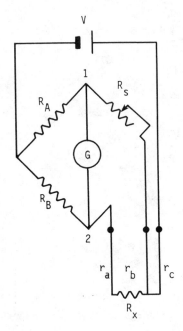

Figure 8.15. Three-wire Callendar bridge.

If $R_A = R_B \approx R_x$, and $\alpha_a = \alpha_b$ so that $r_a = r_b$ despite a thermal gradient along r_a and r_b, the effect of Joule heating is reduced. It is not eliminated, even if the resistances were equal because the heat transfer conditions around the sensor will not be the same as those within the bridge, so the temperature of R_s will not be the same as R_x.

Actually, it is not possible that $r_a = r_b$, but careful matching of the resistances and twisting them together in a cable so that they have the same thermal gradient greatly reduces the error. An improvement that eliminates the connecting wire resistance is the four-wire bridge (Fig. 8.16). This bridge requires two measurements at balance. One is made with wires a, c, and d connected as shown. Then the connections a and c are interchanged and d is shifted to b. Then if $R_A = R_B$,

$$R_{s1} + r_c = R_x + r_a \tag{8.49}$$

and

$$R_{s2} + r_a = R_x + r_b . \tag{8.50}$$

Adding Eqs. (8.49) and (8.50) gives

$$R_{s1} + R_{s2} = 2R_x , \tag{8.51}$$

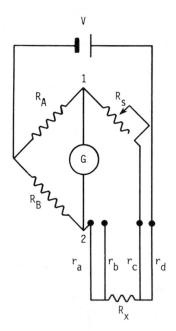

Figure 8.16. Schematic of a four-wire Mueller bridge.

so

$$R_x = \frac{R_{s1} + R_{s2}}{2} \tag{8.52}$$

and the effect of the connecting wires has been eliminated.

The four-wire bridge requires two manual balances. For many decades, a precision four-wire bridge, the Mueller bridge, was used by the U.S. National Bureau of Standards as the resistance thermometer measuring instrument.[11] It has been replaced by ratio transformer bridges (discussed later).

8.10.2 The Kelvin Double Bridge

Lord Kelvin introduced a further modification of the Wheatstone bridge that performs a double balance simultaneously (Fig. 8.17). At balance, the variable resistors are adjusted so that the voltages at the detector contacts are equal. Then some current bypasses both resistors through the external loop. The current flowing through the inner variable resistor (very small because it is in parallel with the shorting connection in parallel with it) flows through both R_s and R_x. Therefore, the same current flows through both R_s and R_x. The contacts for the inner and outer loops slide together, so they stay in the same ratio. The outer loop contacts have resistance r per turn. Then at balance, the IR drops are proportionate, even though $I_1 \neq I_2$. So

$$R_x = R_s \frac{n_1 r}{n_2 r} = R_s \frac{n_1}{n_2}. \tag{8.53}$$

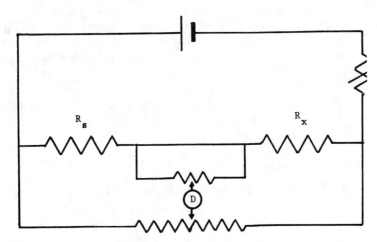

Figure 8.17. Double Kelvin bridge.

We see that the Kelvin double bridge can be a resistance ratio bridge, such that the turns ratio allows one to compare an unknown resistance with a standard resistance when they both have the same current flowing through them. Obviously, since the current flowing around R_s and R_x through the external loop does not flow through the internal loop, both should have high resistance.

8.11 OUT-OF-BALANCE BRIDGES

An out-of-balance bridge is often used when dynamic response is too rapid for null-balancing methods. This is often used as a three-wire bridge, as in Fig. 8.15. The null detector will be replaced with a current measuring device such as a microammeter to measure the deviation from balance as the sensor changes temperature away from the calibration point. The sensitivity can be changed by a variable resistor in series with the current measuring device. The range can be changed by varying R_A and R_B. The zero can be changed by changing R_s. Equations for relating current to resistance are given in Chapter 9. The resistances can be contained in integrated circuits.

8.12 ALTERNATING-CURRENT BRIDGES

In principle, all the dc resistance bridges can also be ac bridges, and are sometimes used as such. For the very high precision required for cryogenic measurements and for standards measurement, where R/R_0 may be measured to 1 part per million or better, dc bridges were used into the 1970s, but were difficult to operate. Bridges like the Mueller bridge had many standard resistors that required periodic recalibration. Massive, mercury-wetted contacts were used to minimize contact resistance. It was necessary to control ambient temperature within narrow limits. But still, the bridges were plagued by thermoelectric and parasitic effects. Both the Seebeck and Thomson effects caused the resistance to have a higher apparent resistance than the true resistance.[*,24,32] Adoption of low-frequency ac methods permitted superior resistance standards to be produced and avoided the thermoelectric errors because current reversal was rapid enough to prevent them. Usually, 200–300 Hz is sufficient to eliminate thermoelectric effects.

8.12.1 Ratio Transformers

The heart of an ac bridge is the ratio transformer.[32] Before it was developed in the early 1960s, transformers did not give output currents in perfect

---

*See Chapter 10 for Seebeck and Thomson effects.

proportion to their turns ratio (the ratio of the number of turns in the secondary to the number in the primary). Improved magnetic materials and insulators made new designs possible. Two important design improvements were to protect the magnetic core from mechanical strain and to use multiple-sector windings on a high-permeability magnetic core to reduce the effect of ambient and stray magnetic fields. Addition of a third winding made it possible to monitor the current in the transformer and allowed primary and secondary currents to be compared. When two ratio windings magnetized the core in opposite directions, the third winding could be used as a current comparator. When the currents were adjusted so that the magnetic flux was zero, as indicated by the current comparator, then

$$N_1 I_1 = N_2 I_2 , \tag{8.54}$$

where N is the number of turns, I is the current, and the subscripts indicate the first and second windings. The comparator coil was shielded for electrostatic fields with a copper shield, and from stray magnetic fields with a high-permeability, toroidal metal shield. In addition, compensation capacitative effects are usually added to ratio transformers. Kusters used a compensation winding in parallel with the ratio winding having the largest number of turns. The compensation winding had the same number of turns and was located inside the magnetic shield.[33]

8.12.2 The Hill and Miller Bridge

The first ac bridge to use ratio transformers was constructed very early in the 1960s. Hill and Miller designed and tested a ratio transformer bridge for platinum resistance thermometry based on the Kelvin double bridge[34] (Fig. 8.18). They used the approximation for Eq. (8.38),

$$R_\omega = R(1 + 2\omega^2 LC - \omega^2 R^2 C^2) \tag{8.55}$$

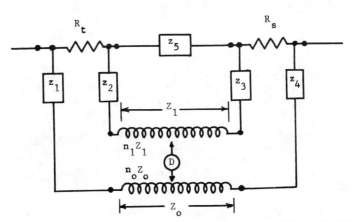

Figure 8.18. Hill and Miller inductively coupled bridge. (After Ref. 34.)

and designed the bridge for a standard 25-Ω wire-wound platinum resistance sensor. By calculating and experimentally verifying the effects of frequency, connecting wire impedances Z_1, Z_2, Z_3, and Z_4, and other inductances and capacitances, they were able to design a bridge in which the effects of the connecting wire impedances could be neglected for $W = R_t/R_0$ measurements. Error was less than 1 part in 10^6 and predicted to be improvable to 5 parts in 10^7, using the simple equation

$$R = R_s \frac{n_0}{1 - n_0}. \tag{8.56}$$

The bridge was designed to operate at 400 Hz with identical inner and outer ratio arms. Each arm consisted of eight decade windings on toroidal transformer cores (Fig. 8.19). The reactance was slightly inductive, so reactance balance was achieved with a ± 2.8-μH air-core mutual inductance and a 25-pF variable capacitor.

Modern versions of this bridge design are commercially available (Fig. 8.20).[35,36] One of Tinsley's manual bridges has the following specifications:[35]

Range: 10 $\mu\Omega$ to 111.1111 Ω
Frequency: 435 Hz $\pm$ 1%
Resolution: 10 $\mu\Omega$
Linearity: better than 1 ppm
Accuracy: $\pm$1 ppm
Readout: seven in-line decade switches
Meter sensitivity: $\pm$20 divisions; 1 division = 1 mK at 25 Ω
Weight: 10 kg
Size: 480 $\times$ 310 $\times$ 100 mm

The internal standard resistor is maintained at 27°C.

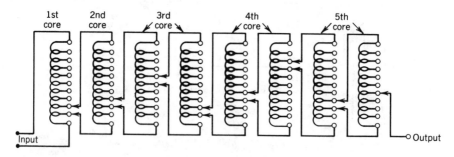

Figure 8.19. Method of connecting eight-decade windings in a five-core inductively coupled ratio arm. (After Ref. 34.)

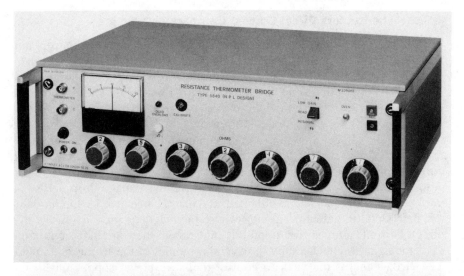

Figure 8.20. Two Tinsley resistance thermometer bridges. (*a*) Original manual NPL design. (*b*) "Senator" automatic model. (Courtesy of H. Tinsley & Co., Ltd.)

Two automatic versions are available from Tinsley, with microcomputer control and an optional IEEE-488 bus. The manual bridge models have either 1- or 0.5-mA sensor current, with $\sqrt{2}$ increase for extrapolation to zero current.

The newest version of the automatic bridge from Tinsley uses 200-μA current with $\sqrt{2}$ steps to 5 mA. The small current is especially useful for germanium thermometers at cryogenic temperatures, where connecting wire resistance may be 10 times the sensor resistance.

8.12.3 The Kusters Direct-Current Comparator

Ratio transformers were used in an ingeneous way by MacMartin and Kusters to make a self-balancing dc comparator for resistance thermometry[37] (Fig. 8.21). The ratio transformer was wound with two identical windings in opposition, one for the standard resistor R_s and one for the unknown R_x. A 700-Hz modulator coil was wound on the bridge as a sensing coil. Two power supplies were provided and they could be adjusted so that the voltage drop across R_x equals the voltage drop across R_s regardless of the current flowing in them. To facilitate balancing, the R_s current supply was designed as the slave current, following R_x. Then adjusting the taps on R_x ratio winding adjusts the voltage drop across R_s until current ratio balance is obtained. Balance is indicated by the modulator, demodulator, detector circuit because the magnetic fluxes from the two direct current sources are opposed, and when they are not equal, the detector signal is enhanced by the even harmonics of the strongest dc field. The error signal is fed back to control the slave power supply and to the detector, keeping the currents equal. Adjusting N_x to balance produces both equal currents and equal voltage drops in R_x and R_s. Then

$$R_x = \frac{N_x}{N_s} R_s .$$

(8.57)

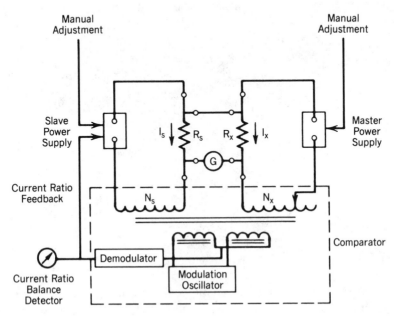

Figure 8.21. Direct-current comparator ratio bridge. [From Ref. 38. © 1970 IEEE, reprinted with permission.)

The bridge is accurate to 1 part per million. An improved design made the bridge more sensitive and allowed measurements to be made for small resistance thermometers $(0–10\,\Omega)$ as well as larger resistance thermometers (up to $0–1100\,\Omega$).[38]

8.12.4   The Cutkosky Bridges

R. D. Cutkosky at the U.S. National Bureau of Standards designed several devices for precision measurement of admittances before designing bridges for resistance thermometers. In that work, he used high-impedance voltage followers to increase the input impedance of ratio transformers.

In this "bootstrap" method, the impedance increase required at least two cores, one of them common to both windings.[39] In 1970 he reported the design of a new ac resistance thermometer of higher input impedance, in order to reduce the importance of the resistance of the voltage wires.[40] This bridge was built around a three-stage transformer, containing three high-permeability cores. A stepped constant-current source was used to excite the bridge (Fig. 8.22) at 400 Hz, and applied equally to R_s and R_T. The voltage across R_t is determined by adjusting the transformer T_1 until the detector coupled by transformer T_2 indicates null. Then

$$R_T = \rho R_s \,, \tag{8.58}$$

where ρ is the turns ratio N_T/N_s.

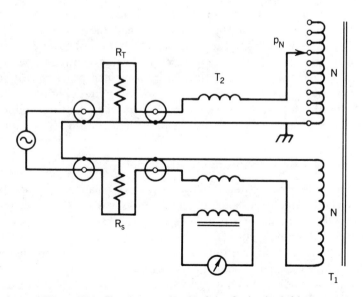

Figure 8.22. Elementary circuit of the Cutkosky bridge.

The actual circuit is much more sophisticated than shown in Fig. 8.22. The transformer T_1 has three windings on three cores. The third stage was wound on the inner core. The second stage was wound around both the inner core and a second core. The first stage was the tapped stage, wound around all three cores. The first stage was tapped and connected to a seven-stage commercial inductive divider. Voltage-follower operational amplifiers were used to couple the second and third stages, with a resistor of about $10^8 \, \Omega$ between the first stage and the first voltage follower. The signal from T_2 was coupled through an impedance-matching transformer to an amplifier and then to two detectors. One detector measured the current in phase with the voltage and the other measured the current out of phase with the voltage. Null was obtained by adjusting T_1 using the resistance detector signal. The quadrature detector signal was fed back to a varacter bridge excited by the voltage follower in the second stage of T_1 to produce a voltage in series with T_2, thus automatically balancing the phase angle. The bridge required only a single balance and produced an error of only 3 μΩ from 0 to 100 Ω, using a 100-Ω standard and 1 mA of current. Automatic balancing and programmed calibration sequences were developed. Various versions have been extensively used at the National Bureau of Standards using standard microcomputers and interfaces to automate the platinum sensor calibration service.

Cutkosky also designed an automatic, microprocessor-controlled bridge for low-resistance sensors because extending platinum resistance thermometry to high temperatures such as the gold point is being investigated with lower resistances at the ice point than the usual 25- or 100-Ω sensors. It is a five-stage bridge patterned after a three-stage bridge designed by Knight.[41] The original version of the bridge used 15- or 30-Hz square waves with 1, 2, 4, or 8 mA of current and had the range of −0.125 to 31.875 Ω.[42] Resolution was 1 μΩ and an accuracy either limited by the resolution or of 1 part in 10^7. Improvements in the operational amplifiers, the detectors, and the guard circuit, and improved software were reported in 1981.[43] It has been used extensively at NBS to study high-temperature platinum thermometers. Cutkosky reported modifications to reduce the currents to $\frac{1}{8}$, $\frac{1}{4}$, $\frac{1}{2}$, and 1 mA, and also its performance with 100-Ω sensors.[44]

8.12.5 The Neil Brown Automatic AC Bridge

Early automatic bridges required relays to achieve balance. The Neil Brown bridge uses field-effect transistors (FET) for ease in microprocessor control. Because of their resistance it was necessary to design the bridges with a special digital-to-analog converter, multitap transformer to limit current and divide the voltage accurately.[45] A 384–Hz sine-wave oscillator provides the primary signal to the inner core of a two-core multitap transformer. One secondary winding provides 10 V to a 10,000-Ω reference resistor. Current through the sensor resistor is held constant at the reference current value of 1 mA by an operational amplifier. The voltage across the sensor is applied to a multistage two-core transformer bridge with the excitation for the second stage "bootstrapped" by a fixed-gain operational amplifier to obtain high input impedances. The output of the bridge transformer is then applied to the multitap special transformer excited by the sine-wave oscillator. The special transformer makes a digital-to-analog converter using a two-stage ratio transformer cascaded with a second two-stage transformer. The first of the two-stage transformers has six taps controlled by FET switches. The second has seven more, making 13 different switchings possible. The output voltage depends on the sum of the first six plus a step-down ratio times the sum of the second seven. Automatic switching of the bridge transformer output to the ratio transformer to produce a digital signal, coupled with automatic logic circuits to adjust the FET switches of the D/A converter, sets the ratio transformer output to the bridge transformer signal and allows the unknown voltage to be measured. A microprocessor is used to control the bridge. Outputs include numeric and alphanumeric displays and can supply external devices through RS 232 or IEEE 488 interfaces.

8.12.6 The Rosemount Temperature Transmitter

The Rosemount temperature transmitter is a small solid-state instrument that uses a feedback system for automatic balance of a dc three-wire Wheatstone bridge circuit, and provides an output signal of 4–20 mA linearly proportional to temperature. Referring to Fig. 8.15, if a constant-current source is used to maintain the current through the sensor wire r_c, the voltage drop across R_x will vary with temperature and cause the null detector to deflect if R_s remains fixed. However, if an additional external current loop is added to R_s, the voltage at point 1 can be maintained equal to point 2. The current through R_x remains constant. The current through R_s is the sum of the R_x current plus the external current. It is necessary to have a way to control the current in the external loop to maintain the bridge in balance.

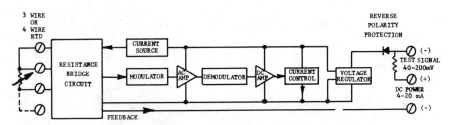

Model 444R Block Diagram

Figure 8.23. Block diagram for Rosemount 444R temperature transmitter. (Courtesy of Rosemount, Inc.)

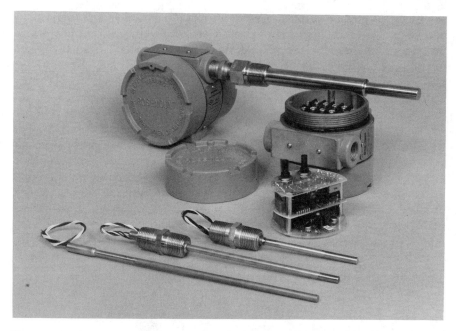

Figure 8.24. Rosemount field-mount temperature transmitter, components, and three resistance thermometer sensors. (Courtesy of Rosemount, Inc.)

The Rosemount temperature transmitter converts the out-of-balance error signal to an ac signal with a solid-state modulator to feed it to a very stable ac amplifier (Fig. 8.23). The ac signal is then demodulated and fed to a dc amplifier to provide the balancing signal. The "constant"-current source for R_s is controlled by the external loop current to compensate for nonlinearity of the sensor. The transmitter is available in three standard ranges: 25–75°C, 70–210°C, and 200–600°C. It requires a dc power supply suitable for the output load. The transmitter is very small, sealed, and can be attached directly to the sensor (Fig. 8.24).

8.13 COMPENSATION FOR SENSOR NONLINEARITY

The resistance thermometer does not produce a linear response to temperature. Some of the methods of dealing with the nonlinearity are:

1. For low accuracy, over a narrow temperature range the error is neglected. The device is calibrated for the average slope in that segment of the resistance–temperature curve.

2. An analog correction can be made before digitizing and displaying the temperature. In one design an operational amplifier was designed

with nonlinear gain, so that as temperature (and sensor resistance) increased, the gain increased just enough to compensate for the negative second-order term in the Callendar equation.[46]

3. Piecewise linearization can be used so that a different linear approximation is used for different sections of the resistance–temperature curve.

4. A read-only memory (ROM) can be used in which every possible digitized resistance is tabulated with its respective temperature. This requires a large ROM but is very fast.

5. The resistance, or its equivalent current or voltage, can be digitized and a computer used to solve the Callendar or Callendar–Van Dusen equation.

6. For platinum the resistance ratio can be used to solve the IPTS polynomial for range I, or the IPTS quadratic and correction term.

7. Experimental calibration plus curve fitting can be used to develop an interpolation for specific applications.

8.14 THE INTERPOLATING EQUATIONS

Because different manufacturers produce resistance thermometers of different total resistance at the ice-point, it is convenient to specify in general terms using the resistance ratio, $W(t) = R_t/R_0$. Therefore, the interpolating equations for IPTS 68 are written in terms of W. At low temperatures the resistance of metals is not a linear function of temperature but varies as T^3 or even as T^5. The resistance approaches zero as the temperature approaches absolute zero. Therefore, curve-fitting techniques have been used to establish the IPTS as nearly as possible to the thermodynamic temperature scale, using the platinum resistance thermometer as the most stable sensing instrument down to about 13.81 K. Here we will discuss metallic temperature sensors in range I and range II, using platinum as the example because of its importance.

8.14.1 Range I Sensors

As explained earlier, the Callendar equation was not accurate below 0°C, and the Callendar–Van Dusen equation, incorporating third- and fourth-power correction terms, was used from 1928 to 1968. Now range I has been divided into four parts and temperature is calculated from computer-developed equations. The original IPTS 68 has a polynomial,

$$T_{68} = A_0 + \sum_{i=1}^{20} A_i (\ln W)^i , \qquad (8.59)$$

where the resistance ratio was calculated from the measured resistance modified by a resistance ratio correction term:

$$W = W_m + \Delta W. \tag{8.60}$$

In 1975 the polynomial equation was replaced with a more efficient form,

$$T_{68} = \sum_{i=0}^{20} B_i \left(\frac{\ln W + 3.28}{3.28} \right)^i. \tag{8.61}$$

An alternative equation that was considered but not adopted was a Chebyshev polynomial. Coefficients for these three polynomial equations are given in Table 8.6.[47]

The reason for changing from Eq. (8.59) to Eq. (8.61) was that even for lesser accuracy than the six significant figures obtained with Eq. (8.59), all 21 terms to 16 significant figures were required. With Eq. (8.61) all 21 terms are required, but the number of significant figures can be truncated to fit lower required accuracy.[48] With the Chebyshev polynomial, both the number of terms and their length can be truncated.

TABLE 8.6 Coefficients for the Polynomial Equations[a]

i	A_i	B_i	C_i
0	0.2731500000000000 E +3	+38.59276	+161.82457
1	0.2508462096788033 E +3	+43.44837	+101.20451
2	0.1350998699649997 E +3	+39.10887	+51.52727
3	0.5278567590085172 E +2	+38.69352	+24.77384
4	0.2767685488541052 E +2	+32.56883	+10.11559
5	0.3910532053766837 E +2	+24.70158	+3.53282
6	0.6556132305780693 E +2	+53.03828	+0.91392
7	0.8080358685598667 E +2	+77.35767	+0.14666
8	0.7052421182340520 E +2	−95.75103	−0.00327
9	0.4478475896389657 E +2	−223.52892	+0.00265
10	0.2125256535560578 E +2	+239.50285	+0.01172
11	0.7679763581708458 E +1	+524.64944	+0.01013
12	0.2136894593828500 E +1	−319.79981	+0.00366
13	0.4598433489280693 E +0	−787.60686	+0.00007
14	0.7636146292316480 E −1	+179.54782	−0.00071
15	0.9693286203731213 E −2	+700.42832	−0.00066
16	0.9230691540070075 E −3	+29.48666	−0.00085
17	0.6381165909526538 E −4	−335.24378	−0.00028
18	0.3022932378746192 E −5	−77.25660	+0.00034
19	0.8775513913037602 E −7	+66.76292	+0.00026
20	0.1177026131254774 E −8	+24.44911	+0.00005

[a]The values for A_i are written in the form: decimal number E signed integer exponent, where E denotes the base 10.

The purpose of the resistance ratio correction term was to make the interpolating equation conform more closely to the temperatures believed to be closest to thermodynamic temperature scale at various fixed points. For this purpose, range I was divided into four parts. The correcting equations for these parts are:

Part 1, range I: 13.81 to 20.28 K

$$\Delta W = A_1 + B_1 T + C_1 T^2 + D_1 T^3 \tag{8.62}$$

Part 2, range I: 20.28 to 54.361 K

$$\Delta W = A_2 + B_2 T + C_2 T^2 + D_2 T^3 \tag{8.63}$$

Part 3, range I: 54.631 to 90.188 K

$$\Delta W = A_3 + B_3 T + C_3 T^2 \tag{8.64}$$

Part 4, range I: 90.188 to 273.15 K

$$\Delta W = A_4 t + C_4 t^3 (t - 100) \tag{8.65}$$

Each of the constants in the foregoing equations for the first three parts is evaluated at specified fixed points at the extremes and within the part. In addition, the first derivatives of resistance corrections with temperature are set equal at 20.28, 54.361, and 90.188 K, working down from part 4 to part 1.

The values of the resistance ratio corresponding to temperatures in range I are given in Table 8.7, and the values for the resistance ratio needed for the constants in the ΔW equations above are given in Table 8.8.[49]

8.14.2 Range II Sensors

In range II the platinum resistance thermometer has interpolating equations according to IPTS 68:

$$t = t' + \Delta t \tag{8.66}$$

$$t' = \frac{1}{\alpha} [W(t') - 1] + \delta \left(\frac{t'}{100} \right) \left(\frac{t'}{100} - 1 \right) \tag{8.67}$$

$$\Delta t = 0.045 \left(\frac{t'}{100} \right) \left(\frac{t'}{100} - 1 \right) \left(\frac{t'}{419.58} - 1 \right) \left(\frac{t'}{630.74} - 1 \right). \tag{8.68}$$

TABLE 8.7 Values of $W_{CCT-68}(T_{68})$ at Integral Values of T_{68}

T_{68} (K)	$W_{CCT-68}(T_{68})$	T_{68} (K)	$W_{CCT-68}(T_{68})$	T_{68} (K)	$W_{CCT-68}(T_{68})$	T_{68} (K)	$W_{CCT-68}(T_{68})$	T_{68} (K)	$W_{CCT-68}(T_{68})$	T_{68} (K)	$W_{CCT-68}(T_{68})$
		50	0.07537756	100	0.28630201	150	0.49861135	200	0.70496694	250	0.90738309
		51	0.07940123	101	0.29062154	151	0.50278707	201	0.70905004	251	0.91139753
		52	0.08287595	102	0.29493841	152	0.50696058	202	0.71313161	252	0.91541074
		53	0.08669859	103	0.29925245	153	0.51113172	203	0.71721174	253	0.91942274
		54	0.09056600	104	0.30356359	154	0.51530065	204	0.72129026	254	0.92313343
		55	0.09447515	105	0.30787183	155	0.51946737	205	0.72536733	255	0.92744283
		56	0.09842336	106	0.31217710	156	0.52363180	206	0.72944288	256	0.93145101
		57	0.10240774	107	0.31647939	157	0.52779409	207	0.73351690	257	0.93545805
		58	0.10642583	108	0.32077856	158	0.53195417	208	0.73758947	258	0.93946371
		59	0.11047506	109	0.32507467	159	0.53611211	209	0.74166059	259	0.94346822
		60	0.11455312	110	0.32936765	160	0.54026792	210	0.74573026	260	0.94747152
		61	0.11865789	111	0.33365751	161	0.54442167	211	0.74979841	261	0.95147352
		62	0.12278722	112	0.33794416	162	0.54857336	212	0.75386518	262	0.95547430
13	0.00123061	63	0.12693914	113	0.34222768	163	0.55272291	213	0.75793043	263	0.95947385
14	0.00145973	64	0.13111189	114	0.34650800	164	0.55687048	214	0.76199430	264	0.96347219
15	0.00174541	65	0.13530363	115	0.35078519	165	0.56101606	215	0.76605672	265	0.96746931
16	0.00209474	66	0.13951284	116	0.35505910	166	0.56515958	216	0.77011770	266	0.97146513
17	0.00251512	67	0.14373800	117	0.35932989	167	0.56930112	217	0.77447730	267	0.97545980
18	0.00301428	68	0.14797773	118	0.36359754	168	0.57344076	218	0.77823545	268	0.97945325
19	0.00359962	69	0.15223058	119	0.36786199	169	0.57757848	219	0.78229223	269	0.98344541
20	0.00427780	70	0.15649541	120	0.37212331	170	0.58171423	220	0.78634756	270	0.98743642
21	0.00505495	71	0.16077108	121	0.37638151	171	0.58584806	221	0.79040151	271	0.99142614
22	0.00593668	72	0.16505643	122	0.38063657	172	0.58997999	222	0.79445409	272	0.99541471
23	0.00692804	73	0.16935049	123	0.38488851	173	0.59411008	223	0.79850523	273	0.99940199

24	0.00803316	74	0.17365240	124	0.38913732	174	0.59823835	224	0.80255506
25	0.00925504	75	0.17796117	125	0.39338316	175	0.60236478	225	0.80660352
26	0.01059585	76	0.18227605	126	0.39762594	176	0.60648931	226	0.81065054
27	0.01205690	77	0.18659628	127	0.40186567	177	0.61061208	227	0.81469625
28	0.01363901	78	0.19092107	128	0.40610242	178	0.61473310	228	0.81874059
29	0.01534261	79	0.19524992	129	0.41033628	179	0.61885229	229	0.82278364
30	0.01716768	80	0.19958212	130	0.41456709	180	0.62296972	230	0.82682531
31	0.01911363	81	0.20391714	131	0.41879507	181	0.62708540	231	0.83086561
32	0.02117944	82	0.20825445	132	0.42302015	182	0.63119939	232	0.83490461
33	0.02336343	83	0.21259344	133	0.42724233	183	0.63531164	233	0.83894224
34	0.02566335	84	0.21693388	134	0.43146169	184	0.63942213	234	0.84297857
35	0.02807645	85	0.22127523	135	0.43567831	185	0.64353094	235	0.84701353
36	0.03059953	86	0.22561712	136	0.43989210	186	0.64763807	236	0.85104726
37	0.03322916	87	0.22995916	137	0.44410322	187	0.65174352	237	0.85507963
38	0.03596155	88	0.23430105	138	0.44831159	188	0.65584730	238	0.85911069
39	0.03879305	89	0.23864248	139	0.45251730	189	0.65994947	239	0.86314046
40	0.04171968	90	0.24298315	140	0.45672033	190	0.66404996	240	0.86716894
41	0.04473760	91	0.24732290	141	0.46092077	191	0.66814886	241	0.87119611
42	0.04784292	92	0.25166128	142	0.46511861	192	0.67224607	242	0.87522199
43	0.05103178	93	0.25599836	143	0.46931387	193	0.67634176	243	0.87924657
44	0.05430036	94	0.26033369	144	0.47350660	194	0.68043577	244	0.88326994
45	0.05764486	95	0.26466718	145	0.47769682	195	0.68452825	245	0.88729200
46	0.06106161	96	0.26899870	146	0.48188459	196	0.68861913	246	0.89131269
47	0.06454679	97	0.27332807	147	0.48606985	197	0.69270841	247	0.89533224
48	0.06809690	98	0.27765516	148	0.49025274	198	0.69679617	248	0.89935049
49	0.07170835	99	0.28197988	149	0.49443319	199	0.70088232	249	0.90336744
50	0.07537756	100	0.28630291	150	0.49861135	200	0.70496694	250	0.90738309

TABLE 8.8   Values of $W_{CCT-68}(T_{68})$ at the Fixed-Point Temperatures

Fixed point	T_{68} (K)	t_{68} (°C)	W_{CCT-68}
e-H_2 triple	13.81	−259.34	0.00141206
e-H_2 17.042	17.042	−256.108	0.00253444
e-H_2 boiling	20.28	−252.87	0.00448517
Ne boiling	27.102	−246.048	0.01221272
O_2 triple	54.361	−218.789	0.09197252
O_2 boiling	90.188	−182.962	0.24379909
Ice point	273.15	0	1
H_2O boiling	373.15	100	1.39259668
Sn freezing	505.1181	231.9681	1.89257086

The procedure is to calculate t' from the resistance ratio using Eq. (8.67), calculate Δt from Eq. (8.68), and then calculate t from Eq. (8.66). Equation (8.67) is the Callendar equation given previously in Eq. (8.1) with α as defined in Eq. (8.11) substituted in the first term, and with the resistance ratio used instead of R_t.

As we learned earlier, the resistance ratio for many metals can be written in the form

$$W = 1 + At' + Bt'^2 . \qquad (8.69)$$

In Eq. (8.67) the relationships of the constants α and δ to the fundamental constants A and B in Eq. (8.69) are given by

$$\alpha = A + 100B \qquad (8.70)$$

and

$$\delta = \frac{10^4 B}{A + 100B} \qquad (8.71)$$

or

$$A = \alpha\left(1 + \frac{\delta}{100}\right) \qquad (8.72)$$

$$B = -10^{-4}\alpha\delta$$

These relationships were originally presented by Callendar in his paper of 1887.[3]

The purpose of the Δt correction term is to make the IPTS 68 temperature scale correspond more closely with the thermodynamic temperature scale. The range II correction term goes to zero at each of the

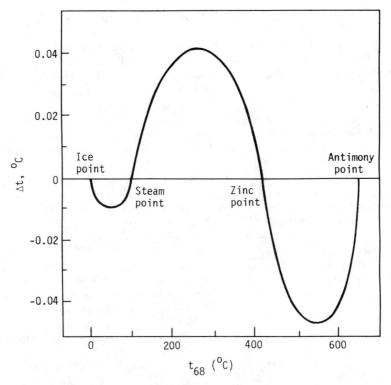

Figure 8.25. Deviations of the Callendar equation in range II from t_{68}.

calibrating points 100, 419.58, and 630.74°C (representing the steam point, the zinc point, and the antimony point, respectively). It is a maximum midway between the points (Fig. 8.25).

REFERENCES

1. Sir Humphrey Davy, Further researches on the magnetic phenomena produced by electricity; with some new experiments on the properties of electrified bodies in their relations to conducting powers and temperature. *Philos. Trans. Royal Soc.* **425**, (1821).

2. W. H. Siemens, On the increase of electrical resistance in conductors with rise of temperature, and its application to the measure of ordinary and furnace temperatures; also on a simple method of measuring electrical resistances. *Proc. Royal. Soc.* London, **19**, 443 (1871).

3. H. L. Callendar, On the Practical measurement of temperature: Experiments made at the Cavendish Laboratory, Cambridge, *Philos. Trans. Royal Soc. London* **178**, 161 (1887).

4. E. Griffiths, "Methods of Measuring Temperature." Griffin, London, 1918.

5. H. L. Callendar, On construction of platinum thermometers. *Philos. Mag.* [4] **32**, 104 (1891). Notes on platinum thermometry. *Ibid.* **47**, 191 (1899).

6. M. S. Van Dusen, Platinum resistance thermometry at low temperatures. *J. Am. Chem. Soc.* **47**, 326 (1925).

7. N. F. Mott and N. Jones, "The Theory of the Properties of Metals and Alloys," p. 245. Oxford Univ. Press (Clarendon), London and New York, 1936.

8. R. Price, The platinum resistance thermometer. *Platinum Met. Rev.* **19**, 78–87.

9. C. R. Barber and J. A. Hall, Progress in platinum resistance thermometry. *Br. J. Appl. Phys.* **13**, 147 (1962).

10. Deutsches Institut für Normen (DIN), "Electrical Temperature Sensors; Reference Tables for Sensing Resistors for Resistance Elements," DIN 43760. Beuth Verlag, Berlin, 1980.

11. J. L. Riddle, G. T. Furukawa, and H. Plumb, Platinum resistance thermometry. *NBS Monogr.* (*U.S.*) **126**, 9–10 April 1973).

12. R. L. Rusby, Towards the absolute zero. *Platinum Met. Rev.* **25**, 57–61 (1981).

13. R. J. Berry, Control of oxygen-activated cycling effects in platinum resistance thermometers. In "Temperature Measurement 1975" (B. F. Billings and T. J. Quinn, eds.), pp. 99–106. Inst. Phys. London, 1975.

14. R. J. Berry, Effect of Pt oxidation on Pt resistance thermometry. *Metrologia* **16**, 117–126 (1980).

15. R. J. Berry, Evaluation and control of platinum oxidation errors in standard platinum resistance thermometers. *Temp.: Its Meas. Control Sci. Ind.* **5** (Part 2), 743–752 (1982).

16. R. J. Berry, Oxidation, stability and insulation characteristics of Rosemount standard platinum resistance thermometers. *Temp.: Its Meas. Control Sci. Ind.* **5** (Part 2), 753–762 (1982).

17. J. S. Johnson, Resistance thermometry. *In* "Temperature Measurement, 1975" (B. F. Billings and T. J. Quinn, eds.), pp. 80–90. Inst. Phys. London, 1975.

18. W. D. J. Evans, Thick film platinum resistance temperature detectors. *Platinum Met. Rev.* **25**, 2–11 (1981).

19. Omega Engineering, Inc., "Temperature Measurement Handbook." Stanford, Connecticut, 1981.

20. J. P. Evans, Experience with high-temperature platinum resistance thermometers. *Temp.: Its Meas. Control Sci. Ind.* **5** (Part 2), 771–781 (1982).

21. J. V. McAllan, Practical high temperature resistance thermometry. *Temp.: Its Meas. Control Sci. Ind.* **5** (Part 2), 789–793 (1982).

22. L. Xumo, Z. Jinde, S. Jinrong and C. Deming, A new high-temperature platinum resistance thermometer. *Metrologia* **18**, 203–208 (1982).

23. L. Guang and T. Hongtu, Stability of precision high temperature platinum resistance thermometers. *Temp.: Its Meas. Control Sci. Ind.* **5** (Part 2), 783–787 (1982).

24. F. J. Wilkins and M. J. Swan, Precision AC and DC resistance standards. *Proc. Inst. Electr. Eng.* **117**, 841–849 (1970).

25. A. M. Thompson and G. W. Small, AC bridge for platinum resistance thermometry. *Proc. Inst. Electr. Eng.* **118**, 1662–1666 (1971).

26. R. F. Coughlin and F. F. Driscoll, "Operational Amplifiers and Linear Integrated Circuits," 2nd ed. Prentice-Hall, Englewood Cliffs, New Jersey, 1982.

27. J. J. Carr, "Elements of Electronic Instrumentation and Measurement." Reston Publ. Co., Reston, Virginia, 1979.

28. Analog Devices, Inc., "Data-Acquisition Datebook 1982," Vols. 1 and 2. Norwood, Massachusetts, 1982.

29. Tegam, Inc. 7230 North Ridge Road, Madison, OH, 44057.

30. Keithley Instruments, Inc., "1987–1988 Catalog and Buyer's Guide." Cleveland, Ohio, 1984.

31. John Fluke Manufacturing Company, Inc., P.O. Box C9090, Everett, Washington 98206.

32. C. G. M. Kirby and M. J. Laubitz, The error due to the Peltier effect in direct-current measurements of resistance. *Metrologia* **9**, 103–106 (1973).

33. N. L. Kusters, The precise measurement of current ratios. *IEEE Trans. Inst. Meas.* **IM-13**, 197–209 (1964).

34. J. J. Hill and A. P. Miller, An AC double bridge with inductively coupled ratio arms for precision platinum resistance thermometry. *Proc. Inst. Electr. Eng.* **110**, 453–458 (1963).

35. Tinsley-Aztec Instruments, P.O. Box 391, Andover, New Jersey 07821.

36. Automatic Systems Laboratories, Ltd., Leighton Buzzard, England L U7 8SX.

37. M. P. MacMartin and N. L. Kusters, A direct-current-comparator ratio bridge for four terminal resistance measurements. *IEEE Trans. Inst. Meas.* **IM-15**, 212–220 (1966).

38. N. L. Kusters and M. P. MacMartin, Direct-current comparator bridge for resistance thermometry. *IEEE Trans. Inst. Meas.* **IM-15**, 212–220 (1966).

39. R. D. Cutkosky, Active and passive direct reading ratio sets for the comparison of audio frequency admittances. *J. Res. Natl. Bur. Stand., Sect.* **68C**, 195–322 (1964).

40. R. D. Cutkosky, An A-C resistance thermometer bridge. *J. Res. Natl. Bur. Stand., Sect. C.* **74C**, 15–18 (1970).

41. R. B. D. Knight, A precision bridge for resistive thermometry using a single inductive current divider. *IEE Conf. Publ.* **152**, 132 (1977).

42. R. D. Cutkosky, An automatic resistance thermometer bridge. *IEEE Trans. Inst. Meas.* **IM-29,** 330–333 (1980).

43. R. D. Cutkosky, Guarding techniques for resistance thermometers. *IEEE Trans. Inst. Meas.* **IM-30**, 217–220 (1981).

44. R. D. Cutkosky, Automatic resistance thermometer bridges for new and special applications. *Temp.: Its Meas. Control Sci. Ind.* **5** (Part 2), 711–715 (1982).

45. N. L. Brown, A. J. Fougère, J. W. McCleod, and R. J. Robbins, An automatic resistance thermometer bridge. *Temp.: Its Meas. Control Sci. Ind.* **5** (Part 2), 719–727 (1982).

46. T. E. Foster, An easily calibrated, versatile platinum resistance thermometer. *Hewlett Packard J.* **25**, 13–17 (1974).

47. T. J. Quinn, Temperature standards. *In* "Temperature Measurement, 1975" (B. F. Billing and T. J. Quinn, eds.), pp. 1–15. Inst. Phys. London, 1975.

48. G. G. M. Kirby and R. E. Bedford, A proposed reformulation of the IPTS-68 reference function. *Metrologia* **8**, 82–84 (1972).

49. H. Preston-Thomas, The international practical temperature scale of 1968 amended edition of 1975. *Metrologia* **12**, 7–17 (1976).

CHAPTER 8 PROBLEMS

8.1 If a platinum resistance thermometer has R_0 equal to $100\,\Omega$, if A is 3.98×10^{-3} and B is $-5.85 \times 10^{-7}/°C$, calculate the resistance at 50, 100, and 200°C. Then calculate α for each of the temperature intervals.

8.2 From Fig. 8.3, estimate the slope and calculate the value of resistance for a platinum sensor ($R_0 = 100\,\Omega$).

8.3 Compare the resistances at 200°C of two metallic temperature sensing elements if they both have $10\,\Omega$ resistance at $0°C$ and their temperature coefficients are as given in Table 8.2.

8.4 Calculate the sensitivity of a high-purity platinum sensor at $-100°C$ if its resistance is given by

$$R = 25.5[1 + 3.98 \times 10^{-3}t - 5.85 \times 10^{-7}t^2 - 4.35 \times 10^{-12}(t - 100)t^3].$$

8.5 Calculate the temperature for a platinum resistance thermometer by the Callendar method if $R_0 = 25.512$, $\alpha = 0.0039265$, and $\delta = 1.493$, and if the indicated resistance at the unknown temperature $R_x = 29.0585$.

CHAPTER 8 ANSWERS

8.1
$$R = R_0(1 + At + Bt^2)$$
$$= 100(1 + 3.98 \times 10^{-3}t - 5.85 \times 10^{-7}t^2)$$
$$R_{50} = 100(1.19754) = 119.75$$
$$R_{100} = 100(1.39215) = 139.22$$
$$R_{200} = 100(1.81940) = 177.26$$
$$\alpha_{0-50} = \frac{R_{50} - R_0}{R_0(50)} = \frac{119.75 - 100}{(100)(50)} = 3.951 \times 10^{-3}$$
$$\alpha_{0-100} = \frac{R_{100} - R_0}{R_0(100)} = \frac{139.22 - 100}{(100)(100)} = 3.92 \times 10^{-3}$$
$$\alpha_{0-200} = \frac{R_{200} - R_0}{R_0(200)} = \frac{177.26 - 100}{(100)(200)} = 3.863 \times 10^{-3}$$

Note that these values for high-purity platinum are slightly higher in resistivity, and also in α, than those for DIN 43760 (Table 8.1).

8.2 The slope of the graph from -100 to $+800°C$ can be estimated:

$$\frac{d^2R}{dt^2} = \frac{0.295 - 0.405}{800 - (-100)} = \frac{0.110}{900}$$
$$= -0.0001222.$$

Integrating gives us

$$\frac{dR}{dt} = \frac{-1.22 \times 10^{-4}t}{2} + C_1.$$

At $t = 0$, $dR/dt = 0.390$, $C_1 = 0.39$, and

$$\frac{dR}{dt} = \frac{-1.22 \times 10^{-4}}{2} t + 0.39 .$$

Integrating, we have

$$R = \frac{-1.22 \times 10^{-4}}{2} t^2 + 0.39t + C_2$$

At $t = 0$, $R = 100$, $C_2 = 100$, and

$$R = 100 + 0.39t - 0.61 \times 10^{-4} t^2 .$$

Testing, at 200°C

$$R_{200} = 100 + 0.39(200) - 0.61 \times 10^{-4}(200)^2$$
$$= 175.56 \ \Omega$$

From Table 8.1, $R_{200} = 175.84 \ \Omega$.

8.3 $R_{100} = R_0(1 + \alpha t)$

Nickel: $\quad R_{100} = 10[1 + 0.0069(100)]$
$$= 16.9 \ \Omega$$

Platinum: $\quad R_{100} = 10[1 + 0.003927(100)]$
$$= 13.9 \ \Omega .$$

8.4 Simplifying the last term, we have

$$R = 25.5(1 + 3.98 \times 10^{-3} t - 5.85 \times 10^{-7} t^2 - 4.35 \times 10^{-12} t^4$$
$$+ 4.35 \times 10^{-9} t^3)$$

$$\frac{dR}{dt} = S = 25.5[3.98 \times 10^{-3} - 2(5.85 \times 10^{-7})t - 4(4.35 \times 10^{-12})t^3$$

$$+ 3(4.35 \times 10^{-9})t^2]$$

$$= 25.5[3.98 \times 10^{-3} + 11.70 \times 10^{-7}(100) - 17.4 \times 10^{-12}(100)^3$$

$$+ 13.05 \times 10^{-9}(100)^2]$$

$$= 25.5(3.98 \times 10^{-3} + 11.7 \times 10^{-5} - 17.4 \times 10^{-6} + 13.05 \times 10^{-5})$$

$$= 25.5(0.00421)$$

$$= 0.107 \ \Omega/°C .$$

8.5   We can solve this problem simply using the Callendar equation in the form of Eq. (8.1):

$$t = \left(\frac{R_t - R_0}{R_{100} - R_0} \right) 100 + \delta \left(\frac{t}{100} \right)\left(\frac{t}{100} - 1 \right).$$

We can calculate R_{100} from R_0 as

$$R_{100} = (1 + 100\alpha)R_0$$
$$= 1.39265(25.512) = 35.529 \, .$$

Then

$$t = \left(\frac{29.0585 - 25.512}{35.529 - 25.512} \right) 100 + \delta \left(\frac{t}{100} \right)\left(\frac{t}{100} - 1 \right)$$

$$= 35.404°C + \delta \left(\frac{t}{100} \right)\left(\frac{t}{100} - 1 \right).$$

As a first approximation we use $t_{pt} = 35.40$ to calculate the δ term:

$$t_1 = 35.404 + 1.493\left(\frac{35.404}{100} \right)\left(\frac{35.404}{100} - 1 \right)$$

$$= 35.404 + (-0.34) = 35.063°C \, .$$

Making a second approximation gives us

$$t_2 = 35.404 + 1.493\left(\frac{35.063}{100} \right)\left(\frac{35.063}{100} - 1 \right)$$

$$= 35.404 + (-0.340)$$

$$= 35.064°C \, .$$

Obviously, a third approximation is unnecessary.

9

THERMISTORS AND SEMICONDUCTORS FOR TEMPERATURE MEASUREMENT

9.1 INTRODUCTION

Semiconductors can be classified into three types; elemental semiconductors, compound semiconductors, and junction semiconductors. All are of recent scientific development. The compound semiconductors used for temperature measurements are usually called thermistors to identify a special resistor that varies in resistance with temperature.

The first compound semiconductor was discovered by Michael Faraday in 1833, who noticed that the resistance of silver sulfide decreased dramatically as temperature increased. The first commercial thermistors were produced in Germany and Holland during the 1930s, based on UO_2, CuO, or SiC. Commercial production of modern compositions has been quite active in the United States since about 1950. Thermistors can have either negative or positive temperature coefficients of resistance. The former are particularly useful for temperature compensation of electronic circuits and are widely used for that purpose. Almost all thermistors produced for temperature measurement have negative coefficients of resistance.[1,2]

Semiconducting resistors and junction semiconductors were developed at Bell Laboratories in the early 1940s and commercial development has been rapid since 1948. However, their use for temperature measurement began about 1955, when germanium resistors were introduced for cryogenic temperature measurement.[2] We will discuss thermistors, resistors, and junction semiconductors separately after describing the properties of semiconductors.

Materials can be classified as electrical conductors, semiconductors, or insulators depending on their electrical conductivity. At room temperature most metals have resistivities around 10^{-3} to 10^{-6} Ω-cm. Semiconductors have about 10 to 10^6 Ω-cm and insulators have greater than 10^{12} Ω-cm. Whether a material is an insulator or a conductor depends on its atomic or ionic composition and its crystalline structure. Metals are conductors because their nuclei approach closely enough in their crystalline structure so that the outer electronic orbits overlap. The repulsion between overlapping electronic charges causes the orbitals to split into bands in accordance with the Pauli exclusion principle. The overlapping of filled (valence band) and unfilled (conduction band) orbitals permits electrons to move under an applied field. In insulators the ions do not approach close enough for splitting to occur. In semiconductors the bands almost overlap, so that thermal energy can excite some electrons into the conduction band (Fig. 9.1).

Conductivity, the reciprocal of resistivity, depends on

$$\sigma = \frac{1}{\rho} = ne'\mu \, , \tag{9.1}$$

where n is the number of charge carriers flowing, e' is the charge on the carrier, and μ is the mobility. For metals n is relatively constant, e' is the electronic charge of one electron, and μ decreases with temperature because lattice vibrations increase and interfere with the electron movement (decrease the mean free path of the electrons). That is the reason the

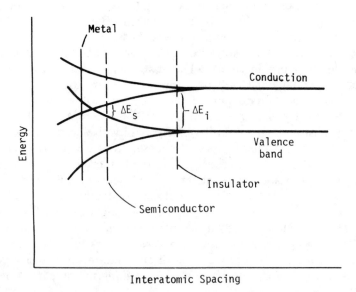

Figure 9.1. Orbital energy versus interatomic spacing. The gap between the valence and conduction bands for semiconductors, $\Delta E_s \approx kT$ at room temperature.

decrease in mobility causes the conductivity of metals to decrease as temperature increases, and the change is only moderately sensitive to temperature. (The resistance increases with temperature.)

In semiconductors the number of carriers can change as the temperature changes. Although more sophisticated equations are required in theory, for many semiconductors the number of charge carriers is given approximately by the Boltzmann equation,

$$n = n_0 e^{-\Delta E/kT} . \tag{9.2}$$

Here n_0 is the total number of electrons that could be excited to the conduction band, ΔE is the energy gap between the valence and the conduction bands, k is Boltzmann's constant, and T is absolute temperature.

In a pure, elemental semiconductor thermal energy causes electrons in the valence band to move to the conduction band in accordance with Eq. (9.2). This puts electrons in the conduction band and creates an equal number of holes in the valence band. Then the conductivity is given by

$$\sigma = n_e e' \mu_e + n_h e' \mu_h , \tag{9.2a}$$

where the subscript e refers to electrons and h refers to holes, and $n_e = n_h$.

At about room temperature the conductivity is quite small and can be increased by adding an impurity (doping). For germanium, adding small concentration acceptors such as of boron or gallium increases the conductivity and makes it p type. (If the major transport species is negative, it is n type; if it appears to be positive, it is p type.) Adding donors such as phosphorus or indium to germanium makes it n type and increases the conductivity. When both donors and acceptors are present, the acceptors trap electrons from the donors, reducing the conductivity, or compensating part of the donor contribution.

The advantage of doping for service near room temperature is that the conductivity can be controlled by the concentration of the dopant, and makes the conductivity less temperature dependent. This is useful for many electronic applications.

In practice, elemental semiconductors such as germanium and silicon are not actually impurity free, even though they can be produced remarkably pure by zone refining. If they could be produced impurity free, the conductivity at low temperatures would be exponentially dependent, as in Eq. (9.2), but would be virtually zero. The resistivity would be too high for a useful device. At low temperatures the conductivity is dominated by the number of charge carriers available from the impurities. Germanium resistors for cryogenic applications are discussed in Section 9.7.

Compound semiconductors also conduct in accordance with Eq. (9.2). But in that case, the conductivity is often described as a hopping mechanism or small polaron mode. The electrons are loosely trapped at a lattice site, but hop with low mobility from site to site. Where, for example, a slightly reduced $BaTiO_3$ crystal is used, the presence of Ti^{3+} instead of Ti^{4+} on some lattice sites allows an electron to move under a field from a Ti^{3+} site to a Ti^{4+} site. In the process the Ti^{3+} site is converted to a Ti^{4+} site and the Ti^{4+} site is converted to a Ti^{3+} site. The oxygen anions surrounding the site respond to the change in valence by moving closer to the quartravalent ion and farther from the trivalent ion. This, then, involves phonon coupling. The physics of hopping and small polaron conduction is not well understood, but is involved at very low temperatures in impure intrinsic semiconductors, and at most temperatures for compound semiconductors.

9.2   THE THERMISTOR AS A SEMICONDUCTOR

A thermistor is a special type of semiconductor that varies in resistance in a predictable way as temperature is changed. Thermistors are usually inexpensive ceramic compounds which are extremely sensitive to temperature change. They may be 1000 to 1 million times more sensitive than a platinum resistance thermometer. However, thermistors are less stable and nonlinear. Despite their high sensitivity, they are most commonly used for less precise temperature control, and for switching, in the temperature range from about −100 to 500°C.

Most thermistors are compound semiconductors that have complex conduction processes. They usually have low mobility in narrow bands which have some temperature dependence. The number of carriers [Eq. (9.2)] is dominant in conduction equation (9.1), but they do not have perfect semilogarithmic behavior as they should if μ were constant.

Compound semiconductors are usually ceramic compositions in which solid solutions of transition metal oxides are used.[1,2] Many have the spinel structure, a close packing of oxygen ions in which the cations can occupy either octahedral (large) or tetrahedral (small) interstitial sites within the oxygen packing. Tremendous changes in electrical properties can be achieved by substituting various transition and nontransition metal cations in these sites. Also, control of the oxygen pressure during the firing process permits still further control by manipulation of the cation oxidation state. The number of charge carriers is controlled by the formulation and firing treatments. It is sensitive to oxidation-reduction processes and annealing or "aging" may be necessary to prevent instability in use. Manufacturers have improved their products so that variation from one thermistor to another can be controlled.

Another class of compound semiconductors is that of the ionic conductors. For this class the ions move and carry charge. They have strong negative coefficients of resistivity because the concentration of lattice vacancies increases with temperature. The ionic semiconductors are subject to dc polarization—so, usually, are used in ac circuits. Their principal appeal is to make devices for temperature sensing at temperatures from 1000 to 2000°C. Thus far they have not been adapted to commercial use.

Compound semiconductors made with transition metal oxides such as Cu_2O, Co_2O_3, SnO, TiO_2, Fe_2O_3, NiO, U_2O_3, and Mn_2O_3 conduct by complex, poorly understood processes. The ions are spaced so far apart that electrons from the cation d orbitals must pass through the oxygen p orbitals when moving from one cation to another. The electron movement can be described as a jumping process, or as a small polaron process. Regardless of the mechanism, the resistivity can, in theory, be described to a first approximation as Arrhenius, dominated by Eq. (9.2). Thus the resistivity,

$$\rho = \frac{1}{\sigma} \approx \frac{1}{n_0 e^{-\Delta E/kT} e' \mu} \approx \text{constant}(e^{\Delta E/kT}). \tag{9.3}$$

Because the exponential usually dominates, the resistivity of a thermistor can often be approximated by

$$\rho = \rho_0 e^{\beta/T}, \tag{9.4}$$

where ρ_0 is the resistivity at infinite temperature and β is a constant for the device. Then the resistance should be a function of absolute temperature, T, on a semilog plot (Fig. 9.2). When plotted as a function of t, log resistance is nonlinear (Fig. 9.3). The linearity of Fig. 9.2 is typical, but empirical equations can be made to fit the curves better than Eq. (9.4). One

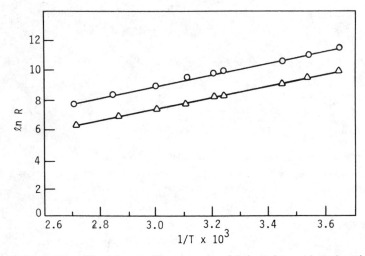

Figure 9.2. Logarithm of resistance of two commercial thermistors plotted against $1/T$. On this scale, the relationship is linear.

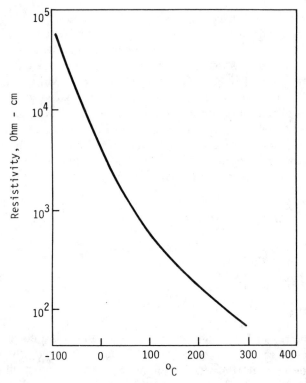

Figure 9.3. Carborundum single-crystal silicon carbide thermistor. Note curvature because it is plotted vs *t*, not 1/ *T*. (Courtesy of Carborundum.)

equation uses a temperature-dependent preexponential

$$\rho = AT^{-C}e^{B/T} \tag{9.5}$$

The constant C is small and may be positive or negative. Another empirical equation is

$$\rho = \rho_0 e^{\dfrac{\Delta E}{R(T+\theta)}}, \tag{9.6}$$

where θ is a small constant with units of kelvin.

Resistance is determined by resistivity and geometry according to the equation

$$R = \frac{\rho L}{A}, \tag{9.7}$$

where L is the length and A is the cross-sectional area of the conductor.

Then the resistance from Eq. (9.4) can be represented by

$$R = R_0 e^{\beta(1/T-1/T_0)}, \tag{9.8}$$

where the resistance R_0 is measured at the absolute temperature T_0. In semiconductor technology T_0 is often taken at 25°C, and is equal to 298 K. For some of our experiments and problems we will measure R_0 at the ice point for use in Eq. (9.8).

An alternative way to represent thermistor response is to expand the exponent to obtain

$$R = \exp\left(A_0 + \frac{A_1}{T} + \frac{A_2}{T^2} + \frac{A_3}{T^3} + \cdots\right). \tag{9.9}$$

This relationship lends itself to computer curve fitting so that the resistance to temperature relationship can be very accurately, albeit empirically, represented.

The temperature coefficient of resistance α is defined as

$$\alpha \equiv \frac{1}{R}\frac{dR}{dT} \tag{9.10}$$

It should not be confused with the mean $\bar{\alpha}$ for metallic resistance sensors which for platinum resistance thermometry is the mean of α for the interval from 0 to 100°C. For metals there is only a small difference between $\bar{\alpha}$ and α because α is not a strong function of temperature. For thermistors α is very sensitive to temperature and the mean for an interval will not be near that for either end of the interval. For thermistors, from Eq. (9.8),

$$\alpha = \frac{1}{R}\frac{dR}{dT} = \frac{-\beta T^{-2}R_0 e^{\beta(1/T-1/T_0)}}{R_0 e^{\beta(1/T-1/T_0)}}$$

$$= -\beta T^{-2}. \tag{9.11}$$

Example 9.1. A thermistor has a resistance at the ice point of 18,080 Ω and at 53.2°C has a resistance of 1855 Ω. Calculate β. Calculate α at each temperature.

Solution

$$R = R_0 e^{\beta(1/T-1/T_0)}$$

$$1855 = 18,080 e^{\beta[1/(273.15+53.2)-1/273.15]}$$

$$0.1025 = e^{\beta(-0.000597)}$$

$$\beta = 3815 \text{ K}$$

At 0°C:
$$\alpha = \frac{3815}{(273.15)^2} = 0.05113/\text{K}$$

At 53.2°C:
$$\alpha = \frac{3815}{(326.35)^2} = 0.03582/\text{K}$$

For platinum, $\bar{\alpha}$ was 0.00392/K, so this thermistor is more than 10 times more sensitive than platinum at 0°C.

Thermistors are used as active units in many electronic devices. They are often used to control the temperature response or to compensate for temperature changes in electronic circuits. For temperature measurement the current flowing through a thermistor should be small, to minimize Joule heating, which would cause the thermistor to be hotter than its surroundings so that the measured temperature would be too hot. For switching purposes, for temperature control, and for measurements of fluid flow or other properties that affect heat transfer from the thermistor, sufficient current to cause Joule heating is desirable. When a small voltage is applied to a thermistor so that its internal heating is negligible, the thermistor temperature will be virtually the same as the temperature of its surroundings. As the current is increased by applying higher voltage, a point will be reached where the temperature of the thermistor exceeds that of its surroundings. Thermistors have negative temperature coefficients of resistance. So the Joule heating causes the resistance to drop and the current to increase. Heat transfer from the thermistor also increases. The volt–current relationship is nonlinear because R is not constant, but comes to a new value after a short time at the new voltage. As the applied voltage is increased, the increase in heat transfer will compensate for the Joule heating, and a point will be reached where a maximum voltage will appear for any particular current source (Figs. 9.4 and 9.5). As the resistance of the thermistor drops, the current–voltage relationship becomes unstable, and

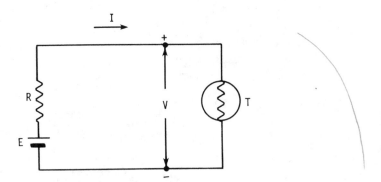

Figure 9.4. The voltage drop V across thermistor T depends on the current I, the source resistance R, and the temperature of T. [From "1946 Transactions of the AIEE," © 1946 AIEE (now IEEE).][3]

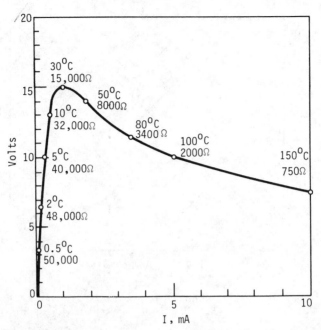

Figure 9.5. Voltage–current response of the thermistor in Fig. 9.4, showing the thermistor resistance and temperature at various current levels. [From "1946 Transactions of the AIEE," © 1946 AIEE (now IEEE).][3]

spontaneous increase in current occurs between points B and C (Fig. 9.6). The electrical power N supplied to the thermistor equals the heat dissipated to the surroundings P along the $V-I$ curves. Above it, $P < N$, and below, $P > N$. This $V-I$ curve is very useful in many thermistor circuit applications.[3]

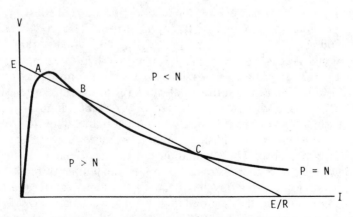

Figure 9.6. Stable (C) and unstagle (B) points of the simple thermistor circuit of Fig. 9.4. [From "1946 Transactions of the AIEE," © 1946 AIEE (now IEEE).][3]

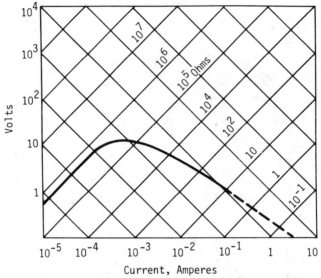

Figure 9.7. Log-log of voltage–current characteristics of a thermistor that has $R_0 = 50,000\,\Omega$ at 300 K, with $\beta = 3600/K$. [From "1946 Transactions of the AIEE," © 1946 AIEE (now IEEE).]

Because thermistor properties have a wide range of values, the V–I characteristics are usually plotted on a log-log plot (Fig. 9.7). On such a plot, a device obeying Ohm's law would plot as a straight line with a slope of +45°. A device having constant power characteristics would plot as a straight line having a slope of −45°. Thermistor manufacturers often present the properties of their products on these graphs for comparison purposes.

For temperature measurement applications the current must be kept as small as possible to minimize self-heating errors. This requires a small applied voltage or current limiting. Errors in voltage and current measurements increase as their value is decreased. In most applications a compromise is required between minimizing self-heating errors and minimizing resistance measuring errors. Selection of the thermistor resistance to obtain the best compromise is facilitated by the wide range of thermistor choices supplied by various manufacturers. Most manufacturers provide the dissipation constant for each shape and resistance value to aid in self-heating error planning. The dissipation constant is defined as the power required to raise the thermistor 1°C above the temperature of the surroundings. Obviously, this depends on both the temperature and the heat transfer conditions of the surroundings. For tiny beads in still air the dissipation factor may be 0.1 mW/°C. For a disk mounted on a copper heat sink it may be as much as 60 mW/°C. The time constant for thermal response of a massive assembly is much greater than that for a tiny bead and must be considered in selecting a suitable thermistor.

9.2.1 Thermistor Properties

Thermistors are available in a wide variety of shapes. They are usually fired ceramics with fired-on platinum, silver, or gold wires for connecting, and are usually encapsulated in glass, epoxy, or a ceramic sheath. They are available as rods, films, flakes, disks, washers, and other forms. Their small size makes them very versatile and they are often included in electronic circuits attached to a circuit board. Temperature measuring probes may be smaller than 0.5 mm (about 20 mils) in diameter by 1.5 mm long ($\frac{1}{16}$ in.). One manufacturer produces thermistors with resistances at 25°C of from $10\,\Omega$ to $5\,M\Omega$. In one size alone there are about 50 different thermistor products having different resistances at 25°C, different "ratios" (the ratio of the resistance at 0°C to that at 50°C), or different tolerances. From the resistance ratio it is possible to calculate β in Eq. (9.8).

When thermistors were first produced, tolerances of ±20% from nominal resistance value at 25°C were common. Because β also varied, tolerances at a higher or lower temperature were even worse. Improved manufacturing methods, including firing atmosphere control and aging, have reduced variations so that manufacturers provide thermistors that can be used interchangeably in many applications. Tolerance specifications of 1, 2, 5 and 20% of nominal resistance at 25°C are provided by several manufacturers. Interchangeable temperature probe thermistors with ±1% tolerance for the range −20 to 200°C are accurate over the entire range, to ±1% of the range. Some manufacturers provide thermistors to meet required tolerances at two or more temperatures. Selection of the best thermistor for a particular application is important and will be discussed later. For temperature measurement, because thermistors are so nonlinear, they are usually recommended to cover a narrow range of temperatures (Fig. 9.8).

In 1978, the National Bureau of Standards evaluated the stability of thermistors from six manufacturers: Fenwal Electronics, Inc.; Gulton Industries, Inc.; Keystone Carbon Company; Thermometrics, Inc.; Victor Engineering Corporation; and Yellow Springs Instrument Company. The tests for 550 to 770 days at 0, 30, and 60°C revealed that the drift rate for beads and disks was fairly uniform for each manufacturer. Beads were more stable than disks. Stability decreased as temperature increased.[4] Improvements in aging prior to shipment probably resulted from this study.

In a more recent study one manufacturer produced glass encapsulated disk $NiMn_2O_4$ thermistors with ±0.1°C interchangeability for 0 to 70°C.[5] Aging studies for 5000 hours at −80, 0, 25, 70, 100, 150, and 200°C revealed remarkable stability at 0, 25, and 70°C. At 100°C and above, negative drift increased with temperature. Glass-encapsulated thermistor disks with dumet wires were 20 times more stable than epoxy-coated

Ranges

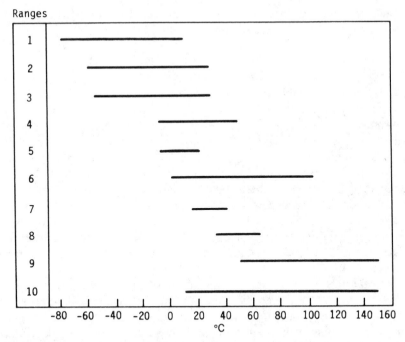

Figure 9.8. Temperature ranges available (Courtesy of the Virtis Company, Inc.)

thermistor disks. Thermal shock could crack the glass and change the calibration. Allowance for drift over the entire range of temperatures could be accomplished by recalibration at any one temperature.

9.3 MODIFICATION OF THE RESISTANCE–TEMPERATURE RELATIONSHIP

The nonlinear nature of Eq. (9.8) is undesirable in many applications. For electronic circuitry as well as for temperature measurement, the resistance is often modified by combining one or more thermistors with other resistances. The other resistances usually have small positive coefficients of resistance. The resulting circuits are less sensitive to temperature change but are more nearly linear.

9.3.1 Simple Series Circuit

The simplest modification is to put a thermistor in series or parallel with a conventional resistor. If the conventional resistor has a small positive coefficient of resistance, the sum of the two resistances in series R' will be dominated by the thermistor at low temperatures and the resistor at high temperatures:

$$R' = R_t + R_1, \tag{9.12}$$

where R_t is the thermistor resistance and R_1 is the added resistance (Fig. 9.9). There will be inflection in the circuit resistance versus temperature curve and Beakley recommends designing temperature measuring equipment to operate around this temperature to minimize departures from linearity.[6] For the simple series circuit, the constant β in Eq. (9.8) is modified:

$$\beta' = \beta \frac{R_t}{R_t + R_1}. \tag{9.13}$$

For a fixed applied voltage the series current is

$$I = \frac{V}{R'} = \frac{V}{R_t + R_1}. \tag{9.14}$$

The temperature term in Eq. (9.8) can be expanded in an infinite Taylor series about a mean temperature T_m, which is chosen to be that of the inflection point on the R' versus T curve. The expansion for a distance ϕ on either side of T_m is

$$f(T_m + \phi) = f(T_m) + \phi f'(T_m) + \phi^2 f'' \frac{(T_m)}{2!} + \cdots + \frac{\phi^n f^{(n)}}{n!}(T_m). \tag{9.15}$$

Here ϕ represents $T - T_m$ and the primes indicate the derivatives of the resistance function, evaluated at T_m. Using R in the form of Eq. (9.8), but evaluating R_0 at T_m, the inflection-point temperature for the circuit, and

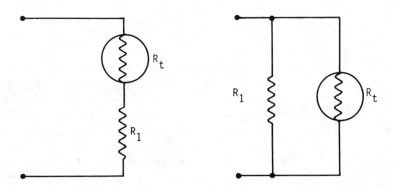

Figure 9.9. Simple smoothing circuits: left, series circuit; right, parallel circuit.

designating it R_m, Eq. (9.15) can be used to calculate the best linear approximation to the resistance versus temperature curve. If optimum linearization is achieved, the third- and higher-degree terms of Eq. (9.15) should be minimized. Ideally, they should be zero, but setting the second term in the expansion of Eq. (9.12) to zero can be used to determine the best circuit parameters to a first approximation. The second derivative is zero if

$$R_1 = \frac{\beta + 2T_m}{\beta - 2T_m} R_0.$$ (9.16)

This establishes the value of R_1 needed to obtain a nearly linear relationship of R' with temperature for the interval $\pm\phi$. Because Eq. (9.8) is not actually linear in T, the higher-order terms are needed to represent the true relationship. For the value of R' determined by the use of R_1 from Eq. (9.16), we can calculate the third- and fourth-order terms and find

$$f'''(T_m) = \frac{-\beta^2}{2T_m^4} f'(T_m)$$ (9.17a)

$$f^{(iv)}(T_m) = \frac{2\beta^2}{T_m^5} f'(T_m).$$ (9.17b)

Substitution in Eq. (9.15) gives

$$R'(T_m + \phi) = R_m + \phi f'(T_m) - \frac{\beta^2 \phi^3 f'(T_m)}{2T_m^4 3!} + \frac{2\beta^2 \phi^4 f'(T_m)}{T_m^5 4!} + \cdots.$$ (9.18)

Then the last two terms on the right give the departure of R' from linear temperature dependence. If we designate that departure from linearity as an error in resistance ε, we see that the ratio of ε_4 to ε_3 is given by

$$\frac{\varepsilon_4}{\varepsilon_3} = \frac{-2\beta^2 \phi^4}{4! \, T_m^5} \frac{3! \, 2T_m^4}{\phi^3 \beta^2} = -\frac{\phi}{T_m}.$$ (9.19)

From Eq. (9.19) it is apparent that the fourth- and higher-order terms will be negligible if ϕ is $\ll T_m$. With Eq. (9.16) we defined ϕ as the temperature interval from T_m. Therefore, ϕ is one-half the temperature range of our thermistor. Then the simple series circuit will approach linearity only for small values of ϕ or very large values of T_m.

Example 9.2. Calculate the departure from linearity for a series circuit containing a thermistor if the thermistor has a resistance of 5000 Ω at 120°C, the minimum in the resistance of the circuit occurs at 120°C, and β

is 3000. Calculate the circuit properties. Are the higher-degree terms negligible?

Solution

$$T_m = 273 + 120 = 393 \ K$$

$$R_1 = 5000 \left[\frac{3000 - 2(393)}{3000 + 2(393)} \right]$$

$$= 5000(0.584) = 2924 \ \Omega$$

$$R_m = 5000 + 2924 = 7924 \ \Omega \ .$$

Then

$$\beta' = 3000 \left(\frac{5000}{5000 + 2924} \right) = 1892 \ K \ .$$

The linear approximation is

$$R(T_m + \phi) = R_m - \phi \frac{\beta'}{T_m^2} R_m$$

$$= 7924 \left(1 - \frac{1892}{(393)^2} \phi \right)$$

or

$$R' = 7924(1 - 0.012\phi).$$

When $\phi = \pm 20°C$,

$$R' = 7924(1 \pm 0.24)$$

$$= 7924 \pm 1941 \ \Omega \ .$$

The departure from linearity is given, to a first approximation, by ε_3:

$$\varepsilon_3 = \frac{(3000)^2(20)^3}{2.3! \ (393)^4} \left[\frac{3000}{(393)^2} \right] (5000)$$

$$= \pm 1.22 \ \Omega \ \text{at} \ \phi = \pm 20°C \ .$$

The ratio of $\varepsilon_4/\varepsilon_3$ is quite small,

$$\frac{\varepsilon_4}{\varepsilon_3} = -\frac{\phi}{T_m} = \frac{-20}{393} = 0.05 \ ,$$

or about $10 \ \Omega$, so the higher terms are negligible.

From Example 9.2 it is apparent that a simple series modification is suitable for high-sensitivity temperature measurements very near a desired temperature, as in calorimetry or changes in body temperature.

An alternative to a simple series circuit is a simple parallel circuit where a fixed resistor is placed in parallel with a thermistor. The results are similar to those for a series circuit, except that the resistance of the circuit is given by

$$R' = \frac{R_t R_1}{R_t + R_1} \qquad (9.20)$$

and

$$\beta' = \frac{R_1}{R_1 + R_t} B \qquad (9.21)$$

and the curve is flattened at low temperatures (Fig. 9.10).

The use of two fixed resistors, in series and parallel combinations (Fig. 9.11), reduces the sensitivity further but extends the range of linearity to about $\pm 40°C$ at $T_m = 300$ K. Both the circuits shown in Fig. 9.11 have resistances in the form

$$R' = \frac{a + bR_t}{c + R_t}, \qquad (9.22)$$

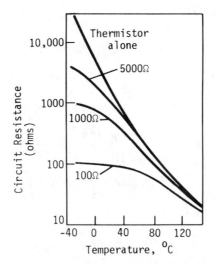

Figure 9.10. Circuit resistance for various values of R_1 in parallel with a thermistor. (After Ref. 1. Reprinted with permission from Electrochemical Publications Ltd., 8 Barns Street, Ayr, Scotland.)

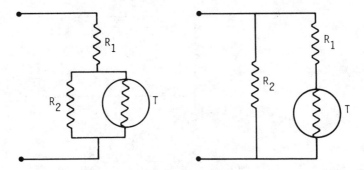

Figure 9.11. Series and parallel shaping networks. (After Ref. 1. Reprinted with permission from Electrochemical Publications Ltd., 8 Barns Street, Ayr, Scotland.)

where a, b, and c are constants whose values depend on R_1 and R_2. More complex circuits are possible, but the combination of two thermistors with fixed resistors gives better linearity and higher sensitivity (Fig. 9.12). These are widely used to produce temperature measuring probes having nearly linear response.

Selection of suitable values for the thermistors and resistors can best be done by computer. One commercial product has a $R_{t_1} = 6000\,\Omega$, $R_{t_2} = 30{,}000\,\Omega$, $R_1 = 6250\,\Omega$, and $R_2 = 3200\,\Omega$. The circuit resistance is given by

$$R' = 2768.23 - 17.151t.$$

The deviation from linearity for the range zero to 100°C is about ±0.5°C for interchangeable temperature probes. The two-thermistor assembly is enclosed in an epoxy bead with 0.080 in. minor diameter and 0.150 in. major diameter.

Thermistor applications in electronic devices often make use of a voltage-dividing circuit (Fig. 9.13). The output voltage is given by

$$V_{O_1} = V_i \left(1 + \frac{R_1}{R_t} \right)^{-1}$$

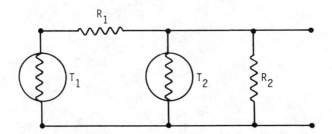

Figure 9.12. Linearized thermistor circuit.

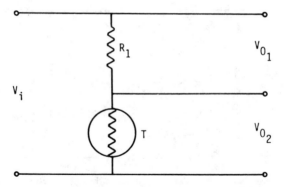

Figure 9.13. Simple voltage-divider circuit. (After Ref. 1. Reprinted with permission from Electrochemical Publications Ltd., 8 Barns Street, Ayr, Scotland.)

The advantage of this circuit is that for R' to be linear with temperature, R_2 needs to have values similar to thermistor properties (Fig. 9.14). For the positive output V_{O_2}, the output can be normalized to obtain

$$F(T) \equiv \frac{V_{O_2}}{V_i} = \frac{1}{1 + R_t/R_1}$$

Some manufacturers provide graphs of $F(T)$ versus temperature for their products. By inspection it is then possible to select the product having

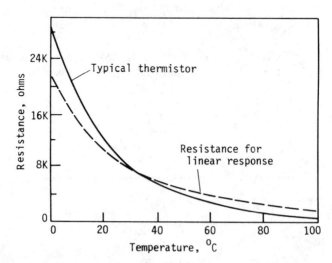

Figure 9.14. Comparison of a typical thermistor response with the resistance needed for a linear output from the circuit in Fig. 9.13. (After Ref. 1. Reprinted with permission from Electrochemical Publications Ltd., 8 Barns Street, Ayr, Scotland.)

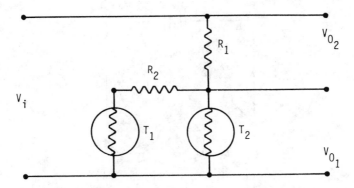

Figure 9.15. Linear divider voltage network. (After Ref. 1. Reprinted with permission from Electrochemical Publications Ltd., 8 Barns Street, Ayr, Scotland.)

linear response for the range of temperatures expected in a particular application.

Combinations of two thermistors and two fixed resistors give better response (Fig. 9.15). The commercial thermistor pair described above, when used in a voltage-dividing circuit as in Fig. 9.15, has output voltages given by

$$V_{O_1} = V_i(0.86507 - 0.0053483t)$$
$$V_{O_2} = V_i(0.13493 + 0.0053483t).$$

If V_i is provided by a stable source such as a zener diode, V_O can be used to indicate temperature electronically. Inexpensive commercial products using thermistors are widely available with narrow range accuracies of 1 or 2°C, and with broader range accuracies of 2 to 5°C. The precision with a particular probe for small temperature changes may be 0.1 and 0.5°C, respectively.

9.4 WHEATSTONE BRIDGE THERMISTOR CIRCUIT

A Wheatstone bridge is a very sensitive resistance measuring instrument in which the resistance of the temperature sensor is compared with that of a standard. Chapter 8 contains a description of the use of bridges for precision resistance measurement. It is also possible to use the out-of-balance current in a Wheatstone bridge for direct indication of temperature, but with less precision. The out-of-balance bridge is often used for thermistor applications. Figure 9.16 shows an out-of-balance bridge in which the null detector current, I_g, causes the null detector to deflect and indicate temperature. We need to analyze this circuit to determine what conditions

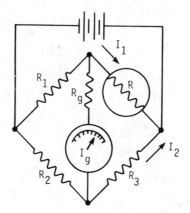

Figure 9.16. Thermistor Wheatstone bridge. (After Ref. 5.)

will give a nearly linear dependence of the null detector current on temperature.

The equations for this out-of-balance condition are

$$V = R_1(I_1 - I_g) + RI_1$$

$$V = (R_2 + R_3)I_2 + R_2I_g \qquad (9.23)$$

$$0 = R_tI_1 - I_2R_3 + R_gI_g .$$

We want to know how I_g depends on the bridge construction. Solving in terms of I_g gives

$$I_g = \frac{V(R_1R_3 - R_2R_t)}{R_t(R_1R_2 + R_1R_3 + R_2R_3 + R_2R_g + R_3R_g) + R_1(R_2R_3 + R_2R_g + R_3R_g)} . \qquad (9.24)$$

This is in the form

$$I_g = \frac{V(l - mR_t)}{PR_t + q} , \qquad (9.25)$$

where l, m, P, and q are constants. Dividing all terms by P gives

$$I_g = \frac{V[l/P - (m/P)R_t]}{R_t + q/P} . \qquad (9.26)$$

Dividing the denominator into the numerator gives

$$I_g = \frac{-V}{P} + \frac{V(l/P + mq/P^2)}{R + q/P}. \tag{9.27}$$

This is the same form as Eq. (9.14) if q/P is equal to R_1, the effective series resistance, and the quantity in parentheses is a multiplying constant. Therefore, the same conditions for linearization of I_g with T should apply that applied to the linearization of R' with T in Section 9.3. These include:

$$R_s = \frac{\beta - 2T_m}{\beta + 2T_m} R_t, \tag{9.28}$$

$$R_s = R_t \frac{R_g + [R_2 R_3/(R_2 + R_3)]}{R_t + [R_2 R_3/(R_2 + R_3)]}, \tag{9.29}$$

and

$$\frac{dI_g}{dT} = \frac{sV(\beta + 2T_m)^2}{4T_m^2 \beta R_m} \left(1 - \frac{\phi^2 \beta^2}{12 T_0^4}\right), \tag{9.30}$$

where R_3 is the effective series resistance and s is a dimensionless sensitivity reduction factor that depends on the resistors,

$$s = \frac{l}{P} + \frac{mq}{P^2}. \tag{9.31}$$

In designing a circuit R_s, the effective series resistance is calculated from Eq. (9.28) and S is calculated for the microammeter sensitivity from Eq. (9.30). McLean recommends setting $R_2 = R_3$ to minimize the effects of connecting wire resistance and of variations in β.[7] At balance $R_1 = R_t$ if $R_2 = R_3$. Then R_1 and R_g can be calculated from Eqs. (9.29) and (9.31) for various possible values of R_t and β. Adjusting the supply voltage to

$$V = \theta R_m T_m \tag{9.32}$$

will reduce the sensitivity of the circuit to small variations in R_m from one thermistor to another. Here θ is the sensitivity, current change per degree.

9.5 ANALOG OUTPUT THERMISTOR CIRCUITS

Thermistors have such high sensitivity that indications of temperature can be obtained with simple instruments. The indicating instrument is often an ammeter or a voltmeter calibrated to read temperatures directly. Some circuits are shown in Figs. 9.17 to 9.20.

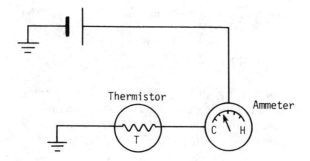

Figure 9.17. Automobile temperature gauge.

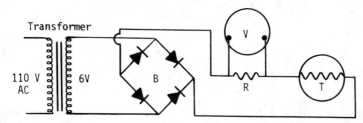

Figure 9.18. Voltmeter used to measure current by measuring the *IR* drop across *R*.

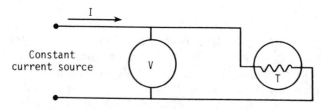

Figure 9.19. Temperature indicated directly by a digital voltmeter calibrated for the constant current *I*.

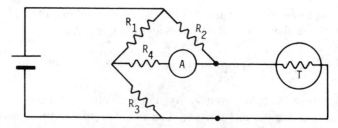

Figure 9.20. Wheatstone bridge calibrated to read temperature directly at the micro-ammeter.

The range of the bridge circuit can be adjusted to semilinear output over a narrow temperature range. For Fig. 9.20 the Wheatstone bridge should have R_2 and R_3 equal, and similar in resistance to R_T at 25°C. Suppose that R_T has resistances 16,325 Ω at 0°C, 5000 Ω at 25°C, and 1801.5 Ω at 50°C. Then R_2 and R_3 should be 5000 Ω. Since the meter should read 0° when the thermistor is at 0°C, R_1 should have a resistance of 16,325 Ω. To minimize errors, the meter resistance, including R_g and R_m, where R_m is the microammeter resistance, should be about 10 times the minimum value of R_T, or about 18,000 Ω. If this is sufficiently high, we can approximate the current flowing through the meter as zero and calculate the voltage drop across R_T and R_1. When $I_{\text{meter}} = 0$, and if the applied voltage is 1.35 V, at 50°C

$$V_T = \frac{V(R_T)}{R_T + R_3} = \frac{1.35(1800)}{1800 + 5000} = 0.3574 \text{ V}$$

$$V_1 = \frac{V(R_3)}{R_2 + R_1} = \frac{1.35(16,325)}{16,325 + 5000} = 1.0335 \text{ V} .$$

$$(9.33)$$

The difference is the potential across the meter leg of the circuit.

$$V_1 - V_T = 1.0335 - 0.3574 = 0.6761 \text{ V} . \tag{9.34}$$

If the meter has 18,000 Ω total resistance, the current flowing in the meter is

$$I_{\text{meter}} = \frac{V_{\text{meter}}}{R_{\text{meter}}} = \frac{0.6761}{18,000} = 37.6 \times 10^{-6} \text{ A} . \tag{9.35}$$

Therefore, a 50-μA meter should work. It will be necessary to adjust R_g downward, however, because we assumed that $I_{\text{meter}} = 0$ when we calculated V_T and V_1. This can be adjusted experimentally and at the same time, the circuit can be calibrated by placing a resistor of the desired values at R_T and adjusting R_g, then marking the meter as the resistor at R_T is stepped through the calibration values of R_T.

9.6   DIGITAL OUTPUT THERMISTOR CIRCUITS

Thermistor probes containing linearizing circuits lend themselves well to solid-state devices. Many commercial digital thermometers are in use because of their small size, low cost, and reliability. One manufacturer recommends the circuit shown in Fig. 9.21.[8] For telemetry purposes the voltage output of a solid-state amplifier can be used to control an oscillator

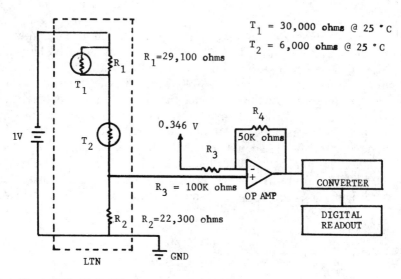

Figure 9.21. Digital thermometer circuit. (Courtesy of Fenwal Electronics.)

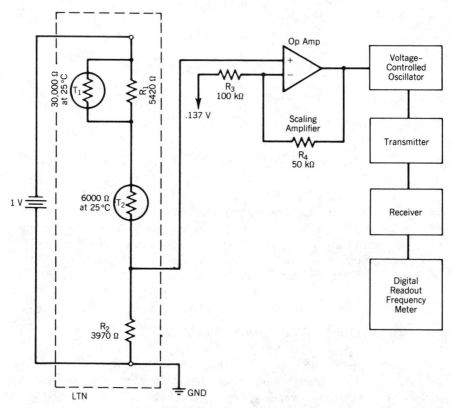

Figure 9.22. Telemetry circuit using frequency proportional to temperature. (Courtesy of Fenwal Electronics.)

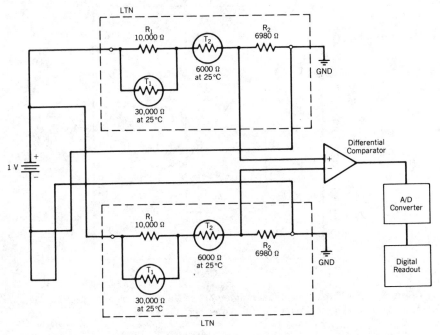

Figure 9.23. Opposing linear circuits that produce a digital comparison output. (Courtesy of Fenwal Electronics.)

frequency. The signal can be transmitted to a receiver and displayed as a digital signal by a digital output frequency meter (Fig. 9.22). Measurement of the difference in temperature between two sites also lends itself well to solid-state components (Fig. 9.23).

9.7 GERMANIUM RESISTANCE THERMOMETERS

Germanium resistance thermometers were developed for use at cryogenic temperatures, especially in the range of about 1–80 K, although a thermometer sensitive at the one end of the scale may not be as sensitive as another designed for the other end of the scale. Carefully calibrated germanium thermometers have been very useful for comparisons of different thermometric devices in the range 1–40 K.[9] Unfortunately, some will produce an unexpected change in calibration during use, and enough cross-checking is required so that such changes can be recognized.

Blakemore designed germanium resistance thermometers by doping them with a donor, arsenic, and a lesser amount of an acceptor, gallium.[10] Bulk crystalline rectangular rods with projections on one side (bridge shaped) were tested with current wires connected to the ends of the rod and

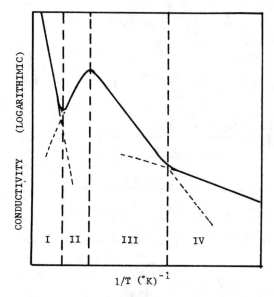

Figure 9.24. Logarithm of electrical conductivity of a doped elemental semiconductor as a function of reciprocal temperature. (After Ref. 9.)

voltage wires connected to the projections. He classified the electrical conduction into four regimes (Fig. 9.24). (Note that temperature increases to the left in the reciprocal temperature plot.) At high temperatures the concentration of electrons and holes dominate in accordance with Eq. (9.2a) and (9.3) (region I). Conductivity rises steeply with temperature and quickly becomes negligible as temperatures decrease along the dashed line. In this region

$$\sigma_I = Ae^{-(\alpha/T)}. \tag{9.36}$$

In regions II and III the conductivity depends on the concentration and mobility of the electrons supplied by the dopant. In region III the conductivity increases with temperature as more and more electrons are released from the donor. However, it is the net concentration, donor minus acceptor, that is important, so the concentration of electrons can be approximated by

$$n \approx \frac{n_D - n_a}{1 + Bn_a \exp(\beta/T)}, \tag{9.37}$$

where B and β depend on the donor. Here the conductivity depends on both the donor and the acceptor. Because the energy required for escape is less than the band gap, the slope in region III is less than that in region I. This is called the extrinsic carrier range because the dopant concentrations fix the concentration of free carriers. Then

$$\sigma_{III} \approx C \frac{n_d - n_a}{n_a} e^{-\beta/T}. \qquad (9.38)$$

The slope β is less than for the intrinsic region. Region II exists because the concentration of carriers released by the impurities is exhausted when all electrons contributed by the impurities are freed. Then, with a constant concentration of charge carriers, the mobility dominates in Eq. (9.1). The

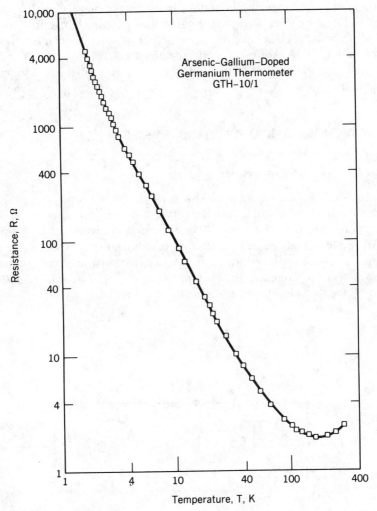

Figure 9.25. Log-log plot of resistance versus temperature for a germanium resistance thermometer. (After Ref. 9.)

decrease in mobility as temperature increases causes a decrease in conductivity.

In region IV thermal energy is too low to excite even the donor electrons into the conduction band. Conductivity should decrease according to the dashed line extending into region IV. However, electron hopping is believed to occur. The electrons travel from donor to donor with difficulty, but the activation energy is less than in region III. The conductivity can be approximated by

$$\sigma_{IV} \approx De^{-(\gamma/T)}.$$ (9.39)

In doped germanium the boundary between regions III and IV is often about 10 K, and can be abrupt, giving an undesirable change in sensitivity. By doping with special concentrations the abrupt changes can be smoothed out as in Fig. 9.25. Different concentrations can be used to control the region of maximum sensitivity and the total resistance at a particular temperature.[10]

9.8 JUNCTION SEMICONDUCTOR DEVICES

9.8.1 Diodes and Transistors

Junction semiconductors are used in several different forms. The simplest is a single diode or transistor where the forward-connected junction resistance is very sensitive to temperature. A more common form is obtained from a pair of diode-connected transistors, especially selected to have the same electrical characteristics, and mounted together in the same package so they will share a common temperature (Fig. 9.26). According to Carr,[11] for any transistor, the base–emitter voltage is given by

$$V_{be} = \frac{kT}{q} \ln \frac{I_c}{I_s},$$ (9.40)

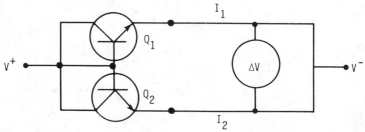

Figure 9.26. Temperature measurement with matched diode-connected transistors. (After Ref. 11).

where k is Boltzmann's constant, T is absolute temperature, q is the electronic charge of an electron, I_c is the collector current, and I_S is the reverse saturation current. Since I_c and I_S can be made constant, the voltage can be made to depend only on absolute temperature. For our circuit, the difference between the base–emitter voltages for Q_1 and Q_2 is given by

$$\Delta V_{\text{be}} = \frac{kT}{q} \ln \frac{I_1}{I_2}. \qquad (9.41)$$

Because the logarithm of 1 is zero, the circuit must be adjusted so that $I_1 \neq I_2$. Sometimes I_1 is set to twice I_2. By supplying I_1 and I_2 from constant-current sources, the difference voltage will depend only on T. By choosing the values of I_1 and I_2 correctly, ΔV_{be} can be set to any desired value, for example, $10\,\mu\text{V/K}$. This makes the system suitable for temperature measurement at temperatures near room temperature, or for temperature compensation for thermocouples. Note that suitable constant-current sources (not shown in the figure) and suitable voltmeters must be provided.

Silicon. The design of a commercial precision transistor thermometer was described at the Sixth International Conference on Temperature Measurement.[12] Verster had previously shown that the collector current for a transistor, mounted to have a base–emitter voltage V_{be} to drive a constant current through the transistor is determined by

$$I_c = (\alpha T^r / \eta) \exp\left(-q V_{g0}/kT\right)[\exp\left(q V_{\text{be}}/kT\right) - 1]. \qquad (9.42)$$

Here α and r are constants, T is temperature, η is the ionization efficiency, q is electron charge, V_{g0} is the extrapolated energy gap of silicon, and k is Boltzmann's constant. Values of V_{g0} and r are often given as 1.205 and 1.5 V, respectively. However, they are not constant, so the collector voltage is not. For temperatures at which the term in the brackets is negligible and letting $\eta = 1$ gives

$$V_{\text{be}} = V_{g0} - (kT/q) \ln\left(\alpha T^r / I_c\right). \qquad (9.43)$$

Then with I_c constant, V_{be} is almost linear in temperature. The value of the ln term, especially T^r, produces slight nonlinearity. Ohte et al.[12] found they could compensate for T^r by controlling I_c to compensate for T^r. For their commercial silicon transistors, r was 3.49. A quadratic approximation of I_c versus T was found to give suitable compensation for nonlinearity. However, a nonlinear digital indicator circuit was found to be less expensive and satisfactory (Fig. 9.27). The block diagram shows the digital output, proportional to E_x/E_s, where E_s is the reference supply voltage, is

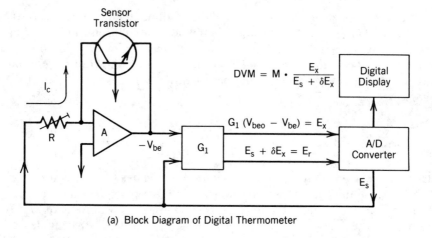

(a) Block Diagram of Digital Thermometer

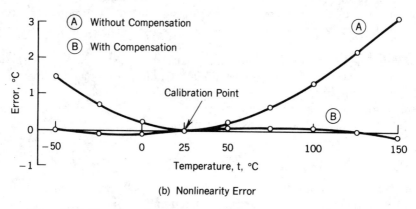

(b) Nonlinearity Error

Figure 9.27. Transistor thermometer with digital indicator. (After Ref. 11.)

modified by adding an increment δE_s to the supply voltage. The value of δ was selected by computer to produce the compensation shown in the lower part of Fig. 9.27. Probes were supplied with the transistor mounted in various sizes of stainless steel sheaths.

9.8.2 Integrated-Circuit Temperature Sensor

A sophisticated modification of the two transistor circuit is available as a two-terminal monolithic integrated circuit (AD590).[13-15] When a dc supply voltage (+4 to +30 V) is provided, the output current is proportional to the temperature of the device with a nominal temperature coefficient of $1 \mu A/K$. The chip is available as a chip, sealed in a flat ceramic package or

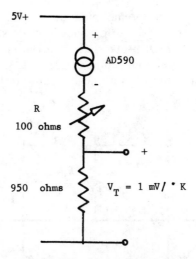

Figure 9.28. AD 590 with one temperature trim.

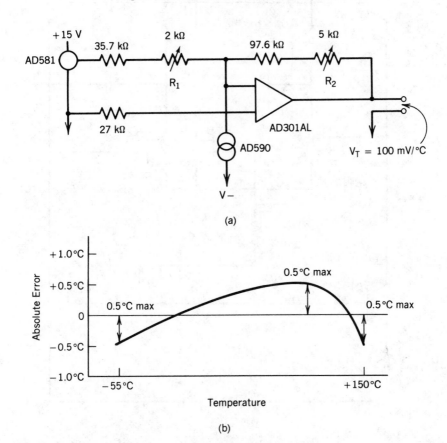

(a)

(b)

Figure 9.29. (a) Two-temperature trim circuit; (b) accuracy after two-temperature trim.

a TO-52 metal can, and premounted in probes. It is rated at −55 to +150°C. It is not very sensitive to changes in dc supply voltage. Its high output impedance, 10 MΩ, protects it from output shorting and makes it insensitive to long output connecting wires.

The simplest application has a one-temperature trim circuit (Fig. 9.28).

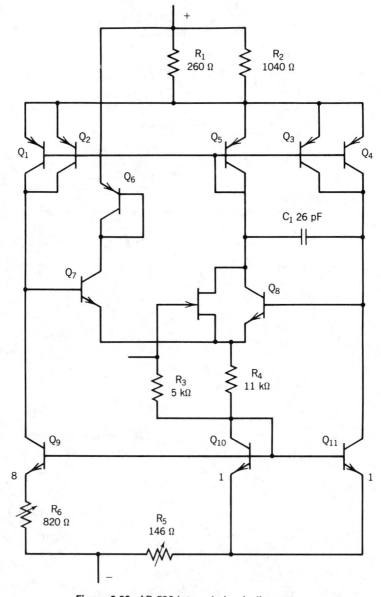

Figure 9.30. AD 590 internal circuit diagram.

The 100-Ω potentiometer is adjusted to give a desired value at a reference temperature. Output current, proportional to temperature, produces an output voltage across the 950-Ω external resistor. A two-trim circuit with better precision is shown in Fig. 9.29.

The internal circuit of the AD590 integrated circuit is shown in Fig. 9.30. The sum of the currents from transistors Q_9, Q_{10}, and Q_{11} make up the total output currents. The voltages supplied are regulated to produce a total current output through resistors R_5 and R_6 proportional to the absolute temperature. Resistors R_5 and R_6 are laser trimmed during manufacture to produce interchangeable devices. The cost is low, and the accuracy sufficient to allow the AD590 to be used for many applications of temperature measurement or for temperature compensation (Fig. 9.31). The circuit shown is for a type J thermocouple. Other thermocouples would require different resistors.

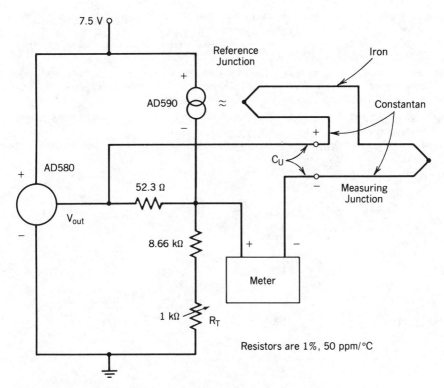

Figure 9.31. Temperature compensation for a thermocouple circuit. (AD 580 is a precision 1.5-V reference supply voltage.)

REFERENCES

1. E. D. Macklin, "Thermistors." Electrochemical Publications, Ltd., Ayn, Scotland, 1979.

2. H. B. Sachse, "Semiconducting Temperature Sensors and Their Application." Wiley, New York, 1975.

3. J. A. Becker, C. B. Green and G. L. Pearson, Properties and uses of thermistors—thermally sensitive resistors. *Trans. AIEE* **65**, 711–725 (1946).

4. S. D. Wood, B. W. Mangum, J. J. Filliben, and S. B. Tillett, An investigation of the stability of thermistors. *J. Res. Natl. Bur. Stand.* **83**, 247–263 (1978).

5. T. H. LaMers, J. M. Zurbuchen, and H. Trolander, Enhanced stability in precision interchangeable thermistors. *Temp.: Its Meas. Control Sci. Ind.* **5** (Part 2), 875–873 (1982).

6. W. R. Beakley, The design of thermistor thermometers with linear calibration. *J. Sci. Instrum.* **28**, 176–179 (1951).

7. J. A. McLean, A method of constructing direct reading thermistor thermometers. *J. Sci. Instrum.* **31**, 455–457 (1954).

8. Fenwal Electronics, 63 Fountain Street, Framingham, Massachusetts 01701.

9. L. M. Beasley and W. R. G. Kemp, An intercomparison of temperature scales in the range 1 to 30 K using germanium resistance thermometry. *Metrologia* **13**, 35–51 (1977).

10. J. S. Blakemore, Design of germanium resistance thermometers for thermometric applications. *Rev. Sci. Instrum.* **33**, 106–112 (1962).

11. J. J. Carr, "Elements of Electronic Instrumentation and Measurement," pp. 295–297. Reston Publ. Co., Reston, Virginia, 1979.

12. A. Ohte, M. Yamagata, and K. Akiyama, Precision silicon transitor thermometer. *Temp.: Its Meas. Control Sci. Ind.* **5** (Part 2), 1197–1203 (1982).

13. Analog Devices, Inc., "Analog Devices Data-Acquisitions Data Book." Norwood, Massachusetts, 1982; National Semiconductors, Inc., "National Semiconductors Data Conversion/Acquisition Data Book." Santa Clara, California, 1984.

14. M. P. Timko and A. P. Brokow, "Integrated Circuit Two Terminal Temperature Transducer", Oct. 31, 1978, U.S. Patent 4,123,698.

15. M. P. Timko, A two-terminal IC temperature transducer. *IEEE J. Solid-State Circuits* **SC-11**, 784–788 (1976).

10

THERMOELECTRIC TEMPERATURE MEASUREMENT

Thermoelectric temperature measurements are made with thermocouples, devices that convert a temperature difference into an electromotive force called the Seebeck voltage. Thermocouples are usually made from two dissimilar metal wires connected so that one junction is held at a reference temperature and the other junction serves as the temperature sensing device. Thermoelectric measurements then require (1) a sensing element connected through a reference junction by (2) electrical lead wires to a (3) voltage measuring instrument. These will be described in detail later.

The thermoelectric method is not as precise as the platinum resistance thermometer in ranges I and II of the IPTS. Despite interest in other methods, it is still the basis for temperature measurement in range III. Except for thermometers in range II, it is the most common method of temperature measurement in ranges II, III, and IV because thermocouples are low cost, moderately reliable, and easily used with recording and controlling instruments. Their widespread use makes it necessary for us to understand the various types of thermocouples, their attributes and limitations. In order to understand the principles by which they indicate temperature, we will trace their historical development and then develop the thermodynamic relationships involved.

10.1 HISTORICAL PERSPECTIVE

In 1821 in Germany, Thomas Johann Seebeck experimented with the effect of temperature on the electrical properties of bismuth, copper, and antimony.[1] He discovered that when two wires of different composition were connected at their ends only, to make a complete electrical circuit, an electrical current would flow in the circuit if one of the connections (junctions) was heated. He reasoned that current could only flow if an electrical potential had been generated between the hot junction and the cold junction. It was convenient to have the cold junction (reference junction) at room temperature. Then the higher the temperature of the hot junction (measuring junction), the higher the current flow would be. The potential responsible for that current flow is now called the Seebeck voltage in his honor.

In 1824 in France, Jean Charles Athanase Peltier experimented with Seebeck's bismuth–antimony thermocouple and applied an external voltage to cause an increased current flow.[2] He discovered that the current caused one of the junctions to be heated when current flowed in one direction, but that it was cooled when the current flowed in the reverse direction. The amount of heating or cooling depended on the current flow, which depended on the external applied EMF. This effect can be used as an energy transducer, converting electrical energy to heat, or vice versa. The Peltier effect is the basis for the Peltier refrigerator, a method of cooling that has no moving parts, just a series of junctions and an electrical current source.

In 1826 in Paris, A. C. Becquerel studied thermocouple outputs as a function of temperature.[3] He proposed that thermocouples be used for thermometer applications. He discovered that the EMF of an iron–copper thermocouple reversed sign and gave zero output at 280°C as the hot junction was heated. This effect was unexpected, and considerable confusion existed concerning the principles that governed thermocouple output. Becquerel believed that there must be a second EMF to balance the Seebeck EMF when conditions were such as to produce zero output.

In 1847 in England, Lord Kelvin (at that time called William Thomson) explained the relationships between the Seebeck, Peltier, and Becquerel discoveries based entirely on thermodynamic arguments.[4] He proposed that an electric current in a homogeneous wire delivered heat from the hot region to the cold region when the current flowed from a hot region to a cold region; and in the opposite direction when the current was reversed. Such an effect would require an EMF to exist between the hot region and the cold region if no current flowed in a homogeneous conductor. That EMF is now called the Thomson voltage.

Various combinations of metals or other electrical conductors were tried as thermometric elements. In 1885 in France, Henri Le Châtelier[5] experimented with platinum and its alloys as a thermocouple system. He proposed a thermocouple with pure platinum as one leg and platinum 90, rhodium 10 as the other. This thermocouple proved to be more stable, particularly at high temperatures, than other combinations. In 1927 it was adopted as the sensing element for range III of the IPTS and is still used for that purpose.

10.2 THEORY OF THERMOELECTRICITY

Most thermocouples are based on metal conductors, although ceramic conductors are also used. The theory of metallic conduction is a reasonable base for qualitative discussions of thermoelectric measurements.

The band theory of metals is based on the concept of metal nuclei arranged in a periodic crystalline array in such a way that the outer electrons of each atom come so close together that the Pauli exclusion principle requires them to be arranged in a quasi-free electron cloud around the nuclei.[6] The energy levels of these electrons are degenerate; that is, they are collected together in bands of energy so that any electron in the band can have a particular allowed energy state. The occupied band of a metallic conductor overlaps an allowed, unoccupied band (Fig. 9.1) so that electrons can flow under an applied field. The distribution of energy among the electrons is called the density of states (Fig. 10.1). The probability of finding an electron at an energy E is given by the Fermi function,

$$P(E) = \frac{1}{e^{(E-E_f)/RT} + 1}.$$ (10.1)

Here E_f is called the Fermi energy and represents the average energy of the electrons at an absolute temperature T. Because every metal or alloy has a unique electronic and crystalline structure, the allowed energy states and their electronic population will also be unique. Therefore, when two metals come in contact, the electrons in the metal with the higher energy will flow into the one with the lower energy. This will occur at the junction of the two metals until the excess electrons in the metal of lower energy builds up a reverse EMF which opposes the flow. This occurs when all the electrons in both energies come to a common Fermi energy intermediate between the two. The potential produced in this way is the Peltier EMF, a reversible thermodynamic EMF which, if current is allowed to flow by making a second connection at a different temperature, can deliver heat at the first junction equal to the product of current and EMF; that is, the heat equals the electrical energy, or

$$dQ = \pi I \, d\theta,$$ (10.2)

where dQ is the heat delivered at a junction in time $d\theta$ by a current I. The Peltier EMF, π, depends only on the temperature and the two junction materials. It is the fundamental potential responsible for the Seebeck voltage. Thus the Peltier voltage for two materials, A and B, with junctions at temperatures T_1 and T_2, when current flow is zero, is

$$\pi_{AB} \atop {T_1 \to T_2} = (\pi_{AB})_{T_1} - (\pi_{AB})_{T_2}$$ (10.3)

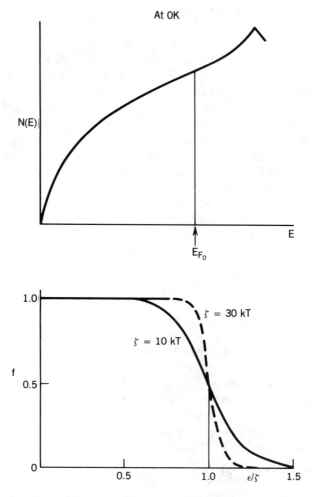

Figure 10.1. Density of states (above) and probability of occupancy of a state (below).

The Seebeck voltage is the net voltage for such a junction. It includes another term, which also depends on the electronic structure of A and B. It depends on the two Thomson voltages, which are also fundamental reversible thermodynamic quantities that can deliver heat if current is allowed to flow. The Thomson voltage depends on the way the Fermi energy of each conductor varies with temperature. In simplest form it can be written

$$V_{\text{Thomson}} = \int_{T_1}^{T_2} \sigma \, dT \,, \tag{10.4}$$

where σ is an empirically determined Thomson coefficient. Then the first law of thermodynamics requires that

$$dQ_T = V_T I \, d\theta = \left(\int_{T_1}^{T_2} \sigma \, dT \right) I \, d\theta . \qquad (10.5)$$

The direction of heat flow depends on the direction of current flow, being delivered in the direction the electrons travel.

The Seebeck voltage, then, is the sum of two reversible thermodynamic electropotentials at open-circuit conditions: the Peltier potential and the Thomson potential. The latter is the net Thomson potential depending on the difference in E_f for the temperature range imposed. In equation form,

$$V_S = V_P + V_T$$

$$= (\pi_{AB})_{T_2} - (\pi_{AB})_{T_1} + \int_{T_1}^{T_2} \sigma_A \, dT + \int_{T_2}^{T_1} \sigma_B \, dT$$

$$= (\pi_{AB})_{T_2} - (\pi_{AB})_{T_1} + \int_{T_1}^{T_2} \sigma_A \, dT - \int_{T_1}^{T_2} \sigma_B \, dT$$

$$= (\pi_{AB})_{T_2} - (\pi_{AB})_{T_1} + \int_{T_1}^{T_2} (\sigma_A - \sigma_B) \, dT . \qquad (10.6)$$

The Seebeck coefficient α_S is often called the thermodynamic power and is defined, for materials A and B, as

$$dV_S = \alpha_{AB} \, dT . \qquad (10.7)$$

This equation holds at a particular temperature. Data are usually tabulated for α at 25°C. The output of the platinum–platinum 90, rhodium 10 thermocouple in IPTS range III is

$$V_S = a + bt + ct^2 . \qquad (10.8)$$

Obviously,

$$\frac{dV_S}{dt} = \alpha_S = b + 2ct \qquad (10.9)$$

and is not constant. Comparison with Eq. (10.6) shows that

$$\alpha_{AB} = \frac{dV}{dT} = d\pi_{AB} + (\sigma_A - \sigma_B) , \qquad (10.10)$$

so that Eq. (10.9) is an empirical approximation of Eq. (10.10).

10.2.1 The Kelvin Relations

There is, however, a more fundamental relationship between α, π, and σ. Consider an AB thermocouple connected as shown in Fig. 10.2 with a steady-state current I flowing. Then the rate of heat absorbed by the thermocouple per unit time is

$$q = \frac{Q_{net}}{\Delta\theta}.$$

Taking the $\Delta\theta$ for time from θ_1 to θ_2 as 1 second is convenient. Then $q = Q_{net} = \sum Q$. The heat absorbed by the thermocouple can be calculated from

$$Q_{net} = \sum Q = \int_{\theta_1}^{\theta_2} dQ_{AB} \bigg|_{T_2} - \int_{\theta_1}^{\theta_2} dQ_{AB} \bigg|_{T_1} + \int_{\theta_1}^{\theta_2} dQ_A + \int_{\theta_2}^{\theta_1} dQ_B .$$

If we neglect thermal conduction along the wire and Joule heating, for $\theta_2 - \theta_1 = 1$,

$$Q_{net} = dQ_{AB} \big|_{T_2} - dQ_{AB} \big|_{T_1} + dQ_A \big|_{T_2 \rightarrow T_1} + dQ_B \big|_{T_1 \rightarrow T_2}$$

$$= V_S I = \pi_{AB} I \big|_{T_2} - \pi_{AB} I \big|_{T_1} + \int_{T_1}^{T_2} \sigma_A I\, dT - \int_{T_1}^{T_2} \sigma_B I\, dT .$$

At steady state I is constant and can be taken outside the integral:

$$V_S I = I \left[\pi_{AB} \big|_{T_2} - \pi_{AB} \big|_{T_1} + \int_{T_1}^{T_2} \sigma_A\, dT - \int_{T_1}^{T_2} \sigma_B\, dT \right]$$

$$= I \left[\pi_{AB} \big|_{T_2} - \pi_{AB} \big|_{T_1} + \int_{T_1}^{T_2} (\sigma_A - \sigma_B)\, dT \right]. \tag{10.11}$$

The term in brackets is the sum of the Peltier and Thomson coefficients. The product of the term in brackets and the current is the reversible

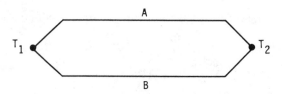

Figure 10.2. Basic thermocouple circuit.

electrical work. This analysis shows that for current to flow, a net absorption of heat must occur sufficient to equal the electrical work. It is
consistent with the first law of thermodynamics,

$$dE = \delta Q - \delta W ,$$

because for the cycle dE is zero and $\delta Q = \delta W$. By canceling the current
from both sides of Eq. (10.11), we can write

$$dV_S = d\pi + (\sigma_A - \sigma_B)\, dT . \tag{10.12}$$

The second law of thermodynamics can also be applied to thermocouples. For a current flowing from the junction at T_1 to T_2 and returning
to T_1, the net change in entropy must be zero because we have returned to
the original state. That is,

$$dS = \frac{\Sigma\, dQ_{\text{rev}}}{T} = 0 .$$

We have already summed dQ reversible. Using Eq. (10.12) and dividing by
T, we have

$$dS = I\left[d\left(\frac{\pi}{T}\right) + \frac{\sigma_A - \sigma_B}{T}\, dT \right] = 0 \tag{10.13}$$

$$= I\left(\frac{T\, d\pi - \pi\, dT}{T^2} + \frac{\sigma_A - \sigma_B}{T}\, dT \right) = 0$$

$$= I\left[\frac{-\pi\, dT}{T^2} + \frac{d\pi + (\sigma_A - \sigma_B)}{T}\, dT \right] = 0 ,$$

but the second term is dV_S/T, so

$$\frac{-\pi\, dT}{T^2} + \frac{dV_S}{T} = 0$$

or

$$\pi = \frac{T^2}{T}\frac{dV_S}{dT} = T\frac{dV_S}{dT} = T\alpha_S . \tag{10.14}$$

Substituting π into Eq. (10.13) gives

$$dS = I\left[d\left(\frac{T\alpha_S}{T}\right) + \frac{\sigma_A - \sigma_B}{T}\, dT \right] = 0 ,$$

so

$$\frac{\sigma_A - \sigma_B}{T} = -\frac{d\alpha_S}{dT},$$

which gives

$$\sigma_A - \sigma_B = -T\frac{d^2 V_S}{dT^2}. \qquad (10.15)$$

Then if we know

$$V_S = a + bt + ct^2 + \cdots,$$

we find that

$$\alpha_S = b + 2ct + \cdots \qquad (10.16)$$

$$\pi = T\alpha_S = T(b + 2ct + \cdots) \qquad (10.17)$$

$$\Delta\sigma = -T(2c + \cdots). \qquad (10.18)$$

10.3   THE CONCEPT OF ABSOLUTE THERMOELECTRIC POWER

As we observed in deriving Eq. (10.10), the Seebeck coefficient for a thermocouple can be considered to be the sum of Peltier and Thomson effects. In Eq. (10.10),

$$\alpha_{AB} = \frac{dV_S}{dt} = d\pi_{AB} + \sigma_A - \sigma_B, \qquad (10.10)$$

it is clear that the Peltier differential depends on the Peltier components of both thermocouples A and B. This must be made up of contributions from both elements of the thermocouple individually. That is, $d\pi_{AB}$ will be different for thermocouples made with different components used for A or B, except at absolute zero, where no potential will exist. However, if component A had the property of not contributing at all to either the Peltier or Thomson potential, Eq. (10.10) could be written as

$$\alpha_B = \frac{dV_S}{dt} = d\pi_B + \sigma_B. \qquad (10.19)$$

Note that conceptually, $d\pi_B = 0$ if there is no second component. Then the output of the thermocouple would depend only on B. This is a way of introducing the concept of *absolute thermoelectric power*.[7] Thus the thermoelectric power of such a thermocouple would depend only on B. A logical extension of this argument is to postulate that the output of an AB thermocouple could depend on the absolute thermoelectric powers of both A and B.

Fortunately, the metal lead has a very low Seebeck voltage. A thermocouple with lead as one element has a Seebeck coefficient that depends almost entirely on the second element. Therefore, it is possible to test the postulate that absolute thermoelectric power exists. And it is possible to compare two different elemental thermocouples, with lead as one leg, and then to join them in a thermocouple to determine if the Seebeck coefficient does depend predictably on the sum of the absolute thermoelectric powers of the individuals. This is found to be true in the form

$$\frac{dV_{AB}}{dT} = \alpha_A - \alpha_B = \int_0^T \frac{\sigma_A}{T} \, dT - \int_0^T \frac{\sigma_B}{T} \, dT . \tag{10.20}$$

From Eq. (10.20) we know the empirical relation

$$\alpha_S = b + 2ct \tag{10.21}$$

holds, so

$$\alpha_S = \alpha_A - \alpha_B .$$

If α_A and α_B are linear functions of absolute temperature,

$$\alpha_A = x_A + m_A T \tag{10.22}$$

and

$$\alpha_B = x_B + m_B T ,$$

then

$$\alpha_S = \alpha_A - \alpha_B = (x_A - x_B) + (m_A - m_B) T . \tag{10.23}$$

If we want the output of a thermocouple to be linear with temperature, the term $2c$ should be as small as possible. Comparing Eqs. (10.21) and (10.23), this occurs when

$$2c = m_A - m_B \tag{10.24}$$

is very small. Linear output only is possible when there is a small difference between the slopes of the absolute thermoelectric power of the two components of a thermocouple. This is the reason the number of thermocouple compositions is so limited. Only a few metal and alloy combinations have been found which have the same slope of their absolute thermoelectric powers with temperature. Other combinations do not have a linear output.

10.4 ELECTRONIC SPECIFIC HEAT

One approach to an understanding of the thermoelectric properties of metals is based on the electronic specific heat of a metal. If we neglect Joule heating, a unit volume of a thermocouple wire in a temperature gradient (i.e., somewhere in the middle of a thermocouple leg) will have a heat evolution rate

$$\dot{Q} = -\sigma\rho_I \frac{dT}{dx},$$
(10.25)

where ρ_I is the current density and dT/dx is the thermal gradient in the wire. The negative sign indicates that the heat is released from the unit volume. Under these conditions Thomson observed that σ was the "specific heat of electricity," the thermal energy delivered by a unit current of electricity.

Well above the Debye temperature the theory of the electronic contribution to the specific heat of classic noble metals gives the equation

$$C_{v_e} = \frac{\pi^2}{2} \frac{nN_A k^2 T}{E_f},$$
(10.26)

where n is the valence of the metal, N_A is Avogadro's number, k is Boltzmann's constant, and E_f is the Fermi energy. This can be compared to theoretical derivations of the thermoelectric power,

$$\alpha_S = \frac{-\pi^2 k^2 T}{2eE_f}.$$
(10.27)

Allowing for the negative sign of e, the electronic charge, it can be seen that Thomson's observation that σ is the equivalent to the electronic specific heat is supported by theory. It should be a linear function of temperature and inversely proportional to E_f. Changes in alloy composition to reduce E_f should increase α.

For the transition element metals, where electrons in d orbitals can be thermally excited to S orbitals, leaving holes behind in the d orbital, the appropriate equations are

$$C_{v_h} = \frac{\pi^2}{6} \frac{nNk^2 T}{E_0 - E_f}$$
(10.28)

and

$$\alpha_S = \frac{-\pi^2}{6} \frac{k^2 T}{e(E_0 - E_f)}, \tag{10.29}$$

where E_0 is the highest allowed d orbital energy. The absolute thermoelectric power for both types of metals is equal to the ratio of the specific heat of the carrier to its electronic charge. Thus control of the electronic specific heat can be used to control the absolute thermoelectric power.

10.5 COMMERCIAL THERMOCOUPLE COMPOSITIONS

Only certain metals and metal alloys are suitable for use as thermocouples because of the requirements described in the preceding sections. Historically, certain proprietary alloys were discovered as being useful in many industrial furnace applications. The manufacturers gave them trade names which were adopted by the industry. When several companies had similar alloys, each with its own trade name, the designation of the correct thermocouple, and the calibration curve for it, became confusing. Therefore, the Instrument Society of America assigned letter designations for various thermocouple pairs to correspond to standard EMF–temperature relationships for each pair. These letter designations were adopted by the American National Standards Institute and by the American Society for Testing and Materials. The nominal compositions for these letter designations are shown in Table 10.1.[8] The Seebeck voltage from each standard thermocouple is tabulated in standard tables as a function of temperature. It is the V_S to t relationship, not the metal composition, that is regarded as standard.

In Table 10.1 the designation of a particular alloy can be made by adding

TABLE 10.1 ISA Designations for Nominal Thermocouple Alloys (wt %)

Letter Designation	Positive Wire	Negative Wire
B	$Pt_{70}Rh_{30}$	$Pt_{94}Rh_6$
E	$Ni_{90}Cr_{10}$	$Ni_{45}Cu_{55}$
J	Fe	$Ni_{45}Cu_{55}$
K	$Ni_{90}Cr_{10}$	$Ni_{95}Mn_2Al_2Si_1$
N	$Ni_{84.5}Cr_{14.2}Si_{1.4}$	$Ni_{95.5}Si_{4.4}Mg_{0.1}$
R	$Pt_{87}Rh_{13}$	Pt
S	$Pt_{90}Rh_{10}$	Pt
T	Cu	$Ni_{45}Cu_{55}$

Source: ASTM. Used by permission.

a P or an N to the letter designation. Thus RP is the positive wire for the type R thermocouple. From the table, it is $Pt_{87}Rh_{13}$.

Some of the trade names for particular alloys are: Constantan, Cupron, and Advance for the EN, JN, and TN composition; Chromel, Tophel, T-1, and ThermoKanthal KP for the KP or EP alloy; and Alumel, Nial, T-2, and ThermoKanthal KN for the KN alloy. The AMAX Specialty Measurements Corp., formerly Wilbur B. Driver Company, manufactures Cupron, Nicrosil, Nisil, Tophel, and Nial. The Driver-Harris Company manufactures Advance, T-1, and T-2. The Hoskins Company manufactures Chromel and Alumel. The Kanthal Corporation manufactures ThermoKanthal JN, KP, and KN. The physical properties of these alloys are shown in Table 10.2.[8]

The type N thermocouple was developed as a replacement for type K thermocouples and is called Nicrosil and Nisil for the positive and negative alloys, respectively, indicating the alloy composition.[9] Nicrosil and Nisil are trademarks owned by the AMEX Specialty Metals Corporation; these products are available from several manufacturers in the United States and Europe. Publication of properties data is scheduled for 1987 for ASTM and ANSI. The density should be similar to that of KP and KN. Experiments on prototype alloys give the thermal expansions:

$$\text{NP:} \quad \alpha = \left(9.44 + 0.00633\,T + \frac{815}{T}\right)10^{-6}/\text{K}$$

$$\text{NN:} \quad \alpha = (12.2 + 0.0034\,T)10^{-6}/\text{K},$$

where $400 < T < 1100$ K. This is about 15% less than type 304 stainless steel, so type N thermocouples in mineral-insulated sheaths should be in sheaths made from Nicrosil/Nisil-type alloys.

The electrical resistivity of NP is almost constant from 80 to 400 K and is given by

$$\rho = (94.79 + 0.010732\,T)10^{-6}\ \Omega\text{-cm}.$$

The resistivity of the NN alloy is much more temperature sensitive, varying from 24×10^{-6} Ω-cm at 100 K to 38×10^{-6} at 400 K.

The thermal conductivity of the NP alloy is not very temperature sensitive, about 0.10 W/cm^2 per kelvin at 100 K, 0.12 at 250 K, and 15.5 at 400 K. The NN alloy is more sensitive and has a small discontinuity at the ferromagnetic Curie temperature just below 300 K. It varies nearly linearly from 0.145 W/cm^2 per kelvin at 100°C to 0.029 at 400 K.

The compositions and properties shown in Tables 10.1 and 10.2 are modified slightly by each manufacturer in order to produce wire of the correct EMF–temperature relationship. These are only nominal compositions. Minor additions, and the carbon content, are unspecified and vary from manufacturer to manufacturer and from lot to lot.

TABLE 10.2 Typical Physical Properties of Thermoelement Materials

Property	Thermoelement Material									
	JP	JN, EN, TN	TP	KP, EP	KN	RP	SP	RN, SN	BP	BN
Melting point (solidus temperatures)										
°C	1490	1220	1083	1427	1399	1860	1850	1769	1927	1826
°F	2715	2228	1981	2600	2550	3380	3362	3216	3501	3319
Resistivity										
μΩ-cm:										
at 0°C	8.57	48.9	1.56	70	28.1	19.0	18.4	9.83	—	—
at 20°C	9.67	48.9	1.724	70.6	29.4	19.6	18.9	10.4	19.0	17.5
Ω cmil/ft:										
At 0°C	51.5	294.2	9.38	421	169	114.3	110.7	59.1	—	—
At 20°C	58.2	294	10.37	425	177	117.7	114.0	62.4	114.5	106
Temperature coefficient of resistance, $\Omega/\Omega°C$ (0 to 100°C)	65×10^{-4}	-0.1×10^{-4}	43×10^{-4}	4.1×10^{-4}	23.9×10^{-4}	15.6×10^{-4}	16.6×10^{-4}	39.2×10^{-4}	13.3×10^{-4}	20.0×10^{-4}
Coefficient of thermal expansion, in./in.·°C (20 to 100°C)	11.7×10^{-6}	14.9×10^{-6}	16.6×10^{-6}	13.1×10^{-6}	12.0×10^{-6}	9.0×10^{-6}	9.0×10^{-8}	9.0×10^{-6}	—	—
Thermal conductivity at 100°C										
Cal-cm/s-cm²-°C	0.162	0.0506	0.901	0.046	0.071	0.088	0.090	0.171	—	—
Btu-ft/h-ft²-°F	39.2	12.2	218	11.1	17.2	21.3	21.8	41.4	—	—
Specific heat at 20°C, cal/g-°C	0.107	0.094	0.092	0.107	0.125	—	—	0.032	—	—
Density:										
g/cm³	7.86	8.92	8.92	8.73	8.60	19.61	19.97	21.45	17.60	20.55
lb/in.³	0.284	0.322	0.322	0.315	0.311	0.708	0.721	0.775	0.636	0.743
Tensile strength (annealed)										
kgf/cm²	3500	5600	2500	6700	6000	3200	3200	1400	4900	2800
psi	50,000	80,000	35,000	95,000	85,000	46,000	45,000	20,000	70,000	40,000
Magnetic attraction	Strong	None	None	None	Moderate	None	None	None	None	None

Source: ASTM. Used by permission.

TABLE 10.3 ANSI Tolerances for Thermocouples[a]

Type[d]	Temperature Range °F	Temperature Range °C	Limits of Error (°F)[b,c] Standard	Limits of Error (°F)[b,c] Special
B	1600 to 3100	871 to 1705	$\pm\frac{1}{2}\%$	
E	32 to 600	0 to 316	$\pm3°F$	$\pm2\frac{1}{4}°F$
	600 to 1600	316 to 871	$\pm\frac{1}{2}\%$	$\pm\frac{3}{8}\%$
J	32 to 530	0 to 277	$\pm4°F$	$\pm2°F$
	530 to 1400	277 to 760	$\pm\frac{3}{4}\%$	$\pm\frac{3}{8}\%$
K, N[e]	32 to 530	0 to 277	$\pm4°F$	$\pm2°F$
	530 to 2300	277 to 1260	$\pm\frac{3}{4}\%$	$\pm\frac{3}{8}\%$
R or S	32 to 1000	0 to 538	$\pm2\frac{1}{2}°F$	
	1000 to 2700	538 to 1482	$\pm\frac{1}{4}\%$	
T	−300 to −150	−184 to −101		$\pm1\%$
	−150 to −75	−101 to −59	$\pm2\%$	$\pm1\%$
	−75 to 200	−59 to 93	$\pm1\frac{1}{2}°F$	$\pm\frac{3}{4}°F$
	200 to 700	93 to 371	$\pm\frac{3}{4}\%$	$\pm\frac{3}{8}\%$

[a]In this table the limits of error for each type of thermocouple apply only over the temperature range for which the wire size in question is recommended. These limits of error should be applied only to standard wire sizes. The same limits may not be obtainable in special sizes.

[b]Where limits of error are given in percent, the percentage applies to the temperature being measured when expressed in degrees Fahrenheit. To determine the limit of error in degrees Celsius, multiply the limit of error in degrees Fahrenheit by 5/9.

[c]Limits of error apply to thermocouples as supplied by the manufacturer. The calibration of a thermocouple may change during use. The magnitude of the change depends on such factors as temperature, the length of time, and the conditions under which it was used.

[d]Type T wire cannot be expected to meet the limits of error at temperatures below the ice point unless so specified at time of purchase. Selection is usually required.

[e]The tolerances for type N were just being adopted as this book was written, and are presumed to be the same as type K.

Source: ASTM. Used by permission.

Manufacturers usually produce two grades of wire, standard and precision. The latter costs more. ANSI tolerances are shown in Table 10.3.[8]

In addition to the compositions given in Table 10.1, there are many other thermocouple materials available.[10] One is a noble metal pair manufactured by Englehard Industries, Inc., designated Platinel II (a replacement for Platinel I). It has the same temperature-to-voltage calibration as that of type K thermocouples. The positive alloy is $Au_{55}Pd_{31}Au_{14}$ (Platinel 1813). The negative alloy is $Au_{35}Pd_{65}$ (Platinel 1503). Another substitute for chromel–alumel thermocouples is the Nicrosil–Nisil thermocouple (discussed later).

Thermocouples used at very high temperatures must be very refractory. For vacuum applications to 2800°C, or in inert atmospheres, the $W_{95}Re_5$ versus $W_{74}Re_{26}$ has been popular. Other tungsten–rhenium couples include W versus $W_{74}Re_{26}$ or $W_{97}Re_3$ versus $W_{75}Re_{25}$. For graphite furnaces and other nonoxidizing situations, a thermocouple of boron-doped graphite versus graphite is commercially available that is stable to 2000°C. The boron-doped graphite is the positive leg.

There are hundreds of other thermocouple pair compositions reported in the literature, but only a few are commercially successful. One should also be aware that thermocouples from one lot or manufacturer's spool may be slightly different from another. Especially, thermocouples intended for use both above and below 0°C may have different characteristics in the different temperature ranges, so that the manufacturer should be informed which regime of temperature will be encountered, and thermocouples intended for use in both regimes should be calibrated in both.

10.6 THE SEEBECK VOLTAGE–TEMPERATURE RELATIONSHIP

If the reference junction of a thermocouple is held at 0°C, the Seebeck voltage produced by the measuring junction can be related to the temperature of the measuring junction. The Seebeck voltage–temperature relationship must be known in order to obtain the temperature of the measuring junction. Early manufacturers of proprietary thermocouples published tables of Seebeck voltage versus temperature. As more manufacturers entered the market and as better control of alloys became possible, it has been necessary to define the voltage–temperature relationship more accurately. This has been done several times by the National Bureau of Standards. The most recent thermocouple Reference Tables, corresponding to IPTS t_{68}, are given in NBS Monograph 125, March 1974,[11] and NBS Monograph 161, March 1978.[12]

The need for such tables can be illustrated by the history of type S and type R thermocouples. The first noble metal thermocouples, which were nominally platinum to platinum, 10% rhodium, had the rhodium leg contaminated by iron, which was introduced with the rhodium. Later, removal of the iron impurity produced thermocouples that had a lower Seebeck voltage for the same measuring junction temperature. Many of the early thermocouples had been sold with instruments calibrated for the higher Seebeck voltage. By increasing the pure rhodium content to 13%, the Seebeck voltage could be made to correspond to that of the impure 10% rhodium thermocouple. The 13% thermocouple was made available by manufacturers so that instruments which read temperature directly would not have to be recalibrated. Thus two similar platinum–rhodium thermocouples came into use, type R, corresponding more nearly to the original EMF–temperature relationship, and type S, corresponding to the pure Pt–$Pt_{90}Rh_{10}$ relationship.

For practical reasons the Bureau of Standards uses high-purity platinum, designated Pt_{67}, as a standard reference material. The Seebeck voltage from each type of thermocouple wire joined to Pt_{67} is determined as a function of the measuring junction temperature, with the reference junction at the ice

TABLE 10.4 Coefficients for determining $V_s(\mu V)$ or t(°C) from the power series approximation. Use Eq. (10.30) for type S and K thermocouples. Use Eq. (10.31) for type N thermocouples.

Temperature Range (°C)	a_0 Argument	Exp.	a_1 Argument	Exp.	a_2 Argument	Exp.	a_3 Argument	Exp.	a_4 Argument	Exp.	Error Range (μV) Exact-Approx.
Type S Thermocouples											
I. Quartic equation											
−50 to 900	· · · · ·	·	5.5439639	+0	1.0103667	−2	−1.0944499	−5	4.9628963	−9	−7 to 14
0 to 1100	· · · · ·	·	5.8791282	+0	7.9098118	−3	−6.7450002	−6	2.5247577	−9	−16 to 12
0 to 1400	· · · · ·	·	6.2516859	+0	5.8347856	−3	−3.4351369	−6	9.4022202	−10	−35 to 25
0 to 1650	· · · · ·	·	6.5554932	+0	4.4519908	−3	−1.6378513	−6	2.4140360	−10	−55 to 35
0 to 1768	· · · · ·	·	6.6834421	+0	3.9334084	−3	−1.0384046	−6	3.4244511	−11	−60 to 35
400 to 1100	−3.8051591	+2	8.7228147	+0	6.2984807	−4	9.0526670	−7	−2.9241601	−10	−0.7 to 0.5
400 to 1400	−5.2412524	+2	9.5827994	+0	−1.2077351	−3	2.5723104	−6	−8.3681057	−10	−1.6 to 1.5
400 to 1650	−5.0061921	+2	9.4591354	+0	−9.7986687	−4	2.3967559	−6	−7.8867971	−10	−1.8 to 1.9
1050 to 1400	1.4352322	+3	2.9873073	+0	6.9951678	−3	−1.8986036	−6	6.5006637	−11	−0.05 to 0.05
1050 to 1650	1.3054176	+3	3.4129348	+0	6.4741403	−3	−1.6163524	−6	7.9103746	−12	−0.05 to 0.05
1400 to 1550	1.8695098	+2	6.4091373	+0	3.4664812	−3	−2.7553724	−7	−2.1606150	−10	−0.05 to 0.05
1400 to 1650	1.0863331	+3	3.9952876	+0	5.8933317	−3	−1.3595782	−6	−3.4675031	−11	−0.05 to 0.05
1400 to 1768	−7.4180405	+4	2.0043202	+2	−1.8607781	−1	8.1899566	−5	−1.3556030	−8	−1.0 to 1.3
1666 to 1768	8.2703440	+4	−1.3532278	+2	8.0243878	−2	−1.0633404	−5	−1.7212343	−9	−0.05 to 0.05
II. Cubic equation											
−50 to 900	· · · · ·	·	6.1727088	+0	5.7640155	−3	−2.4622638	−6	· · · · ·	·	−30 to 60
0 to 1100	· · · · ·	·	6.5318477	+0	4.4314159	−3	−1.3617466	−6	· · · · ·	·	−51 to 31
0 to 1400	· · · · ·	·	6.7532549	+0	3.7355199	−3	−8.8324648	−7	· · · · ·	·	−70 to 35
0 to 1650	· · · · ·	·	6.7665573	+0	3.7028557	−3	−8.6548628	−7	· · · · ·	·	−70 to 35
0 to 1768	· · · · ·	·	6.7203042	+0	3.8113477	−3	−9.2098883	−7	· · · · ·	·	−65 to 35
400 to 1100	−3.0763512	+2	8.2834531	+0	1.5802448	−3	2.8242989	−8	· · · · ·	·	−0.8 to 1.0
400 to 1400	−1.3792205	+2	7.5202132	+0	2.6472719	−3	−4.3957897	−7	· · · · ·	·	−6 to 10
400 to 1650	6.2024804	+1	6.6930160	+0	3.6814923	−3	−8.3654340	−7	· · · · ·	·	−16 to 17
1050 to 1400	1.2918771	+3	3.4602987	+0	6.4119568	−3	−1.5800960	−6	· · · · ·	·	−0.05 to 0.05
1050 to 1650	1.2804330	+3	3.4888591	+0	6.3883551	−3	−1.5736364	−6	· · · · ·	·	−0.05 to 0.05

Range											Error range
1400 to 1550	1.2058568	+3	3.6416331	+3	6.2841184	−3	−1.5499505	−6			−0.05 to 0.05
1400 to 1650	1.2722747	+3	3.5056963	+3	6.3768029	−3	−1.5710024	−6			−0.05 to 0.05
1400 to 1768	9.9067391	+3	−1.3424041	+3	1.7418790	−2	−3.966469	−6			−2 to 4
1666 to 1768	9.7633500	+4	−1.7012950	+4	1.1066893	+2	−2.2451634	−5			−0.05 to 0.05

Variable reference junction correction

Range											Error range
0 to 50			5.3994446		1.2467754	+0	−1.9934168	−5			−0.01 to +0.01

III. Quadratic equation

Range											Error range
−50 to 900			7.0208414		2.7110736	−3					−90 to 110
0 to 1100			7.2387148		2.3629498	−3					−110 to 50
0 to 1400			7.4959163		2.0280528	−3					−140 to 70
0 to 1650			7.7773856		1.7310445	−3					−190 to 130
0 to 1768			7.9553026		1.5630968	−3					−260 to 180
400 to 1100	−2.9751812	+2	8.2381836		1.6438011	−3					−1.1 to 1.2
400 to 1400	−3.8977488	+2	8.5122920		1.4602609	−3					−22 to 19
400 to 1650	−6.0634145	+2	9.1029993		1.1089823	−3					−80 to 60
1050 to 1400	−1.5719586	+3	1.0540713		6.0483629	−4					−2.6 to 2.7
1050 to 1650	−2.4585270	+3	1.1994364		1.5127873	−5					−13 to 13
1400 to 1550	−3.7608363	+3	1.3754357		−5.7526222	−4					−0.2 to 0.2
1400 to 1650	−4.2764947	+3	1.4453542		−8.1196867	−4					−0.9 to 1.0
1400 to 1768	−5.7168249	+3	1.6348865		−1.4335467	−3					−9 to 7
1666 to 1768	−1.5978814	+4	2.8439331		−4.916234	−3					−0.9 to 1.0

Variable reference junction correction

Range											Error range
0 to 50			5.4231535		1.0940881	+0					−0.1 to +0.2

Type K Thermocouples

I. Quartic equation

Range											Error range
−270 to 0			3.9575518	+1	3.1063355	−2	−9.1607995	−5	3.0006628	−8	−1.1 to 1.2
−200 to 0			3.9478446	+1	2.8256412	−2	−1.1488433	−4	−2.8153447	−8	−0.08 to 0.05
−200 to 800			3.6762217	+1	2.4544587	−2	−4.3081993	−5	2.5127588	−8	−180 to 200
−20 to 500			4.0999640	+1	−3.2619221	−3	8.5714137	−6	−1.6912373	−9	−25 to 45
0 to 400			4.0981103	+1	−1.5992510	−4	−1.2525700	−5	3.2784725	−8	−25 to 20
0 to 1370			3.9443859	+1	5.8953822	−3	−4.2015132	−6	1.3917059	−10	−60 to 110

TABLE 10.4 (cont.)

Temperature Range (°C)	a_0 Argument	Exp.	a_1 Argument	Exp.	a_2 Argument	Exp.	a_3 Argument	Exp.	a_4 Argument	Exp.	Error Range (μV) Exact–Approx.
400 to 1000	1.3223524	+3	3.0191663	+1	2.7508912	−2	−2.4734437	−5	6.9799332	−9	−0.9 to 1.4
600 to 1370	−3.5456236	+1	3.8349319	+1	9.9993329	−3	−8.7444446	−6	1.7108618	−9	−12 to 11
600 to 800	2.1326086	+3	2.5608012	+1	3.7091744	−2	−3.3517324	−5	9.9607405	−9	−0.05 to 0.07
850 to 1000	−9.0373549	+2	4.0577145	+1	9.5092149	−3	−1.0989249	−5	3.0753213	−9	−0.05 to 0.03
1050 to 1150	−2.5972816	+3	5.2075276	+1	−1.4576419	−2	9.4854151	−6	−3.1178779	−9	−0.05 to 0.05
II. Cubic equation											
−270 to 0	· · · · · ·	·	3.9458846	+1	2.8553429	−2	−1.0733611	−4	· · · · · ·		−3 to 2
−200 to 0	· · · · · ·	·	3.9523013	+1	2.9550516	−2	−1.0394181	−4	· · · · · ·		−0.4 to 0.5
−200 to 800	· · · · · ·	·	3.5636003	+1	1.9650485	−2	−1.5602926	−5	· · · · · ·		−250 to 400
−20 to 500	· · · · · ·	·	4.0960115	+1	−2.7895730	−3	6.9465824	−6	· · · · · ·		−25 to 45
0 to 400	· · · · · ·	·	4.1405218	+1	−6.2431359	−3	1.3032352	−5	· · · · · ·		−35 to 25
0 to 1370	· · · · · ·	·	3.9514688	+1	5.5954388	−3	−3.8312918	−6	· · · · · ·		−65 to 110
400 to 1000	−5.6641071	+1	3.8903709	+1	7.6011228	−3	−5.1867039	−6	· · · · · ·		−8 to 8
400 to 1370	−7.8572408	+2	4.2404631	+1	2.3321601	−3	−2.6773698	−6	· · · · · ·		−25 to 18
600 to 800	−2.1824276	+2	3.9168943	+1	7.8581167	−3	−5.6043952	−6	· · · · · ·		−0.14 to 0.15
850 to 1000	−3.1416568	+3	5.0285062	+1	−6.2656016	−3	3.9088120	−7	· · · · · ·		−0.04 to 0.03
1050 to 1150	1.9581711	+3	3.5492848	+1	8.0515773	−3	−4.2332477	−6	· · · · · ·		−0.05 to 0.05
Variable reference junction											
0 to 50	· · · · · ·	·	3.9448872	+1	2.4548362	−2	−9.0918433	−5	· · · · · ·		−0.06 to +0.14
III. Quadratic equation											
−270 to 0	· · · · · ·	·	4.2845408	+1	6.8763665	−2	· · · · · ·		· · · · · ·		−80 to 100
−200 to 0	· · · · · ·	·	4.1360116	+1	5.8725165	−2	· · · · · ·		· · · · · ·		−30 to 35
−200 to 800	· · · · · ·	·	3.7665390	+1	5.7515653	−3	· · · · · ·		· · · · · ·		−600 to 1500
−20 to 500	· · · · · ·	·	4.0201888	+1	2.0594884	−3	· · · · · ·		· · · · · ·		−40 to 70
0 to 400	· · · · · ·	·	4.0501566	+1	9.8960972	−4	· · · · · ·		· · · · · ·		−40 to 45
0 to 1370	· · · · · ·	·	4.2625178	+1	−1.6847755	−3	· · · · · ·		· · · · · ·		−450 to 350

Temperature Range (°C)	a_1' Argument	Exp.	a_2' Argument	Exp.	a_3' Argument	Exp.	a_4' Argument	Exp.	Error Range (°C)
400 to 1000	−1.6032853	+3	4.6193308	+1	−3.2869106	−3	. . . . . .	—	−40 to 50
400 to 1370	−2.2148965	+3	4.8187566	+1	−4.7474730	−3	. . . . . .	—	−90 to 100
600 to 800	−2.1041329	+3	4.7343554	+1	−3.8942290	−3	. . . . . .	—	−1.6 to 1.6
850 to 1000	−2.8338815	+3	4.9283561	+1	−5.1810406	−3	. . . . . .	—	−0.06 to 0.06
1050 to 1150	−3.6676410	+3	5.0851682	+1	−5.9181402	−3	. . . . . .	—	−0.17 to 0.15

Variable reference junction correction

Temperature Range (°C)	a_1' Argument	Exp.	a_2' Argument	Exp.	a_3' Argument	Exp.	a_4' Argument	Exp.	Error Range (°C)
0 to 50	3.9557007	+	1.7584397	−2	. . . . . .	—	. . . . . .	—	−0.4 to +1.0

Type N Thermocouples (AWG 28)

I. Quartic equations

Temperature Range (°C)	a_1' Argument	Exp.	a_2' Argument	Exp.	a_3' Argument	Exp.	a_4' Argument	Exp.	Error Range (°C)
−200 to 0	3.143455	−2	−4.375508	−6	—	—	—	—	−5 to 4
0 to 50	3.833763	−2	−8.255147	−7	—	—	—	—	<0.01
0 to 400	3.645264	−2	−4.480891	−7	—	—	—	—	−2 to 3

II. Cubic equations

Temperature Range (°C)	a_1' Argument	Exp.	a_2' Argument	Exp.	a_3' Argument	Exp.	a_4' Argument	Exp.	Error Range (°C)
−200 to 0	4.262622	−2	4.276051	−6	1.520918	−9	—	—	−1.5 to 1.3
−200 to 400	4.074915	−2	−1.503463	−6	5.816619	−11	—	—	−10 to 8
0 to 50	3.826460	−2	−6.512970	−7	−9.331708	−11	—	—	<0.01
0 to 400	3.812049	−2	−8.711717	−7	2.396423	−11	—	—	−0.3 to 0.3

III. Quartic equations

Temperature Range (°C)	a_1' Argument	Exp.	a_2' Argument	Exp.	a_3' Argument	Exp.	a_4' Argument	Exp.	Error Range (°C)
−200 to 0	3.523634	−2	−6.080743	−6	−2.748231	−9	−5.396201	−13	−0.4 to 0.7
−200 to 400	3.920032	−2	−1.650547	−6	1.469131	−10	−5.394067	−15	−7 to 6
0 to 50	3.823891	−2	−5.403608	−7	−2.328326	−10	5.340190	−14	<0.01
0 to 400	3.857711	−2	−1.080822	−6	5.182122	−11	−1.122986	−15	−0.06 to 0.05

Type N Thermocouples (AWG 14)

I. Quartic equation

Temperature Range (°C)	a_1' Argument	Exp.	a_2' Argument	Exp.	a_3' Argument	Exp.	a_4' Argument	Exp.	Error Range (°C)
0 to 50	3.860831	−2	−9.570146	−7	—	—	—	—	<0.01
0 to 400	3.656723	−2	−4.566896	−7	—	—	—	—	−1.7 to 2.5

TABLE 10.4 (cont.)

Temperature Range (°C)	a'_1		a'_2		a'_3		a'_4		Error Range (°C)
	Argument	Exp.	Argument	Exp.	Argument	Exp.	Argument	Exp.	
450 to 1000	3.141413	−2	−1.105032	−7	—	—	—	—	−5 to 8
850 to 1000	2.873831	−2	−3.054420	−8	—	—	—	—	−3 to 4
1000 to 1300	2.820624	−2	−1.847854	−8	—	—	—	—	−1 to 2
II. Cubic equation									
0 to 50	3.861843	−2	−9.812299	−7	1.300932	−11	—	—	<0.01
0 to 400	3.825962	−2	−8.862736	−7	2.434387	−11	—	—	−0.3 to 0.3
450 to 1000	3.445072	−2	−3.482337	−7	4.397892	−12	—	—	−0.6 to 1.0
850 to 1000	3.240401	−2	−2.192067	−7	2.382986	−12	—	—	−0.1 to 0.1
1000 to 1300	3.226729	−2	−2.127340	−7	2.304693	−12	—	—	−0.01 to 0.01
III. Quartic equation									
0 to 50	3.861153	−2	−9.513633	−7	−2.466762	−11	1.446443	−14	<0.01
0 to 400	3.868881	−2	−1.083431	−6	5.055108	−11	−1.056823	−15	−0.02 to 0.02
450 to 1000	3.593932	−2	−5.313789	−7	1.158359	−11	−9.053322	−17	0.1 to 0.1
850 to 1000	3.308804	−2	−2.726068	−7	3.756099	−12	−1.163691	−17	−0.05 to 0.04
1000 to 1300	3.184042	−2	−1.819054	−7	1.570465	−12	5.823682	−18	<0.01

point. The tables in the NBS Monographs give the Seebeck voltage relative to platinum, measured with an accuracy of 0.01 μV. Also presented are the Seebeck coefficient and its derivative. From these data reference tables for types S, R, B, E, J, K, N, and T thermocouples are presented at 1° intervals for the Seebeck voltage, Seebeck coefficient, and the first derivative of the Seebeck coefficient. Compact tables for the Seebeck voltage of the standard thermocouples are also presented for both °C and °F, both as voltage versus temperature and temperature versus voltage. Approximation formulas are presented in the form

$$V_S = a_0 + a_1 t + a_2 t^2 + a_3 t^3 + a_4 t^4 \tag{10.30}$$

or

$$t = a_0' + a_1' V_S + a_2' V_S^2 + a_3' V_S^3 + a_4' V_S^4. \tag{10.31}$$

The formulas for types K, S, and N, and their coefficients over selected intervals of temperature or voltage, are given in Table 10.4.[11,12]

The research producing type N thermocouples revealed that the Seebeck voltage relationships depend on wire size, presumably because cold-working and oxidation processes depend on wire size.[12] (This is believed to be true of all commercial base-metal thermocouples, but has not been documented.) Therefore, the power series for type N thermocouples, Nicrosil versus Nisil, was given for 28-gauge thermocouples at low temperatures, where oxidation is not very important, and for 14-gauge thermocouples for the higher temperatures. Also, type N thermocouples that perform satisfactorily at temperatures above 0°C may not do so below 0°C, and vice versa. Thermocouples intended for both cryogenic and high-temperature regions must be calibrated across the entire temperature span.

The power series for highest accuracy presented in the Bureau of Standards monographs gives values to 0.01 μV. They require series expansions as high as thirteenth degree and very powerful computing facilities. Errors are estimated for 12-, 16-, 24-, 27-, and 36-bit calculations. Errors are often excessive at less than 24-bit processing. The equations for Table 10.4 are the reduced-accuracy equations that typically give errors of less than 0.05% of the absolute temperature, but no error estimates for reduced bit processing are presented. The user is cautioned to test for rounding-off errors in using the equations. Tables for the Seebeck voltages of ANSI thermocouples are given in the Appendix.

The thermoelectric voltage of each thermocouple element referred to platinum is shown in Figure 10.3.[8] From this figure it is possible to estimate the Seebeck voltage to be obtained from any combination of the various thermoelectric materials. It is also apparent that both positive and negative deviations from linearity exist. Figure 10.4, constructed from Fig. 10.3, gives the thermoelectric voltages from each of the standard types of thermocouples.

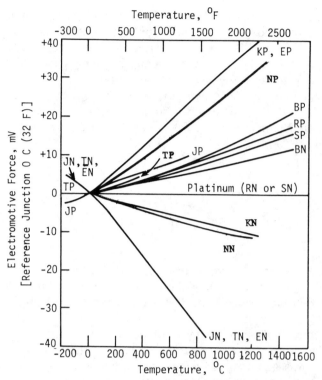

Figure 10.3. Thermal EMF of thermoelements relative to platinum. (Adapted from ASTM Committee E20, "Manual on the Use of Thermocouples in Temperature Measurement," STP470A, 1974, with permission from the American Society for Testing and Materials, 1916 Race Street, Philadelphia, PA 19103.)

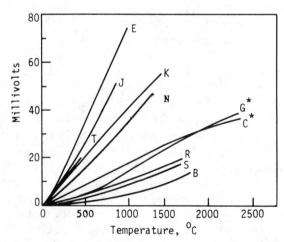

Figure 10.4. Seebeck voltage of various thermocouples. $G = W - W_{74}Re_{26}$; $C = W_{95}Re_5 - W_{74}Re_{26}$ (not ANSI symbols).

10.7 ABSOLUTE THERMOELECTRIC POWER OF COMMERCIAL THERMOCOUPLE ALLOYS

When we consider the commercial thermocouples, represented by types B, S, R, J, T, K, N, and E, we find that all except one composition can be classified as dilute transition elements, for which Eq. (10.29) gives the thermoelectric power.[7] The lone exception is the JN, TN, EN alloy, which is considered later.

For the other commercial compositions the term $(E_0 - E_f)$ in the denominator of Eq. (10.28) is controlled by alloying. Typical densities of state for the dilute noble metal series are as shown in Fig. 10.5, where alloying changes E_f, but not E_0. Conduction in the d states by hole conduction can be modified by alloying with a higher or lower valence metal. Adding a metal with more electrons per atom than the solvent decreases $(E_f - E_0)$ and makes the absolute thermoelectric power more positive. Adding a metal with fewer electrons per atom makes the absolute thermoelectric power more negative. The addition of Rh to Pt (types B, S, R), of Mn, Si, and Al to Ni (type KN), and of Cr to Ni (type KP) increases absolute thermoelectric power (Fig. 10.6).[7] Note that all have about the same slope as that required for nearly linear response.

Concentrated solid solutions of transition element thermocouples alloyed with classical metals, such as copper, have more complex density of states because both E_0 and E_f are modified as E_f approaches E_0. For the addition of copper to nickel, the densities of state are as shown in Fig. 10.7.[7]

Additions of Cu to Ni contribute electrons to fill the d orbitals until, at about 55 to 60% Cu, all d states are filled, and the absolute thermoelectric power reaches a maximum [when $(E_f - E_0)$ reaches a minimum]. This is a broad maximum, so that precise control of the Cu–Ni ratio is not required (Fig. 10.8). When $(E_f - E_0)$ is small, thermal energy can excite an electron from the d to the S bands [when $(E_f - E_0) \approx kT$], where k is Boltzmann's constant. Then $E_0 - E_f$ becomes a linear function of kT,

$$E_0 - E_f = B + m(E)kT . \tag{10.32}$$

Experimentally, the slope $m(E)$ is found to be 1.44 to 1.46, close to the theoretical value of $\frac{3}{2}kT$. Addition of a third alloying element has the effect of increasing $E_0 - E_f$ and reducing the temperature dependence of $E_0 - E_f$. Empirically, adding small amounts (usually <0.5 atomic percent) of one or more alloying elements to copper–nickel alloys is an effective way of controlling the absolute thermoelectric power (Fig. 10.9).[7]

Thermocouple alloy compositions are selected on the basis of other physical properties after their Seebeck performance considerations are met. The ideal thermocouple would have an infinite melting temperature and would be completely inert. Real alloys have many limitations. Maximum use temperature, ease of wire drawing, oxidation resistance, freedom from

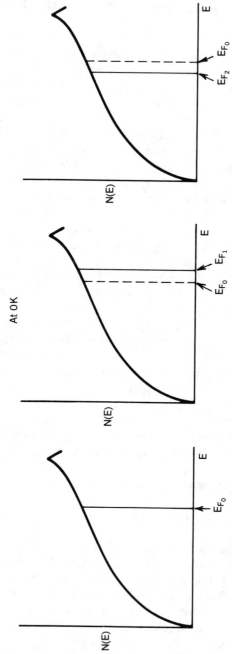

Figure 10.5. Schematic representation of the densities of states for dilute alloys of noble metals. (From D. D. Pollock, "Theory and Properties of Thermocouple Elements," STP492, 1971, with permission from the American Society for Testing and Materials, 1916 Race Street, Philadelphia, PA 19103.)

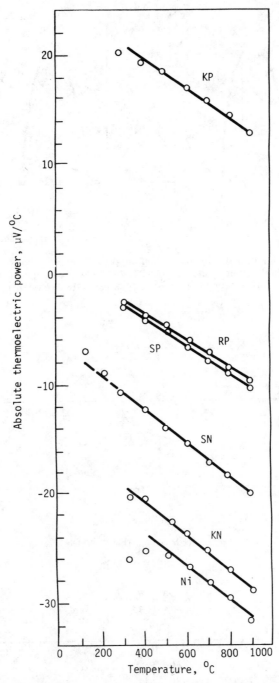

Figure 10.6. Absolute thermoelectric power for nickel and types S, R, and K thermoelements. (From D. D. Pollock, "Theory and Properties of Thermocouple Elements," STP492, 1971, with permission from the American Society for Testing and Materials, 1916 Race Street, Philadelphia, PA 19103.)

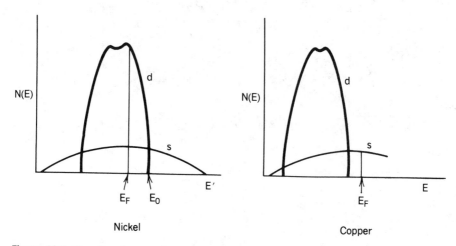

Figure 10.7. Density of states for nickel and copper. (From D. D. Pollock, "Theory and Properties of Thermocouple Elements," STP492, 1971, with permission from the American Society for Testing and Materials, 1916 Race Street, Philadelphia, PA 19103.)

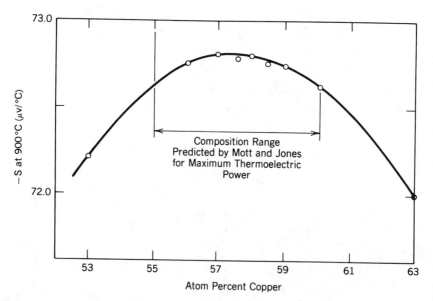

Figure 10.8. Absolute thermoelectric power of copper–nickel alloys at 900°C. (From D. D. Pollock, "Theory and Properties of Thermocouple Elements," STP492, 1971, with permission from the American Society for Testing and Materials, 1916 Race Street, Philadelphia, PA 19103.)

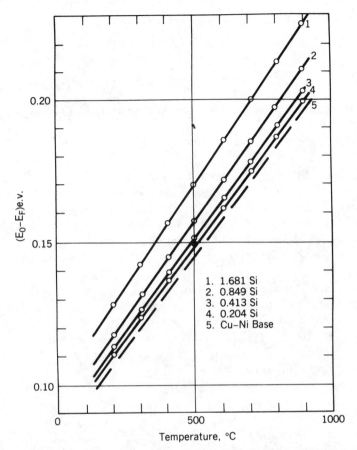

Figure 10.9. Effect of adding a typical third element, Si, to a $Cu_{60}Ni_{40}$ base alloy. (From D. D. Pollock, "Theory and Properties of Thermocouple Elements," STP492, 1971, with permission from the American Society for Testing and Materials, 1916 Race Street, Philadelphia, PA 19103.)

chemical reactions with common furnace materials and atmospheres, freedom from phase changes, and so on, are important considerations. The established compositions are ones that have proved useful in real industrial and laboratory situations. Many nonstandard alloys are available. Kinzie gives 194 different thermocouples and their EMF characteristics.[10] Many are available as proprietary alloys. New alloys are continuously being developed.

10.8 PRACTICAL THERMOCOUPLE LAWS

In the use of thermocouples, certain "laws" must be considered in making measurements. There are three such laws which are very helpful in real situations.

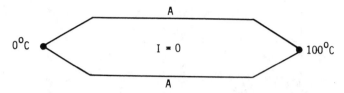

Figure 10.10. If both legs are the same composition, no net Seebeck potential exists and no current will flow.

1. The Law of the Homogeneous Circuit. This is an empirical law which says that no temperature variation in a homogeneous circuit will produce a net voltage, that is, will sustain a current (Fig. 10.10). Changes in cross-sectional area and temperature distribution have no effect. Any voltage that does appear, as the result of a temperature variation in a wire that is supposed to be homogeneous, is evidence that the wire is actually in-homogeneous.

2. The Law of Intermediate Conductors at Constant Temperature. This is a fundamental law that says no matter what dissimilar conductors are in contact, no net voltage will appear if the temperature of all the junctions is the same (Fig. 10.11). This is a consequence of the second law of ther-modynamics because, if such a potential existed, work could be extracted without a temperature difference.

3. The Law of Successive Temperature. This is a fundamental law which says that the Seebeck voltage from a lower temperature to a higher temperature is equal to the sum of the Seebeck voltage from the lower temperature to any intermediate temperature and the Seebeck voltage from that intermediate temperature to the higher temperature (Fig. 10.12). Mathematically, we have

$$\int_{T_1}^{T_3} \alpha_S \, dT = \int_{T_1}^{T_2} \alpha_S \, dT + \int_{T_2}^{T_3} \alpha_S \, dT , \qquad (10.33)$$

where T_2 is any intermediate temperature. The law of successive tem-

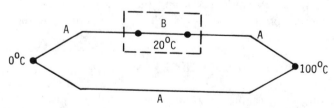

Figure 10.11. Both *AB* junctions at the same temperature. No Seebeck voltage. No current.

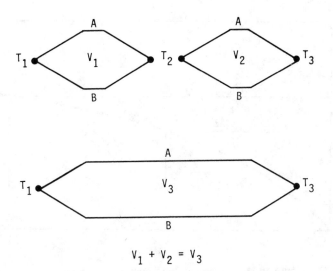

$$V_1 + V_2 = V_3$$

Figure 10.12. Illustration of the law of successive temperatures.

peratures is very important because α_S is a function of T and we can use this law to compensate for changes in reference junction potentials.

10.8.1 Reference Junction Corrections

For convenience, most thermocouples are calibrated using the ice point as the reference junction. In practice, it may be more convenient to use room temperature, or some other temperature, as the reference junction. By applying the law of successive temperatures we can use the standard calibration tables even though the reference junction was not at the ice point. An example will make this clear. Consider a copper–constantan calibration curve shown in Fig. 10.13 and with the curvature exaggerated for clarity.

From this calibration curve it is easy to see that if the measuring junction were at 300°C and the reference junction were at 20°C, the thermocouple voltage would be less than if the reference junction were at 0°C. In fact, V_{measured} would be $V_{300} - V_{20}$. Therefore, using 20°C as a reference junction is permissible because the Seebeck voltage for a reference junction of 0°C can be calculated by adding $V_{20} - V_0$ to it. Thus

$$V_{300} - V_0 = (V_{300} - V_{20}) + (V_{20} - V_0). \tag{10.34}$$

Now if we measure V_x and want to calculate the temperature, using a 20°C reference junction, we can determine t_x from the calibration curve by

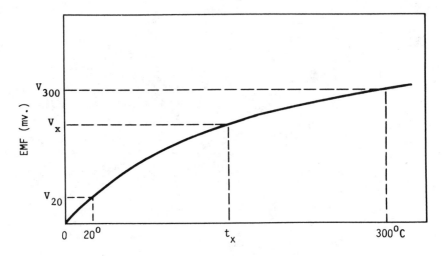

Figure 10.13. Schematic thermocouple calibration curve.

the calculation

$$V_x - V_0 = (V_x - V_{20}) + (V_{20} - V_0) .   (10.35)$$

The first term on the right side is the experimentally measured potential. The second we can look up in the calibration tables (or from our calibration curve). Note that because V_S is not linear in t, we must make *voltage corrections. Never add or subtract temperatures* in making reference junction corrections.

Most calibration curves are provided by the Bureau of Standards in table form. For example, consider the copper–constantan table (see Appendix). Suppose that t_x were 300°C. Then

$$V_{measured} = V_x - V_{20} = 14.06 \text{ mV}$$

Adding the correction $V_{20} - V_0 = 0.789 \text{ mV}$, we obtain $V_x - V_0 = 14.78 \text{ mV}$. From the table this is 298.5°C.

Suppose, instead, that we had looked up the temperature corresponding to 14.06 mV without allowing the cold-junction correction. This is 286°C. Now suppose that we added 20°C to this, to obtain 306°C. This answer is obviously wrong. But it is the most common error made in using thermocouples. Again: In making reference junction corrections, *never add or subtract temperatures.*

10.9 ANALYSIS OF THERMOELECTRIC POTENTIALS

The practical thermocouple laws can be used to analyze the voltages appearing in a thermocouple circuit. This is done most conveniently using Kirchhoff's first law of electrical circuits, which states: "The summation of EMFs around an electric circuit is equal to zero." When current flows, this includes the voltage drops for all the impedances. When current does not flow and the voltages are all thermoelectric,

$$\sum_i V_{S_i} + V_{oc} = 0, \tag{10.36}$$

where V_{S_i} is the Seebeck voltage at each junction and V_{oc} is the open-circuit voltage. For example, in the typical thermocouple circuit (Fig. 10.14) there are three junctions, each of which could have a Seebeck voltage. These are Cu–KP at 0°C, KP–KN at 100°C and KN–Cu at 0°C. Then if we sum the EMFs clockwise around the circuit, we obtain

$$V_{oc} + V_{\underset{0°C}{\text{Cu-KP}}} + V_{\underset{100°C}{\text{KP-KN}}} + V_{\underset{0°C}{\text{KN-Cu}}} = 0,$$

so

$$-V_{oc} = V_{\underset{0°C}{\text{Cu-KN}}} + V_{\underset{100°C}{\text{KN-KP}}} + V_{\underset{0°C}{\text{KP-Cu}}}.$$

Each of the thermocouple junctions produces a thermoelectric voltage. However, if the reference junction is at zero degrees, then having the measuring junction at zero degrees allows us to apply the law of intermediate conductors to

$$V_{\underset{0°C}{\text{Cu-KN}}} \quad \text{and} \quad V_{\underset{0°C}{\text{KP-Cu}}},$$

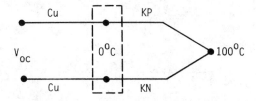

Figure 10.14. Typical thermocouple circuit.

and these two Seebeck voltages are zero. Then

$$-V_{oc} = V_{\substack{KP-KN, \\ 100°C}},$$

and the right side is tabulated in standard thermocouple tables. We see that for this circuit, the open circuit voltage must equal the value 4.095 mV found in the type K thermocouple tables (see the Appendix).

In this example let us assume that we connected a source of an opposing voltage at V_{oc} and that it is made of copper. Then, even if the temperature is not 0°C at the potentiometer terminals, since we have Cu connected to Cu, the law of homogeneous circuits tells us that no thermoelectric EMF will be generated at the Cu–Cu terminals. If we adjust the voltage source to produce a potential equal and opposite to the sum of the other sources of potential, current will not flow. Then we can measure the applied voltage necessary to prevent current flow. We will call this V_m. Obviously, because our source of voltage is $E_{KP-KN,100°C}$ the potential that we must apply to prevent current flow must be equal but opposite in polarity to $V_{KP-KN,100°C}$; that is, $V_m = -V_{oc} = 4.095$ mV. (A millivolt potentiometer can be used to produce this opposing voltage. It will be described later.)

Suppose, now that our circuit is modified as shown in Fig. 10.15. If we again connect a source of opposing voltage, we want to know what value it must have to prevent current flow. Numbering the junctions as shown, we see that junctions 1 and 5 for Cu–Cu at 20°C are not really junctions. They produce no Seebeck voltage V. Then the real junctions are 2, 3, and 4. They must be balanced by an applied voltage, V_m, so applying Kirchhoff's law in a clockwise circuit gives

$$V_m = V_{\substack{Cu-KP \\ 20°C}} + V_{\substack{KP-KN \\ 100°C}} + V_{\substack{KN-Cu}{20°C}}.$$

We notice that the junctions at 2 and 4 are at the same temperature, and they have a common connecting wire, copper, so the law of intermediate conductors at a common temperature tells us that the copper component should not produce a voltage. Yet we see that V_{Cu-KP} and V_{KN-Cu}, both at 20°C, are real junctions at temperatures above the reference temperature,

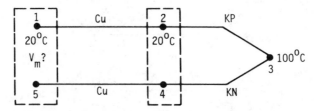

Figure 10.15. Thermocouple circuit with reference junction at 20°C.

$0°C$. A potential must exist because of the fact that KP and KN are different, that

$$V_{\substack{Cu-KP \\ 20°C}} \neq V_{\substack{KN-Cu \\ 20°C}}.$$

We have been using a shorthand notation to specify each junction. We see that we could combine the junction potentials if we write them in this order,

$$V_{\substack{KN-Cu \\ 20°C}} + V_{\substack{Cu-KP \\ 20°C}} = V_{\substack{KN-KP \\ 20°C}}.$$

We can combine the potentials to get a single potential because the law of intermediate conductors tells us that the copper does not contribute to the Seebeck voltage. It is as though the circuit were actually as shown in Fig. 10.16.

In summing potentials around a circuit, potentials are reversed if the direction is changed. That is,

$$V_{\substack{KP-KN \\ 100°C}} = -V_{\substack{KN-KP \\ 100°C}}.$$

In our circuit, then,

$$V_m = V_{\substack{KN-KP \\ 20°}} + V_{\substack{KP-KN \\ 100°C}}$$

$$= -V_{\substack{KP-KN \\ 20°C}} + V_{\substack{KP-KN \\ 100°C}}.$$

We can look up V_{KP-KN} at $20°C$ in the tables. It is $1.203\,mV$. So

$$V_m = -1.203 + 4.095$$

$$= 2.892\,mV.$$

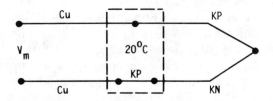

Figure 10.16. Alternative circuit for Fig. 10.15.

We see that our analysis agrees with what we learned in the preceding section. We have verified the law of intermediate temperatures and have learned how to analyze circuits. This method of analysis is extremely useful in many real situations. By applying it, we can often determine the correct temperature from unusual circuits and sometimes can use standard tables to determine the output of nonstandard thermocouples.

10.10   THE IMPORTANCE OF TEMPERATURE GRADIENTS

Because the Seebeck voltage depends on the temperature difference between the measuring and reference junctions, it is apparent that the Seebeck voltage depends on the gradients in the absolute thermoelectric powers of the two metals making up the thermocouple.[13] The net Seebeck voltage for a thermocouple of metal A with metal B, with the reference junction at the ice point, can be determined from the Seebeck coefficients of the individuals. With the measuring junction at t,

$$V_S = \int_0^t \alpha_A + \int_t^0 \alpha_B .$$ (10.37)

In Section 10.3 we wrote this in the equivalent form,

$$V_S = \int_0^t (\alpha_A - \alpha_B) .$$

Equation (10.37) emphasizes the fact that it is the temperature difference for each material which produces the Seebeck voltage. That voltage depends on the absolute thermoelectric power of the two materials making up our voltage when subjected to the thermal gradient between zero and t. We can write Eq. (10.37) in an alternative form, for wires of length L_A and L_B,

$$V_S = \int_0^{L_A} \alpha_A \frac{dt}{dx} dx + \int_{L_B}^0 \alpha_B \frac{dt}{dx} dx .$$ (10.38)

This alternative form emphasized the importance of the temperature gradient on the Seebeck voltage. Suppose that $L_A = L_B = 10$ ft. But suppose that both wires are fed through a furnace wall, so that the entire temperature change takes place over a distance of only $\frac{1}{10}$ ft. Inside the furnace, the temperature is uniformly hot at temperature t, and outside the furnace the temperature is uniform at the ice point. Then for each wire, the entire temperature change takes place in only $\frac{1}{10}$ ft. Because no Seebeck voltage is produced if no gradient exists, the uniform-temperature regions inside and outside the furnace do not participate in generating the Seebeck voltage (because in those regions dt/dx is zero). We see from Eq. (10.38) that the Seebeck voltage depends only on the absolute thermoelectric power of materials A and B in that short, $\frac{1}{10}$-ft section, not on α_A and α_B for the whole length L. This is important because it shows us that the homogeneity of the wire *in the thermal gradient* is what is important. Any inhomogeneity, any contamination or any physical structure changes (e.g., order–disorder transformations) that occur in the parts of the thermocouples exposed to the thermal gradient will affect the Seebeck voltage. This is the reason that a type K thermocouple often gives different EMFs if, after use for a period of time, the thermocouple is moved farther into, or out of, the furnace. Then the thermal gradient in the furnace wall comes at a new position, and if the gradient has modified the structure, the values of α_A or α_B will be different.

For precision work, one must be very careful about inhomogeneities within the thermal gradient. Even if a measuring junction is located in a uniform-temperature region of a calibration furnace, the calibration may not be accurate if the thermocouple, when used to measure temperature in another furnace, does not have the temperature gradient at the same place on the thermocouple. No thermocouple wire is perfectly homogeneous, and local variations in homogeneity will affect the Seebeck voltage if they occur within the thermal gradient. One way to minimize the effect of variations in homogeneity over short distances is to deliberately increase the distance over which dt/dx occurs. This can be accomplished with thermal insulation and by binding a good thermal conductor into the thermocouple cable. When precision is required, annealing, tests for inhomogeneity, and prevention of contamination become important.

10.11 THE CONNECTING WIRES

Thermocouples may be connected directly to the measuring instrument by extending them from the furnace to the instrument. This is the usual case for inexpensive thermocouple materials such as types J, K, T, and E. Sometimes the thermocouple in the furnace is many feet from the instrument, so the manufacturers sell less expensive grades of thermocouple wire to be used to extend from the thermocouple to the instrument. These grades have less precise EMF characteristics, but because the temperature is near room temperature the error may not be significant. These extension wires have an X added to their letter designation. Thus type JPX would be the designation for extension wire for the positive component of a type J thermocouple.

Noble metals and other expensive or easily oxidized alloys, such as tungsten–rhenium compositions, are often used with extension connecting wires. Special wire compositions are used which, at the low temperatures expected at the terminals of the thermocouple, have approximately the

TABLE 10.5 Extension Wires for Common Thermocouples[a]

Thermocouple Type	Extension Type	Alloy Type Positive Element	Alloy Type Negative Element	Temperature Range (°F)	Limits of Error (°F) Standard	Limits of Error (°F) Special	Magnetic Response[b] P	Magnetic Response[b] N
Base metal	Category 1							
E	Ex	NiCr (Chromel)[c]	Constantan	32–400	±3	—	O	O
J	JX	Iron	Constantan	32–400	±4	±2	M	O
K	KX	NiCr (Chromel)[c]	NiAl (Alumel)[c]	32–400	±4	—	O	M
T	TX	Copper	Constantan	−75–200	±1$\frac{1}{2}$	±$\frac{3}{4}$	O	O
Noble metal	Category 2							
R	SX	Copper	Copper alloy	75–400	±12	—	O	O
S	SX	Copper	Copper alloy	75–400	±12	—	O	O
B	BX	Copper	Copper	32–200	—	—	O	O

[a]Type N thermocouples require NPX and NNX elements, presumably with the same limits as type K. NNX is ferromagnetic.
[b]M denotes ferromagnetic alloy: O denotes nonferromagnetic alloy.
[c]Former (proprietary) designation.
Source: ASTM. Used by permission.

same EMF–temperature characteristics as the thermocouple. They are often described as compensating extension wires. They, too, are given the X designation to indicate their extension function; for example, SPX is the designation for the positive compensating extension wire for a type S thermocouple. If the measuring junction is at 1300°C, and outside the furnace the thermocouple is connected to compensating extension wire at 150°C, which connects it to the instrument at 20°C, the Seebeck voltage can be made the same as it would be if the thermocouple itself were long enough to reach the measuring instrument.

Because their Seebeck voltages are almost the same at low temperatures, type S and R use the same compensating extension wires, copper for the positive component and alloy No. 11 for the negative component. Compositions of some common extension wires are shown in Table 10.5.[8]

10.11.1 EMF Relationships

Actually, the Seebeck characteristics of compensating extension wires are never exactly the same as those of the thermocouple. The most common situation in which compensating extension wires are used is that of a type S thermocouple, connected just outside the furnace to SNX and SPX wires which are connected to the measuring instrument at a remote location. Usually, the SPX wire is pure copper and the SNX wire is an alloy of copper and a small amount of nickel. By adjusting the amount of nickel in the negative leg, it is possible to produce nearly exact compensation at a single temperature, usually 100 to 150°C. If both the positive and negative junctions of the compensating wire to the thermocouple are at that temperature, compensation is achieved. If either or both junctions are at a different temperature, an error, usually small, will exist. If the junction between the extension wires and the thermocouple is unusually hot, considerable error may exist and excessive oxidation of the compensating wire may introduce additional error.

The EMF relationships can be shown schematically. Suppose that we have a hot junction at a temperature t_1 (Fig. 10.17). If we imagine that we connect a potentiometer to the thermocouple right at the hot junction, both the hot junction and the reference junction will be at the same temperature, and the voltage we measure will be zero. Now suppose that we connect the potentiometer at t_2. Each of the thermocouple legs will have some absolute thermoelectric power, and the difference will give the Seebeck voltage that we will measure. Now if we move the potentiometer to the ice-point temperature, t_0, we will measure the total Seebeck voltage, $V_{AO} + V_{BO}$, expected from such a thermocouple and tabulated in standard thermocouple reference tables with reference junction at zero degrees. By connecting compensating extension wire to the thermocouple at t_2, even though we have different absolute thermoelectric powers for AX and BX, the Seebeck voltage will be the same if $V_{AXO} + V_{BXO} = V_{AO} + V_{BO}$. If we

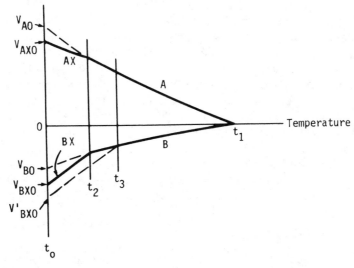

Figure 10.17. Compensating extension wire Seebeck relationships.

have adjusted the alloy to achieve that effect, the measured voltage will allow us to determine the temperature, t_1. Then if the temperature of the B–BX junction shifts to t_3, an error will occur because the BXO voltage will shift from V_{BXO} to V'_{BXO}. The error will be the difference.

10.11.2   Three-Leg Compensating Extension Wires

An ingeneous method of adjusting the compensating extension wire to improve compensation has been devised.[14] It makes use of two wires connected in parallel as the extension wires for the negative leg of a type S or type R thermocouple, plus a single wire for the positive leg to produce a "three-leg" system (Fig. 10.18). This system requires that at any tem-

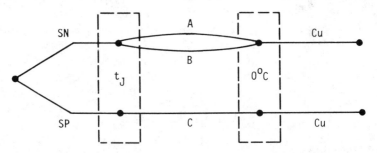

Figure 10.18. Typical three-leg compensating extension wire system.

TABLE 10.6 Composition of Three-Wire Extensions

Type	Leg	Chrome	Nickel	Iron
S	Negative, A	16.5 ± 0.5	20 ± 1.0	Balance
	Negative, B	23.5 ± 1.0	20 ± 1.0	Balance
R	Negative, A	15.0 ± 0.5	12.0 ± 0.5	Balance
		22.0 ± 1.0	12.0 ± 0.5	Balance
R, S	Positive, C	22.0 ± 1.0	Balance	10.0 ± 0.5

perature t_j for which compensation is to be made,

$$V_{AC} \gtrless V_S$$

$$V_{BC} \lessgtr V_S$$

and the difference between V_S and V_{AC} or V_{BC} should be as small as possible. It also requires that the ratio of the resistances

$$\frac{R_A}{R_B} = \frac{V_S - V_{AC}}{V_S - V_{BC}}. \tag{10.39}$$

When the latter requirement is met, the current in the AB loop will produce IR drops which will automatically compensate for the error between V_S and $V_{AC} - V_{AB}$. Because the value of R_A/R_B must change with temperature, the inventors recommend multiple-strand extension wires, so that the resistance ratio can be adjusted by choosing the number of AX wires and BX wires that are connected. The alloys they recommend for this application are shown in Table 10.6.[14]

10.11.3 Polarity and Color Code

It is important in using extension wire that its polarity be connected properly, and that it not reach too high a temperature. The extension wire often is more easily oxidized than the thermocouple to which it is attached.

Manufacturers of thermocouple assemblies sell special terminals that have different size connectors for each leg. When properly installed, these make it possible to plug thermocouples to extension wires, or into terminal blocks, with polarity automatically assured.

It is unfortunate that the color code adopted by the Instrument Society of America for thermocouples has the negative wire colored red. In most other types of dc circuitry, red is positive. One of the very common mistakes with compensating wire is to connect the positive compensating wire to the negative thermocouple wire. Then if the connecting wires are also interchanged at the instrument, the voltage produced is less than it should be by twice the voltage error that would occur if copper connection wires had been used. The following example is typical.

Example 10.1. A type S thermocouple at 1300°C is connected with compensating extension wire to a millivoltmeter. The temperature where the thermocouple wires are connected to the extension wire is 150°C. The red connecting wire is connected to the positive thermocouple wire and to the positive millivoltmeter terminal. If the millivoltmeter terminals are 20°C, and if the millivoltmeter has cold-junction compensation for 20°C built into it, what is the error caused by the improper extension wire connections?

Solution. The circuit diagram is shown in Fig. 10.19. Summing the voltages clockwise gives us

$$\underset{20°C}{V_{Cu-SN}} + \underset{150°C}{V_{SN-SP}} + \underset{1300°C}{V_{SP-SN}} + \underset{150°C}{V_{SN-SP}} + \underset{20°C}{V_{SP-Cu}} = V_m$$

$$V_m = \underset{1300°C}{V_{SP-SN}} - 2\underset{150°C}{V_{SP-SN}} + \underset{20°C}{V_{SP-SN}} = V_m .$$

The compensation at the instrument also adds $V_{SP-SN,20°}$C. So the total measured potential

$$V_m = \underset{1300°C}{V_{SP-SN}} - 2\underset{150°C}{V_{SP-SN}} + 2\underset{20°C}{V_{SP-SN}} .$$

From the tables,

$$V_m = 13.16 - 2(1.03) + 2(0.11) = 11.32 \text{ mV} .$$

The apparent temperature, associated with 11.32 mV, from the tables, is 1148°C. Thus an error of 152°C was caused by the improper connections.

It is also important not to use the wrong extension wire. Sometimes

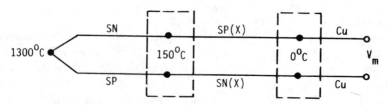

Figure 10.19. Circuit diagram, type S extension wire example.

people accustomed to type S thermocouples will use the same compensating extension wire for type B thermocouples. This causes serious errors because the type B thermocouple produces nearly zero EMF at temperatures up to 150°C.

Example 10.2. Calculate the temperature error resulting when a type B thermocouple with a measuring junction at 1500° inside a furnace is connected just outside the furnace, where the temperature is 150°C, to type S compensating connecting wires leading to an ice-point unit and a potentiometer.

Solution. This circuit can be represented by Fig. 10.20.

$$V_m = \underset{150°C}{V_{SP-BP}} + \underset{1500°C}{V_{BP-BN}} + \underset{150°C}{V_{BN-SN}} \cdot$$

The standard reference tables do not include Seebeck voltages for the two 150° junctions. However, it is helpful to refer both to platinum. This can be done because $V_{SP-BP} = V_{SP-SN} + V_{SN-BP}$. With this substitution, we have

$$V_m = \underset{150°C}{V_{SP-SN}} + \underset{150°C}{V_{SN-BP}} + \underset{1500°C}{V_{BP-BN}} + \underset{150°C}{V_{BN-SN}} \cdot$$

The second and fourth terms can be combined because

$$\underset{150°C}{V_{BN-SN}} + \underset{150°C}{V_{SN-BP}} = \underset{150°C}{V_{BN-BP}} = -\underset{150°C}{V_{BP-BN}} \cdot$$

Then

$$V_m = \underset{150°C}{V_{SP-SN}} - \underset{150°C}{V_{BP-BN}} + \underset{1500°C}{V_{BP-BN}} \cdot$$

From the thermocouple reference tables the Seebeck voltages can be evaluated. $V_m = 1.028 - 0.092 + 10.094 = 10.930 \, \text{mV}$. The temperature for a type B thermocouple at this voltage is 1572°C. Thus the temperature indicated is 72°C too high.

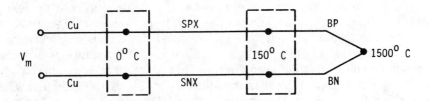

Figure 10.20. Circuit diagram, type B thermocouple with type S extension wires.

10.12 METHODS OF MAKING THERMOCOUPLE JUNCTIONS

Thermocouple junctions can be made by welding, crimping, soft soldering, hard soldering, bolting, or simply twisting the wires together.[15] At low temperatures where adverse chemical reactions are not expected to occur, bolting or soft soldering are often used. Often a terminal block at an instrument has either screw or solder connections. For high temperatures, welding is usually preferred, although one manufacturer makes a connector and crimping tool widely used for type K thermocouples used to about 1050°C.

Bare thermocouple wire can be purchased from one of the companies specializing in thermocouple wire manufacture. Bare or electrically insulated wire, or completed thermocouples, can be purchased from one of the companies specializing in thermocouple supply. Completed thermocouples are often supplied as integral parts of furnaces or other apparatus by the manufacturer of the apparatus. The cost increases with the number of manufacturers involved, although it is convenient to buy insulated wire for most applications.

Welded junctions can be made by (1) stripping the insulation from about 1 in. of the end of each wire to be joined, (2) cleaning the stripped area with sandpaper or emery paper (not silicon carbide), (3) dipping the cleaned ends in "crystallized" boric acid flux ("powdered" is too fine), (4) supporting the ends so that they will stay in position when the molten bead is formed, (5) heating the ends of the wires until the flux melts, then very rapidly until a bead of molten metal of both legs is formed, and (6) allowing it to cool quickly without disturbing the molten bead.

The flux and the heating method depend on the type of metal being welded. Type K should be welded under reducing conditions, type J under neutral conditions, and type T should be silver soldered. Types J and K can be heated with an oxyacetylene flame or with an electric arc. Types S, R, and B should be welded in the oxidizing flame or with an electric arc. Tungsten will burn in a welding arc, so the welds must be made in an arc shielded by an inert gas.

The flux to be used also depends on the metals to be welded. For types E, J, K, and T, boric acid is suitable. The noble metals do not require a flux, nor does the shielded arc. Fluxes that might remain and react with the insulators should be avoided. Boric acid must be washed away.

The easiest way to support the wire during welding is to twist the wires together about three turns, leaving the ends to be welded extended about 5–10 mm side by side beyond the twist so that when the ends are melted, the bead will be held to both wires by surface tension. The bead forms better if the welded end is pointed down during welding. One wire sometimes has a higher melting temperature than the other. Then it is best to head that wire to melting first. This is true for type JP and KP wires. Wires to be butt welded or heated in a shielded arc should be held in a fixture during welding.

Electric arc welding can be accomplished with one or two graphite electrodes. One thermocouple arc welder commercially available consists of graphite electrode mounted in an insulating cup (the cup is filled with flux for convenience), a current-limiting wire-wound resistor, an insulated alligator clip for grasping the thermocouple, and a glass optical filter for eye protection. To weld the thermocouple it is only necessary to insert the twisted pair into the alligator clip, touch the graphite electrode and draw back to start the arc, and pull farther away until the bead is formed. Fine wires of the less refractory metals are more difficult to weld without overheating in an electric arc. They are easily vaporized.

The double carbon arc is recommended for types B, R, and S thermocouples. The arc is established between two soft-carbon electrodes and the junction is moved into the arc just long enough to form the bead. The ends should not be twisted before welding.

The purpose of the welded junction is to establish permanent, low-resistance electrical contact. Twisted noble metals will contact weld at high temperatures, but this is not usually a reliable method. Either soft or hard solders can be used where chemical reactions will not occur in use if, at the junction, both legs and the solder are at the same temperature.

10.13 THERMOCOUPLE INSULATION

Thermocouple wires are often used in pairs, insulated electrically from each other. They are sometimes supplied with the insulated wire bound together in a cable. Ceramic fiber and fiberglass fabrics, impregnated with ceramic powders and strengthened with organic fibers, are often used for types E, J, T, and K thermocouples. The organic portion burns out at high temperatures, but the ceramic materials remain to provide insulation. Manufacturers recommend the ceramic fiber insulation for use up to 1425°C, but it will usually crack if bent after the organic is burned out (after heating to about 400°C).

For higher temperatures, hollow ceramic tubes are usually used as electrical insulation. These tubes (sometimes called spaghetti) have one, two, or four longitudinal holes through which the wires are threaded. Spaghetti is available from ceramic manufacturers in many different outside diameters and internal bores. They are often supplied in 30-in. lengths and can be broken to smaller lengths if needed. They are also available in standard short lengths or as "fish spine" beads so that the insulated assembly can be bent.

The quality of the ceramic insulation is important. Electrical insulation can be provided by porous ceramic materials. Then, in situations were fluids may penetrate the insulation, they may have less electrical resistance than required. Porous insulators are less expensive and can be used if suitable protection from harmful fluids is provided, by a protection tube if necessary.

Better insulation resistance for the same diameter is obtained from dense ceramics. Dense aluminum oxide (Al_2O_3) has the highest electrical resistance and is almost always used at temperatures above about 1250°C.

Sometimes the electrical insulation is provided by ceramic powders surrounding the thermocouple wires and enclosed in a metal outer sheath. This powder is usually magnesium oxide, although Al_2O_3 is also used. Because powdered MgO reacts with water to produce magnesium hydroxide, it is essential that the ends of the metal sheath be sealed. Any water absorbed by the magnesia (MgO) when the ends are temporarily open must be driven off by heating. The heat should be applied with a torch a few inches from the open end first, then moving it to the end. This is necessary to prevent driving the water vapor farther into the tube.

At low temperatures moisture penetration can also harm the electrical resistance. Wires shielded with Teflon, polyvinyl chloride, and polyimide are very useful where moisture penetration is to be prevented.[8] Wax, silicones, and resins are also used to impregnate fabrics and inhibit moisture penetration. Flurocarbons (Teflon) and organic chlorides produce dangerous fumes when heated too hot. Also, when the organic is burned out, no moisture resistance remains.

The importance of electrical insulation resistance depends on the application. The insulation provides a shunt path that allows the current to short-circuit the measuring instrument (Fig. 10.21). The effect on the open-circuit voltage, V_{oc}, of the insulation leakage through R_i is to reduce V_{oc}. Then the leakage current, I_l, will be given by

$$I_L = \frac{V_{AB}}{R_i + R_{TC}}. \tag{10.40}$$

The open-circuit voltage will depend on the relative values of R_i and R_{TC}.

$$V_{oc} = I_l R_i = V_{AB} \frac{R_i}{R_i + R_{TC}}. \tag{10.41}$$

As R_i approaches zero, so does the open-circuit voltage.

The situation is more complex when a measuring instrument is connected

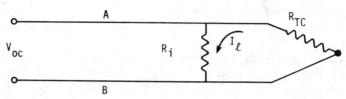

Figure 10.21. Equivalent circuit showing electrical shunt path R_i through electrical insulator.

to the terminals. Then the voltage measured by the instrument depends on the relative resistance of the thermocouple, the shunt path, the leads, and the measuring instrument. It is clear, however, that the instrument will read low. Whenever thermocouples are used, the limitations imposed by the electrical insulation should be recognized.

10.14 THERMOCOUPLE PROTECTION

Thermocouples are rarely used with the measuring junction bare. Usually, the thermocouple and insulation is inserted into some kind of protection tube. When this is done, the thermocouple indicates the temperature inside the tube. So it is important to consider the effects of the tube on the indicated temperature and on its response time. This is discussed in Chapter 18.

Protection tubes are needed at low temperatures to prevent moisture absorption, especially at temperatures below the dew point. Glass, stainless steel, and organic protective envelopes are available in a wide variety of forms. These are often insulated inside with Teflon or other organic materials. The entire system should be hermetically sealed whenever moisture penetration or condensation can occur.

At moderate temperatures metal sheaths with ceramic powder insulation are very effective. These are made by inserting the thermocouple wires and powder into a larger tube, then drawing (or swaging) the tube to elongate it, reduce it in diameter, and compact the ceramic powder around the thermocouple wires. Subsequent cycles of annealing and drawing elongate the tubes and the thermocouple wires until the final dimensions are achieved. The final length is proportional to the square of the ratio of the original diameter to the final diameter. Sheathed thermocouples are available in lengths up to about 250 ft.

The sheathed thermocouple should have an insulation resistance of at least $100\,M\Omega$ at $500\,V$ dc for diameters greater than 1/16 in.[8] Magnesia (MgO), alumina (Al_2O_3), zirconia (ZrO_2), beryllia (BeO), and thoria (ThO_2) have been used as insulating powder. Usually, magnesia is used because it is more ductile, has a thermal expansion only slightly less than stainless steel, and is less expensive.

The hot junction of a sheathed thermocouple is usually sealed in the tube for protection from atmospheric reactions. If the tube is left open at the hot junction so that the thermocouple bead is exposed, it will respond more quickly to temperature changes but will not have the protection of the metal sheath. Therefore, sheathed thermocouples usually have their hot junction ends welded closed. The junction is often formed by welding the thermocouple wires and the sheath together under an inert atmosphere. Then the external sheath is in better thermal contact with the thermocouple, and more rapid response will result. However, mechanical strain

TABLE 10.7 Sheath Materials of Ceramic-Packed Thermocouple Stock and Some of Their Properties

				Recommended	Tensile Strength (psi)[a]	
		Recommended		Continuous		
	Melting	Maximum	Operating	Maximum	At	At
Material	Point (°F)	in Air (°F)	Atmosphere[b]	Temperature (°F)	200°F	1600°F
Stainless steel						
304	2560	1920	ORNV	1650	68,000	
309			ORNV	2000		
310	2560	2000	ORNV	2100	87,000	23,000
316	2500	1650	ORNV	1700	75,000	23,000
321	2550	1650	ORNV	1600	70,000	17,000
347	2600	1680	ORNV	1600	75,000	
430	2700	1550	ORNV	1200		
446	2700	2000	ORNV	2000		
Inconel	2550	2000	ONV[c]	2100	93,000	5,000
Inconel X	2620	1500	ONV[c]	2200	150,000	11,000
Incoloy	2500	1640			77,000	3,000
Hastelloy X	2350	2300			106,000	7,000
Hastelloy C	2310	1820			136,000	64,000
Haynes 25	2425	1820			147,000	13,000
Hastelloy B	2375	1400			125,000	51,000
Monel	2460	1640				
Chromel	2600	2100	ONV		90,000	21,000
Copper	1980	600	O[d]RNV	600		
Brass	1850	700				
Aluminum	1220	800	ORNV	700		
Nichrome	2550	2200		2000		
Alumel	2550	2100	ONV		82,000	19,000
Nickel	2647	1100				
Iron	2798	600				
Zircaloy	3350	1400				
Platinum	3217	3000	ON[c]	3000		
Pt–Rh 10%	3362	3100	ON	3100		
Columbium	4474	1600	VN	3800	110,000	
Molybdenum	4730	400	VNR		137,000	30,000
Molybdenum disilicized		3100	ON	3000		
Molybdenum chromalized		3100	ON	3000		
Tantalum	5425	750	V	5000	96,000	22,000
Titanium	3035	6000	VN	2000		

[a]After exposure to temperature for 100 h except for stainless steels; Haynes 25, W, Mo, Ta, and Cb.
[b]O, Oxidizing; R, reducing; N, neutral; V, vacuum.
[c]Very sensitive to sulfur corrosion.
[d]Scales readily in oxidizing atmosphere.
Source: ASTM. Used by permission.

and breakage may occur if the thermal expansion of the sheath is much different from that of the thermocouple. The stability of type N thermocouples is degraded if used in a stainless steel sheath instead of a NP/NN alloy. If the thermocouple junction is welded first and the sheath welded afterward, and not in contact with the thermocouple bead, the response will be slower but more dissimilar thermal expansions can be accommodated. Sheathed thermocouples should be fully annealed by the manufacturer.

Sheathed thermocouples are shipped in coils. The allowable coil diameter depends on the sheath material and its diameter and wall thickness. They should not be coiled too tight, or permanent set will occur. Then annealing may be required. Flexing, particularly repeated flexing, should be avoided because changes in calibration may result. The maximum use temperature depends on the sheath material, the furnace atmosphere, and the sheath dimensions (Tables 10.7 and 10.8).[8]

At higher temperatures metal or ceramic protection tubes are used. They must withstand the application conditions for a reasonable period. Metal tubes can be used to temperatures of 1150°C, although some furnace gases can diffuse through metals at temperatures above about 800°C. High-carbon steels can be used to about 700°C without excessive oxidation. Austenitic stainless steels can be used in oxidizing atmospheres to about 870°C. Ferritic stainless steels can be used to about 1150°C and high-nickel alloys can be used to about 1150°C. Manufacturers' literature should be consulted to determine the conditions for which particular alloys can be used.

Ceramic protection tubes are used for more severe conditions of temperature and atmosphere. Mullite protection tubes have the best thermal shock resistance and the lowest permeability. They can be provided vacuum tight to about 1350°C. Each ceramic tube manufacturer provides a slightly different product. Tubes sold as mullite tubes usually contain the mineral mullite $Al_6Si_2O_{13}$, bonded together with a high-alumina glass. The glass helps reduce the permeability. They are stable in oxidizing and reducing atmospheres to 1350°C. Most start to soften at about 1450°C, although if supported they may be used to higher temperatures.

Alumina (Al_2O_3) tubes can be used to higher temperatures than mullite.

TABLE 10.8 Recommended Maximum Long-Term Service Temperature for Various Sheath Dimensions

Nominal diameter (in.)	0.040	0.062	0.125	0.188	0.250
Nominal wall (in.)	0.007	0.010	0.020	0.025	0.032
Type K (°F)	1400	1600	1600	1600	1800
Type J (°F)	1000	1200	1400	1400	1600
Type E (°F)	1200	1400	1400	1600	1700

Source: ASTM. Used by permission.

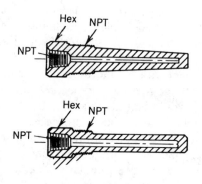

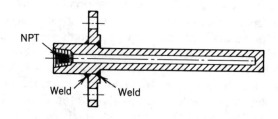

Straight and Tapered
Drilled Thermowells

Flanged Thermowell

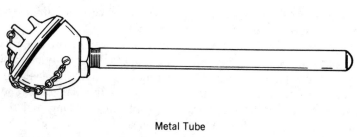

Metal Tube

Ceramic Tube

Figure 10.22. Typical protection tubes. Top: for metal vessels and pipes.[8] Bottom: for insertion through furnace walls. (From ASTM Committee E20, "Manual on the Use of Thermocouples in Temperature Measurement," STP470A, 1974, with permission from the American Society for Testing and Materials, 1916 Race Street, Philadelphia, PA 19103.)

They may be very porous, or they may be vacuum tight. Each manufacturer usually has several different products. They are usually not 100% alumina, so selection of a tube will depend on the requirements in service. Some can be used to 1750°C.

Silicon carbide has high thermal conductivity. The best grades of silicon nitride–bonded silicon carbide are relatively impermeable, but usually furnace gases penetrate. The principal advantage of this material is its thermal shock resistance and its resistance to the slags often associated with coal-fired furnaces.

Zirconia (stabilized with CaO, Y_2O_3, or MgO), berylia, and thoria protection tubes are sometimes used for extremely high temperatures. However, ceramic products have properties that depend on the method of manufacture. It is dangerous to make decisions based on general information such as the melting temperatures of pure oxides. Manufacturers' specifications should be consulted to determine the properties of each product and the conditions for which it is suited. It is better to purchase directly from the tube manufacturer than from someone else distributing the tubes.

The configuration of protection tubes depends on the application. Two commercial examples are shown in Fig. 10.22.[16]

10.15   DETERIORATION OF THERMOCOUPLES

The usual mechanisms of destruction of thermocouples are either mechanical breakage or chemical deterioration. When thermocouple wire is received, fully annealed and ready for application, it is at its best. Any mechanical flexing, contamination in handling, or chemical reaction with the environment may cause deterioration. In assembling thermocouples great care should be exercised to minimize cold working and to prevent contamination.

The noble metal thermocouples are particularly susceptible to damage from cold working. Flexing the wire, as in bending in while inserting into spaghetti, will introduce dislocations. Then, when heated, recrystallization accompanied by massive grain growth may occur. Platinum wire from type R and S thermocouples, after use at 1400°C, will often be recrystallized to the point that an entire segment of the wire may be a single crystal. Then any mechanical strain can cause massive slip on a single slip plane, and the offset accompanying it may drastically reduce the cross section of the wire. Creep also occurs under these conditions.

Mechanical working, whether during assembly or while in use, introduces dislocations that raise the electrical resistivity and lower the Seebeck voltage. Because the Seebeck voltage depends on the temperature gradient, all thermocouples are sensitive to mechanical bending in that portion of the thermocouple subjected to the thermal gradient. This is often in the portion within a furnace wall.

Dislocations can be consolidated into grain boundaries by annealing. When large amounts of type B, R, or S wire are used, it is often practical to cut out deteriorated sections, weld and anneal, to reduce thermocouple costs.[17] The repaired thermocouple wire can be annealed by passing an electric current through it. The Pt wire should be heated to 1100°C. The PtRh wire should be heated to at least 1450°C. Usually, $\frac{1}{2}$ hour is sufficient. Repaired thermocouples should be checked against secondary standards before being returned to use.

Chemical deterioration can be accelerated by salt from perspiration or other impurities introduced in handling. It is more common, however, for deterioration to occur slowly by contact with the atmosphere, the working materials in the furnace, or the refractories. Atmospheric deterioration of base metal thermocouples is usually caused by oxidation. As the surface is oxidized the metallic, current-carrying cross-sectional area is reduced. That is the reason that the recommended service life and maximum temperature depend on the original wire size. As the metal progressively oxidized, types J, K, and N thermocouples usually develop adhering oxide coatings which inhibit further oxidation. The rate of oxidation is controlled by the rate of diffusion of the oxygen through the coating. This gives parabolic oxidation kinetics. The rate is inversely proportional to the square root of time. The diffusion is proportional to

$$D = D_0 e^{-Q/RT} , \qquad (10.42)$$

where D_0 is a constant, Q is the activation energy for diffusion, R is the gas constant, and T is absolute temperature. If the temperature is too high, the rate of diffusion, and therefore the rate of oxidation, will be rapid. Then the service life will be short.

When the oxygen partial pressure is moderately low and the temperature is in the range 700–1000°C, type KP thermocouple wire develops "green rot," a nonadherent oxide coating that does not protect the wire. At 1200°C the KN alloy oxidizes more rapidly. The development of the Nicrosil–Nisil alloys produced less external oxidation and eliminated internal oxidation, resulting in improved thermal EMF stability.

Noble metal deterioration by chemical reaction often comes from atmospheric transport of alloying metals. Contact with fumes containing volatile transition metals, such as iron, is especially harmful. At extremely high temperatures, the pure platinum element of types S and R thermocouples is susceptible to Rh vaporized from the other leg. That is the reason type B thermocouples are superior above about 1400°C. At low oxygen partial pressure SiO_2 may be reduced to SiO, which has a high vapor pressure. Platinum thermocouples in contact with gaseous SiO produce a platinum silicide eutectic at the grain boundaries and the platinum falls apart. Therefore, types S, R, and B thermocouples should never be used at temperatures above 1350°C in oxygen pressures less than 10^{-4} atm unless all sources of SiO_2 (including the refractories) are excluded.

Deterioration by any of the mechanisms described above may not cause sudden failure. However, oxidation and embrittlement do make thermocouples much more sensitive to breakage caused by bending. The usual result of deterioration is gradual reduction of the Seebeck voltage, often extended over several weeks. This is often not detected. If the Seebeck voltage is low, the temperature is believed to be low. So the actual temperature may be increased to produce the desired Seebeck voltage. The net result can be excessive temperature, with resulting damage to materials and processes. Those who use thermocouples should be aware of the possibility of slow deterioration and its consequences.

10.16 THERMOELECTRIC VOLTAGE MEASUREMENTS

The Seebeck voltage is a small direct-current voltage that must be measured accurately to be able to add the reference junction voltage and compute temperature. These millivolt dc voltages can be measured with a millivoltmeter, with a millivolt potentiometer, and with a solid-state voltmeter. The type of instrument that is used depends on the application. In recent years solid-state voltmeters, calibrated to read temperature directly, have become available. All three classes of instruments are in use and will be described separately.

10.16.1 The Millivoltmeter

Millivoltmeters are in wide use as temperature-indicating instruments. Most of them are actually d'Arsonval milliammeters with a resistance in series to limit the current. The milliammeter has a lightweight coil suspended between the poles of a magnet so that when a dc current flows in the coil, the coil is rotated by the induced magnetic field in the coil (Fig. 10.23).[18] The coil is usually suspended between the poles of the magnet using jeweled bearings or a taut band suspension. A spiral spring holds the coil at the zero position until the current in the coil produces a rotating force. The indicator attached to the coil indicates either temperature or voltage. Movement of the coil is directly proportional to the current flowing, which depends on the applied voltage. Millivoltmeters of this type are mass produced and are available for all types of thermocouples. They usually indicate temperature to about ±10°C, so they are not accurate enough for precise work and usually have a linear scale.

Milliammeters draw current. Therefore, they are only accurate when the proper external resistance is included in the circuit. Most direct-indicating instruments have the required external resistance printed on their scale. If

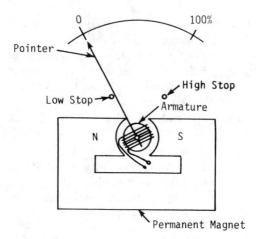

Figure 10.23. Usual configuration for a d'Arsonval meter movement.

the thermocouple resistances increases, as often happens over a long period of time as a base metal thermocouple oxidizes, the temperature indicated will be low.

Example 10.3. A milliammeter is calibrated for an external resistance of 12.5 Ω but is connected to a thermocouple having a resistance of 125 Ω. If its internal resistance is 500 Ω, what voltage will it indicate when a voltage of 16 mV is generated by the thermocouple?

Solution. The current that should flow in the coil of the millivoltmeter can be calculated:

$$I_c = \frac{16.0}{500 + 12.5} = \frac{16}{512.5} = 0.0312 \text{ mA} .$$

The actual current is

$$I_A = \frac{16.0}{500 + 125} = 0.0256 \text{ mA} .$$

Therefore, the indicated voltage is

$$V = 16\left(\frac{0.0256}{0.0312}\right) = 13.1 \text{ mV} .$$

Millivoltmeters are often equipped for control functions. One design has an indicator for the set point which supports a photoelectric cell. When the temperature indicator reaches the set-point temperature, an attached vane shuts off the light path to the photoelectric cell. The photoelectric cell output controls a relay for the furnace power so that the power is interrupted when the temperature reaches the set point. On–off controllers are discussed in Chapter 20.

Another design has a set of flat-wound coils attached to the set point. The metal flag on the temperature indicator changes the inductance of the coils and causes a solid-state amplifier providing power to a solenoid to discontinue output, turning off the solenoid power (Fig. 10.24).

Millivoltmeter controllers often have an open-circuit protection device built into them. Without such protection, when a thermocouple is broken, the voltage will drop to zero. Then the temperature-indicating vane will never reach the set point. The power will be on all the time and the furnace will overheat, burning out elements and overheating the product. Protection is achieved by providing a source of direct current through a very large resistor connected to the thermocouple terminals (Fig. 10.24).[19] When the

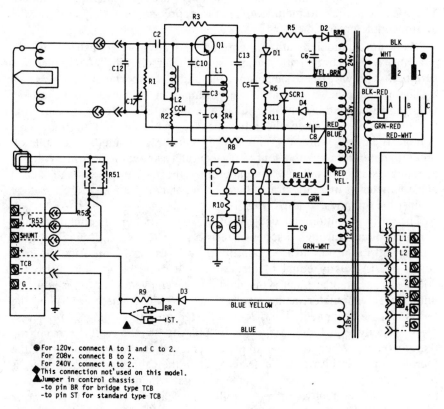

Figure 10.24. Schematic of an on–off controller operated by a millivoltmeter. (Courtesy of Capacitrol.)

thermocouple is intact, the voltage leaks off through the resistor as such a small current that it does not affect the indication significantly. When the thermocouple is broken, the current cannot leak off in the thermocouple circuit, causing the applied potential to appear at the voltmeter terminals. This drives the voltmeter up scale and shuts off the furnace. The example in Fig. 10.24 has an 18-V secondary transformer suppling a dc potential through diode A_3 and resistor R_9 to the thermocouple break (TCB) protection terminals. These are connected with copper wires to the thermocouple terminals at the furnace.

10.16.2 The Millivolt Potentiometer

Millivolt potentiometers have been used for decades for the measurement of Seebeck voltages. They are available in simple portable forms capable of measuring to three significant figures, and in very sophisticated forms capable of five to seven significant figures. With precision potentiometers the Seebeck voltage can be measured more accurately than is necessary for the accuracy of the voltage-to-temperature interpolation formulas. The electrical measurement can be more precise than required by the thermocouple errors.

Many automatic indicating recording and controlling instruments are available which contain the basic potentiometer circuitry. A detailed description is provided here because of the circuit's importance.

The potentiometer circuit is very simple in principle. A measuring voltage source is adjusted to be equal, but opposite in polarity, to that from the thermocouple. When the adjustment is achieved, the thermocouple voltage is known. Then the temperature can be computed. The millivolt potentiometer contains the source of accurately known measuring voltage, a voltage standard suitable for calibrating the measuring voltage, and a galvanometer or null detector to be used in balancing the measured voltage against the unknown voltage (Fig. 10.25).

The basic potentiometer consists of a working battery B connected in a complete circuit through a series of resistors R_1, R_2, R_3, R_4, and R_5. In use the current in this circuit is standardized by adjusting R_2 until the IR drop across R_1 is exactly equal in magnitude but opposite in polarity to the voltage of the standard cell. This condition is achieved when R_2 is adjusted until the galvanometer in the standard cell circuit reaches zero. Then the thermocouple is connected and R_5 is adjusted until the voltage drop across R_5 is exactly equal and opposite in polarity to the voltage from the thermocouple. When this condition is achieved the galvanometer in the thermocouple circuit will indicate zero. Then the calibrated resistance R_5 is read to determine the thermocouple voltage, which can be translated to temperature. (Actually, there is only one galvanometer. It is merely switched to the standard cell circuit when adjusting R_2, and to the thermocouple circuit when adjusting R_5.) The purpose of R_3 and R_5 is to

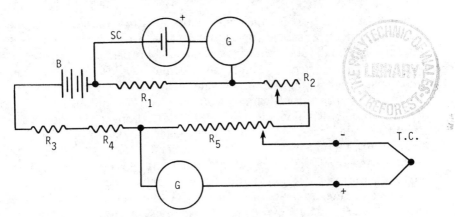

Figure 10.25. Simple millivolt potentiometer circuit.

adjust the range of R_5 by limiting the current. For multiple-range instruments the IR drop across both R_4 and R_5, or $R_3 + R_4 + R_5$, can also be used.

There are three types of standard voltage sources that can be used for standardizing the current. Portable potentiometers usually contain an unsaturated Weston standard cell because it is less sensitive to damage by bumping than is the Weston saturated cell. A new unsaturated cell voltage is 1.019 V, but the voltage decreases about 0.01% per year. Cells should be replaced about every 10 years. For higher precision, a saturated Weston cell is used. It has a potential of 1.01864 absolute volts at 20°C, reproducible within a few microvolts. Room-temperature changes produce only about a $-40\,\mu V/°C$ change. Many solid-state temperature recorders containing automatic balancing potentiometers have zener diode circuits to produce a standard voltage. The precision of the zener diode circuit depends on its complexity, because ambient temperature variations cause variation in the simpler circuits. These can be compensated by building more sophisticated, more expensive circuits.

Weston cells deliver the correct voltage only at low current levels. Recovery after heavy current draw depends on the time and amount of current, and may take hours. Therefore, Weston cells should never be short circuited, and balancing should be achieved with a minimum of current use. Depress the galvanometer key only long enough to determine whether it deflects left or right. Make the appropriate adjustment of the R_2 resistance and depress momentarily again. Repeat until balance is achieved. When at balance the key may be held or locked down and minor variations compensated.

For precision work several modifications of the basic potentiometer circuit can be used.[20] Usually, this includes massive contacts to reduce resistance variation on switching, substitution of an electronic null detector (an amplifier plus a sensitive meter) for the galvanometer, substituting a series of fixed resistors for the coarse adjustments of R_5, using a precision slidewire of low resistance for the final balance adjustment, and using a saturated cell as the standard cell. Commercial precision potentiometers for voltage and resistance measurements are quite sophisticated and can be purchased with computer-controlled automatic balancing features.

One of the principal advantages of a millivolt potentiometer is that the current is zero at balance. Therefore, no IR drop in the thermocouple circuit will alter the voltage. In principle, then, oxidation of the thermo-couple, poor contacts, long, fine wire, and other factors that cause higher-than-normal resistance are not important. However, if the resistance in the thermocouple circuit is too large, at off-balance the galvanometer will respond sluggishly because current does flow in the off-balance con-dition. If the galvanometer appears insensitive to R_5 adjustments, or if balance appears to permit wide variation in R_5, excessive external resis-tance should be suspected.

A portable potentiometer is very convenient in troubleshooting. It can be used to check the output of thermocouples manually to determine if the voltage they produce is that indicated by another instrument. It can also be used as a low-impedence voltage source. When connected to an instrument instead of the thermocouple, the instrument can be tested for voltage response as the potentiometer is adjusted to various values.

10.16.3 Manual Reference Junction Compensation

Manual reference junction compensation is often provided in commercial portable potentiometers. By adjusting the reference junction compensation dial to the correct voltage for the thermocouple reference junction, the instrument will automatically add that value to all readings so that the voltage, or temperature, indication will be corrected to the ice-point value. This may be a source of error, however. If the reference junction was set by a previous user to an inappropriate value (and not reset to the correct value), the value obtained at balance will be in error. Some commercial instruments always add the reference junction compensation voltage—and it must be reset to zero if that voltage is not wanted. Other commercial instruments have a switch which when adjusted to "thermocouple" applies the reference junction voltage, but when adjusted to "voltage" does not. A reference junction compensation circuit adds considerable expense to a potentiometer because two precision slidewire circuits are required (Fig. 10.26).

Example 10.4. If the measuring junction of a type K thermocouple is inserted into a furnace at 1000°C and the reference junction is connected to a potentiometer at 25°C, the voltage measured without compensation will be too low. The voltage with the measuring junction at 25°C and the reference junction at 0°C is 1.000 mV. The indicated voltage will be $V_{1000} - V_{25}$, or

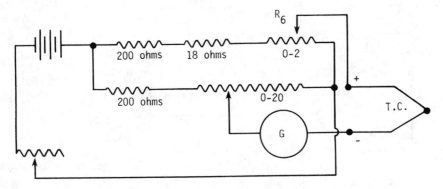

Figure 10.26. Reference junction compensation.

$41.269 - 1.000 = 40.269$ mV. To make the potentiometer read correctly, R_6 is adjusted until its IR drop is 1.000 mV. Then, at balance, that potential is added to 40.269 mV to get the correct indication corresponding to $1000°$C.

10.16.4 Automatic Reference Junction Compensation

Many commercial potentiometers, recorders, and controllers have built-in reference junction compensation that automatically compensates for a change in temperature of the measuring instrument.[20] This is convenient but is limited to a particular thermocouple type, as we will demonstrate. Never use the wrong type of thermocouple when using an instrument with automatic reference junction compensation, and never use an ice-point reference junction connected with copper wires to the potentiometer. The automatic compensation will be applied when unneeded, and an error will result.

An automatic reference junction circuit can be constructed as shown in Fig. 10.27. The resistor R_7 serves as the reference junction compensating resistor. If the temperature of the instrument is $20°$C, the current flowing through R_3, R_4, and R_5 should equal that flowing through R_6 and R_7. If 10 mA flows through R_2 when it is standardized, 5 mA will flow through each of the two branches. If the thermocouple is also at $20°$C, the EMF produced by the thermocouple will be zero. Then R_5 will balance at zero and the value of R_7 should equal R_4. As the temperature of the instrument is increased, the Seebeck voltage will decrease. If the resistance of R_7 increases the same amount that the Seebeck voltage decreases, the higher IR drop through R_7 will require a displacement of the set point of R_5 to compensate. If $R_3 + R_4 \gg R_5$, to a first approximation the compensation can be achieved by making $\Delta IR_7 = \Delta V_S$. The Seebeck voltage can be

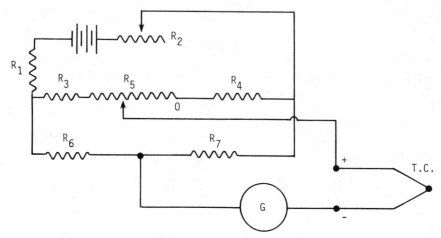

Figure 10.27. Automatic reference junction compensation.

represented by

$$V = a_1 t + b_1 t^2$$

and the resistance by

$$R_7 = R_{20}(1 + a_2 t + b_2 t^2).$$

Then R_7 can be adjusted so that $\Delta I R_7 = \Delta V_S$ if $a_1 = R_{20}a_2$, that is, if $\alpha = a_2 R_{20}$.

The terms b_1 and b_2 represent the curvature from linearity. By choosing the correct metals R_7 can be wound with two metals so that $b_1 = R_{20}b_2$. Usually, this is nickel and copper wire, for which b_2 is large and small, respectively. Then compensation is achieved by adjusting the total resistance of R_7 at a fixed reference temperature and using the proportions of the two metals so that the curvature is also corrected. Then, at 20°C, $R_6 = R_3 + R_4 + R_5 - R_7$.

Automatic reference junction compensation can also be provided by incorporating a bimetallic strip in the measuring instrument. For example, a millivoltmeter can be constructed with the scale supported by a bimetallic strip which will move the zero of the scale as the ambient temperature of the meter changes. This will move the entire scale by the amount necessary to make it indicate correctly at a particular temperature. Because of the nonlinearity of the EMF temperature function, an error will exist at all other temperatures, but it will not be noticed if the millivoltmeter scale can only be read to about ±5°C.

Another ingenious method for automatic reference junction compen-

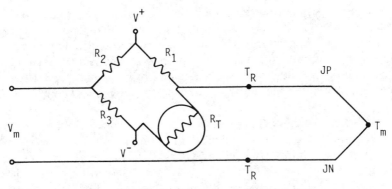

Figure 10.28. Bridge compensating circuit.

sation is the double-oven method. It requires two ovens controlled at constant temperatures T_1 and T_2. The reference junction of the measuring thermocouple is connected in the first oven at T_1 to wires of the same thermocouple stock, but of opposite polarity. They are extended to the second oven, where they are connected to copper wires connected to the measuring instrument. By adjusting the temperature of the two ovens to the correct T_1 and T_2, the output can be made equal to that of a thermocouple with its reference junction at the ice point.

A bridge circuit has been found to be very useful in some applications where solid-state devices are needed because the application imposes high inertial forces on the system (Fig. 10.28). In this circuit the thermistor assembly R_T is mounted in thermal contact with the reference junction at T_R. When $T_R \neq 0°C$, the bridge is unbalanced because $R_1 \neq R_T$. The applied voltage and the resistors are chosen so that the bridge unbalance produces a compensating voltage which is added to the Seebeck voltage to make V_m equal to the voltage that would have been produced if T_R were zero. Commercial instruments, complete with power supplies, are available.

10.16.5 Solid-State Voltmeters

Solid-state voltmeters are available in a termendous range of performance standards. Some are simple, inexpensive, and not very accurate. Others are complex, expensive, and moderately accurate. The best solid-state milli-voltmeters at the present time will indicate to $5\frac{1}{2}$ digits, more accurate than is needed for almost all industrial temperature measuring applications. Such meters are available for particular types of thermocouples, calibrated to indicate temperature directly, with indications of $5\frac{1}{2}$ digits and resolution of 0.01°C.[21] These sophisticated meters include internal standardization coupled to an ac line supply frequency of 60 Hz, usually $16\frac{2}{3}$ ms. Their circuitry is much too complex to explain here.

Inexpensive solid-state voltmeters indicating thermocouple temperatures are becoming popular because they provide adequate control of many industrial processes at low cost. Some are made to indicate temperature with a LED or LC display visible from considerable distance; some have analog signal outputs for recording purposes; and some have control function outputs. The more sophisticated forms may have IEEE 448 connectors for interfacing with microcomputers for process control.

Example 10.5: Ambient Temperature Compensation. Ambient temperature compensation can be provided with a simple combination of monolithic integrated circuits (IC) and resistors. One of the ICs serves as the reference junction temperature sensor, so that it should be mounted to be at the same temperature as the reference junction. This means that the reference junction terminals must be mounted close to the appropriate IC. A suitable reference temperature sensor could be the LM 35 from National Semiconductor Corporation. It produces a signal of −10 mV/K when adjusted to −2.73 V at 0°C. The National LM 135, LM 235 and LM 335 ICs have outputs in °C instead of K (Fig. 10.29).[22] These circuits only

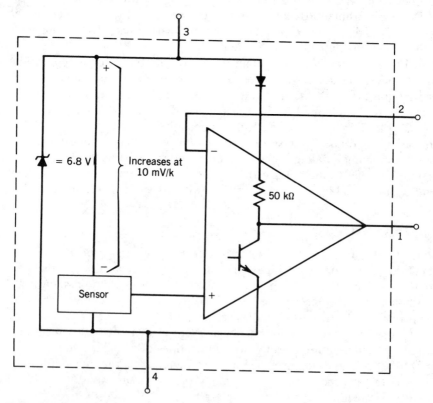

Figure 10.29. Block diagram of National M35. (Courtesy of National Semiconductor.)

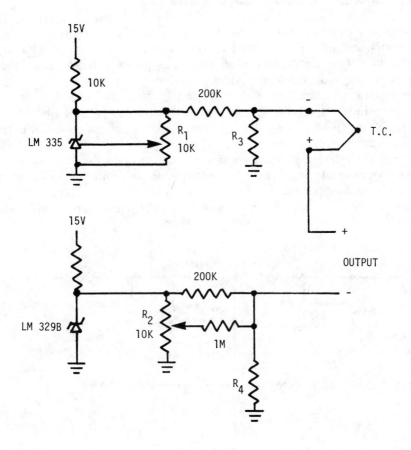

SELECT R_3 AND R_4 FOR THERMOCOUPLE TYPE

	R_3	R_4	Seebeck Coefficient
J	1.05 KΩ	385 Ω	52.3 μV/°C
T	856 Ω	315 Ω	42.8 μV/°C
K	816 Ω	300 Ω	40.8 μV/°C
S	128 Ω	46.3 Ω	6.4 μV/°C

Adjustments

1. Adjust R_1 for the voltage across R_3 equals the Seebeck coefficient times ambient temperature in K.

2. Adjust R_2 for voltage across R_4 corresponding to T.C.

J	14.32 mV	K	11.17 mV
T	11.79 mV	S	1.768 mV

Figure 10.30. A solid-state circuit to add a reference junction correction to a thermo-couple signal. (From Ref. 22.)

require four external connections. The sensor in the IC has two matched transistors with different collector currents and with their bases in parallel so that the difference in the emitter voltage responds to temperature. (See Section 9.8 on Junction Semiconductor Devices.)

A circuit suggested by the manufacturer uses the LM 335 as the reference junction sensor. When combined with a LM 329B, the circuit shown in Fig. 10.30 provides the reference junction correction added to the thermocouple signal.

When the reference correction signal is used to bias the input of a LM 308A operational amplifier, the output can be adjusted to produce 10 mV/°C and supplied to any suitable display (Fig. 10.31). The accuracy is limited to the linear approximation to the type K thermocouple output, and

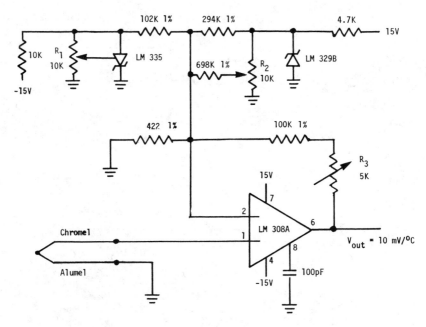

Terminate thermocouple reference junction in close proximity to LM 335.

Adjustments:

1. Apply signal in place of thermocouple and adjust R_3 for a gain of 245.7.
2. Short noninverting input of LM 308A and output of 329B to ground.
3. Adjust R_1 so that $V_{out} = 2.982$, V @ 25°C.
4. Remove short across LB 329B and adjust R_2 so that $V_{out} = 246$ mV at 25°C.
5. Remove short across thermocouple.

Figure 10.31. A solid-state thermocouple thermometer based on a linear approximation for the thermocouple response, and with a reference junction correction. (From Ref. 22.)

the resistors and circuit design must be carefully constructed to minimize thermal effects within the instrument.

10.17 REFERENCE JUNCTIONS

The simplest thermocouple circuit is a thermocouple with its measuring junction in a desired location and its reference junction at the potentiometer (Fig. 10.32). Such a circuit will produce a Seebeck voltage which depends on both the measured temperature t_m and the potentiometer temperature t_p. By making the reference temperature at the ice point, the voltage will depend only on t_m, and standard tables can be used to convert the Seebeck voltage to temperature (Fig. 10.33). This arrangement is desirable, but it was used primarily for laboratory work because an ice–water mixture required continual maintenance. In the past, reference junction compensation, rather than ice-point maintenance, was required. About 1960 commercial ice-point units became available which automatically provide good ice–water equilibrium. They are convenient and accurate but somewhat expensive. Both methods of obtaining ice-point temperatures will be described.

A recommended ice bath unit is shown in Fig. 10.34.[8] It has the thermocouple reference junctions, each connected to copper leads, inserted into mercury held in glass tubes to prevent shunt path leakage. The ice is shaved and is wet on the surface to maintain the thermodynamic equilibrium, but a drain is provided at the bottom to prevent the ice from floating on top of water. (If that occurs, an error of 2 or 3° can easily result.) The ice is contained in a dewar flask held in a Styrofoam block to reduce heat loss. The thermocouple junctions and the mercury vials are inserted deep enough so that no immersion error exists.

A typical commercial ice point consists of a polyethylene container into which a copper block is inserted (Fig. 10.35). Inside the block the soldered junctions of the thermocouple wires to copper extension wires are held in glass vials filled with diffusion pump oil. The purpose of the copper block is to ensure good thermal contact with the ice–water mixture. The polyethylene container is filled with pure water which is partially frozen by a Peltier refrigerator. As the ice is frozen, it expands and increases the size of the

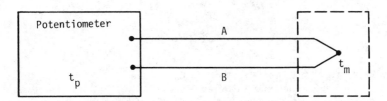

Figure 10.32. Simple thermocouple circuit.

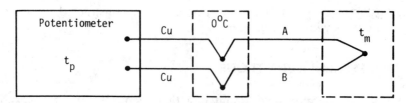

Figure 10.33. Typical ice-point reference circuit.

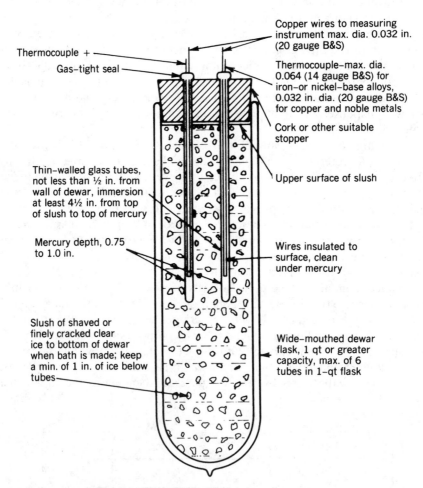

Figure 10.34. Ice bath Dewar. (From ASTM Committee E20, "Manual on the Use of Thermocouples in Temperature Measurement," STP470A, 1974, with permission from the American Society for Testing and Materials, 1916 Race Street, Philadelphia, PA 19103.)

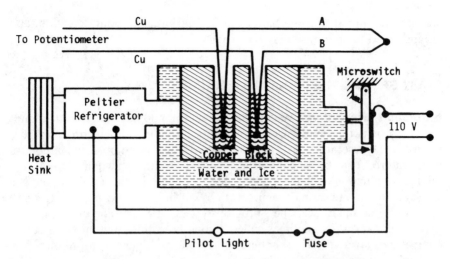

Figure 10.35. Block diagram of a typical ice-point unit.

polyethylene container. That expansion is used to regulate a microswitch controlling the refrigerator, so that the water is never completely frozen or the ice completely melted, providing ice-point reference temperatures. Of course, the power supply for the Peltier refrigerator is incorporated in the unit so that it can operate on 110 V ac.

10.17.1 Auxiliary Voltage Source

Reference junction compensation can also be provided by an external battery. All that is necessary is to provide a circuit that adds a voltage equal to the desired reference junction voltage to the voltage from the thermocouple. The circuit described previously for potentiometers is an example of the use of a battery for this purpose. Commercial reference junction compensation circuits that can be attached with connectors to copper wires are available (Fig. 10.36).[23] The principal disadvantage of the battery-

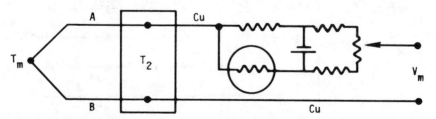

Figure 10.36. Schematic diagram of a battery-operated reference junction compensation. The thermistor is in thermal contact with the junction board T_2 so that the bridge unbalance gives an added voltage to the thermocouple.

operated circuits is that the battery may deteriorate without detection, and the indicated temperature may be wrong.

10.18 SPECIAL THERMOCOUPLE CIRCUITS

Multipoint Recording. When many thermocouples are used simultaneously, it is frequently useful to have a multiple-point recorder make a record of temperature at each location. This, or simple indication of temperature, can be accomplished with a rotating switch. Often, the reference junction will be located deep enough in the earth to have a constant temperature, and compensation will be provided for that through one of the methods described previously (Fig. 10.37). The constant-temperature location is sometimes called a zone box. Note that microprocessor control can be used with this type of circuit, multiplexing to measure the Seebeck voltage sequentially in time.

Thermopile. A thermopile is a group of thermocouples connected in series. If each has the same measuring and reference junction temperatures, the output voltage of N thermocouples will be N times the output of a single thermocouple. If not, the output will be the sum of the individual thermocouples.

Parallel Thermocouples. When a group of thermocouples are connected in parallel with a common reference junction, the voltage produced will be approximately the average of the voltages corresponding to the tem-

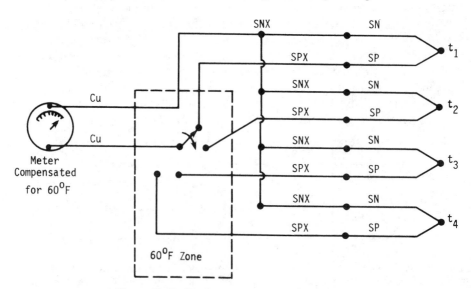

Figure 10.37. Multipoint recorder with zone box.

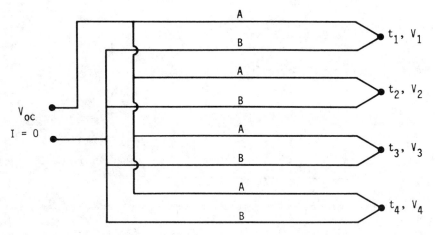

Figure 10.38. Parallel *AB* thermocouples.

peratures of the measuring junctions (Fig. 10.38). This is sometimes useful if rapid fluctuations in temperature from place to place within a furnace are expected. The thermocouples should have about the same electrical resistances.

$$I_1 = \frac{V_{oc} - V_1}{R_1} \qquad I_2 = \frac{V_{oc} - V_2}{R_2}$$

$$I_3 = \frac{V_{oc} - V_3}{R_3} \qquad I_4 = \frac{V_{oc} - V_4}{R_4}$$

If $R_1 = R_2 = R_3 = R_4 = R$,

$$R(I_1 + I_2 + I_3 + I_4) = 4 V_{oc} - V_4 - V_3 - V_2 - V_1 = 0$$

$$V_{oc} = \frac{V_1 + V_2 + V_3 + V_4}{4}$$

Difference Thermocouples. When two thermocouples are mounted with their Seebeck voltages in opposition, their output will depend on the difference in temperature between the measuring junctions (Fig. 10.39). This is a configuration used to make differential thermal analysis. That method of analysis is to heat a material to be analyzed, and also a standard inert material, at a constant rate. One thermocouple is mounted in the unknown and one is mounted in the standard. If an endothermic or exothermic reaction occurs in the unknown as it is heated, $T_1 \neq T_2$. Then the difference $\Delta V_S \neq 0$ and can be used to analyze the exothermic and endothermic reactions occurring as the material is heated.

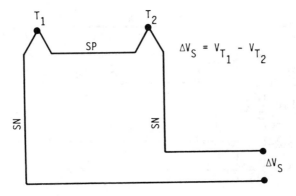

Figure 10.39. Two type S thermocouples to measure the difference in T_1 and T_2.

REFERENCES

1. T. J. Seebeck, Evidence for the thermal current from the combination Bi–Cu by its action on the magnetic needle. *Ab. K. Akad. Wiss. Berlin* **265** (1822–1823).

2. J. C. A. Peltier, Nouvelles expériences sur la caloricité des courans électriques. *Ann. Chim. Phys.* [2] **56**, 371 (1834).

3. A. C. Becquerel, Procédé à l'aide duquel on peut mesurer l'intensité d'un courant électrique, *Ann. Chim. Phys.* [2] **31**, 371 (1826). (According to Le Chateier,[5] p. 130, it was Becquerel who first proposed in 1830 the use of the Seebeck voltage to measure temperature)

4. W. Thomson, Theory of thermoelectricity . . . in crystals. *Trans. Edinburgh Soc.* **21**, 153 (1847).

5. H. Le Chatelier, Sur la variation produite par une élevation de température, dans la force électromotice des couples thermo-électrique. *C. R. Hebd. Seances Acad. Sci.* **102**, 819–822 (1886); De la mesure des température elévées par couples thermo-électrique. *J. Phys.* **6**, 23–31 (1987); see also H. Le Chatelier and O. Boudouard, "High Temperature Measurements" (Transl. by G. K. Burgess), 2nd ed. Wiley, New York, 23–31, 1907.

6. A. J. Dekker, "Solid State Physics," pp. 212–217. Prentice-Hall, Englewood Cliffs, New Jersey, 1957.

7. D. D. Pollock, "The Theory and Properties of Thermocouple Elements," Publ. No. 492. Am. Soc. Test. Mater., Philadelphia, Pennsylvania, 1971.

8. "Manual on the Use of Thermocouples in Temperature Measurement," Publ. No. 470A. Am. Soc. Test. Mater., Philadelphia, Pennsylvania, 1974. (Type N compositions have been added)

9. N. A. Burley and T. P. Jones, Practical performance of nicrosil-nisil thermocouples. In "Temperature Measurement 1975" (B. F. Billings and T. J. Quinn eds.), pp. 172–80. Inst. Phys., London, 1975.

G. W. Burns, The nicrosil versus nisil thermocouple: Recent developments and present status. *Temp.: Its Meas. Control Sci. Ind.* **5** (Part 2), 1121–1127 (1982).

T. P. Wang and C. D. Starr, Nicrosil-nisil thermocouples in production furnaces in the 538°C (1000°F) to 1177°C (2150°F) range. *ISA Trans.* **18**, 83–99 (1979).

G. Majesko, W. R. Brady, and N. A. Burley, Development and application of high performance Ni-Cr-Si/Ni-Si-Mg thermocouple. *Ind. Heat.* **53**, 44–46 (1985).

10. P. A. Kinzie, "Thermocouple Temperature Measurement." Wiley, New York, 1973; see also U.S. Pats. 2,946,835 and 3,060,251 for Boron-graphite thermocouples.

11. U.S. Department of Commerce, Thermocouple reference tables. *NBS Monogr. (U.S.)* **125** (1974).

12. N. A. Burley, R. L. Powell, G. W. Burns, and M. G. Scroger, The nicrosil versus nisil thermocouple: Properties and thermoelectric reference data. *NBS Monogr. (U.S.)* **161** (1978).

13. R. J. Moffat, Understanding thermocouple behavior: The key to precision. *Advances in Metrology* **5**, Annual ISA Test Measurement Symposium, New York, Oct. 28–31, 1968.

14. Johnson Matthey & Co., Ltd. "Thermocouple Compensating Leads" Jan 1972 British Pat. 1,379,546; W. G. Bugden, J. A. Tomlinson, and G. L. Selman, Improved compensation lead systems for platinum base thermocouples. In "Temperature Measurement 1975" (B. F. Billings and T. J. Quinn, eds.), pp. 181–187. Inst. Phys., London, 1975.

15. Leeds and Northrup Company, "Thermocouple Welding," Booklet No. 077987. Philadelphia, Pennsylvania, 1966.

16. General Catalog, Honeywell Process Control Division, Fort Washington, Pennsylvania.

17. R. S. Bradley, "Methods for Servicing Noble Metal Thermocouples," Bull. No. 39. Am. Refract. Inst., Pittsburgh, Pennsylvania, 1933.

18. J. J. Carr, "Elements of Electronic Instrumentation and Measurement," Reston Publ. Co., Reston, Virginia, 1979.

19. Barber-Colman Co., "Instruction Manual," Model 471H Capacitrol, Rockford, Illinois, 1966.

20. P. H. Dike, "Thermoelectric Thermometry," Leeds and Northrup Company, Philadelphia, Pennsylvania, 1954.

21. 2180A and 2190A Digital Thermometers, John Fluke Manufacturing Company, Inc., Everett, Washington, 1980.

22. National Semiconductors, Inc., "National Semiconductors Data/Conversion Acquisition Book, 1984," Santa Clara, California, 1984.

23. Leeds and Northrup Company, "Automatic Thermocouple Reference Systems," Booklet No. 177521. Philadelphia, Pennsylvania, 1968.

CHAPTER 10 PROBLEMS

10.1 A type K thermocouple with a resistance of 1.5 Ω is inserted into a furnace at the zinc point. Its cold junction is connected to a millivoltmeter that has an internal resistance of 500 Ω and is calibrated to read temperature directly from 30 to 600°C when the external resistance is 1.5 Ω and the cold junction is at 30°C.

(a) What temperature will the millivoltmeter indicate if room temperature is 30°C?

(b) How much current will flow in the thermocouple?

(c) If the cold junctions were placed in an ice bath and copper leads were connected from the ice bath to the millivoltmeter, what

temperature would it indicate? Would it be the correct temperature?

(d) If the thermocouple is oxidized until its resistance is 26 Ω, what
will the millivoltmeter indicate? What will be the thermocouple
current? What will be the indicated temperature?

10.2 What will a potentiometer read for each of the circuits shown in Fig.
10.2P?

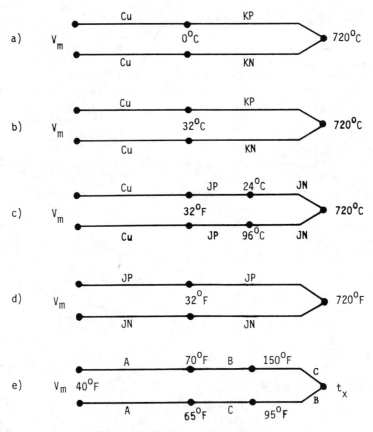

Figure 10.2P. Circuits, Problem 10.2.

10.3 Kirchhoff's law   For the circuit shown in Fig. 10.3P, what is V_m?

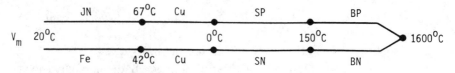

Figure 10.3P. Circuit, Problem 10.3.

10.4 A chromel–alumel thermocouple with a cold junction at 42°C produces an EMF of 14.91 mV. What EMF would a type S thermocouple produce in the same furnace if its cold junction were at 17°C?

10.5 If the Seebeck voltage is given by $V_s = a + bt + \cdots$, what is the Seebeck coefficient, α, for a copper–constantan thermocouple based on the following data for the individual metals referred to platinum?

	a (mV)	b (mV/°C)
Copper	+2.7	+0.0079
Constantan	−35.6	−0.0558

10.6 A type R thermocouple is inserted in a furnace and compensating lead wires are connected from the thermocouple to a potentiometer. The positive terminal at the potentiometer is connected to the red compensating lead wire. When the potentiometer was used, it deflected with reverse polarity, so the connections at the type R thermocouple were reversed to make it read on scale. A thermometer near the furnace wall where the compensating leads join the thermocouple reads 175°C. The potentiometer reading is 6.741 mV. What is the furnace temperature?

10.7 For the circuit shown in Fig. 10.7P, calculate V_M.

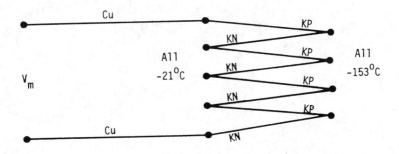

Figure 10.7P. Circuit, Problem 10.7.

10.8 From Fig. 10.8P, calculate V_M. (*Note*: t_c and t_n are the same for each pair.)

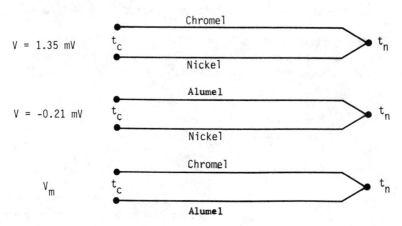

Figure 10.8P. Circuits, Problem 10.8.

CHAPTER 10 ANSWERS

10.1 (a) 419.6°C

(b) $I = \dfrac{17.225 - 1.203 \text{ mV}}{501.5} = \dfrac{16.022 \text{ mV}}{501.5} = 3.19 \times 10^{-5} \text{ A}$

(c) $V = 17.225 + 1.203 = 18.428 \text{ mV}$

$t = 449°C$, not the correct t

(d) $I = \dfrac{16.022}{526} = 3.05 \times 10^{-5} \text{ A}$

$V_{error} = IR_{drop} = [26(3.05) - 1.5(3.19)]10^{-5} = 0.744 \text{ mV}$

$V_{indicated} = 16.022 - 0.744 = 15.278 \text{ mV}$

$$+ V_{30} = \underline{1.203 \text{ mV}}$$
$$16.481 \text{ mV}$$

$t = 402.0°C$

10.2 (a) 29.965 mV

(b) $29.965 - 1.285 = 28.680 \text{ mV}$

(c) $1.225 - 5.050 = -3.825 \text{ mV}$

(d) Cannot answer; depends on t_x.

$V_m = V_{J,720°F} - V_{J,t_x} = 20.866 - V_{t_x}$

(e) $V_m = V_{\underset{70°F}{AB}} + V_{\underset{150°F}{BC}} + V_{\underset{t_x}{CB}} + V_{\underset{95°F}{BC}} + V_{\underset{65°F}{CA}}$

10.3 $V_m = \underset{20°C}{V_{Cu-const}} + \underset{67°C}{V_{const-Cu}} + \underset{0°}{\cancel{V}^{\,0}_{Cu-SP}} + \underset{150°C}{V_{SP-BP}} + \underset{1600°C}{V_{BP-BN}}$

$\qquad + \underset{150°C}{V_{BN-SN}} + \underset{0°}{\cancel{V}^{\,0}_{SN-Cu}} + \underset{42°C}{V_{Cu-Fe}} + \underset{20°C}{V_{Fe-Cu}}$

but

$$\underset{20°C}{V_{Cu-const}} + \underset{20°C}{V_{Fe-Cu}} = \underset{20°C}{V_{Fe-const}},$$

$$\underset{150°C}{V_{SP-BP}} = \underset{150°C}{V_{SP-SN}} + \underset{150°C}{V_{SN-BP}},$$

and

$$\underset{150°C}{V_{BN-SN}} + \underset{150°C}{V_{SN-BP}} = -\underset{150°C}{V_{BP-BN}}.$$

Also,

$$\underset{42°C}{V_{Cu-Fe}} = \underset{42°C}{V_{Cu-const}} + \underset{42°C}{V_{const-Fe}}.$$

So

$$V_m = \underset{20°C}{V_{Fe-const}} + \underset{67°C}{(-V_{Cu-const})} + \underset{150°C}{V_{SP-SN}} + \underset{150°C}{(-V_{BP-BN})} + \underset{1600°C}{V_{BP-BN}}$$

$$\qquad + \underset{42°C}{V_{Cu-const}} + \underset{42°C}{(-V_{Fe-const})}$$

$$= 1.019 - 2.775 + 1.029 - 0.092 + 11.257 + 1.695 - 2.163$$

$$= 9.970 \text{ mV}.$$

10.4 $V_K(t_h) = V_{K,42°C} + V_{observed} = 1.693 + 14.910 = 16.603 \text{ mV}$

$t_h = 405°C$

$V_{obs_S} = V_x = V_{S_{405°C}} - V_{S_{17°C}} = 3.308 - 0.095 = 3.213 \text{ mV}$

10.5 $V_{Cu-const} = V_{Cu-pt} + V_{pt-const} = V_{Cu-pt} - V_{const-pt}$

$\qquad = [a + bt]_{Cu-pt} - [a + bt]_{const-pt}$

$\qquad = 2.7 + 0.0079t - (-35.6 - 0.00558t)$

$\qquad = 38.3 + 0.0637t$

$\qquad \alpha = \dfrac{\partial V_{Cu-const}}{\partial t} = 0.0637$

10.6 Type R

$$V_m = 6.741 = - \underset{175°C}{V_{PtRh-Pt}} + \underset{t_x}{V_{PtRh-Pt}} - \underset{175°C}{V_{PtRh-Pt}}$$

$$= V_{PtRh-Pt} - 2(1.251)$$

$$V_x = 6.741 + 2(1.251) = 9.243 \text{ mV}$$

$$t_x = 903°C$$

10.7 $$V_m = \underset{-21°C}{V_{Cu-Cr}} + \underset{-153°C}{V_{Cr-Al}} + \underset{-21°C}{V_{Al-Cr}} + \underset{-153°C}{V_{Cr-Al}} + \underset{-21°C}{V_{Al-Cr}} + \underset{-153°C}{V_{Cr-Al}}$$

$$+ \underset{-21°C}{V_{Al-Cr}} + \underset{-153°C}{V_{Cr-Al}} + \underset{-21°C}{V_{Al-Cu}}. \tag{1}$$

The first and last terms can be combined:

$$\underset{-21°C}{V_{Al-Cu}} + \underset{-21°C}{V_{Cu-Cr}} = \underset{-21°C}{V_{Al-Cr}}. \tag{2}$$

Substituting (2) into (1) and collecting terms gives us

$$V_m = 4\underset{-21°C}{V_{Al-Cr}} + 4\underset{-153°C}{V_{Cr-Al}} = 4(\underset{-153°C}{V_{Cr-Al}} - \underset{-21°C}{V_{Cr-Al}})$$

$$= 4(-7.387 + 1.208)$$

$$= 4(-6.179) = -24.716 \text{ mV}.$$

10.8 $$V_m = V_{KN,KP} = - (V_{KP,Ni} + V_{Ni,KP}) = -(1.35 - 0.21) = -1.14 \text{ mV}$$

11

THEORY OF RADIANT HEAT TRANSFER AS A BASIS FOR TEMPERATURE MEASUREMENT BY RADIANT TECHNIQUES

11.1 INTRODUCTION

In range IV of IPTS, temperature is defined using Planck's law and the measuring instrument is not specified. Originally, IPTS used Wien's law and the disappearing filament optical pyrometer was the only suitable measuring instrument. Now, photoelectric instruments are more powerful and used by metrology laboratories to implement IPTS in range IV. Both spectral and total radiation instruments have become more powerful, so that measurements can be made in ranges III, range II, and even, experimentally, in range I. In this chapter we present the theory of radiant heat transfer as a necessary background for the instruments described in Chapters 12, 13, and 14.[1,2,3,4] Additional theory is included in Chapters 12, 13, and 14 as necessary to the concepts in those chapters.

11.2 KIRCHHOFF'S LAW OF OPTICS

When visible light or any other form of electromagnetic radiation approaches the surface of a body, conservation of energy requires that it be either transmitted, reflected, or absorbed. The simplest case is for near-normal incidence (Fig. 11.1). If a given intensity of light I_0 impinges on the surface, a portion can be reflected by the surface, I_r, a portion can be transmitted, I_t, and a portion can be absorbed at the surface, I_a (not shown). We can calculate the absorbed portion,

$$I_a = I_0 - I_r - I_t ,$$

(11.1)

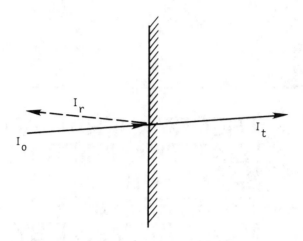

Figure 11.1. Surface irradiated by I_0, showing the reflected I_r and transmitted I_t components.

because to conserve energy,

$$I_0 = I_r + I_t + I_a . \tag{11.2}$$

Dividing through by I_0, we obtain

$$1 = \frac{I_r}{I_0} + \frac{I_t}{I_0} + \frac{I_a}{I_0} . \tag{11.3}$$

Letting the fraction reflected, transmitted, and absorbed be indicated by r, t, and a, respectively, this can be written as Kirchhoff's law of optics:

$$1 = r + t + a . \tag{11.4}$$

We will discuss each term of Eq. (11.4) separately.

11.3 REFLECTANCE AND REFLECTIVITY

Reflection of electromagnetic waves is the result of the interaction of the oscillating electric field of the wave with the electrons of the solid. Solids are generally classified as dielectrics or as electronic conductors in deriving the equations for reflection from the Maxwell relationships. However, many real materials are not well represented by the theoretical relationships because of surface roughness, scattering in transparent or semitransparent materials, electronic structure differing from classical theory, inhomogeneity in composition, grain boundaries, porosity, oxide coatings, and other considerations. The property of reflection from a homogeneous, ideally smooth surface is called reflectivity. The reflection from a non-optically smooth or inhomogeneous material is called reflectance. Understanding the reflectivity of a body will help us understand the thermal emission properties of real bodies. This is true because reflection prevents absorption or transmission, and absorptivity is directly related to emissivity. As we will see later, for an opaque body, absorptivity is equal to emissivity when an energy balance is properly constructed.

Reflection from a surface at any angle not normal to the surface produces polarization of the reflected beam. The general Fresnel equations for reflectivity from a homogeneous, ideally smooth surface, for a beam incident from a vacuum, are, for the reflection polarized perpendicular to the surface,

$$r_{s,\lambda} = \frac{a^2 + b^2 - 2a \cos \theta + \cos^2 \theta}{a^2 + b^2 + 2a \cos \theta + \cos^2 \theta}. \tag{11.5}$$

For the beam polarized parallel to the surface,

$$r_{p,\lambda} = r_{s,\lambda} \frac{a^2 + b^2 - 2a \sin \theta \tan \theta + \sin^2 \theta \tan^2 \theta}{a^2 + b^2 + 2a \sin \theta \tan \theta + \sin^2 \theta \tan^2 \theta} \tag{11.6}$$

where

$$2a^2 = [(n^2 - k^2 - \sin^2 \theta)^2 + 4n^2 k^2]^{1/2} + (n^2 - k^2 - \sin^2 \theta) \tag{11.7}$$

$$2b^2 = [(n^2 - k^2 - \sin^2 \theta)^2 + 4n^2 k^2]^{1/2} - (n^2 - k^2 - \sin^2 \theta) \tag{11.8}$$

and θ is the angle from the normal, n the index of refraction, and k the extinction coefficient. Both n and k depend somewhat on the temperature of the reflector and are very sensitive to the wavelength of the incident radiation. The light reflected is the sum of the light with electrical vector perpendicular, $r_{s,\lambda}$, and parallel, $r_{p,\lambda}$, to the surface. When the incident beam is not polarized, the light reflected is usually estimated from the mean reflectivity, $(r_s + r_p)/2$.

If the incident beam has a frequency associated with the energy ΔE necessary to excite an electron from a low level to a higher level, strong resonant absorption occurs, and the beam is strongly absorbed, not reflected. Absorption is a maximum when

$$\Delta E = h\nu, \tag{11.9}$$

where h is Planck's constant and ν is the frequency. Thus semiconductors such as germanium (Fig. 9.1) are opaque in the visible region but transparent at the lower frequencies because there are no electrons excited from the valence band to the conduction band by the lower-frequency radiation. Therefore, the values of n and k in Eqs. (11.5) to (11.8) are very sensitive to both frequency and the electronic structure of the materials, especially for metals and semiconductors.

Values of n are usually in the range 1.3–1.7 for dielectrics, but can be as low as 0.1 and as high as 35 for metals. For dielectrics, the value of k is effectively zero. For metals it can be greater than 500. The net result is that the reflectance of metals tends to be elliptically polarized if smooth and clean enough for the Fresnel equations to apply.

Many furnace components, and many materials of interest, are dielectrics. Typical dielectric materials are electrical insulators and have a wide band gap. They have much lower reflectivity than metals at most wavelengths and are often transparent where Eq. (11.9) does not permit resonance. For transparent insulators where the absorption index, k, is negligible, an incident beam at angle θ_1 to the normal, from a vacuum, is refracted at the surface to an angle θ_2 in accordance with Snell's law (Fig. 11.2):

$$\sin \theta_2 = \frac{\sin \theta_1}{n}. \tag{11.10}$$

The reflected ray will be polarized and shifted in phase. The reflectivity of the ray polarized with its electric vector perpendicular to the surface is given by

$$r_s = \frac{\sin(\theta_1 - \theta_2)^2}{\sin(\theta_1 + \theta_2)}. \tag{11.11}$$

The ray polarized parallel to the surface has a reflectance given by

$$r_p = \frac{\tan(\theta_1 - \theta_2)^2}{\tan(\theta_1 + \theta_2)}. \tag{11.12}$$

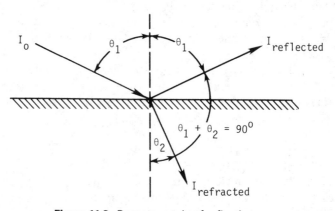

Figure 11.2. Brewster angle of reflection.

When a wave goes from a low-index medium to a high-index medium, there will be an angle θ_1 at which $\theta_1 + \theta_2 = 90°$. This is called the Brewster angle. At that angle, $\tan(\theta_1 + \theta_2)$ is infinite and the parallel ray is not reflected. The reflected ray is composed entirely of the perpendicular component. It is completely polarized (Fig. 11.2). Polarization by reflection from metals is much less specific and tends to higher angles of reflection.

Reflectance depends not only on composition, but also on the molecular structure. For example, fused silica is isotropic but quartz is not. Both have the composition SiO_2, but in fused silica the molecular structure is random and in quartz the molecular structure is crystalline. Quartz has hexagonal periodicity with a different density of electrons in its hexagonal basal plane than it does perpendicular to that plane. Therefore, both the transmissivity and the reflectivity of quartz are anisotropic. The reflectivity depends on the angle of incidence relative to the crystallographic axis. (This is in accord with the Fresnel equations because the index of refraction also depends on the electron packing density.) A 4×4 matrix is needed to describe the polarized reflected rays. Fortunately, most temperature measurement situations have thermal radiation emitted by polycrystalline diffuse surfaces so that only the average reflectance need be considered.

At normal incidence, the angular dependence of the Fresnel equations (11.5) to (11.8) disappears, the reflectivity is unpolarized, and the equations for a wave approaching from and reflected into a vacuum become

$$r = \frac{(n_2 - 1)^2 + n_2 k^2}{(n_2 + 1)^2 + n_2 k^2} . \tag{11.13}$$

For metals, the values of n and k are very wavelength sensitive (Fig. 11.3).[1] Since metals are opaque, this means that transmissivity is zero. Then, in accordance with Eq. (11.4), the absorptivity is very small when the reflectance is very high. This also means that the emissivity is small, as will be discussed later.

For insulators k in Eq. (11.13) is negligible. The index of refraction of a vacuum is 1. For optic systems in air, the index of refraction of air is close enough to 1 that the difference usually can be neglected. When the wave travels in one medium of index n_1 to approach another medium of index n_2, the reflectivity at normal incidence is given by

$$r = \left(\frac{n_2 - n_1}{n_2 + n_1}\right)^2 . \tag{11.14}$$

For example, for sodium light entering a glass of index 1.75 from a liquid of index 1.22, the reflectivity at normal incidence is

$$r = \left(\frac{1.78 - 1.22}{1.78 + 1.22}\right)^2 = 0.0348 . \tag{11.15}$$

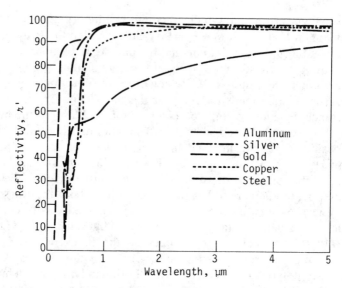

Figure 11.3. Reflectivity of selected metals. (From T. R. Harrison, "Radiation Pyrometry and Its Underlying Principle of Heat Transfer." Copyright 1960, John Wiley & Sons, Inc., New York. Reprinted by permission.)

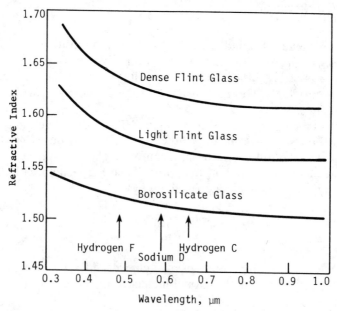

Figure 11.4. Index of refraction of certain glasses as a function of wavelength. (After Ref. 5.)

The index of refraction is important in calculating the reflectivity of an insulator and also, of course, in designing lenses for optical systems often used in radiation temperature measurement. Most lens materials have indices dominated by the tale of the resonant frequency curve of Eq. (11.9). Because the wavelength depends on the frequency,

$$\lambda = \frac{c}{\nu},$$
(11.16)

where c is the velocity of light in the body and ν is the frequency, the reflectivity at normal incidence depends on an index of refraction that decreases with wavelength (Fig. 11.4).[5] As we will see later, the wavelength depends on temperature, so that the fraction reflected depends on temperature. Note also that many metals have oxide layers. Whether Eq. (11.13) or (11.14) applies depends on the oxide properties and thickness.

11.4 TRANSMITTANCE OR TRANSMISSIVITY

Transmissivity is a property of a pure, optically perfect body. Transmittance is what is observed because no real body has perfect optical properties. However, the transmissivity property is approached in optical glasses, fiber optics, infrared transmitting semiconductor lenses, filters, and so on. Real bodies that are the targets for temperature measurement by radiant measuring techniques are usually opaque, or nearly so, but we should be very cautious when transparency can occur because the target will not emit as much as an opaque target would. (See Section 12.8 for the effect of transparency on temperature measurement.)

The transmittance of a body depends on its composition. Glass and crystalline lenses are often intended to be completely transparent. Filters are intended to be partially transparent. However, even transparent materials may be virtually opaque if they have enough grain boundaries and pores distributed through them. That opacity requires considerable thickness and is caused by multiple refraction, reflection, and nonspecific absorption. If no refraction or internal reflection occurs, the transmittance depends on the concentration of absorbing ions through the medium. For highly transparent materials the transmittance is dominated by impurities, and since perfect purity is never obtained, the transmittivity of the pure body is never obtained. Rather, for a uniform distribution of impurities, the reduction in transmission for intensity I going through an infinitesimal thickness dx is given by Beer's law,

$$dI_\lambda = -I_\lambda C_i q_i dx,$$
(11.17)

where C_i is the concentration of element i, and q_i is its molar absorption coefficient for the wavelength of the light of intensity I. The minus sign indicates a reduction in intensity. Separating variables and integrating from an initial intensity I_0 at thickness zero to a final intensity I at thickness ℓ gives

$$I_{\lambda,\ell} = I_0 e^{-\Sigma C_i q_i \ell}. \tag{11.18}$$

The summation is for all species i causing absorption. The value of q_i must be for the wavelength or wavelengths of light being considered. For optical filters, Eq. (11.18) is often written

$$\frac{I}{I_0} = e^{-C\ell} \tag{11.19}$$

where C is a constant representing the summation of the $C_i q_i$ terms.

Example 11.1. Calculate the transmission of a telescope lens for sodium light if the optical path length is 9 mm, its density is 2.50 g/cc, its index of refraction is 1.55, and its impurity properties are as follows:

	C_i (wt%)	q_i (cm^2/mol)
Fe	1.50	1,122
Mn	0.70	2,065
Co	0.01	113,026

Solution

$$I_t = I_0(1-\rho)(e^{-\Sigma C_i q_i \ell})(1-r)$$

$$r = \left(\frac{1.55-1}{1.55+1}\right)^2 = 0.0465 \qquad (1-\rho)^2 = (1-0.0465)^2 = 0.909$$

Fe: $C_i q_i \ell = \dfrac{0.015\,\text{g}}{\text{g}} \;\left|\; \dfrac{2.5\,\text{g}}{\text{cm}^3} \;\right|\; \dfrac{\text{mol}}{55.847\,\text{g}} \;\left|\; \dfrac{1122\,\text{cm}^2}{\text{mol}} \;\right|\; 0.9\,\text{cm}$

$$= 0.678$$

Mn: $C_i q_i \ell = \dfrac{(0.007)(2.5)(2065)(0.9)}{54.938} = 0.594$

Co: $C_i q_i \ell = \dfrac{(0.0001)(2.5)(113026)(0.9)}{58.933} = 0.432$

$$\Sigma\, C_i q_i \ell = 1.703$$

$$\frac{I_t}{I_0} = (1 - r)^2(e^{-\Sigma C_i q_i \ell})$$

$$= 0.909(e^{-1.703}) = 0.166.$$

11.5 ABSORBTANCE AND ABSORPTIVITY

Absorptivity is the property of a pure homogeneous body and either takes place at the surface because of fundamental electronic absorption, or in depth in accordance with Beer's law. Although absorptivity of high-absorbtance materials such as metals is usually considered a surface phenomenon related to the value of k in Eqs. (11.5) to (11.8), reflection requires that the electromagnetic wave penetrate the electron cloud of a metal before it is reflected. But with large values of k, the penetration distance is extremely small. It appears to be a surface phenomenon. When there are energy states available for electronic transitions, Eq. (11.9) applies and strong absorption occurs. If a surface is rough, so that multiple reflections can occur, the absorption is increased because a fraction is absorbed at each reflection. Then the angular dependence of Eqs. (11.5) to (11.8) also is not clearly defined because surface topography causes variation in the angle between the incident beam and the normal to the surface. Absorptance will be discussed further after blackbody concepts are introduced. However, it is clear that radiation not reflected or transmitted is absorbed.

11.6 THE CONCEPT OF A BLACKBODY

If a body is opaque, the transmittance is zero. Black objects have high absorbtivity. In 1860, Kirchhoff studied the effect of geometry on the apparent brightness of a body heated to incandescence. He discovered that when it was suddenly removed from a red-hot furnace, a metal specimen with a hole drilled in it appeared to be hotter inside the hole than elsewhere on the body. His reasoning based on this led to the concept of a blackbody.

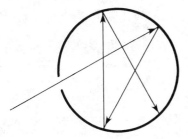

Figure 11.5. Planar representation of a blackbody by a small hole in a hollow sphere as a perfect absorber.

First he said that a blackbody is one that absorbs all radiation that falls on it. None is reflected or transmitted. This could be represented by a tiny hole into a large hollow sphere, with the interior surface of the sphere painted black. Any light entering the hole would be absorbed inside the sphere. It would not be reflected back out the hole because of the multiple reflection needed to have the light beam return to the exit, and because of the very high absorption at each reflection (Fig. 11.5).

11.7   THE BLACKBODY AS A PERFECT EMITTER

If a hollow blackbody sphere is heated to incandescence, every part of the interior can emit radiation out through the hole (Fig. 11.6). The apparent brightness of a hole drilled in a metal bar when suddenly removed from the furnace can be attributed to this affect. Kirchhoff recognized the difference between the brightness of the metal surface and the hole drilled in the bar as a difference in emittance. The effect was so rapid that the metal surface had not had enough time to cool. There was a difference between the emittance of the surface and the emittance of the hole.

11.8   EMITTANCE AND ABSORBANCE OF A BLACKBODY

If the ratio of the diameter to the length of a hole drilled in a real body is reduced, the brightness of the hole changes. This can be shown by drilling a series of holes in a metal rod, varying the length-to-diameter ratio of the holes. At a ratio of 1:1, the hole and the surface, when suddenly removed from a furnace at incandescent temperatures, appear to be of nearly the same brightness. As the ratio is changed from 1:1 to 1:2 to 1:4 to 1:8 to 1:10 to 1:20 by drilling successively smaller-diameter holes all the same depth and repeating the experiment with a bar containing all the holes, the smaller holes appear brighter than the surface. However, there does not seem to be any difference between the brightness of the three smallest holes.

These facts led Kirchhoff to recognize that a true blackbody would have

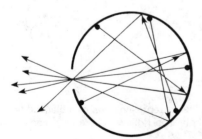

Figure 11.6. Planar representation of a small hole in a hollow sphere as a perfect emitter.

a certain brightness that depended on temperature, and that all blackbodies would have the same brightness if they were at the same temperature. He recognized also that heat at incandescent temperatures is transferred by radiation, and conducted a remarkable thought experiment that led to a relationship between emittance and absorbance.

Consider a black body as a circular surface heated by a hemisphere above it. Heat would flow from the hemisphere to the body until they both came to the same temperature. Suppose, then, that the absorbance of the blackbody were greater than its emittance. The body would reject less heat than it received, so it would continue to become hotter, to a temperature even greater than that of the source. This is impossible. Therefore, for a blackbody absorbance must be equal to emittance. For a blackbody, because t and r are zero in Eq. (11.4), it must be true that

$$a = e = 1 , \tag{11.20}$$

where e is the emittance. This is a remarkable result because it was discovered by Kirchhoff in 1860, long before the fundamentals of radiant heat transfer were understood.

11.9 EMITTANCE FROM REAL BODIES: BLACKBODY CONDITIONS

We now recognize that real bodies are not blackbodies, although some, such as rough graphite, approach blackbody properties. When a shiny metal such as platinum is held in an incandescent furnace at 1500°C, both the furnace and the metal appear to be equally bright when viewed through a small hole in the furnace wall, that is, when viewed under blackbody conditions. If the metal is close to the door so that it can radiate through a large area when the door is opened, it will appear to be darker than the furnace. This is true because its emittance is less than that of the refractories in the furnace. For the same experiment with an easily oxidized metal, the emittance becomes higher as the oxide layer is formed. If the surface is roughened by sanding, the roughened portion appears brighter. Thus we see that opaque objects will have emittance that depends on surface roughness, oxide films, or scales, and shape.

11.10 DEFINITIONS AND NOMENCLATURE FOR BLACKBODY RADIATION

There are a number of different ways to describe blackbody radiation.[3] We will explain them with the basic premises that radiation is emitted from the blackbody solely as the result of its temperature, that it is emitted into free space in all directions, and that it travels in straight lines once emitted. A blackbody is one that, as the result of its thermal energy level (i.e., its temperature), emits radiant energy at the maximum power possible. That rate is given by Planck's law at all temperatures above absolute zero. Planck's law also describes the frequencies (wavelengths) at which the radiant energy is emitted. Let us start by considering a blackbody as a flat plane, select on its top surface an infinitesimal area, dA, and consider the radiation emitting from dA using a hemisphere of unit radius to measure the angular relationships (Fig. 11.7).

We start with the definition of intensity. The intensity, I, of a blackbody is the radiant energy (total in the sense that it includes all frequencies emitted) emitted within an infinitesimal solid angle $d\omega$ normal to the surface of the blackbody from an infinitesimal area dA in the time interval dt.* Because it is normal to the surface, it is a vector quantity. It has units

$$I = \frac{\text{energy}}{\text{time} \times \text{area} \times \text{steradian}}.$$ (11.21)

The intensity of a blackbody in free space is everywhere invariant. It describes the blackbody source.

When viewed at an angle to the surface as shown in Fig. 11.7, the intensity is reduced according to the Lambert cosine law for black bodies:

$$I_\theta = I \cos \theta.$$ (11.22)

However, the projected area dA in the direction θ is given by $dA \cos \theta$, so

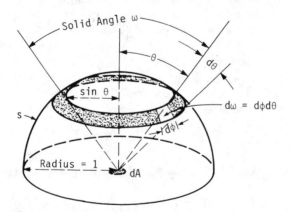

Figure 11.7. Geometrical relationships for an emitting blackbody.

* dt is used here for time to avoid confusion with $d\theta$ as an angular measurement.

the radiant emission per unit of projected area, known as the radiance L, is given by

$$L = \frac{I \cos \theta}{\cos \theta} = I. \tag{11.23}$$

Thus the radiance of a blackbody is independent of angle and equal in magnitude to the intensity. This is only true where blackbody properties are exhibited. The units of L are also

$$L = \frac{\text{energy}}{\text{area} \times \text{time} \times \text{steradian}}. \tag{11.24}$$

The intensity and radiance described above are for all the energy emitted by the body at all frequencies. For temperature measurements we are particularly interested in monochromatic (or nearly monochromatic) radiation. The monochromatic intensity is defined as radiant energy emitted by a body between frequencies v and $v + dv$ within an infinitesimal solid angle $d\omega$ normal to the blackbody surface from an infinitesimal area dA in the time interval dt. The monochromatic intensity is also a vector quantity. It has units

$$I_v = \frac{\text{energy}}{\text{area} \times \text{time} \times \text{steradian} \times 1/\text{time}}$$

$$= \frac{\text{energy}}{\text{area} \times \text{steradian}}. \tag{11.25}$$

For a blackbody the spectral radiance is also equal to the monochromatic intensity

$$\text{spectral radiance} \equiv L_v = I_v. \tag{11.26}$$

This is true only for blackbody behavior.

For most purposes, L_v or I_v is better than L_λ or I_λ because the frequency of a radiant beam is invariant. If the beam passes through a liquid of different index of refraction, the wavelength and the velocity will change. For many temperature measuring applications, wavelength is used by convention or for convenience. The International Practical Temperature Scale specifies monochromatic radiance L_λ instead of L_v. Usually, this causes no inconvenience, as

$$\lambda = \frac{c}{v}, \tag{11.27}$$

where c is the velocity of light in the medium and v is the frequency.

Another way of describing blackbody radiation is to express definitions in terms of hemispherical properties. This is useful in heat transfer situations because hemispherical measurements measure the loss in all directions available from a plane surface. Much of the literature on total radiation pyrometry is written in hemispherical radiant terms, so they are described here.

For the blackbody in Fig. 11.7, the radiant energy escapes from the area dA in all directions above the plane. The entire loss from the surface is given by integrating the intensity over the entire solid angle of the hemisphere:

$$\text{hemispherical total radiant intensity} \equiv W = \int_0^{\pi} I_\theta \, d\omega \, . \qquad (11.28)$$

By Eq. (11.17),

$$I_\theta = I \cos \theta$$

$$d\omega = \sin \theta \, d\theta \, d\phi$$

$$W = \int_0^{2\pi} \int_0^{\pi/2} I \cos \theta \sin \theta \, d\theta \, d\phi \qquad (11.29)$$

$$= \pi I \, . \qquad (11.30)$$

Hemispherical radiation is an energy flux. It is a vector quantity with units,

$$W = \frac{\text{energy}}{\text{area} \times \text{time}} \, . \qquad (11.31)$$

By restricting our measurement to a frequency between the interval ν and $\nu + d\nu$ as before, we can make a monochromatic hemispherical radiant intensity measurement. This radiant flux has dimensions

$$W_\nu = \frac{\text{energy}}{\text{area} \times \text{time} \times 1/\text{time}}$$

$$= \frac{\text{energy}}{\text{area}} \, . \qquad (11.32)$$

Instead of integrating I_θ over the entire hemisphere, we could integrate over any angle desired. When the solid angle is taken as 1 steradian about the normal,

$$W_{\omega=1} = \int_0^{\theta} 2\pi \cos \theta = 2\pi(1 - \cos \theta) = 1$$

$$\cos \theta = 1 - \frac{1}{2\pi} = 0.8409$$

$$\sin^2 \theta = 1 - \cos^2 \theta = 0.2929$$

$$W_{\omega=1} = \pi I \sin^2 \theta = 0.9201 I. \tag{11.33}$$

Here we see the interesting result that the rate of energy emission for a solid angle of 1 steradian about the normal is only 92% of the rate of emission normal to the surface.

If the radiant energy release rate from a surface is desired, we can integrate over the surface. The entire thermal loss, J, from the surface is given by

$$J = \int W \, dA. \tag{11.34}$$

The spectral loss is

$$J_\nu = \int W_\nu \, dA. \tag{11.35}$$

The units for the thermal energy loss are

$$J = \frac{\text{energy}}{\text{time}} \tag{11.36}$$

and

$$J_\nu = \frac{\text{energy}}{\text{time} \times 1/\text{time}} = \text{energy}. \tag{11.37}$$

11.11 PLANCK'S LAW OF BLACKBODY RADIATION

Prior to 1900 the wave theory of electromagnetic radiation had been used successfully for many purposes. The Rayleigh–Jeans equation was successful in explaining radiant emission at long wavelengths. Wien's law had been derived in 1891 and fit the short-wavelength portion of radiant energy emission. Neither was satisfactory over the entire range of temperature in representing experimental data. Max Planck considered the models used by Rayleigh and Jeans and by Wien. He used a similar model, a closed container at an equilibrium temperature containing a constant field density of radiant energy. He assumed the radiation to be the result of a field of ideal linear oscillators. He devised a phase-plane representation to describe the momentum versus displacement of any individual oscillator, and he concluded that the area of the phase plane under the curve of possible microstates must be a constant, an element of action we know as Planck's constant, h. Then the oscillator could only contribute energy to the radiation field in multiples of $h\nu$ because the quantitization of the oscillator prohibited intermediate-energy states. Based on this assumption, he was able to derive the famous law, which does appear to represent experimental data over the entire range of experimental temperatures:

$$I_\nu = \frac{2h\nu^3 n^2}{c^2} \frac{1}{\exp(h\nu/kT) - 1},\qquad(11.38)$$

where h is Planck's constant, ν is frequency, n is the index of refraction of the medium into which the radiation is emitted, c is the velocity of light, k is Boltzmann's constant, and T is absolute temperature. For a vacuum $n = 1$; for air, $n = 1.0028$.

The intensity for temperature measuring purposes is frequently expressed in terms of wavelength λ. Then Planck's law is

$$I_\lambda = \frac{2hc^2}{n^2\lambda^5} \frac{1}{\exp[(hc/nk)(1/\lambda T)] - 1}\qquad(11.39)$$

or

$$I_\lambda = C_1\lambda^{-5}(e^{C_2/\lambda T} - 1)^{-1}.\qquad(11.40)$$

Here the first radiation constant C_1 has a value $C_1 = 2hc^2/n^2$, which, for n is 1 and c is the velocity of light in a vacuum, has a value of 1.19089×10^{-16} J·m²/s. The second radiation constant $C_2 = hc/nk$, which for vacuum properties has the value 0.0143883 m-K.

Planck's law can also be written for hemispherical spectral radiant intensity.

$$\begin{aligned} W_\lambda &= \pi I_\lambda \\ &= \frac{\pi hc^2}{n^2\lambda^5} \frac{1}{\exp[(hc/nk)(1/\lambda T)] - 1}. \end{aligned}\qquad(11.41)$$

Then C_1 above must be multiplied by π for radiation into a vacuum and has the value 3.7413×10^{-16} J-m²/s.

Plank's law gives the intensity of radiation per unit wavelength interval per unit of area at absolute temperature. It is a theoretical relationship that agrees with experiment within experimental limits, and for temperature measurements in region IV of the IPTS, it is assumed to be exact. The intensity is a function of both λ and T (Fig. 11.8). At low temperatures the

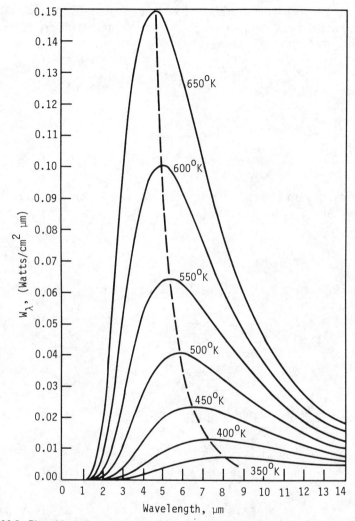

Figure 11.8. Planck's law intensity I_v at various temperatures. (From T. R. Harrison, "Radiation Pyrometry and Its Underlying Principle of Heat Transfer." Copyright 1960, John Wiley & Sons, Inc., New York. Reprinted by permission.)

intensity is very weak and the maximum occurs at very long wavelengths. As the temperature increases, the intensity rises and the maximum shifts to shorter frequencies. This is apparent to an observer. At 1000°C the color of a furnace enclosure viewed through a small hole is a pleasing red. At 1600°C it is a brilliant painful white. The sun's surface temperature, estimated at 5500°C, has an intensity maximum in the visible spectrum. In contrast, the maximum intensity does not fall in the visible spectrum at most furnace temperatures.

Planck's law can be differentiated with respect to λT and set equal to zero to obtain the equation for the maximum. Starting with Eq. (11.40) and writing it with C_1 and C_2, we have

$$I_\lambda = \frac{C_1}{\lambda^5}(e^{C_2/\lambda T} - 1)^{-1}. \tag{11.42}$$

Multiplying by $(T^5/C_2^5)(C_2^5/T^5)$ gives

$$I_\lambda = \frac{C_1 T^5}{C_2^5}\left(\frac{C_2}{\lambda T}\right)^5 (e^{C_2/\lambda T} - 1)^{-1}. \tag{11.43}$$

Differentiating with respect to λT can be, for convenience, with respect to $C_2/\lambda T$ because C_2 is a constant. Let $C_2/\lambda T = x$. Then

$$I_\lambda = \frac{C_1 T^5}{C_2^5} x^5 (e^x - 1)^{-1}. \tag{11.44}$$

For any temperature we can hold T constant. Then

$$\frac{dI_\lambda}{dx} = -x^5(e^x - 1)^{-2}e^x + (e^x - 1)^{-1}(5x^4) \tag{11.45}$$

$$= \frac{C_1 T^5}{C_2^5} \frac{x^4}{e^x - 1}\left(\frac{-xe^x}{e^x - 1} + 5\right) = 0. \tag{11.46}$$

Then x_{max} can be calculated from the term in parentheses:

$$\frac{x_m e^{x_m}}{e^{x_m} - 1} = 5 \tag{11.47}$$

$$x_m = \frac{5(e^{x_m} - 1)}{e^{x_m}} \tag{11.48}$$

$$= 5(1 - e^{-x_m}) \tag{11.49}$$

$$\frac{C_2}{\lambda_m T} = 5(1 - e^{C_2/\lambda_m T}). \tag{11.50}$$

Equation (11.50) cannot be solved directly, but taking $C_2/\lambda_m T$ with an initial value of 5, successive approximations gives a value of 4.9651142. Then

$$\lambda_m T = \frac{C_2}{4.9651142} = 0.00289787 \text{ m-K}$$

$$= 2897.87 \text{ } \mu\text{m-K}. \tag{11.51}$$

Note that C_1 was eliminated by differentiation, so this also applies to W_λ. This result is plotted in Fig. 11.8 as a dashed line.

11.12 WIEN'S LAW OF BLACKBODY RADIATION

Wilhelm Wien, in 1891, derived an equation for radiant intensity from a blackbody that was later replaced by Planck's law. Wien's law differs only by the subtraction of 1 from the denominator.

$$I_\lambda = C_1 \lambda^{-5} e^{-C_2/\lambda T}. \tag{11.52}$$

At short wavelengths, when the disappearing filament optical pyrometer is used, Wien's law departs only slightly from Planck's law (Fig. 11.9). The

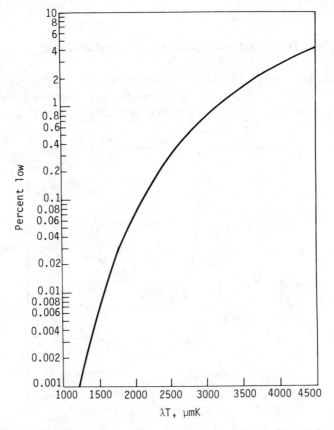

Figure 11.9. Comparison of Wien's law and Planck's law. (From T. R. Harrison, "Radiation Pyrometry and Its Underlying Principle of Heat Transfer." Copyright 1960, John Wiley & Sons, Inc., New York. Reprinted by permission.)

error in using Wien's law is

$$E = \frac{W_\lambda(\text{Planck}) - W_\lambda(\text{Wien})}{W_\lambda(\text{Planck})}$$

$$= \frac{[e^{C_2/\lambda T} - 1]^{-1} - e^{-C_2/\lambda T}}{[e^{C_2/\lambda T} - 1]^{-1}} = e^{-C_2\lambda T} . \tag{11.53}$$

If a radiation pyrometer is used with a wavelength of 10μm (about the largest now in use), the error is

$$E = e^{-C_2/\lambda T} = e^{-1438.3/\lambda T} . \tag{11.54}$$

Taking $T = 1000\,\text{K}$, this gives an error of 0.237 or 23.7%. In the visible spectrum, using $0.655\,\mu\text{m}$ as the wavelength, the error would have been only $2 \times 10^{-8}\,\%$.

11.13 DEVIATIONS FROM BLACKBODY CONDITIONS

Real bodies are not blackbodies. Only under special conditions can blackbody properties be approached. These are used for calibrating devices and require extreme care in preparation. We must therefore expect nonblack-

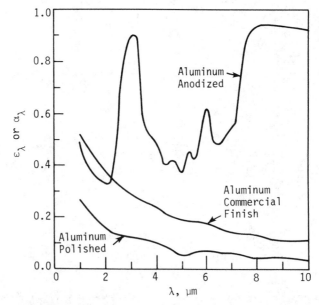

Figure 11.10. Emittance versus wavelength for aluminum. For a blackbody, $\epsilon = \alpha = 1$. (After Ref. 2. Reprinted by permission McGraw-Hill.)

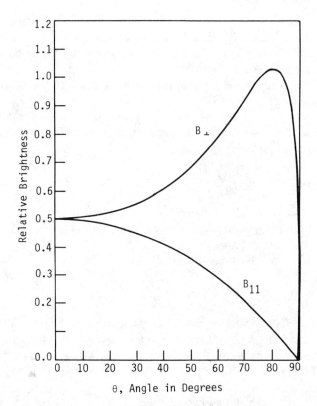

Figure 11.11. Perpendicular and parallel components ($B_\perp$ and $B_\parallel$) emitted by incandescent tungsten at 0.665 μm. (From T. R. Harrison, "Radiation Pyrometry and Its Underlying Principle of Heat Transfer." Copyright 1960, John Wiley & Sons, Inc., New York. Reprinted by permission.)

body conditions in almost every real radiant energy temperature measurement situation. The deviations from blackbody behavior are often extreme (Fig. 11.10).[2] We need to define special terms for deviation from blackbody conditions.

Nonblackbodies do not obey the radiance conditions that $I = L$ or $I_\nu = L_\nu$. Emittance from a nonblackbody may be very irregular with angle, depend on surface conditions, and produce polarized radiation (Fig. 11.11).[1]

11.14 EMITTANCE AND EMISSIVITY DEFINED

In previous sections we said that a blackbody had an emittance of 1. It absorbed or emitted without reflectance or transmission. Real bodies have emittances of less than 1, but the emittance depends on shape, surface roughness, and oxidation or other surface effects. We need to define emittance more precisely and to introduce the concept of emissivity. Emissivity is often used in the literature, erroneously, to describe emittance. It is important to make a clear distinction.

Emissivity is a fundamental property of a material. It is used to describe the emittance from it under the conditions that it is perfectly flat, even on a molecular scale, and is free from porosity, surface scale, oxide coating, or anything else that would alter its properties. For such a perfect material emissivity ε is defined as

$$\varepsilon \equiv \frac{J_p}{J_{BB}}, \tag{11.55}$$

where J_p is the radiant flux density coming from a given surface of the perfect body and J_{BB} is the radiant flux density coming from a blackbody under the same conditions.

Emittance has exactly the same form,

$$e \equiv \frac{J_p}{J_{BB}} \tag{11.56}$$

but the radiant flux density J_R is that coming from the real body as it exists in service with whatever surface roughness, oxide coatings, curvature, or other conditions exist. In most situations we deal with emittance rather than emissivity, and many of the data tabulated in the literature as emissivity are really emittance.

For metals at high temperatures the fresh, clean surface has a low emissivity. Often oxidation is so rapid that true emissivity measurements are very difficult. As the oxide film builds up, the emittance may change rapidly, especially if the oxide is black, such as mill scale on iron.

Ceramic materials at high temperatures are often polycrystalline porous materials. Their emittance will depend on the method of manufacture as well as the composition. For example, aluminum oxide, Al_2O_3, is available as single-crystal sapphire, as dense translucent aluminum oxide, as dense white opaque fine-grained polycrystalline aluminum oxide, as porous insulating brick, and as fiber blanket for high-temperature insulation. All forms are pure aluminum oxide, but the emittance varies from 0.15 for the single-crystal sapphire to 0.55 for the fiber blanket.

Emissivity and emittance, as defined above, can be taken at a single wavelength; for radiation between two wavelength limits, λ_1 and λ_2; or for all wavelengths. These are designated spectral e_λ, partial $e_{\lambda_1-\lambda_2}$, or total e_t.

Because emittance must be known to convert radiant heat transfer measurements to temperature, it is essential that emittance be properly determined for the real material being studied in order to obtain accurate temperature measurements. Some emittances are given in Table 11.1, but *they must be used with caution.*[6] It is not possible to tabulate emittances that are accurate for general use.

TABLE 11.1 Monochromatic Emittance Values, e_λ

Surface	Wavelength (μm)									
	0.50	0.60	0.95	1.8	2.1	3.6	4.4	5.4	8.8	9.3
Aluminum										
Polished		0.26			0.17	0.08		0.05		0.04
Oxidized						0.18		0.12		0.11
Duralumin		0.53								
Chromium		0.49	0.43			0.36	0.26	0.17		0.08
Copper										
Polished			0.26		0.17	0.18		0.05		0.04
Oxidized						0.77		0.83		0.87
Gold, polished					0.03	0.03		0.02		0.02
Steel, polished		0.45	0.37		0.23	0.14		0.10		0.07
Iron										
Polished		0.45	0.35		0.22	0.13		0.08		0.06
Cast, oxidized						0.76		0.66		0.63
Galvanized, new	0.66	0.66	0.67	0.42						0.23
Galvanized, dirty	0.89	0.89	0.89	0.90						0.28
Galvanized, whitewashed	0.24	0.21	0.22	0.37						
Magnesium		0.30	0.26		0.23	0.18		0.13		0.07
Monel		0.43	0.29		0.16	0.10				
Molybdenum	0.55		0.43		0.18	0.11		0.08		0.06
Brass										
Polished										0.05
Oxidized										0.61
Paint										
Lampblack		0.97	0.97				0.97		0.96	
Red		0.74	0.59				0.70		0.96	
Yellow	0.39	0.30					0.59		0.95	
White	0.18	0.14	0.16				0.77		0.95	
Aluminum		0.45								
Silver, polished	0.11		0.04		0.03	0.03		0.02		0.01
Platinum black	0.97		0.97		0.97	0.97	0.96	0.96	0.93	0.93
White paper		0.28	0.25			0.82				0.95
White marble		0.47				0.93				0.95
Graphite	0.78		0.73		0.64	0.54		0.49		0.41

For an opaque nonblackbody, Eq. (11.5) can be written

$$e_\lambda = 1 - r_\lambda .\tag{11.57}$$

In using this equation it is important that e_λ relate properly to r_λ, that is, that the proper energy balance be used. Because r_λ is a function of θ, e_λ must also be a function of θ. Most temperature measurement problems require that e_λ be known at or near the normal. But because an incident beam can be reflected hemispherically, this requires that r_λ be for hemispherical reflectance in calculating e_λ for a particular direction.

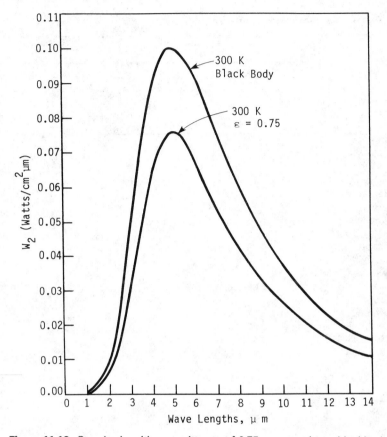

Figure 11.12. Gray body with an emittance of 0.75 compared to a blackbody.

11.15 GRAY BODY

When a body has a constant emittance over the whole range of λT, it is known as a gray body (Fig. 11.12). Gray bodies probably do not actually exist, but some materials and some measuring conditions where a blackbody cavity is approximated permit us to use gray body calculations.

11.16 THE STEFAN–BOLTZMANN LAW

In 1879, Stefan analyzed John Tyndal's data for heat emitted from incandescent metal and concluded that the heat transfer was proportional to the fourth power of the absolute temperature. In 1844, Boltzmann calculated fourth-power dependence from a thermodynamic argument. Planck's law in 1900 confirmed the relationship and gave a sound basis for the Stefan–Boltzmann constant. The Stefan–Boltzmann law is obtained by integrating Planck's law over all wavelengths. Starting with Eq. (11.41),

$$W_\lambda = \pi C_1 \lambda^{-5}(e^{C_2/\lambda T} - 1)^{-1}.$$

We want to find

$$W = \pi \int_0^\infty W_\lambda \, d\lambda = \pi \int_0^\infty C_1 \lambda^{-3}(e^{C_2/\lambda T} - 1)^{-1}\lambda^{-2} \, d\lambda. \qquad (11.58)$$

Since $d(\lambda^{-1}) = -\lambda^{-2} \, d\lambda$,

$$W = \pi C_1 \int_0^\infty \frac{\lambda^{-3} d(\lambda^{-1})}{e^{C_2/\lambda T} - 1}. \qquad (11.59)$$

Multiplying by $(T^4/C_2^4)(C_2^3/T^3)(C_2/T)$ gives us

$$W = \pi \frac{C_1 T^4}{C_2^4} \int_0^\infty \frac{(C_2/\lambda T)^3 d(C_2/\lambda T)}{e^{C_2/\lambda T} - 1}. \qquad (11.60)$$

Now let $x = C_2/\lambda T$. The equation then becomes

$$W = \pi \frac{C_1 T^4}{C_2^4} \int_0^\infty \frac{x^3 \, dx}{e^x - 1}. \qquad (11.61)$$

Because

$$\frac{1}{e^x - 1} = \frac{1}{e^x - 1} \frac{e^{-x}}{e^{-x}} = \frac{e^{-x}}{1 - e^{-x}} = e^{-x} + e^{-2x} + e^{-3x} + \cdots + e^{-\infty},$$

we can write

$$W = \pi \frac{C_1 T^4}{C_2^4} \left(\int_0^\infty x^3 e^{-x} \, dx + \int_0^\infty x^3 e^{-2x} \, dx + \int_0^\infty x^3 e^{-3x} \, dx + \cdots \right). \qquad (11.62)$$

Each of the integrals has a solution in the form

$$\int_0^\infty x^\eta e^{-ax} \, dx = \frac{\eta!}{a^{\eta+1}}.$$

With $\eta = 3$,

$$W = \pi \frac{C_1 T^4}{C_2^4} \left(\frac{3!}{1^4} + \frac{3!}{2^4} + \frac{3!}{3^4} + \cdots + \frac{3!}{\infty^4} \right) \qquad (11.63)$$

$$= \frac{3! \, C_1 T^4}{C_2^4} \sum_{\eta=1}^\infty \frac{1}{\eta^4}. \qquad (11.64)$$

The summation has a value $\pi^4/90$. Substitution yields

$$W = \pi\left(\frac{C_1}{C_2^4}\frac{6\pi^4}{90}\right) T^4. \tag{11.65}$$

The term in parentheses is the Stefan–Boltzmann constant σ. In vacuum

$$\sigma = \frac{C_1\pi^4}{C_2^4(15)} = 5.6687 \times 10^{-8} \text{ J/K}^4\text{-m}^2\text{-s}$$

Then, in air,

$$W = \sigma T^4 = 5.6718 \times 10^{-8} T^4. \tag{11.66}$$

This equation applies if all wavelengths are received by the measuring instrument, and is approximated by the total radiation pyrometer discussed in Chapter 14. Some optical pyrometers use a very narrow band of wavelengths, so Planck's law applies. However, many of the new instruments measure a portion of the wavelengths in the near infrared, and total radiation pyrometers having lenses or mirrors do not accept all wavelengths. Therefore, we need to know what the integrated intensity will be for various regions of the spectrum. This requires us to integrate between certain wavelengths, as discussed in the next section.

11.17 BLACKBODY RADIATION WITHIN WAVELENGTH INTERVALS

Calculation of a portion of the Planck's law distribution could be attempted for the appropriate wavelength interval. Fortunately, this is not necessary because tables are available for various intervals of λT so that the fraction F of the integrated intensity between zero and a given value of λT is tabulated.[1] Then

$$W_{0\rightarrow\lambda T} = F_{0\rightarrow\lambda T}\,\sigma T^4. \tag{11.67}$$

To measure an interval $(\lambda T)_1$ to $(\lambda T)_2$,

$$W = [F_{(\lambda T)_2} - F_{(\lambda T)_1}]\sigma T^4. \tag{11.68}$$

A graph of F versus λT is given in Fig. 11.13.

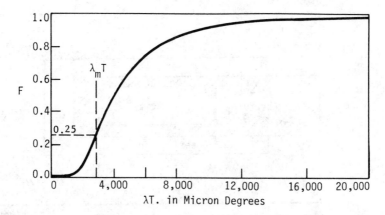

Figure 11.13. Fraction of integrated Planck's law from 0 to λT compared to integration from 0 to ∞. (From T. R. Harrison, "Radiation Pyrometry and Its Underlying Principle of Heat Transfer." Copyright 1960, John Wiley & Sons, Inc., New York. Reprinted by permission.)

11.18 BLACKBODY RADIATION SOURCES

To calibrate instruments it is necessary to have blackbody radiation sources. A blackbody source radiates all radiation possible for a given absolute temperature in accordance with Planck's law. Experimentally, true blackbody behavior is unattainable, but the difference between true blackbody behavior and actual behavior can be made essentially negligible. To test a source, it is necessary to have a cavity at constant temperature. This is done by surrounding the cavity with a high-temperature thermodynamic invariant point. For example, various cavity configurations can be compared by surrounding each with a mixture of solid and liquid pure gold at the freezing point of gold. Different materials or different surface roughnesses could also be compared for a given cavity design at a gold point isothermal arrest. Examples of actual blackbody sources are discussed in Chapter 17. However, a schematic of the requirements is given in Fig. 11.14.

Referring to the figure, we see a long, narrow hole drilled into a block surrounded by a constant temperature produced by the liquid–solid invariant point. Usually, the isothermal block is made of graphite or similar material of high emittance. There is a well-defined baffle and exit aperture, with a planar exit angle of about 20° for the exit cone. Many different designs for blackbody sources are found in the literature. Various radiant energy pyrometers can be used to compare sources. For a particular pyrometer, the source with the highest output for as-given temperature should be considered closest to blackbody behavior.

It is not actually possible for the cavity to be everywhere at the melting point of the medium surrounding it. The opening through which radiation

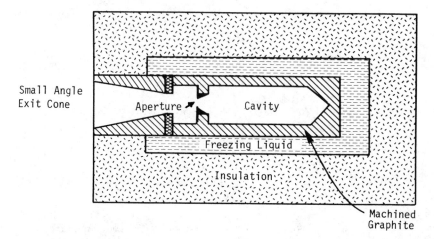

Figure 11.14. Schematic essentials of a blackbody source (heating elements, liquid container, thermocouples, etc., not shown).

escapes for pyrometry measurements allows heat to escape in accordance with the Stefan–Boltzmann equation (11.66). This cools the interior wall of the cavity by some temperature interval, ΔT, and reduces its emittance. In principle ΔT will be a minimum when the thickness of the cavity wall

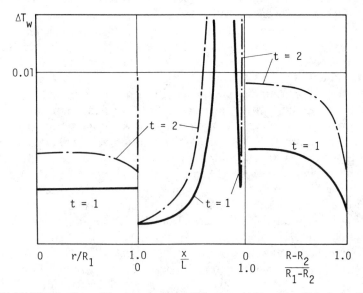

Figure 11.15. Distribution of the temperature difference ($\Delta T_w = T_0 - T_w$) of a cylindrical cavity with $L/R_1 = 8$, $R_2/R_1 = 0.15$, $\lambda = 0.9$, and $e = 0.25$ W-cm^{-1}-K^{-1}, where R_2 is operative radius, R_1 is cavity radius, and L is length, r_{ls}/r_l is the fraction of spectral surface reflection from a 120° end cone. [From Chen Hongpan, Chen Showren, and Chu Zaixiong. The evaluation of the emissivity and the temperature of cavities at the gold freezing point, *Metrologia* **17**, 59–63 (1981). By permission of Springer-Verlag.]

approaches zero, when the thermal conductivity of the cavity material approaches infinity, when the ratio of the radius of the aperture hole to the internal radius of the cavity approaches zero, when the emittance of the cavity wall is 1, and when there is no thermal contact resistance between the external surface of the cavity and the surrounding liquid–metal iso-thermal media.

It is not possible to measure precisely the radiant features inside a cavity.

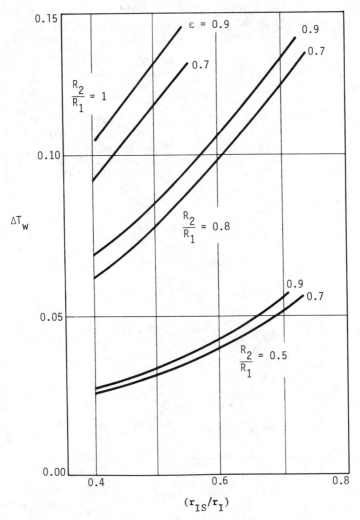

Figure 11.16. Relation between R_w and ΔT_w on the bottom of a cylindrical cavity with $L/R_1 = 8$, where R_w is the thermal resistance of the cavity wall. [From Chen Hongpan, Chen Showren, and Chu Zaixiong, The evaluation of the emissivity and the temperature of cavities at the gold freezing point, *Metrologia* **17** 59–63 (1981). By permission of Springer-Verlag.]

However, computer solutions to mathematical models have shown that existing cavity designs can produce temperature errors in achieving isothermal blackbody behavior only slightly more than the errors of the finest measuring devices, usually taken as about 0.01 K at the gold point (Figs. 11.15 and 11.16).[7]

Computer modeling to determine the effects of specular reflection show that for a flat bottom, error is least when specular reflection is zero ($e = 1$). When the included angle is 120°, a bright reflective bottom views more of the surface of the black body cavity and can actually give higher effective emittance than the same cavity with a diffuse bottom (Fig. 11.17).[8]

If a real cavity has an aperture temperature less than desired, and a

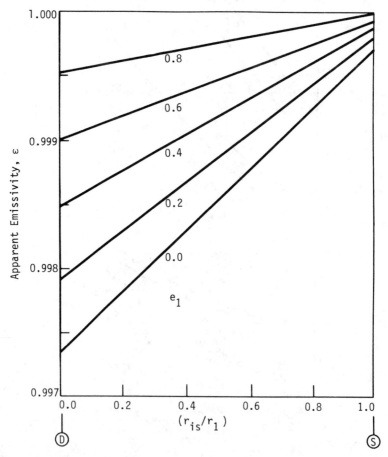

Figure 11.17. Apparent emissivities versus the degree of specularity for a cavity of $L/d = 10$, $e_0 = 0.8$, and $\theta = 120°$. (From A. Ono, "Temperature: Its Measurement and Control in Science and Industry," Vol. 1, Part 1, National Research Laboratory of Metrology, Ibaroki, Japan, 1982.)

temperature gradient exists along its length, it will have an effective emittance of less than 1. In effect, it will have an effective temperature (equivalent to a brightness temperature) less than that of the invarient point of the surrounding media. Then the effective temperature must be used when calibrating a device.

11.19 TUNGSTEN STRIP LAMPS

Wide-filament tungsten strip lamps are useful as calibration sources and to transfer calibration from one pyrometer to another. The U.S. National Bureau of Standards will calibrate such a lamp and issue a certificate of calibration relating brightness temperature to lamp current. In the United States, one manufacturer makes two models of these large lamps.[9] One is the 30AT24/6 lamp, rated at 6 V, 30 A. The other is model 20AT24/2. The 30 A model has a tungsten ribbon filament 3 mm wide, 0.075 mm thick, and 50 mm long. The filament has a tiny notch near its middle to permit the pyrometer to be focused at the calibrated position adjacent to it. One end has a V bend near its support so that changes in length from thermal expansion can be accommodated. The glass envelope is large and has a plane window for sighting. It is filled with argon at about 1/3 atm at room temperature. The lamp must be operated in a vertical position so that any tungsten evaporated from the filament will be carried upward by convection currents and not condense on the window. The pyrometer should be sighted at the center of the filament adjacent to the notch. The lamp should be enclosed in a black enclosure to prevent reflection from adjacent lights.

A great deal of research on tungsten lamps has been conducted at the National Physical Laboratory, London.[10] Tungsten exhibits grain growth on initial heating to 1200°C, secondary recrystallization and faceting at 1900°C, and facet modification with current reversal. Quinn recommends 100 to 200 hours at 1900°C to stabilize a new lamp, and that current reversal be avoided. Vaporization of tungsten onto the lamp envelope takes place by oxidation of tungsten by displacement of oxygen from trace amounts of water vapor, deposition of the tungsten oxide on the lamp surface, reduction of the oxide by hydrogen to reform water vapor, and adherence of the tungsten on the interior surface of the glass lamp envelope. Therefore, residual water vapor must be reduced to an acceptable level during manufacture. The resistance of tungsten is changed by the presence of CO, N_2, and H_2, and can change if the gases are emitted at 1900°C but slowly resorbed at a lower temperature. Therefore, effective degassing is also required during manufacture. If a tube filled with tungsten wires is used as a filament, a blackbody lamp is produced. Sighting into the tiny opening at the end of the tube gives an emittance of about 0.99, but varies with wavelength and angle of observation. This research resulted in the development of massive special lamp designs (Fig. 11.18).

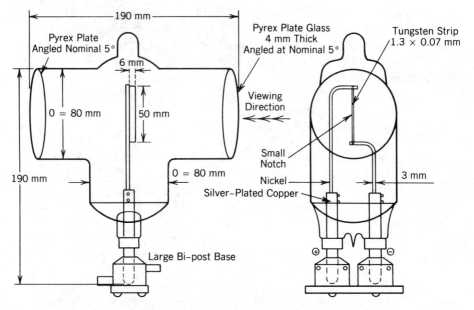

Figure 11.18. Tungsten strip lamp. (Courtesy of The General Electric Co., p.l.c. of England, GEC Hirst Research Centre, London.)

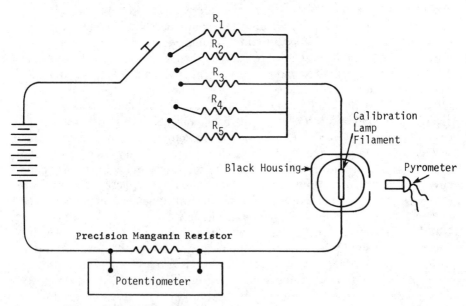

Figure 11.19. Schematic of a tungsten strip lamp calibrating system.

Tungsten strip lamps are annealed prior to use and are reasonably stable. Some are very sensitive to vibrations produced by bumping, and many change radiating characteristics if bumped. Operating at excess temperature can change the calibration due to further recrystallization of the tungsten or evaporation. The maximum temperature of the 30-A lamp should be 2300°C, and for the 75-A lamp should be 2500°C. However, the lamps are more stable if the time above 1900°C is kept to a minimum. The direction of current flow is important because of the Thomson effect, although this does not appear to affect the temperature near the middle (at the notch). Care should be taken to sight the pyrometer at a slight angle through the plane window so that the ray reflected from the inside does not return to the filament to be reflected again into the pyrometer. Polarization of the emitted radiation will occur if the angle is too large.

Usually, dc power is used so that a precision resistor and potentiometer can be used to measure the current accurately. Dc welding power supplies or storage batteries may be used. Any supply must be stable to provide constant current flow. A suitable circuit is shown in Fig. 11.19.

The emittance of tungsten is very sensitive to wavelength. This is one of the reasons that pyrometers of different designs cannot be calibrated against each other using a strip lamp as a transfer device.

11.20 ABSOLUTE INTENSITY MEASUREMENTS

Both Planck's law and Wien's law predict the relationship between absolute radiant intensity and wavelength (frequency) and temperature. Absolute temperature measurements at a single frequency are extremely difficult. They require a method of photon counting at a particular frequency, knowledge of the losses inherent in the counting instrument, and knowledge of all losses between the source and the instrument. Schematically this can be shown as in Fig. 11.20, and mathematically by

$$N_{S,\lambda} = F_{l,\lambda} F_{D,\lambda} I_{D,\lambda} , \tag{11.69}$$

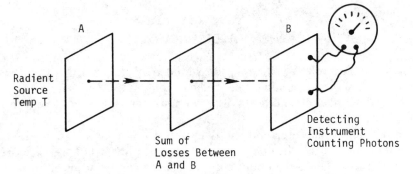

Figure 11.20. Schematic of a photon counting system.

where $N_{S,\lambda}$ is the number of photons per second with wavelength λ leaving the radiant source, $F_{l,\lambda}$ is the relationship between the number of photons leaving the source and arriving at the detector, $F_{D,\lambda}$ is the relationship between the number of photons arriving at the detector and the number of photons actually counted by it, and $I_{D,\lambda}$ is the output signal of the detector produced by the number of photons actually counted by it. For example, if the detector current is 6.342×10^{-3} A for 10^4 photons per second actually counted, 83% of those entering the detector are counted, and 42% of the photons leaving the source actually reach the detector, then

$$N_{S,\lambda} = I_D\left(\frac{10^4}{6.342 \times 10^{-3}}\right)\left(\frac{1}{0.83}\right)\left(\frac{1}{0.42}\right) \qquad (11.70)$$

$$= 4.523 \times 10^6 I_D \text{ photons/s}$$

where I_D is the detector current. Note that this assumes a constant temperature of the source, detector response at a single wavelength, uniformity of source emission in space and time, and stability in all other factors. The number of photons can be converted to intensity through the relationship

$$I_S = N_{S,\lambda}\left(\frac{h_c}{\lambda}\right), \qquad (11.71)$$

where the units are consistent to produce intensity in J/(sec-m^2-sr).

There are many serious difficulties in such measurements. Probably the most serious is that all detectors respond to a profile of photon energy, so that it is not possible to measure energy at a particular wavelength. Counting devices with narrow energy acceptance properties, or narrow bandpass filters, can limit the wavelengths, but then Planck's law (or Wien's law) must be integrated over the range of energy accepted. Detectors must also be linear over the range of intensities measured. This is also a serious limitation for photon counting devices.

11.21 RELATIVE INTENSITY MEASUREMENTS

Experimentally, relative intensity measurements are much more practical. Instead of photon counting, the detector output of one source can be compared to that of another source without knowing all of the various transfer functions.

In a relative measurement, the intensity of one source is compared to that of another. If all the conditions giving rise to the relationships in Eqs. (11.69) to (11.71) are held constant, one source can be adjusted to be the same as another source without knowing any of the absolute values of the measurement. If this is done several times using different thermodynamic invarient points as the primary source, the instrument can be calibrated at each invarient point and a curve can be passed through those points calibrating the instrument. Then the instrument can be used to measure the temperature of a secondary source if the conditions of the measurement are the same as when originally calibrated. This is the basis for all commercial instruments. It is essential, then, that the conditions of measurement must either be identical to those during calibration, or the deviations from calibration conditions must be known and corrections applied.

As an example, suppose that an optical pyrometer has been calibrated using blackbody sources at the gold point, the silver point, and the copper point. Later it is sighted into a hole in a furnace under blackbody conditions (a tiny hole into a massive closed furnace). The temperature indicated should need no corrections if all conditions of operation of the pyrometer are unchanged from the conditions during calibration. However, if the hole is covered by a window that transmits only 92% of the radiation emitted in the region of the spectrum used by the pyrometer, a correction must be applied. If that window is gradually coated with evaporated products from inside the furnace, its transmittance will change with time. Then it must be cleaned or a measurement of the change must be computed. If the furnace sight path contains combustion gases, dust, or other interferences, errors will occur if the measurement is not properly corrected.

It is possible to determine if the inside of a window is coated from outside the furnace. If an incandescent wide-filament bulb is placed near the window and the pyrometer is used to measure the apparent temperature of the reflection from the outside and inside surfaces, the presence of a coating on the inside can be detected by the brighter reflection from the inside surface (Fig. 11.21). The outside surface can be cleaned, if necessary, before making the measurement.

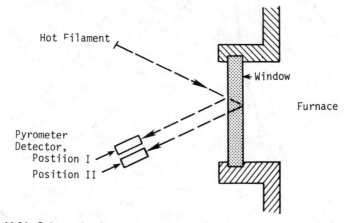

Figure 11.21. Schematic of a measurement of the reflectance of a furnace window.

11.22   THE CONCEPT OF A BRIGHTNESS TEMPERATURE

If a source does not have blackbody emittance, the intensity radiated by that source will be less than that of a blackbody at the same temperature. Suppose, for example, that a source has an emittance of 0.43, a true blackbody temperature of 1500 K, and the intensity is measured at a wavelength of 655×10^{-9} m. Then, by definition,

$$e = 0.43 = \frac{I_A}{I_{BB}}, \qquad (11.72)$$

where I_A is the actual intensity and I_{BB} is the blackbody intensity. There will be some lower temperature, however, that a blackbody could exist and

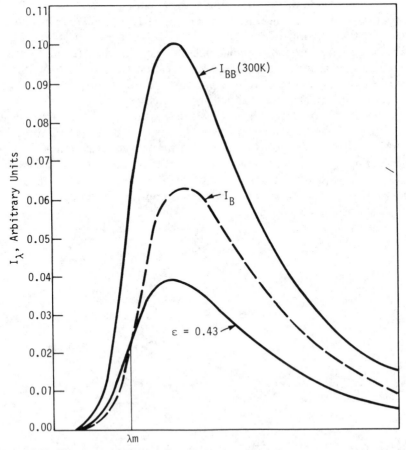

Figure 11.22. The concept of a brightness temperature for measurement of intensity at λ_m.

have the same intensity emitted as that of the actual emittance from our nonblackbody source. We call this a "brightness" temperature. It is an apparent temperature, one that a blackbody would have if we assume it has the actual intensity I_A (Fig. 11.22).

We can represent I_A and I_{BB} by Planck's or Wien's law. Using Wien's law for convenience in Eq. (11.72), we have

$$e = \frac{C_1 \lambda^{-5} \exp(-C_2/\lambda T_\beta)}{C_1 \lambda^{-5} \exp(-C_2/\lambda T_{BB})} \tag{11.73}$$

$$= \exp\left[\frac{C_2}{\lambda}\left(\frac{1}{T_{BB}} - \frac{1}{T_\beta}\right)\right]. \tag{11.74}$$

Taking logarithms gives as

$$\ln e = \frac{C_2}{\lambda}\left(\frac{1}{T_{BB}} - \frac{1}{T_\beta}\right) \tag{11.75}$$

Substituting the values in our example yields

$$\ln 0.43 = -0.844 = \frac{0.0143883}{655 \times 10^{-9}}\left(\frac{1}{1500} - \frac{1}{T_\beta}\right). \tag{11.76}$$

Solving for T_β, we have

$$T_\beta = 1418 \text{ K}$$

$$= 1145°C .$$

Note that as we learned in Section 11.14, the error from using Wien's law instead of Planck's law is given by

$$E = \exp(-C_2/\lambda T) \tag{11.54}\square$$

$$= \exp[-0.0143883/(655 \times 10^{-9})(1500)]$$

$$= e^{-14.6} = 4.36 \times 10^{-7} .$$

This is a negligible error.

11.23 SEGMENTED DISKS

Often it is desirable to have made a known reduction in intensity of a radiant source.[7] This can be done with a rotating segmented disk. Suppose that a disk has windows cut into it as shown in Fig. 11.23. Because, in the

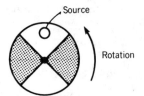

Figure 11.23. Segmented disk used to reduce intensity from a radiant source by 50%.

illustration 50% of the disk is open, only half of the intensity from a source is transmitted through it as it is rotated in front of the source. This makes it possible to reduce the intensity very precisely. Disks can be constructed with 25%, 10%, or as little as 4% open area. A single source can be used with several different disks to reduce the intensity for calibration or measurement purposes. They are useful to extend the range of a calibrated instrument to higher temperatures.

An example of a segmented disk system might be as shown in Fig. 11.24. The intensity reaching the detector is only 25% of that from the blackbody source. Then

$$I_D = 0.25 I_{BB} \tag{11.77}$$

and the detector will indicate a brightness temperature that can be converted by Wien's law or Planck's law to the true blackbody temperature. Using Wien's law gives us

$$C_1 \lambda^{-5} e^{-C_2/\lambda T_\beta} = 0.25 C_1 \lambda^{-5} e^{-C_2/\lambda T_{BB}}. \tag{11.78}$$

Canceling $C_1 \lambda^{-5}$ and taking logarithms yields

$$-\frac{C_2}{\lambda T_\beta} = \ln 0.25 - \frac{C_2}{\lambda T_{BB}} \tag{11.79}$$

$$\frac{1}{T_{BB}} = \frac{1}{T_\beta} + \frac{\lambda}{C_2} \ln 0.25$$

$$T_{BB} = \left(\frac{1}{T_\beta} + \frac{\lambda}{C_2} \ln 0.25 \right)^{-1}.$$

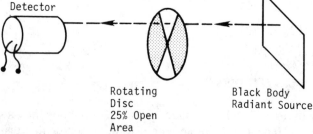

Detector

Rotating
Disc
25% Open
Area

Black Body
Radiant Source

Figure 11.24. Extending the range of a calibrated pyrometer with a segmented disk.

Segmented disks are very useful for calibration purposes but they are not convenient for general measurements because they need an oscillator or a motorized drive. Obviously, the speed of rotation must be rapid enough so that the detector does not oscillate in response to the rotation. For most purposes, the on–off cycle of the disk should be less than one-tenth of the time constant of the detector system. For measuring instruments, filters are usually preferred to extend the range.

A rotating disk does not change the wavelength distribution, only the intensity. The color of the source is unchanged, so λ appears in our example, Eq. (11.79).

11.24 FILTERS OR ABSORBING SCREENS: PYROMETER BROWN GLASS FILTER

Excellent-quality neutral gray glass filters are available for reducing the intensity in the visible region of the spectrum. Filters are also called absorbing screens. Neutral gray filters perform very much like rotating disks, absorbing uniformly across the visible spectrum and not changing the color, only the intensity. As explained previously, the absorption depends on the concentrations of the absorbing ions in the glass and on its thickness. For a particle desired transmittance t, it is convenient to grind the thickness of the filter to satisfy the equation explained previously,

$$t = (1-a)(1-r)(1-r) . \tag{11.80}$$

The desired thickness must allow for both absorbance a and reflection r. It is convenient to express the transmittance as

$$t = e^{-k/\lambda} \tag{11.81}$$

where k is a constant for that particular filter known as the extinction coefficient. When substituted into Wien's law, we have

$$t = e^{-k/\lambda} = \frac{I_2}{I_1} = e^{(C_2/\lambda)(1/T_1 - 1/T_2)} . \tag{11.82}$$

Taking logarithms gives as

$$-\frac{k}{\lambda} = \frac{C_2}{\lambda}\left(\frac{1}{T_1} - \frac{1}{T_2}\right) \tag{11.83}$$

or

$$\frac{k}{C_2} = \frac{1}{T_2} - \frac{1}{T_1} = A, \tag{11.84}$$

where A is a constant describing the performance of the filter. Notice, however, that for t to be constant, k must vary linearly with λ, so that A is not constant with wavelength. It will have a particular value at any wavelength λ and varies linearly with λ for a neutral gray filter. This is identical in behavior to a rotating disk. Because color changes with temperature, it becomes very difficult to match intensity through a narrow-wavelength color filter if a neutral gray filter is used to extend the range. It is better to have a special filter, not a neutral gray filter. For the disappearing filament optical pyrometer, the effective wavelength of the red filter used to provide a "single" wavelength (actually a narrow band of about $630-700 \times 10^{-9}$ m for one particular commercial instrument) is often taken as 655×10^{-9} m. The special filter should have a transmittance such that A is constant. This will occur because

$$\ln t = -\frac{k}{\lambda}. \tag{11.85}$$

when k is constant over the narrow bands of wavelengths. Then the transmittance will be greater at the longer wavelength. A suitable glass is produced commercially and is often called "pyrometer brown" because of its brown appearance with transmitted white light. Use of a filter with a constant A value has the advantage that a blackbody at a particular high temperature viewed through such a filter will have the color of a blackbody at a lower temperature, its brightness temperature.

Interference filters are more important than absorption filters for many applications. They are discussed in Section 13.6.

Optical filters can be used in series. Then the transmittance of the series is the product of the transmittance of the individuals, and the constant A for the series is the sum of the constants for the individual filters. Introducing a second, a third, and a fourth filter would provide a series of extended ranges.

Use of filters or rotating disks to extend the range results in reduced accuracy of the extended range because the error in measurement is increased in proportion to the increase in range.

11.25 SUMMARY

In this chapter we have presented the principles of radiant heat transfer that are essential to radiant methods for temperature measurement. We have learned that radiant heat transfer is complex and that, if accurate measurement is to be achieved, each particular situation must be approached with prudent consideration of the scientific principles. We have learned that absolute temperature measurements based on Planck's law can be calculated, in principle, from the radiant intensity and the three fundamental constants c, h, and k (the velocity of light in a vacuum, Planck's constant, and Boltzmann's constant). Because direct measurement of spectral radiant intensity is impractical, the ratio of intensities at a temperature to that at the gold point has been adopted for the IPTS. That greatly relaxes the measurement requirements because it is only necessary to ensure that the conditions of measurement (e.g., solid angle, attenuation, etc.) are kept constant. Even so, many errors are easily introduced and an understanding of the principles involved is essential.

Radiant methods have traditionally been applied at temperatures above the gold point and called pyrometry. Optical pyrometry and total radiation pyrometry are examples. The development of solid-state detectors and circuits, and of microprocessors, has extended radiation methods to much lower temperatures, down to the ice point and below. The trend to greater use of radiant methods is expected to continue. The next three chapters explain three important classifications of radiant methods: optical pyrometry, photoelectric pyrometry (infrared temperature measurement), and total radiation pyrometry.

REFERENCES

1. T. R. Harrison, "Radiation Pyrometry and Its Underlying Principles of Heat Transfer." Wiley, New York, 1960.

2. E. M. Sparrow and R. D. Cess, "Radiation Heat Transfer." McGraw-Hill, New York. 1978.

3. T. J. Love, "Radiative Heat Transfer." Merrill, Columbus, Ohio, 1968.

4. J. C. Richmond and D. P. Dewitt, eds., "Applications of Radiation Thermometry." Am. Soc. Test. Mater., Philadelphia, Pennsylvania 1985.

5. W. D. Kingery, H. K. Bowen, and D. R. Uhlmann, "Introduction to Ceramics," p. 651. Wiley, New York, 1976.

6. G. G. Gubareff, J. E. Janssen, and R. H. Torborg, "Thermal Radiation Properties Survey," 2nd ed. Minneapolis-Honeywell Regulator Co., Minneapolis, Minnesota, 1960.

7. C. Hongpan, C. Shouren, and C. Zaixiang, The evaluation of emissivity and the temperature of cavities at the gold freezing point. *Metrologia* **17**, 59–63 (1981).

8. A. Ono, Apparent emissivities of cylindrical cavities with partially specular conical bottoms. *Temp.: Its Meas. Control Sci. Ind.* **5** (Part 1), 513–519 (1982).

9. H. J. Kostkowski and R. D. Lee, Theory and methods of optical pyrometry. *NBS Monogr.* **41** (1962).

10. T. J. Quinn, "Temperature," pp. 320–324. Academic Press, New York, 1983.

12

THE DISAPPEARING FILAMENT OPTICAL PYROMETER

12.1 INTRODUCTION

The disappearing filament optical pyrometer is a temperature measuring instrument in which the intensity of radiant energy coming from an incandescent source is matched to the intensity of a calibrated filament when both the source and the filament are viewed through a red filter. When the filament and the source intensities are the same, the image of the filament disappears as it is superimposed on the image of the source. The red filter bandpass is narrow enough so that the intensity match is usually considered to be monochromatic. The method is limited to temperatures that produce visible incandescence, above about 750°C.

To use the optical pyrometer a telescope containing the filament is sighted on a target so that the images of target and filament are in sharp focus and superimposed. Then the intensity of the filament, or the target, is adjusted until the image of the filament disappears in the image of the target. With the intensity matched, the temperature is read from the instrument, allowance being made for target emittance if necessary.

This type of optical pyrometer has the advantages that the distance from the target to the source is not important (it is only necessary to have a clear image to compare with the filament image), various places on a target can be examined for temperature distribution, temperatures can be measured to very high temperatures, and the visual image ensures that the instrument is measuring the temperature of the desired portion of the target. Disadvantages are that the instrument either must be sighted under blackbody conditions or the reading corrected for emittance; that absorption by dust, windows, flame, and other optical interferences can produce errors; and that the usual method of application requires manual adjustment, so that measuring and recording are slow and manual. Automatic pyrometers are available and are described separately in Chapter 13. The typical disappearing filament optical pyrometer is manually operated and therefore cannot be used for automatic control.

Since the discovery of fire, early man must have been aware that both the color and the intensity of light emitted from incandescent object changes with temperature. At about 600°C, looking into a furnace from a very dark room, a dull, barely visible red glow can be detected. As the temperature is increased to 1650°C, the intensity increases and the color shifts from dark red to bright red to orange to yellow to a dazzling intense white. Color has long been used by ceramists and metallurgists to estimate the temperature of heated objects. Some people can estimate temperature to ±25°C or better by looking into an incandescent furnace. However, for precise measurements the shift in dominant wavelength according to Planck's law (see Chapter 11) is confusing, so that intensity measurements at a single wavelength have advantages. The disappearing filament optical pyrometer eliminates the effect of the shift in wavelengths by measuring intensity through a red filter.

Henri Le Châtelier invented the first optical pyrometer in 1892.[1] He used an oil lamp, a red filter, and an iris diaphragm, adjusting the diaphragm opening to achieve a match between the intensity of the furnace and that of the lamp. This was not a practical way to measure temperature at that time because there was no way to convert intensity to temperature.

When Wilhelm Wien derived his equation in 1896, an interpolating equation became available but it required an absolute intensity measurement.[2] Wien's law, Eq. (12.1), is discussed in detail in Chapter 11.

$$I_\lambda = C_1 \lambda^{-5} \exp(-C_2/n\lambda T) \qquad (12.1)$$

Note that absolute temperature T is used to calculate intensity I at a particular wavelength λ. The best values for the constants C_1 and C_2, in vacuum, which depend on other fundamental constants, are

$$C_1 = 1.9089 \times 10^{-16} \text{ J-m}^2/\text{s} \qquad (12.2)$$

$$C_2 = 0.0143883 \text{ m-K} \qquad (12.3)$$

The index of refraction, n, for air is 1.0028 and is used in Eq. (12.1) for precise measurements.

In 1899, Samuel F. B. Morse patented the first disappearing filament optical pyrometer.[3] His patent describes observing, through a red filter, the image of a calibrated filament superimposed on the image of the unknown, adjusting the electrical current heating the calibrated filament until the

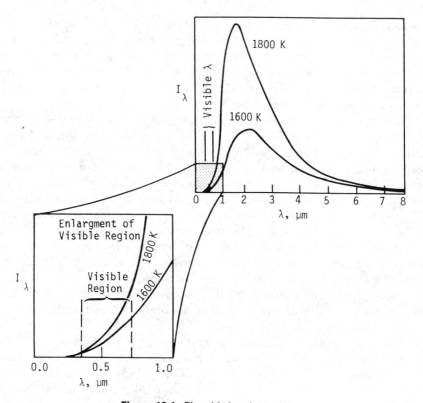

Figure 12.1. Planck's law intensities.

image of the filament disappeared in the image of the unknown, and reading the electrical current from an ammeter to determine temperature.

As we learned in Chapter 11, the intensity maximum, using Planck's law, occurs in the infrared region of the spectrum, even to temperatures of 3000°C. Only a small fraction of the intensity occurs in the visible region of the spectrum (Fig. 12.1). The choice of a red filter gave the maximum intensity in the visible region and permitted measurements to be made to a lower temperature than would be possible with a yellow or a blue filter.

The development of Planck's law in 1900 completed the theoretical background for the disappearing filament optical pyrometer.[4] When the first IPTS was adopted in 1927, an optical pyrometer was specified to determine temperature in range IV, using Wien's law as a defining equation, and based on the ratio of the intensity to that of a blackbody at the gold point. The disappearing filament optical pyrometer was the instrument used for most such measurements for about three decades.[5,6] In 1948, the ninth International Conference on Weights and Measures changed the defining equation to Planck's law [Eq. (12.4) and (12.5)]. The type of measuring instrument is no longer specified.

$$\text{Wien's law ratio: } R = \frac{I_x}{I_{Au}} = \frac{e^{-C_2/\lambda T_x}}{e^{-C_2/\lambda T_{Au}}} \tag{12.4}$$

$$\text{Planck's law ratio: } R = \frac{I_x}{I_{Au}} = \frac{(e^{C_2/\lambda T_x} - 1)^{-1}}{(e^{C_2/\lambda T_{Au}} - 1)^{-1}} \tag{12.5}$$

Note that the definition of temperature is based on an extrapolation for all temperatures above the gold point, and that Planck's law for blackbodies is presumed to be exact. When the IPTS was revised in 1968, the gold point was changed from 1063°C to 1064.43°C. Because of the definition of temperatures in range IV through Eq. (12.5), this changed all temperatures above the gold point by a greater amount. For example, 3000°C increased to about 3005.8°C.[7] Future changes are expected to be smaller, but it is apparent that the accuracy of determinations based on a theoretical extrapolation will be diminished with increasing temperature. If the gold point is not used in Eq. (12.5), the reference temperature must be specified, or ambiguity will result.[8]

To match the intensity of the filament to that of the target, it is apparent that either the intensity of the filament or the intensity from the target could be adjusted. Both methods are used in commercial instruments, although adjustment of the filament intensity is more common. Both types of instruments will be described.

12.2 ACCURACY OF MEASUREMENT

Commercial disappearing filament optical pyrometers have precision that depends on their design. The best portable units can be read to ±0.2° at 775°C, reducing to about ±1° at 1225°C. Actual experience is usually not as good because of errors in matching filament and target intensity, so that ±1°C at 775°C and ±5°C at 1225°C is more common. However, emittance corrections for nonblackbody conditions determine the accuracy of the measurement, as constrasted to precision. Each situation must be evaluated separately. The lower-precision commercial instruments can be read to about ±3°C on the low-temperature scale. Dual- or triple-range instruments are available, where one or more filters are added to reduce target intensity and extend the range. The precision drops off as filters are added. A good instrument may read to ±25°C at 5000°C. The use of filters is discussed in Section 12.4. We emphasize, again, that either blackbody conditions must be used to obtain the possible precision indicated above, or the deviations from blackbody conditions must be known and a correction applied.

12.3 LEEDS AND NORTHRUP MODEL 8630 OPTICAL PYROMETER

The 8600 models of optical pyrometers are based on the Morse method of controlling the temperature of a calibrated filament to achieve an intensity match between the filament and the target.[9] Several models have been available (Fig. 12.2). Although manufacture the model 8630 has been discontinued, many are still in service. It is discussed as a basis for the other optical pyrometers.

In use, the telescope is focused on the target and the intensity of the filament is adjusted until its image disappears in the image of the target (Fig. 12.3). If the target is a blackbody, the indicated temperature at filament disappearance will be the actual temperature of the body. If it is not a blackbody, either the instrument must be calibrated for the emittance of the target or the true temperature must be calculated using Wien's or Planck's law. The essential elements of this L & N optical pyrometer are shown in Fig. 12.4.

Inside the telescope and objective aperture C limits the field of view so that changes in focus of the objective lense B do not change the solid angle of entrance. One or more absorption filters D can be used to extend the range, with a separate measuring scale for each range, or to compensate for a particular target emittance. A clear glass disk is used instead of a filter for temperatures below 1225°C, so that introducing absorption filters for higher ranges does not change the telescope focus, and the glass compensates for surface reflectance of the filters.

The most critical part of the pyrometer is a very stable calibrated pyrometer lamp E. It contains a pure tungsten flat strip filament 2 mils

Figure 12.2. Model 8630 Leeds & Northrup optical pyrometer. (Courtesy of Leeds & Northrup Company.)

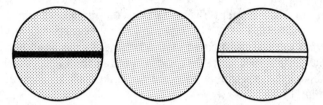

Figure 12.3. Filament disappearance. If the intensity of the filament is too low, it will appear dark against the target image. If it is too high, it will appear bright against the target image. At perfect match the filament will not be visible. It will disappear in the target image. (Courtesy of Leeds & Northrup Company.)

wide, 0.2 mil thick, and 1 in. long. The lamp housing has plane-parallel windows. The width of the filament observed through the windows is perpendicular to the telescope axis. The filament is never heated above 1235°C, a temperature low enough to prevent rapid tungsten evaporation. Although the lamp is extremely stable, recalibration is recommended after each 200 hours of use.

An erecting lens F is used so that the telescope is easy to orient. It erects the target image. The filament image is inverted. The red filter G is described in the next section.

The lenses F, H, and J are contained in a Huygenian eyepiece that can be adjusted to obtain a sharp image of the filament. The objective lens can then be adjusted to obtain a sharp image of the target in the plane of the filament.

In practice the temperature of any part of the target can be obtained if it is large enough to superimpose it on the image of the filament. However, for small targets certain geometric limitations exist that are discussed separately in Section 12.5.

Intensity match is shown schematically by a variable resistor R that has been adjusted to vary the current flowing through the lamp filament so that

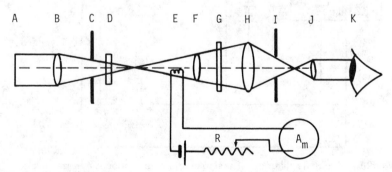

Figure 12.4. Schematic of an optical pyrometer. (Courtesy of Leeds & Northrup Company.)

the image of the filament disappears in the image of the target. The ammeter is calibrated to read temperature of a blackbody. Modern instruments have a more sophisticated method of measuring lamp current, that of reading the setting of a variable resistor limiting the lamp current when the voltage across the resistor is adjusted automatically to a precise standard voltage (Fig. 12.5).[9]

The current flowing through the pyrometer lamp is divided by the parallel precision resistors B, B', and B'' so that only a small fixed fraction flows through the precision variable resistor B, which has a scale calibrated to read temperature directly on the same shaft with the contact P. The amplifier is designed so that the voltage drop e across resistor C and that portion of the resistor B included by the contact P is always equal to the standard cell voltage of 1.019 V. As contact P is moved to adjust the temperature of the filament, the amplifier responds by adjusting its output, the lamp current, to maintain voltage e constant. Since B is a precision resistor, the current flowing in B can be calculated precisely. The scale on B is calibrated to read temperature directly.

The 8630 instrument was available with the following ranges: low, 775–1225°C; medium, 1075–1750°C; high, 1500–2800°C; and extra high, 1950–6200°C. Scales for 0.4 emittance or other special emittances are available. The Fahrenheit equivalent scales are available. Different objective lenses are also available. The minimum target diameter can be as small as 0.5 mm but depends on distance from the target.

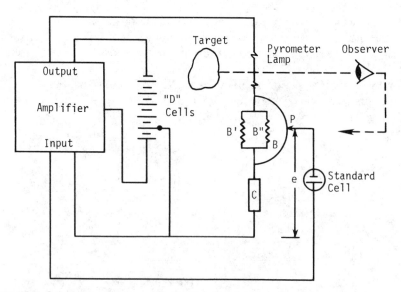

Figure 12.5. Schematic diagram of a single-adjustment optical pyrometer circuit. The observer moves contact P to match filament brightness to target brightness. (Courtesy of Leeds & Northrup Company.)

12.4 EFFECTIVE WAVELENGTH OF PYROMETER RED FILTERS

To use Planck's law (or Wien's law) for temperature determinations, measurement of intensity at a single wavelength is desirable. The disappearing filament optical pyrometer encorporates a red filter that has some transmittance above about 630 nm, but the human eye limits the maximum wavelength to about 700 nm. The weighted average transmittance is often taken as an effective monochromatic wavelength of 655 nm, but the factors affecting wavelength are explained below. It is apparent that some variation will occur between different individuals. However, usually this produces less error than other factors related to the measurement.

The pyrometer red filter used frequently in optical pyrometers has a sharp cutoff at short wavelengths (Fig. 12.6). The reflectance of the two glass surfaces and absorption, combined, give transmittances for the filters between 80 and 90% at wavelengths longer than the cutoff. The visual accuity of the human eye is much more sensitive in the green region of the spectrum (Fig. 12.7). The intensity preceived by the eye as the detector is wavelength sensitive and provides the long-wavelength cutoff. The intensity of the incident radiation being viewed through the optical system follows Planck's law and is highly nonlinear, so the effective wavelength of the system depends on the spectral response of the eye, the transmittance of the filter, the transmittance of other optic elements in the observation if the transmittance varies with wavelength, and the temperature of the observed radiating body. A blackbody at the gold point has a spectral intensity in the visible region given by Fig. 12.8. Then, assuming that all other optic elements have constant transmittance with wavelength, the relative brightness B is given by the product of intensity $I_{\lambda T}$, transmittance t_R, and visual accuity V:

$$B_{\lambda T} = I_{\lambda T} t_R V. \qquad (12.6)$$

This is shown in Fig. 12.9.

Because the relative brightness of the detector depends on the spectral intensity of the source, the mean effective wavelength depends on the temperature of the target. According to a 1961 monograph, a National Bureau of Standards pyrometer using a Corning pyrometer red filter had mean effective wavelengths shown in Fig. 12.10.[6] The mean effective wavelength λ_e is the weighted center of the brightness curve (Fig. 12.9), and is given by

$$\frac{1}{\lambda_e} = \frac{\int_0^\infty (B/\lambda)\, d\lambda}{\int_0^\infty B\, d\lambda}. \qquad (12.7)$$

The integral is really only taken over the range of wavelengths appropriate to the relative brightness curve and must be evaluated experimentally. The

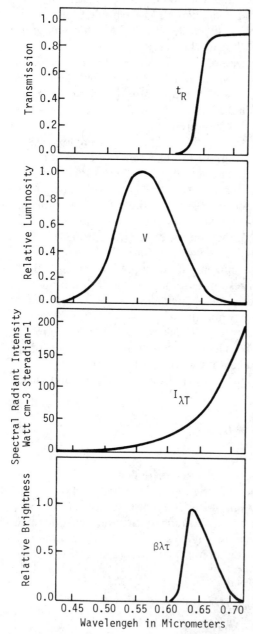

Figure 12.6. Glass filter transmittance. (Courtesy of Leeds & Northrup Company.)

Figure 12.7. Visual acuity. (Courtesy of Leeds & Northrup Company.)

Figure 12.8. Typical Planck's law intensity. (Courtesy of Leeds & Northrup Company.)

Figure 12.9. Brightness, product of τ, V and $I_{\lambda T}$. (Courtesy of Leeds & Northrup Company.)

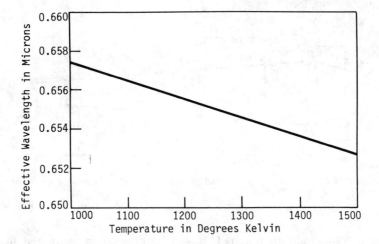

Figure 12.10. Mean effective wavelength as a function of temperature.

NBS research indicated that $1/\lambda_e$ is linear with $1/T$. Then

$$\frac{1}{\lambda_e} \approx \frac{1}{2}\left(\frac{1}{\lambda_{T_1}} + \frac{1}{\lambda_{T_2}}\right),$$

(12.8)

so that the mean effective wavelength for any two temperatures T_1 and T_2 can be calculated. It is helpful to have the experimental measurements at several temperatures.

12.5 SMALL TARGET OPTICAL PYROMETERS

Small target optical pyrometers use shorter-focal-length lenses in the telescope. Some instruments could be considered long-working-distance microscopes, but the principles of operation are the same as those described above. There are two problems frequently encountered with small target devices: vignetting and emittance determination.

Vignetting can occur when sighting into a hole or cavity intended to produce blackbody conditions (Fig. 12.11).[10] When vignetting exists due to nonuniform intensity of the target, the geometry of the target, the size of the objective aperture, and the distance from the target become important. Using Fig. 12.11 as an example, where the target is presumed bright at the bottom of a hole of length L, the diameter D of that hole required to produce a minimum target diameter W can be calculated:

$$\frac{D - W}{L} = \frac{d - D}{l}.$$

(12.9)

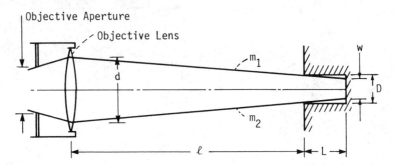

Figure 12.11. Schematic of a small target with a hole to increase emittance.

If these limitations are not met, it will be difficult to make the filament disappear because of nonuniformity of target brightness. The minimum target diameter for error-free measurement depends on the filament diameter, the mean effective wavelength, and the target-to-objective lens distance. For square targets, Buchele has recommended that the minimum target be 10 times the theoretical resolution angle, and for circular targets 15 times the resolution angle.[10] The theoretical resolution angle is given by

$$\theta_C = \frac{\lambda_e}{d}. \tag{12.10}$$

Then W is found, for circular targets, as

$$W = 15 \frac{\lambda_e}{d} l. \tag{12.11}$$

Small targets are often enclosed in vacuum chambers and have radiant intensity losses caused by target emittance, window reflection and absorption losses, and so on. Each situation must be evaluated separately to evaluate these losses. Sometimes a known source can be inserted into the specimen chamber to determine window losses, or sometimes the window can be temporarily demounted for that purpose.

12.6 THE PYRO OPTICAL PYROMETER

The Pyro optical pyrometer contains a lamp the current of which is standardized so that it is always at a constant temperature. Varying the thickness of a calibrated absorption wedge to reduce the intensity of the target image is used to obtain a match between the intensity of the lamp and the intensity of the target[11] (Fig. 12.12). The calibrated wedge is manufac-

Figure 12.12. Picture of pyro optical pyrometer. (Courtesy of the Pyrometer Instrument Co., Inc., Northvale, N.J.)

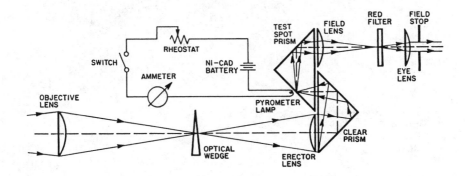

Figure 12.13. Pyro schematic. (Courtesy of the Pyrometer Instrument Co., Inc., Northvale, N.J.)

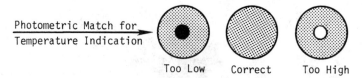

Figure 12.14. Pyro intensity adjustments. (Courtesy of the Pyrometer Instrument Co., Inc., Northvale, N.J.)

tured in a ring shape so that the thickness can be read from the scale accompanying the ring. Additional scales are provided for added absorption filters or for emittance corrections. The most sensitive scale (760–1100°C) can be read to ±2°C at 1100°C.

The optical system contains prisms to provide images of both the target and the lamp (Fig. 12.13). An intensity match is achieved when the image of the lamp (center circle) disappears against the image of the target (outer ring), as shown in Fig. 12.14. The telescope objective lens and the eyepiece lenses are adjustable to facilitate focusing on the target and the filament. An adjusting screw rheostat and an ammeter are provided. Each lamp has a current rating assigned to it so that the adjusting screw can be turned to obtain the current necessary for the lamp to be at its required standard temperature.

12.7 EFFECT OF EMITTANCE ERRORS

As explained earlier, emittance errors must be considered when making temperature measurements with the optical pyrometer. It will be helpful to have some perspective for the amount of error that will appear from an emittance value. Using Wien's law, for convience, we can write

$$I = C_1 \lambda^{-5} e^{-C_2/\lambda T} . \tag{12.12}$$

Taking the derivative, at constant λ, we obtain

$$dI = - C_1 \lambda^{-5} \frac{C_2}{\lambda} e^{-C_2/\lambda T} \frac{dT}{T^2} . \tag{12.13}$$

Dividing Eq. (12.12) by (12.13) gives us

$$\frac{dI}{I} = \frac{-C_2}{\lambda} \frac{dT}{T^2} . \tag{12.14}$$

Then the relative error in temperature, representing the differentials by increments, is

$$\frac{\Delta T}{T} = \frac{-\lambda T}{C_2}\frac{\Delta I}{I}. \tag{12.15}$$

If we substitute for λ and C_2, Eq. (12.15) becomes

$$\frac{\Delta T}{T} = -\frac{(655 \times 10^{-9})T}{0.0143883}\frac{\Delta I}{I} \tag{12.16}$$

$$= 4.552 \times 10^{-5} T \frac{\Delta I}{I}. \tag{12.17}$$

If $T = 1000$ K and e is 0.90, the fractional error is

$$\begin{aligned}\frac{\Delta T}{T} &= (-4.552 \times 10^{-5})(1000)\left(\frac{1-0.9}{1}\right)\\ &= -4.55 \times 10^{-3} = 0.455\%\end{aligned} \tag{12.18}$$

and $\Delta T \approx 4.55°$. Therefore, small errors in emittance do not produce large errors in T.

Representing ΔT by the differential is not justified for large increments of ΔI. It is better to solve for the reciprocal temperatures. This can be done from the definition of emittance. Again using Wien's law as an approximation,

$$e = \frac{C_1\lambda^{-5}e^{-(C_2/\lambda T_\beta)}}{C_1\lambda^{-5}e^{-(C_2/\lambda T_{BB})}}. \tag{12.19}$$

TABLE 12.1 Brightness Temperatures Corresponding to Three Blackbody Temperatures at Various Emittances

	T_β		
Emittance	$T_{BB} = 1000$ K	$T_{BB} = 1500$ K	$T_{BB} = 2000$ K
0.9	995	1489	1981
0.8	989	1478	1960
0.7	984	1464	1937
0.6	977	1449	1911
0.5	969	1432	1880
0.4	960	1412	1846
0.3	948	1386	1802
0.2	932	1352	1744
0.1	905	1296	1653

Taking logarithms, we obtain

$$\ln e = \frac{C_2}{\lambda}\left(\frac{1}{T_{BB}} - \frac{1}{\beta}\right) \tag{12.20}$$

so

$$\frac{1}{T_\beta} = \frac{1}{T_{BB}} - \frac{\lambda}{C_2}\ln e. \tag{12.21}$$

Substituting for λ and C_2 yields

$$\frac{1}{T_\beta} = \frac{1}{T_{BB}} - 4.552 \times 10^{-5}\ln e. \tag{12.22}$$

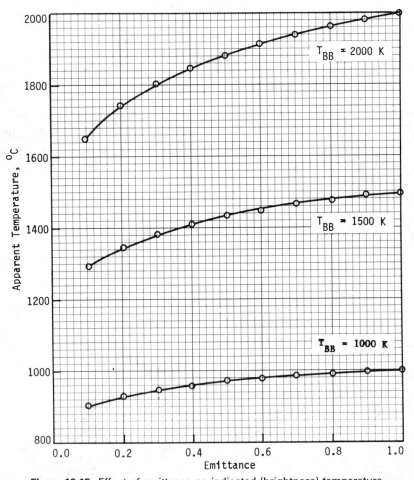

Figure 12.15. Effect of emittance on indicated (brightness) temperature.

Letting e have the values shown in Table 12.1, we can calculate the reciprocal temperature correction term in Eq. (12.22) and the brightness temperatures corresponding to blackbody temperatures of 1000, 1500, and 2000 K. The results are plotted in Fig. 12.15.

12.8 PRACTICAL APPLICATIONS OF DISAPPEARING FILAMENT OPTICAL PYROMETERS

Because they are portable and can be sighted on any incandescent target, these optical pyrometers are remarkably versatile. Their principal limitation is the requirement for either blackbody conditions or emittance corrections. However, they are very useful for many industrial processes if those limitations are observed.

Examples of applications include the following:

1. Adjusting a total radiation pyrometer to indicate correct temperature
2. Determining the temperature of a thermocouple sheath through a window in a furnace to determine if the thermocouple has deteriorated
3. Determining the temperature distribution at various places in a furnace to improve ware quality, prevent overheating refractories, balance furnace power input, and so on.
4. Determining if an incandescent target is uniform enough in temperature for use of a photoelectric optical pyrometer
5. Determining the time required for a metal ingot to cool to a particular temperature
6. Determining the temperature of an enclosed furnace
7. Determining the emittance of an object, including changes in emittance accompanying processes, at the effective wavelength of the pyrometer
8. Determining the reflectance of a surface under controlled illumination conditions

These are a few examples to show the versatility of the instrument. However, it is so useful and portable that every industrial process and every laboratory using incandescent temperatures should have access to a disappearing filament optical pyrometer.

In making measurements with an optical pyrometer, the limitations imposed by emittance and blackbody requirements must be recognized. Sighting into a uniformly heated furnace through a small hole is often sufficient to obtain good blackbody behavior. However, if, when sighting into the furnace, various objects, such as ware, furnace walls, and thermocouple sheaths can be seen, then blackbody conditions do not exist. Only when the furnace temperature is uniform and all objects disappear in the furnace glow has a true blackbody situation been established. When objects can be clearly seen, the difference in temperature between objects can be estimated if a reasonable estimate of emittance is possible. The information obtained may be useful in controlling a process even though it is an approximation.

The effect of reflectance must also be observed. The instrument measures brightness, so that if a shiny metal of low emittance and high reflectance is measured under the conditions that a hot furnace reflects into the instrument, the measurement may be erroneously high. This is especially true if the emittance corrections are applied without regard for reflectance. As an extreme example, a novice with the optical pyrometer is usually surprised to find that the temperature of the snow outside on a cold wintry day has a brightness temperature of, say, 1037°C. Obviously, the snow has a true temperature less than 0°C. It is the reflected light of the sun that gives a high apparent temperature.

Some industrial flames are nearly transparent. A flame for a process heated with natural gas will usually have negligible absorption (and emittance) at pyrometer red wavelengths. Changing to fuel oil, with incandescent carbon particles, or introduction of process dust, and so on, can change the emittance. Usually, a flame opaque enough to obscure the features of a cold object when it is introduced will give erroneous temperatures. Under these conditions, and for controlled atmosphere and vacuum furnaces, a sight tube can be very helpful. If a closed-end tube is inserted into the furnace far enough so that its closed end is heated to the temperature to be measured, a clear sight path can be obtained. It is necessary that the end of the tube represent the correct temperature. If it has low thermal mass in high-velocity flame, it will indicate flame temperature, not ware temperature. Each application is unique and some experimentation may be necessary to achieve the desired result. This is discussed in more detail in Chapter 18.

Many industrial applications involve process measurements on nonopaque materials. Then both transmission and reflection enter into the temperature measurement problem. For example, in producing glass the raw materials are melted to produce a transparent liquid before forming glass articles from it. Different glasses have tremendously different transparencies at pyrometer red wavelengths. Radiation emitted from hot glasses will be from near the surface if the absorption coefficient is large, but from the depth if it is small. If a temperature gradient exists in the glass (if the glass is cooler with depth, as usually occurs in glass tanks), then when the temperature is measured from above, the apparent temperature will be weighted to the lower temperatures. A simple explanation follows.[1]

Suppose that the glass has uniform composition and absorption coefficient, k_a. Consider an element of glass of unit thickness somewhere within its depth (Fig. 12.16). Then the layer will emit radiation in all directions. The intensity normal to the surface will be equal to $I(1 - e^{-k_a})$, where I is the intensity normal to a blackbody surface at that temperature. As that radiation approaches the surface, it will be attenuated all along its path until as it reaches the surface, it is reduced by a factor $e^{-k_a x}$. Then the intensity from the element of unit thickness just as it reaches the surface will be

$$I_x = I(1 - e^{-k_a})(e^{-k_a x}) . \tag{12.23}$$

High values of k_a cause greater emittance from the layer, but also greater attenuation during transmission. Consequently, high values of k_a cause the radiation to eminate from near the surface. Low values, if constant-temperature conditions exist, cause the radiation to eminate from deeper within the surface. The total rate that radiation reaches the surface is the sum of all the radiation from all the layers of unit thickness having integral values of x:

$$I_t = I(1 - e^{-k})(1 + e^{-k} + e^{-2k} + \cdots + (e^{-(x-1)k}) \tag{12.24}$$

$$= I(1 - e^{-k})\left(1 + \frac{e^{-k} - e^{-kx}}{1 - e^{-k}}\right)$$

$$= I(1 - e^{-k_a x}) . \tag{12.25}$$

This shows that the intensity reaching the surface from small values of x is always dominant, but that for small values of k greater contributions from

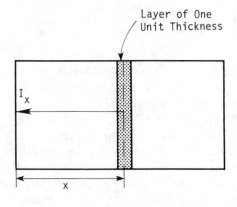

Layer of One
Unit Thickness

I_x

x

Figure 12.16. Schematic of a glass tank with a layer of unit thickness within the melted glass. (From T. R. Harrison, "Radiation Pyrometry and Its Underlying Principle of Heat Transfer." Copyright 1960, John Wiley & Sons, Inc., New York. Reprinted by permission.)

depth occur. Then temperature measurements will give brightnesses, after allowing for reflection, that depend on the absorption of the radiating medium. If I is a function of x, that relationship must be included in the summation. For the usual situation where glass is heated from the top in a melting furnace, the temperature is lower at greater depths. The brightness temperature measured will be reduced by the reduction in I from Planck's law. Such a measurement may indicate temperatures much lower than the actual temperatures of the glass surface. Conversely, if a radiant source is located behind a partially transparent medium, the apparent temperature may be higher than the actual temperature. The actual situation is more complex than explained here, but the general conclusions are the same.[12-14]

12.9   TWO-COLOR PYROMETERS

For some applications a two-color pyrometer may be better. If two wavelengths are used, and if the emittance is the same at both wavelengths, the absolute value of emittance need not be known because it cancels out of the equation for the spectral intensity ratio [Eq. (12.12)]. Using Wien's law, we have

$$R = \frac{eC_1\lambda_1^{-5}\exp(-C_2/\lambda T)}{eC_1\lambda_2^{-5}\exp(-C_2/\lambda_2 T)}. \qquad (12.26)$$

Taking logarithms gives us

$$\ln\left[R\left(\frac{\lambda_1}{\lambda_2}\right)^5\right] = \frac{C_2}{T}\left(\frac{1}{\lambda_1} - \frac{1}{\lambda_2}\right), \qquad (12.27)$$

so

$$T = \left(\frac{1}{\lambda_2} - \frac{1}{\lambda_1}\right)\frac{C_2}{\ln[R(\lambda_1/\lambda_2)^5]}. \qquad (12.28)$$

From this we see that the greater the difference in wavelength, the greater will be the observed intensity ratio, R, and the better the precision of measurement. Although superior in principle, this method often is not as precise as other methods because of the requirement that the emittance be constant as widely separated wavelengths. If there is reason to question the absolute value of emittance, there is usually uncertainty at two wavelengths also. For visible light, the range of available wavelengths is narrow, so that visible two-color pyrometers are best in principle where the temperature is high enough to give good intensity at short wavelengths. Two-color pyrometers are also discussed in Chapter 13, because the longer wavelength can be utilized more effectively in photoelectric optical pyrometers.

REFERENCES

1. H. Le Châtelier, Sur le mesure optique des temperature elevées. *C.R. Hebd. Seances Acad. Sci.* **114**, 214–216 (1892).
2. W. Wien, Ueber die Energievertheilung im Emissionsspectrum eines schwarzen Körpers. *Ann. Phys.* **58**, 662 (1896).
3. S. F. B. Morse, "The Morse ThermoGage" *Am. Mach.* **26**, 1514–1515 (1903).
4. M. Planck, Ueber irreversible Strahlungsvorgänge. *Ann. Phys. (Leipzig)* [4] **1**, 69 (1900).
5. H. F. Stimson, The international practical temperature scale of 1948. *J. Res. Natl. Bur. Stand., Sect. A* **65A**, 139 (1961).
6. H. J. Kostkowski and R. D. Lee, Theory and methods of optical pyrometry. *NBS Monogr.* **41** (1962).
7. C. R. Barber, International practical temperature scale of 1968. *Nature (London)* **222**, 929–931 (1969).
8. R. E. Bedford, The reporting of thermodynamic temperature intervals measured with an optical pyrometer. *Metrologia* **18**, 169–170 (1982).
9. Leeds and Northrup Company, "Optical Pyrometry," Tech. Publ. A1.4000/1966. Philadelphia, Pennsylvania, 1966.
10. J. R. Branstetter, Some practical aspects of surface temperature measurement by optical and ratio pyrometers. *NASA Tech. Note* **NASA TN D-3604** (1966).
11. Pyrometer Instrument Company, "Pyro Optical Pyrometer," Catalog No. 86. Northvale, New Jersey, 1984.
12. R. Gardon, The emissivity of transparent materials. *J. Am. Ceram. Soc.* **39**, 278–287 (1956).
13. J. C. Richmond, Relation of emittance to other optical properties. *J. Res. Natl. Bur. Stand., Sect. C* **67C**, 217–226 (1963).
14. R. C. Folweiler and J. Mallio, "Thermal Radiation Characteristics of Transparent, Semi-transparent and Translucent Materials," OTS Tech. Doc. ASD-TDR-62-719, Part 2, U.S. Dept. of Commerce, Washington, D.C., 1964.

CHAPTER 12 PROBLEMS

12.1 You measure the temperature of a furnace through a small hole and determine a temperature of 1620°C with a disappearing filament optical pyrometer. A specimen removed from the furnace before it has a chance to cool has an apparent temperature of 1540°C. What is the emittance of the specimen? Use Wien's law.

12.2 If you use the same pyrometer to measure the temperature of a platinum wire and obtain an apparent temperature of 1100°C, what is its actual temperature if the wire emittance is 0.37?

12.3 You measure the temperature of a furnace as 1450°C with an optical pyrometer. Then you take the same temperature with your sunglasses in front of the telescope objective and get a temperature of 1307°C. What is the transmittance of the filter? What is the filter constant, A? (*Note*: Use 0.640 μm as λ for this particular optical pyrometer and 14380 μm-K for C_2 in air.)

CHAPTER 12 ANSWERS

12.1   1620°C + 273 = 1893 K
1540°C + 273 = 1813 K

$$\epsilon \equiv \frac{J}{J_{BB}} = \frac{C_1\lambda^{-5}e^{-C_2/\lambda T_\beta}}{C_1\lambda^{-5}e^{-C_2/\lambda T_{BB}}} = e^{(C_2/\lambda)(1/T_{BB}-1/T_\beta)}$$

$$= e^{[(14.380/0.640)(1/1893-1/1813)]} = 0.59$$

12.2   110°C + 273 = 1373 K

$$0.37 = e^{(14,380/0.640)(1/T_{BB}-1/1373)}$$

$$\ln 0.37 = \frac{14,380}{0.640}\left(\frac{1}{T_{BB}} - \frac{1}{1373}\right)$$

$$\frac{1}{T_{BB}} = \frac{1}{1373} - \frac{(\ln 0.37)(14,380)}{0.640} = 0.000684$$

$$T_{BB} = 1462 \text{ K} = 1189°C$$

12.3   1450°C + 273 = 1723 K
1307°C + 273 = 1580 K

$$\tau = \frac{C_1^{-5}e^{-C_2/T_{out}}}{C_1^{-5}e^{-C_2/T_{in}}} = \frac{J_{out}}{J_{in}}$$

$$\tau = e^{-(14,380/0.640)(1/1723-1/1580)} = e^{-1.1802}$$

$$= 0.307 = 30.7\%$$

$$A = \frac{1}{T_{out}} - \frac{1}{T_{in}}$$

$$= \frac{1}{1580} - \frac{1}{1723}$$

$$= 5.25 \times 10^{-5}$$

13

PHOTOELECTRIC OPTICAL PYROMETERS (AUTOMATIC AND INFRARED)

13.1 INTRODUCTION

Development of photoelectric detectors such as photomultipliers and solid-state junction devices made it possible to replace the human eye as the detector in radiant temperature measurements. This made it possible to develop automatic instruments to replace the manual disappearing filament optical pyrometer and also to made it possible to shift the wavelength of measurement into the infrared region where Planck's law radiation is more intense.

Automatic optical pyrometers operating in the visible wavelengths were developed first beginning in the 1950s.[1] The traditional wavelength at about 650 nm was used, partly because there was a tremendous volume of experience in standards laboratories for that wavelength for the IPTS.[2] This included strip lamp calibrations against the gold point with the disappearing filament optical pyrometer. There was also a tremendous body of data on emittance which favored the traditional wavelength for practical applications. When silicon diode photodetectors became available with much higher intensity at 1000 nm and longer wavelengths, the emphasis shifted to the longer-wavelength region. But the visible and near-visible region is still the basis for international temperature standards.

Replacing the human eye with a photoelectric detector provided improved sensitivity, produced an electrical signal for indicating, recording, and controlling, and eliminated variation between individuals. Concomitant with eliminating human eyes, it was necessary to use filters to limit the long-wavelength cutoff, and this gave the opportunity for improved optical design. The new instruments permitted measurements at the gold point to be improved by at least one order of magnitude. It is now possible to use digital computers to expedite the measurements, even to perform calibration sequences automatically at standards laboratories such as the National Bureau of Standards.

Photoelectric pyrometers can be classified by their wavelengths as visible or infrared. They can also be classified by their wavelengths as very narrow bandpass (spectral, according to Planck's law), as broadband (where the wavelengths admitted to the detector are too broad to be characterized by a single effective wavelength), and as total (where all wavelengths are said to be admitted, although both short- and long-wavelength cutoffs exist). Total radiation pyrometers are so described because all important wavelengths according to Planck's law are admitted. The intensity regions beyond the cutoffs are deemed insignificant. For total radiation instruments, see Chapter 14.

Photoelectric pyrometers can also be classified as having stable standardization, quasi-stable standardizations, or internal standardization; and as slow response or rapid response. The diversity of instruments available commercially and in standards laboratories around the world is the result of improvements in detectors, optical systems (especially narrow bandpass interference filters), high-stability operational amplifiers, digital computers, and electronic instrumentation. Major improvements in the application of fundamental physics to radiation methods have accompanied these new instruments. The result has been greatly improved precision and accuracy of temperature measurement, to extending the range of applicability to higher and lower temperatures, even below the ice point, and to a variety of new, inexpensive commercial instruments for temperature measurement.

The development of automatic optical pyrometers was the first commercial application of photoelectric detectors. Naturally, these instruments, which appeared in the early 1960s, were based on the disappearing filament optical pyrometer. This was a transition phase but produced fine commercial instruments with improved performance over the disappearing filament optical pyrometer.

13.2 BASIC ELEMENTS OF PHOTOELECTRIC TEMPERATURE MEASUREMENT

Any photoelectric temperature measuring system must have (1) an optical system capable of focusing radiant energy from the target on a suitable detector, (2) filters or other means to select the wavelengths to be used in the measurement, (3) one or more detectors to convert the radiant intensity to an appropriate electrical signal, (4) amplifiers and other components to convert the detector signal to a useable signal, and (5) a method of calibration or standardization. We will discuss these individually.

13.3 THE OPTICAL SYSTEM

The purpose of the optical system is to focus radiation from the target on the detector. Optical system design depends tremendously on the wavelength of interest. Selection of lenses is easiest at visible and near-infrared wavelengths because conventional ceramic glasses and single or polycrystalline oxides can be used. These have the distinct advantages that they are strong, hard, and stable in most environments. Lenses of crown glass, borosilicate glass, fused silica, and calcium-aluminosilicate glass have long-wavelength cutoffs of about 1, 2.6, 3.5 and 4.8 μm, respectively. The fused silica must be water-free or the cutoff will be similar to that of borosilicate glass. Special low-water-content glasses are available. Both single-crystal (sapphire) and polycrystalline alumina can be used as windows but are not practical as lenses (Fig. 13.1).[3]

Semiconductors and alkaline earth halides and chalcogen glasses or crystals can be used at longer wavelengths. Commercial single-crystal and polycrystalline lenses and windows can be made from CaF_2, MgF_2, ZnS, ZnSe, MgO, CdTe, SrF_2, and PbS. Some of these are opaque to visible light. So is germanium, for which the short-wavelength cutoff is 1.8 μm, and silicon, for which it is 1.2 μm.

Simple lens systems are often used for visible and near-infrared purposes. At wavelengths greater than about 2.6 μm, first-surface mirrors are often used instead of lenses. One or two in a Cassegrange arrangement are usually employed. Sometimes lens and mirror combinations are used. Alu-

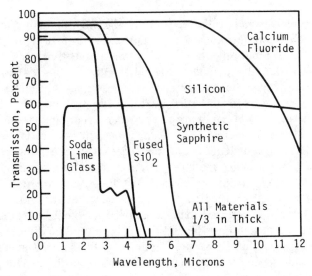

Figure 13.1. Transparency of glasses and crystals used for infrared optics. (Courtesy of the Williamson Corporation.)

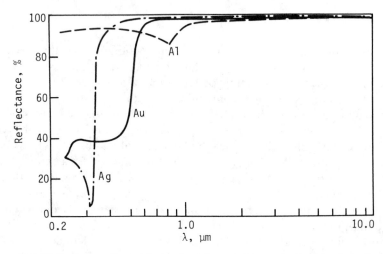

Figure 13.2. Reflectance of metals in the infrared. [From G. Hass and A. F. Turner, Coatings for infrared optics, in "Ergebnisse der Hochvakuumtechnik und der Physik Dünner Schicten" (M. Auwarter, Ed.), Wissenschaftliche, Verlagsgesellschaft mbH, Stuttgart, 1957.]

minum is frequently used as the reflector (Fig. 13.2).[4] Mirror systems must be sealed to prevent dust. Many detectors and lens components react with atmospheric moisture, so those systems also need to be sealed. The choice of window to seal the system is important.

The design of infrared optical systems is similar to that of conventional optics.[5] However, indices of refraction are higher and the chromatic aberration of lenses is usually greater. Some of the most difficult considerations are the background introduced by the thermal emissions of the optical components, unwanted reflections, and transmissions through the target. At the longer wavelengths, incandescent light, with its Planck's law maximum of about 6 μm, must be prevented from reaching the detector. Longer wavelengths are used if the surroundings can produce excessive background radiation. At room temperature, the maximum is about 9.6 μm. Therefore infrared systems making measurements at room temperature must employ long wavelengths and have detectors operating at cryogenic temperatures. But background and reflection contributions must be reduced to acceptable levels. Both optic shielding and filters are used to control background and reflection. But every measurement situation is different. For example, if an infrared heater is used to heat polyethelene film, reflection from that film must be avoided. This would be difficult to do with a broadband instrument. But a filter eliminating the heater wavelengths may make it possible to measure temperature with some other wavelength.

The sensitivity of any detector is proportional to the reciprocal of the square root of the area for the same radiant flux. Therefore, concentrating radiation from a large solid angle from the target onto a small receptor produces maximum sensitivity. One way to enhance the concentration of the radiant energy onto the detector is to use a germanium hemisphere as the receptor for the optical system, and put the detector at the focal point of the hemisphere.[6] Because germanium has an index of refraction of 4, this increases the detector flux by a factor of 4 over that of the optic system without the hemisphere. Such detectors are characterized as hemisphere immersed.

A number of different infrared pyrometers will be discussed later. They have been selected, in part, for their variety of optic systems to illustrate some of the methods used in optic system design.

13.4 OPTICAL FILTERS

With the disappearing filament optical pyrometer the combination of a pyrometer red filter and the visual acuity of the eye gave an asymmetrical bandpass. Because of the steep variation of intensity with wavelength, the product of Planck's law intensity, filter transmittance, and detector sensitivity gave an effective wavelength that depends on temperature (Chapter 12). Adding range change filters then required special absorption properties so that the low- and high-temperature scales could be used to read apparent blackbody temperatures without problems of intensity matching (to avoid problems with the filament not completely disappearing). The first automatic pyrometers used pyrometer brown filters to extend the range even though their use was not justified. Later instruments use neutral density filters (absorption uniform at all the wavelengths admitted) or rotating sectored disks. Filters are more useful than rotating disks because they can have any amount of attenuation desired, whereas rotating disks can be used only at attenuations down to about 1.7%.[7]

The ideal filter would have a square bandpass (Fig. 13.3). With that

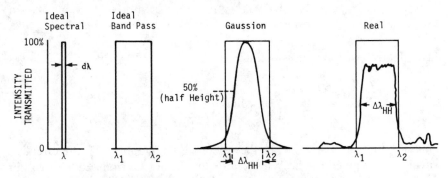

Figure 13.3. Several bandpass possibilities.

shape, for Planck's law spectral intensity determinations, the width would be infinitesimally narrow. A wider bandpass provides more photons for the detector. Real filters do not transmit a single wavelength and do not have a square bandpass. They are somewhere between Gaussian and square in transmittance (Fig. 13.3).

High-quality interference filters are now available that produce transmittance as desired with narrow bandpass or with broader bandpass. The selection of the bandpass width depends in part on the application. For example, high-speed pyrometers sample over such short periods of time that the intensity indicated depends on the total number of photons received. (The rise time is photon dependent and shorter at high photon input rates.) Then a wide band path assists in detector performance. On the other hand, precision standards laboratories will prefer narrow-bandpass filters for better definition of temperature by Planck's law.

In order to solve Planck's law intensity ratios to compute absolute temperature, it is necessary to know the wavelength of the radiation. If a single wavelength were used, there would be no problem. But with finite bandpass width a wavelength must be used that will give an accurate representation of temperature, and the proper wavelength to use depends on temperature because of the shift in intensity with wavelength that accompanies a change in temperature. This subject has received a great deal of attention. Unfortunately, there is no agreement on terminology, although there is agreement on concepts.[8,9] Using the terminology of Kostkowski and Lee, the effective wavelength λ based on Wien's law ratios for two temperatures T_1 and T_2 is

$$R = \frac{\int_0^\infty C_1 \lambda^{-5} e^{-C_2/\lambda T_1}}{\int_0^\infty C_1 \lambda^{-5} e^{-C_2/\lambda T_2}} = e^{(C_2/\lambda)(1/T_2 - 1/T_1)} . \tag{13.1}$$

If the wavelength passing the filter is different at T_1 and T_2, solving for $1/\lambda$ gives

$$\frac{1}{\lambda_{T_1 \to T_2}} = \frac{\ln R}{C_2(1/T_2 - 1/T_1)} . \tag{13.2}$$

The effective wavelength is defined as

$$\frac{1}{\lambda} = \lim_{T_2 \to T} \frac{1}{\lambda_{T \to T_2}} . \tag{13.3}$$

This says that as T approaches the reference temperature T_2, the effective wavelength is given by the Wien's law ratio R and C_2 from Eq. (13.2). For the finite bandwidth represented by Eq. (13.2), we need the mean effective wavelength for the interval $1/T_1$ to $1/T_2$, or

$$\int_{1/T_1}^{1/T_2} \frac{1}{\lambda} d\left(\frac{1}{T}\right) = \frac{\ln R}{C_2}. \tag{13.4}$$

Substituting Eq. (13.4) into (13.2) gives

$$\frac{1}{\lambda_{T_1 \to T_2}} = \frac{1}{1/T_1 - 1/T_2} \int_{1/T_1}^{1/T_2} \frac{1}{\lambda} d\left(\frac{1}{T}\right). \tag{13.5}$$

This shows that the mean effective wavelength between two temperatures is the center of gravity of the reciprocal of the effective wavelength with respect to and between the reciprocal temperatures. This can be determined from experimental data using

$$\frac{1}{\lambda_e} = \frac{\int_0^\infty (B/\lambda)\, d\lambda}{\int_0^\infty B\, d\lambda}. \tag{12.7}$$

According to Kostkowski and Lee, the reciprocal of the mean effective wavelength can be represented adequately as a linear function of reciprocal absolute temperature, and Eq. (12.8) can be used[9]:

$$\frac{1}{\lambda_e}_{T_1 \to T_2} \approx \frac{1}{2}\left(\frac{1}{\lambda_{T_1}} + \frac{1}{\lambda_{T_2}}\right). \tag{12.8}$$

This requires a separate calculation for the mean effective wavelength to be used in Wien's or Planck's laws at every temperature of measurement. An improved method of analysis was proposed by Bezemer.[8] It was improved further by Coates, who recognized that interference filters that have symmetrical band pass properties can be analyzed most effectively if the reference wavelength suggested by Bezemer be taken as the mean wavelength, and the center of gravity of the reciprocal wavelengths for the reciprocal temperature interval be calculated by taking moments about its reciprocal.[10] When this is done, the first term vanishes because there are an equal number of plus and minus terms. Only the square and higher terms appear, because of the symmetry (Fig. 13.4).

Coates begins with the center-of-gravity expressions,

$$\int_0^\infty (\lambda - \lambda_0)^n \tau(\lambda)\, d\lambda = a_n \tau_0 (\Delta\lambda)^{n+1}, \tag{13.6}$$

where $\Delta\lambda$ is the full width of the bandpass at half height, $\tau(\lambda)$ is the filter transmittance, and the subscript zero indicates the mean. The moments a_n depend on the shape of the transmittance curve, and a_1 is zero when λ_0 is

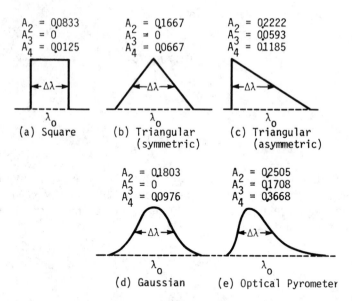

Figure 13.4. Values of A_2, A_3, and A_4 for different optical fiber shapes. (From Ref. 10, reprinted with permission.)

the mean of the transmittance curve. The result of this analysis is that the mean effective wavelength is the reference wavelength λ_0 plus a correction term. To arrive at that correction term, he uses the reference wavelength method of Besemer and expands all terms contributing to the pyrometer signal with a Taylor series about the signal at the reference wavelength. The output signal of the pyrometer $S(\lambda)$ depends on the filter transmittance $\tau(\lambda)$, the spectral sensitivity of the rest of the pyrometer $S(\lambda)$, the emittance of the source $e(\lambda, T)$, and the Planck's law function $P(\lambda, T)$. That signal can be found as an approximation at λ_0 times a correction factor $C(T)$. The approximation is

$$S_0(T) = k a_0 \tau_0(\Delta\lambda) S(\lambda_0) e(\lambda_0) P(\lambda_0, T) \qquad (13.7)$$

where k is an instrument constant not dependent on λ or T. The correction factor is

$$C(T) = \frac{S(T)}{S_0(T)} = \int_0^\infty \frac{\tau(\lambda) S(\lambda) e(\lambda) P(\lambda)\, d\lambda}{S_0(T)}. \qquad (13.8)$$

Expanding all except $\tau(\lambda)$ with a Taylor's series gives

$$\frac{S(\lambda)}{S(\lambda_0)} = 1 + S_1(\lambda - \lambda_0) + S_2(\lambda - \lambda_0)^2 + \cdots \qquad (13.9)$$

$$\frac{e(\lambda)}{e(\lambda_0)} = 1 + e_1(\lambda - \lambda_0) + e_2(\lambda - \lambda_0)^2 + \cdots, \tag{13.10}$$

where the coefficients are found by the Taylor expansion, for example,

$$S_n = \frac{d^n S(\lambda)}{d\lambda^n}\bigg|_{\lambda_0} \frac{1}{n!(\lambda_0)} \tag{13.11}$$

and

$$\frac{P(\lambda, T)}{P(\lambda_0, T)} = 1 + P_1 \frac{\lambda - \lambda_0}{\lambda_0} + P_2 \frac{(\lambda - \lambda_0)^2}{\lambda_0^2} + \cdots, \tag{13.12}$$

where

$$P_1 = \frac{p e^p}{e^p - 1} + 5 \tag{13.13}$$

$$P_2 = \frac{(3p^2/2 - 6p)e^p}{e^p - 1} + 15 \tag{13.14}$$

and p is defined as $p \equiv C_2/\lambda_0 T$. Coates uses Wien's approximation for the higher terms,

$$P_3 = p^3/6 - 7p^2/2 - 19p - 35 \tag{13.15}$$

and

$$P_4 = p^4/24 - 4p^3/3 + 14p^2 - 46p + 70. \tag{13.16}$$

He gives the values of the Planck's function expansion polynomials for two common wavelengths (Table 13.1).

The correction term is computed,

$$C(T) = 1 + \sum_1 \int_0^\infty \frac{(\lambda - \lambda_0)\tau(\lambda)\, d\lambda}{a_0 \tau_0 \lambda_0 \Delta \lambda} + \sum_2 \int_0^\infty \frac{(\lambda - \lambda_0)^2 \tau(\lambda)\, d\lambda}{a_0 \tau_0 \lambda_0^2 \Delta \lambda} + \cdots, \tag{13.17}$$

where $\sum_n$ represents the sum of the coefficients of degree n after collecting like terms in the Taylor expansion. Letting $A_n = a_n/a_0$ gives

$$C(T) = 1 + \sum_2 A_2\left(\frac{\Delta \lambda}{\lambda_0}\right)^2 + \sum_3 A_3\left(\frac{\Delta \lambda}{\lambda_0}\right)^3 + \cdots. \tag{13.18}$$

Note that the first moment vanishes because λ_0 was taken as the mean. The values of A_n for some possible filter characteristics are given in Fig. 13.4. It

TABLE 13.1 Values for the Polynomials P_n

p	P_1	P_2	P_3	P_4	T (°C) 654 nm	T (°C) 1 μm
6	1.01	− 2.96	−11.0	64.0	3394	2125
7	2.01	−2.47	−16.3	76.7	2870	1782
8	3.00	−0.98	−21.7	86.0	2477	1525
9	4.00	1.51	−26.0	91.4	2171	1326
10	5.00	5.00	−28.3	93.0	1927	1166
11	6.00	9.50	−27.7	93.4	1727	1035
12	7.00	15.0	−23.0	94.0	1560	926
13	8.00	21.5	−13.3	98.7	1419	834
14	9.00	29.0	2.3	112	1298	755
15	10.0	37.5	25.0	139	1194	686
16	11.0	47.0	55.7	187	1102	626
17	12.0	57.5	95.3	263	1021	573
18	13.0	69.0	145	376	949	526
19	14.0	81.5	206	535	885	484
20	15.0	95.0	278	750	827	446

Source: From Ref. 10, reprinted with permission.

is necessary, then, to determine the filter transmission characteristics, $\tau(\lambda)$, to measure $S(\lambda)$, and to find $e(\lambda)$. The values of A_n can be determined by numerical integration of $\tau(\lambda)$, or, if lesser accuracy is needed, from Fig. 13.4.

For narrow-bandpass filters, the correction term with excellent accuracy, if $\Delta\lambda/\lambda_0 < 2/p$, is

$$C(T) = 1 + \sum_2 A_2 \left(\frac{\Delta\lambda}{\lambda_0}\right)^2 \qquad (13.19)$$

and

$$\sum_2 = P_2 + P_1(S_1 + e_1)\lambda_0 + (S_2 + e_2 + S_1 e_1)\lambda_0^2. \qquad (13.20)$$

For filters often used in photoelectric standards work, where $\Delta\lambda \approx 10$ nm and $600 \le \lambda_0 \le 900$ nm, Eq. (13.19) is accurate to 0.01% if $S(\lambda)$ is known within 10% and $\tau(\lambda)$ has been measured precisely.

For intensity ratio calculations where $S(T)$ is compared to $S(T_1)$,

$$\frac{S(T)}{S(T_1)} = \frac{S_0(T)}{S_0(T_1)}\left\{1 + \left[A_2\left(\frac{\Delta\lambda}{\lambda_0}\right)^2\right]\left[\sum_2(T) - \sum_1(T)\right]\right\}. \qquad (13.21)$$

For optical pyrometers, using $\lambda_0 = 653.5$ nm and $\Delta\lambda = 33$ nm, Coates found values of the correction terms to be

$$A_2\left(\frac{\Delta\lambda}{\lambda_0}\right)^2 = 6.39 \times 10^{-4}$$

$$A_3\left(\frac{\Delta\lambda}{\lambda_0}\right)^3 = 2.20 \times 10^{-5}$$

$$A_4\left(\frac{\Delta\lambda}{\lambda}\right)^4 = 2.39 \times 10^{-6}.$$

The application of this method to broad-bandpass filters requires determination of the parameters $S(\lambda)$, $\tau(\lambda)$, and $e(\lambda)$. But it is then possible to calculate the mean effective wavelength without complex integration from the equation

$$\frac{1}{\lambda} = \frac{1}{\lambda_0}\left[1 - \left(b_1 + 2\frac{b_2 C_2}{\lambda_0 T}\right)A_2\left(\frac{\Delta\lambda}{\lambda_0}\right)^2\right] \tag{13.22}$$

where

$$b_1 = -6 + (S_1 + e_1)\lambda_0 \tag{13.23}$$

and $b_2 = 0.5$.

In the utilization of filters careful design of the optic system is essential. Improper cleaning of any glass surface can leave residues that affect the transmittance.[11] Filters are normally mounted at a slight angle to the optic axis, but care must be taken that multiple reflections cannot reach the detector or that uncompensated polarization cannot occur. It is also essential to know the transmittance versus wavelength curve. The wings of the curve are important to the mean effective wavelength. Any harmonic transmittance must be excluded.

Selection of narrow, broad, or total radiation filters depends on the requirements for the application. High temperatures and slow response permit narrower bandwidths. Low temperatures and rapid response require greater bandwidth.

13.5 PHOTOELECTRIC DETECTORS

Photoelectric detectors are classified as thermal detectors or quantum detectors.[12] Thermal detectors absorb the incident radiation, producing a temperature rise that is used to measure the radiant intensity. Quantum detectors measure the radiant intensity by producing an electrical signal by absorption of individual photons in an electronic transition that responds specifically to the photon energy. Quantum detectors may be photoconductive or photovoltaic, as in a silicon photodiode, where the incident photon is absorbed and produces an electron transition within the target material. This is an internal device; or it can be a photomultiplier, an external device, where the photon causes ejection of an electron from the target. Both types of photoelectric detectors are rated by the minimum power they can detect.

13.5.1 Noise Equivalent Power

The noise equivalent power (NEP) is a measure of the minimum power required to produce an output signal.[13] The incident radiation in most infrared detecting systems is chopped, or modulated, to allow the use of low-noise ac amplification of the signal. The noise equivalent power is the average power of the modulated incident radiation that will give an electrical signal equal to the average root-mean-square (rms) value of the detector noise voltage,

$$\text{NEP} = P(V_n/V_s), \tag{13.24}$$

where P is the radiant power in watts, V_n is the noise voltage, and V_s is the signal voltage developed by the detector. NEP is the power divided by the signal-to-noise ratio. It depends on the experimental conditions, so it should be determined under the conditions to be used in the application. It is necessary to specify the type of source (a blackbody will give a different result than radiation filtered to be at the most sensitive wavelength). Modulation frequency, response-time constant, detector area, temperature, solid angle, and noise bandwidth should be specified. The NEP is the radiant intensity that will give a signal of the same magnitude as the noise, so it is just barely detectable above the noise.

13.5.2 Detectivity

The detector industry has adopted a modification of NEP that makes it possible to compare different detectors more easily. It is the reciprocal of NEP normalized for detector area and bandwidth and called detectivity:

$$D^* = [A_d(\Delta\lambda)]^{1/2}/\text{NEP}, \tag{13.25}$$

where A_d is the detector area and $(\Delta\lambda)$ is the noise bandwidth. High values of D^* are very desirable. Thermal detectors often have D^* in the neighborhood of 10^8 to 10^9 over the range 1.5–20 μm. Quantum detectors often have D^* values of 10^{10}–10^{11} but are very wavelength sensitive and are very sensitive to operating temperature. Performances of commercial photodetectors are well documented.[12]

13.5.3 Absolute Responsivity

The absolute responsivity (AR) of the detector is the ratio of the output electrical signal to the incident radiant power. It, too, is usually defined for a periodic input signal. Then it is the rms value of the fundamental component of the output signal divided by the rms value of the input radiant power and has units of volts per watt or amperes per watt. The spectral absolute responsivity is

$$AR_\lambda = \frac{V_s}{P(\Delta\lambda)}. \tag{13.26}$$

For a blackbody, it is

$$AR_{BB} = \frac{V_s}{P_{BB}}. \tag{13.27}$$

AR is a function of the temperature, the frequency, the wavelength, and detector electrical parameters such as bias voltage. High values are desirable.

13.6 SOLID-STATE PHOTODIODE DETECTORS

Solid-state photodetectors used for intensity measurements can be classified as either photoconductive or photovoltaic devices. Photoconductive devices produce a current proportional to the incident radiation. Photovoltaics produce a voltage proportional to the incident radiation. Some of the more sophisticated (and expensive) devices must be operated at liquid-nitrogen temperatures to minimize noise and enhance sensitivity. Commercial instruments use conventional solid-state detectors.

One of the most commonly used detectors is the planar-diffused silicon PIN diode.[14] It can be used in either photoconductive or photovoltaic modes and has a spectral response curve for wavelengths of 0.3–1.08 μm (Fig. 13.5). Silicon photodiodes are popular because they are linear over many decades of intensity and do not require cooling to cryogenic temperatures for most applications. Other detectors include:

PbS	2.5 μm
Ge	1.6 μm
PbTe	4.0 μm
InSb	7.0 μm
PbSnTe	13.7 μm
PbHgTe	16.0 μm

TABLE 13.2 Comparison of the Properties of Various Types of Detectors

Detector	Spectral Range (μm)	Rise Time (ns)	Responsivity (A/W)	Dark Current (nA)	Noise equivalent Power (W/$\sqrt{Hz}$)
Vacuum photodiode					
Photocathode Cs_3Sb (S4)	0.3–0.6	0.2	0.035 (at 0.437 μm)	0.5	
Photocathode Na_2KCsSb (S20)	0.3–0.8	0.4	0.01 (at 0.698 μm)	0.2	
Photomultiplier					
Multialkali, side-on type	0.16–0.95	1.2	6.4×10^5 (at 0.45 μm)	2	1.4×10^{-16}
Photocathode AgOCs (S1)	0.3–1.1	1.2	360 (at 0.8 μm)	50	3.1×10^{-13}
Photocathode Na_2KCsSb (S20) extended red	0.32–0.92	12	1.2×10^4 (at 0.63 μm)	10	
Germanium photodiode					
Standard	0.5–1.8	0.3	0.97 (at 1.5 μm)		10^{-12}
Ultrafast	0.5–1.8	<0.08	0.15 (at 1.5 μm)		10^{-10}
Silicon photodiode					
Ultrafast	0.3–1.1	<0.035	0.2 (at 0.72 μm)		10^{-10}
Standard	0.4–1.1	1	0.5 (at 0.8 μm)	2.5	5.7×10^{-14}
Avalanche type	0.4–1.1	2	75 (at 0.9 μm)	100	1.5×10^{-14}

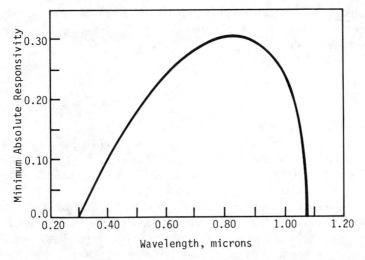

Figure 13.5. Spectral response of a PIN planar-diffused silicon photodiode. (Courtesy of United Technology.)

A recent comparison of photomultiplier and photodiode characteristics is given in Table 13.2.[15]

Advantages of photodiode detectors are that they are stable and have high quantum efficiency, low thermal mass, high sensitivity, and rapid rise time. They are compatible with modern solid-state electronics, micro-processors, and digital computers. They must be kept at a constant temperature for highest precision.

Photodiodes are available in combination with transistors for greater output. The phototransistor and the field-effect phototransistor have similar spectral response to that shown in Fig. 13.5. Preamplifiers are usually employed with photodiodes and phototransistors to enhance their signals. The long-wavelength detectors are more efficient if cooled to cryogenic temperatures. Some instruments operate with a liquid-nitrogen supply (77 K) to have higher signal-to-noise ratios, extending their use to lower temperatures.

13.7 PHOTOMULTIPLIER DETECTORS

Although not useful for practical measuring instruments, photomultipliers have been very useful in the standards laboratories as photon counting devices.

Photomultiplier tubes are electron tubes with a special cathode coated with compounds having a small work function for electron emission. Typically, they are alkali metal compounds with first ionization potentials small enough so that an incoming photon can eject an electron from the cathodes when it is absorbed. The photon energy ΔE depends on frequency ν (or on wavelength),

$$\Delta E = h\nu = h\frac{C}{\lambda} \tag{13.28}$$

where h is Planck's constant, C is the velocity of light in a vacuum, and λ is wavelength. The cathode is mounted inside an evacuated enclosure with a window transparent to the incoming light. The tube may be magnetically shielded except for the window. When an electron is ejected from the cathode, it is accelerated by an applied field to a "dynode," where its energy is sufficient to displace many electrons as the result of the acceleration. The usual tube has many dynodes in series (5 to 10 are common). An avalanche of electrons is emitted from the last stage to the anode, large enough to provide a measurable current pulse[16] (Fig. 13.6).

Photomultipliers require a stable dc high-voltage supply of about 1000 V to provide the accelerating voltage. In the voltage mode they also need a low-voltage dc supply. Integrated photodetection assemblies are available with the power supplies included.

The quantum efficiency (the number of pulses produced per hundred incident photons) may be quite low, perhaps 6%. As the number of photons increases, the output current should increase linearly. The purpose of the

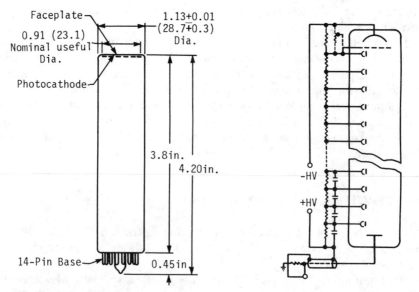

Figure 13.6. Typical photomultiplier configuration. (Courtesy of RCA Corporation, New Products Division, Lancaster, Pa.)

capacitors in the high voltage-divider circuit of the last three or four dynodes in Fig. 13.2 is to prevent significant voltage change by the incoming electron cascade. This helps to increase the linear region to higher photon intensities and to improve the current response when subjected to pulsed light. Linearity of output, stability, and lack of drift at different light levels is critical to some types of photoelectric temperature measurement.

When a single alkali compound is used on the cathode, a rather narrow range of photon energies will cause the cathode to emit. Adding another (bialkali) and another (multialkali) increases the range of photon energies accepted. One of the most popular tubes for near-infrared wavelengths and for optical wavelengths has a multialkali cathode and its JEDEC spectral response is designated as S-20 (Fig. 13.7).

The response curve depends on the tube design and on the type of window. At 650 nm the 50 AT tube has a responsivity of about 20 mA/W. This falls to only 6 mA/W at 750 nm, so this particular photomultiplier tube is best suited to the red or very near infrared wavelengths. Other tubes with special windows and cathodes have usable response to about 1100 nm. However, linearity and drift problems increase as the wavelength increases.

The lowest light level that can be detected depends on the dark current. That current, the leakage current when there is no light, sets the lower limit of use because the dark current must always be less than the lowest practical reading by whatever limit of resolution is desired. For example, if the current must be measured to 0.1% and the dark current is 10^{-12} A, the minimum usable current is

$$I_m = \frac{10^{-12}}{0.001} = 10^{-9}\,\text{A}\,. \qquad (13.29)$$

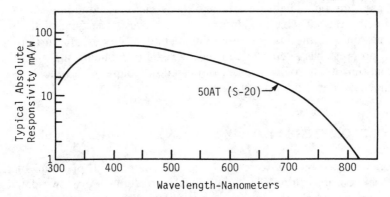

Figure 13.7. Typical response curve for an RCA 50 AT photomultiplier tube. (Courtesy of RCA Corporation, New Products Division, Lancaster, Pa.)

If the dark current changes under different measuring conditions, the highest dark current values must be used in the computation.

When photomultipliers were first used as detectors, considerable variation in linearity and a tendency to drift was recognized. The applications were limited to devices incorporating a radiant intensity standard to nullify that variation. An early study evaluating photomultipliers concluded that they were especially subject to drift when the illumination was decreased. Drift was reduced by maintaining the tube at constant temperature. By proper instrumentation, testing with sectored disks, Kunz found satisfactory linearity if long-term drift were known and compensated, and if the operating conditions were properly chosen.[7] Recently, however, the photomultiplier in that laboratory was replaced with a special vacuum photocell with S-20 spectral response. For this cell, the stability and linearity is so good that single-point calibration is practical. The photocell is linear from 10^{-14} to 10^{-8} A and the estimated total error of the pyrometer was 0.1 to 0.2% of the temperature.[17] It is apparent, then, that photomultipliers are very useful for precision instruments operating at pyrometer red wavelengths, but that improved sensors are now available for higher sensitivity, for longer wavelengths, and for applications where internal calibration is not desirable.

13.8 THERMAL DETECTORS

Thermal detectors can be classified as bolometers, where the resistance of the receptor is used to measure its change in temperature caused by the incident radiation; as thermopiles, where the Seebeck voltage is used; and as pyroelectrics, where the change in surface charge accompanies a change in temperature of the receptor.[12] In each case, the incident beam is absorbed, and this causes a temperature change of the receptor. Because the energy is converted to heat, the sensor is not sensitive to the wavelength received. The smaller the thermal mass of the detector, the more rapid its response can be. The supports and heat sinks surrounding the sensor control the rate at which a sensor cools. The response time of the sensor depends on the thermal mass and the conductivity to its surroundings. Much progress has been made in recent years in reducing thermal mass of thermal sensors. Pyroelectrics are discussed in Chapter 15. Bolometers and thermocouples will be discussed next.

13.9 BOLOMETERS

Historically, bolometers were platinum foil coated with platinum black to increase their absorption. The rough black surface had very low reflectance, so emittance and absorptance were very high. The absorbed energy raised the temperature and increased the electrical resistance. Modern bolometers usually have thermistors substituted for the platinum foil. Thermistors have a negative temperature of resistance, α, and are much more sensitive than metal resistors (Chapter 9).

Bolometers are most effective when the incoming radiant beam is chopped to produce a low-frequency output signal, about 10 Hz. Two common detector circuits are used. The thermistor detector R_d can be mounted in series with a very stable resistor R_s, as shown in Fig. 13.8. The resistance of the detector changes in response to the chopped incoming radiation, J. If R_s is very stable, for example a manganin resistor, then variations in its temperature from either heat conducted from R_d, or from variations in ambient temperature, will not change its resistance. Then the voltage drop from the current flowing through both R_s and R_d will produce an ac voltage signal V_s through the blocking capacitor,

$$V_s = \frac{R_s V (\Delta R_d)}{(R_d + R_s)^2}.$$ (13.30)

The change in the resistance of the thermistor depends on the temperature change ΔT_d:

$$\Delta R_d = R_d \alpha (\Delta T_d),$$ (13.31)

So

$$V_s = \frac{R_s R_d V \alpha (\Delta T_d)}{(R_d + R_s)^2}.$$ (13.32)

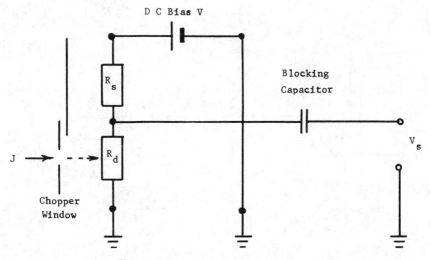

Figure 13.8. Biased thermistor detector.

The responsivity depends on the effective heat conductance q_e for the detector (conductance away from the resistor by all modes of heat transfer), on the thermal mass, and the frequency of chopping the radiation. The thermal mass, mC_p, where m is mass and C_p is specific heat, slows down temperature response. The effective conductance q_e enhances it. A time constant,

$$\tau = \frac{mC_p}{q_e} \tag{13.33}$$

is determined by their ratio (Chapter 19). Then the absolute responsivity is given by

$$AR = \frac{V_s}{J} = \frac{R_s R_d \varepsilon \alpha}{(R_s + R_d)^2 q_e (1 + \omega^2 \tau^2)^{1/2}}. \tag{13.34}$$

where ω is the angular frequency and ε is the emittance for the detector.

The bridge circuit (Fig. 13.9) is usually made with R_d and R_s carefully matched in size, mass, and resistance. Flake thermistors are often used. When $R_1 = R_3$ and $R_s = R_d$, the bridge current I through the detector depends on the applied voltage and $R_s + R_d$. No current flows through R_2. When the radiant flux strikes the thermistor, R_d changes and current flows through R_2 giving an output signal V_s. Then

$$V_s = \frac{I(\Delta R_d) R_2}{R_1 + 2R_2 + R_3}. \tag{13.35}$$

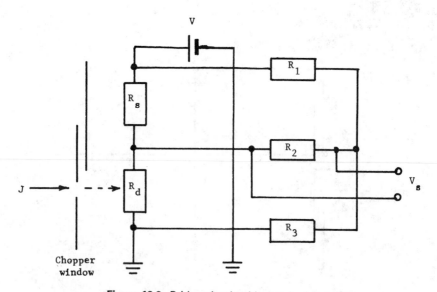

Figure 13.9. Bridge circuit with matched resistors.

The absolute responsivity is

$$AR = \frac{\varepsilon I R_d \alpha}{2}.$$ (13.36)

Commercial instruments are available employing these arrangements.

13.10 THERMOPILE DETECTORS

Thermopile detectors produce a signal without an applied voltage (Chapter 10). They are rugged, stable, and reliable. For this reason, thermopiles are frequently used for unchopped radiation in spectral or in total radiation pyrometers, where the wavelength is short enough (temperature high enough) for their use. Schematically, the device consists of a low-thermal-mass blackened surface in good thermal contact with the measuring junctions of a thermopile or other detector (Fig. 13.10). When a thermopile is used, the reference junctions of the thermopile are held in a high-thermal-mass constant-temperature region and the thermocouples are made from very fine wire to restrict heat flow. The thermocouples are electrically insulated from each other so that the output of the n thermocouples in series is n times that of the individual thermocouples. Type E thermocouples give the highest output. It is necessary either to control the reference junction temperature or to measure it by an auxiliary means and compensate for it. Thermopiles are discussed in Chapter 10. The output of the thermopile is given by

$$V_{\text{out}} = n(V_{t_x} - V_r),$$ (13.37)

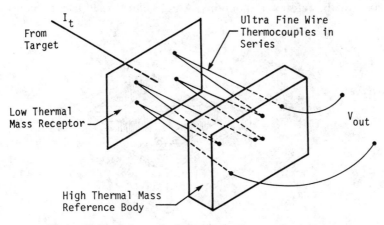

From Target

I_t

Ultra Fine Wire Thermocouples in Series

Low Thermal Mass Receptor

V_{out}

High Thermal Mass Reference Body

Figure 13.10. Schematic diagram of a thermopile detector.

where V_{t_x} is the Seebeck voltage for an individual thermocouple at temperature t_x and V_r is the Seebeck voltage of an individual thermocouple at the reference temperature t_r. The thermopile receptor accepts and responds to all thermal radiation coming to it, so it is in principle a total radiation receptor. Because of optical absorptions at long wavelengths for gases in the air, such as water vapor and carbon dioxide, filters are used to avoid those regions of absorption. Therefore, the thermopile response depends on the design of the optical system for the incident radiation. Although not shown in Fig. 13.10, the receptor should be thermally insulated from the reference body so that its temperature change does not change the reference body temperature.

The principal advantage of a thermopile is its high sensitivity for radiation at low temperatures. Its disadvantages, nonlinear output and the required reference junction compensating, can be accommodated by modern microprocessor and solid-state logic circuitry. The heat delivered to the radiation sensor should be prevented from changing the ambient temperature by shielding to reduce direct radiation and by long, thin wires to reduce conduction. Linearization and compensation circuits are discussed in Chapter 8.

Modern thermopiles are usually made by thick-film photolithography techniques. One example is a thermopile assembly printed on a thin film of Mylar that serves as an insulator and suspends the measuring junctions over a channel in a heat sink, with the reference junctions in good thermal contact with the heat sink (Fig. 13.11).[18]

Another example of a thin-film thermopile is one where the density of junctions is increased by putting the legs of each thermocouple at right angles (Fig. 13.12).[19] Junctions are arranged in four quadrants using thick-film thermocouples of Constantan–Nichrome, or other composition deposited on a film of borosilicate glass suspended over a hole in an oriented silicon crystal. The silicon crystal serves as a support frame and heat sink for the reference junctions. A gold film is used as a reflector over

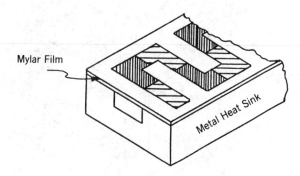

Figure 13.11. Thermopile of low thermal mass.

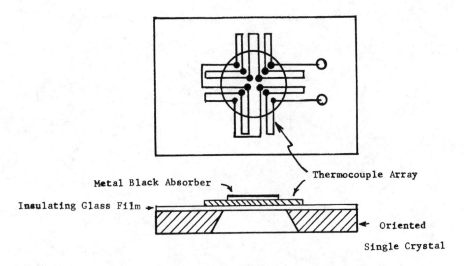

(Gold deposits for reference junction connections   not shown.)

Figure 13.12. Leeds and Northrup thin-film thermopile.

reference junctions areas of the heat sink and for electrical connections. The measuring junctions are covered with a thin glass film, which in turn is covered with metal black to increase emittance.

13.11   INFRARED SIGHTING SYSTEMS

Infrared optical systems are always a compromise between cost and accuracy. Expensive germanium lenses can be used to make optical systems useful at 8 to 14 μm, using a detector cooled in liquid nitrogen, and this may be justified for some applications. Although there are important exceptions, many commercial instruments use lenses for the short wavelengths, in the region 0.7–2.0 μm. Many commercial instruments use mirror optical systems for the 2-μm and longer wavelengths, with windows suitable for the wavelengths used.

Because of its invisibility, an auxiliary visual sighting system is often provided. It is essential to avoid adjustment and parallax errors for such systems. For small targets, sometimes a light beam is projected onto the target to ensure alignment. Mounting the instrument on a sight tube is another way of sighting that is sometimes suitable.

Auxiliary visual focusing lenses are often incorporated into the infrared optical system at short wavelengths. This helps eliminate parallax. Separate focusing telescopes are often used for long-wavelength optical systems, and the degree of compensation for parallax is important in using them (Figs. 13.13 and 13.14).[20]

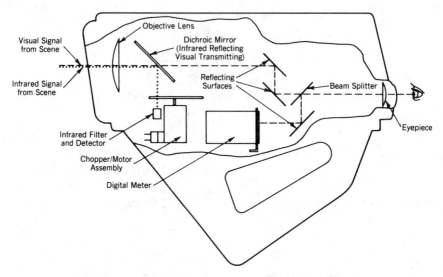

Figure 13.13. Example of an infrared optical system with auxiliary sighting incorporated into the infrared optics. (Courtesy of the Willimson Corporation.)

Infrared sighting systems must accommodate the emittance of the target. Emittances are less well known than for visible and very sensitive to wavelength. Applications where a blackbody target tube can be incorporated into the system without disturbing the thermal balance of the system can be accommodated readily. Emittance of objects can sometimes be enhanced by painting them "black." The paint must be carefully selected because some paints that are black in the visible are white in infrared, and vice versa. Commercial paints are available.[12] However, painting a target to increase the emittance also changes the target's response to the surroundings. Targets of low thermal mass may be cooled or heated if the surroundings are at a lower or higher temperature. So one must be careful

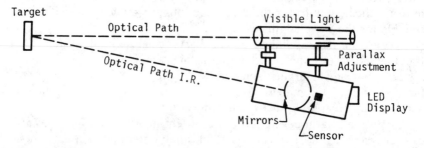

Figure 13.14. Example of an auxiliary telescope used as a sighting aid. The parallax error is reduced by adjusting the telescope line of site for target distance. (Courtesy of the Williamson Corporation.)

that infrared paint does not change the target temperature in an undesirable way.

Infrared optical systems must include elements to shield unwanted radiation. Often, radiation reflected from a target, or transmitted through a target, will give erroneous readings.

The emittance in the infrared spectrum is usually different from the emittance in the visible spectrum. This can sometimes be very useful. Thin sheets of plastic that are almost completely transparent to visible light may be opaque at certain infrared wavelengths. Measuring the temperature of a plastic film during manufacture may be impossible with visible light and entirely practical with infrared radiation.

Absorption by gaseous components in the atmosphere often takes place in the infrared. Then the absorption wavelengths must be avoided if reliable results are to be obtained. For this reason, broadband infrared detectors are often combined with filters to select the band path to be used. For objects requiring long wavelengths, mirrors or other special optical elements are needed.

Water vapor and carbon dioxide are often present in the atmosphere and in flames (Fig. 13.15).[21] When the source has a complex absorption spectrum, the instrument should be operated where the absorption maximum (hence, emission maximum) occurs (Fig. 13.16).[21] Commercial instruments are available to take advantage of absorption by the target and transparency of the atmosphere. For example, one manufacturer has the following selection of instruments available (Figs. 13.17 to 13.23).[22]

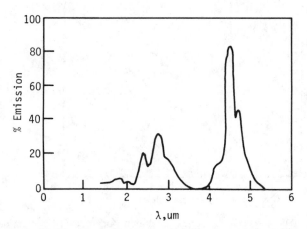

Figure 13.15. Emission spectrum of a bunsen burner (natural gas) flame. Absorption occurs at the same wavelength as emission. Therefore, detector wavelengths should correspond to the valleys of the emission curve. The principal H_2O absorption is 2.7 μm. For CO_2, it is 4.45 μm. (From Burton Bernard, Infrared techniques for temperature measurement of transparent objects, *Industrial Heating*, Dec. 1966, p. 2366.)

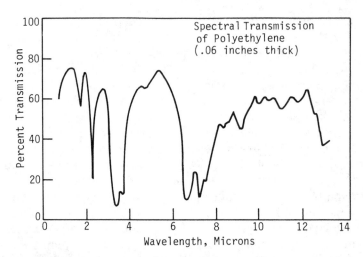

Figure 13.16. Transmission spectrum of polyethylene film. The detector should be sensitive at wavelengths corresponding to valleys in the transmission curve when measuring the temperature of polyethylene. The valleys represent absorption maxima. (From Burton Bernard, Infrared techniques for temperature measurement of transparent objects. *Industrial Heating*, Dec. 1966, p. 2366.)

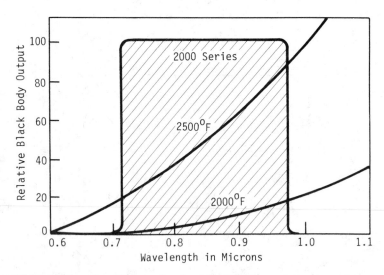

Figure 13.17. 2000 series. This series is suited for measuring incandescent temperatures, with a minimum of extraneous influences. Its short-wavelength response provides a large output change for small temperature changes, thus reducing the effects of unknown or varying emittances. The instrument is not affected by changes in humidity or in carbon dioxide levels and can operate near open flames. (Courtesy of IRCON.)

398

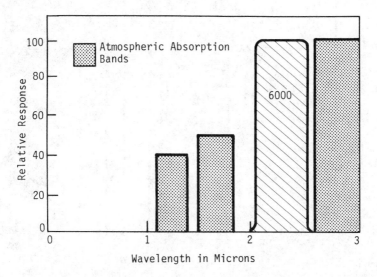

Figure 13.18. 6000 series. The several models cover a temperature span of 150–5000°F and is particularly suited to the measurement of subincandescent temperatures. These instruments can measure plastic temperatures at 300°F as well as steel at 1500°F. (Courtesy of IRCON.)

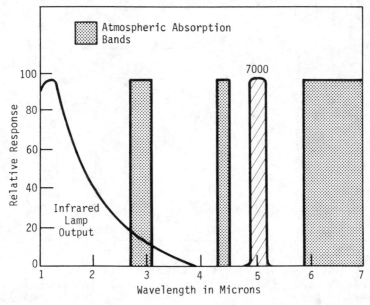

Figure 13.19. The 7000 series receives a longer wavelength than infrared lamps produce, so there is no interference from IR lamp reflections. (Courtesy of IRCON.)

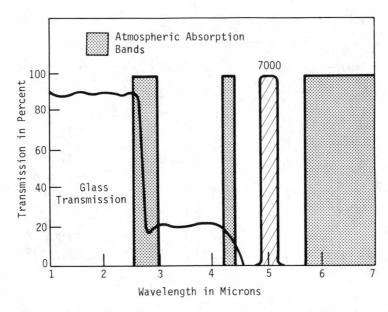

Figure 13.20. 7000 series. This series of instruments has three capabilities which outline its main areas of application: (1) the ability to measure temperatures accurately as low as 100°F; (2) the ability to measure true product temperature in the presence of intense infrared heater radiation; and (3) the ability to measure true glass surface temperature. (Courtesy of IRCON.)

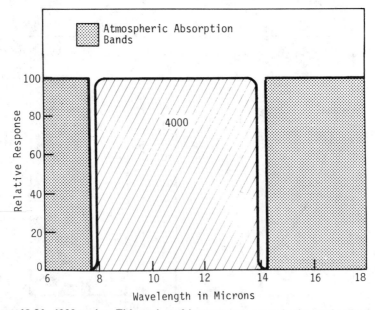

Figure 13.21. 4000 series. This series of instruments extends the performance to very low temperatures, down to 0°C or 0°F. The units operate in the atmospheric window between 8 and 14 μm and are not affected by atmospheric absorption. (Courtesy of IRCON.)

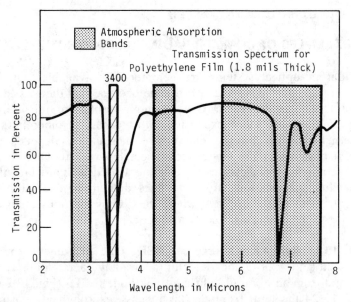

Figure 13.22. 3400 series. The 3400 series instruments are designed to measure temperature of certain thin-film plastics down to 1/4 mil thick and down to 100°F. The instruments are sensitive only to a very narrow band of wavelengths centered at 3.43 μm. This band corresponds with the fundamental carbon–hydrogen absorption band present in many polymers. The instruments are also suited to waxes, oils, organic resins, paints, varnishes, paper, textiles, and so on. (Courtesy of IRCON.)

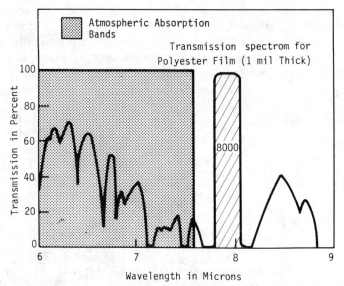

Figure 13.23. 8000 series. The 8000 series instruments are sensitive only to a narrow band of wavelengths centered at 7.9 μm. This corresponds to the fundamental absorption band found in many plastics, particularly polyesters and fluorcarbons. This region also avoids all areas of atmospheric absorption. Temperature ranges down to 50°F are available. (Courtesy of IRCON.)

13.12   EXAMPLES OF COMMERCIAL INFRARED PYROMETERS

Many different optical systems are commercially available. The design depends, in part, on the wavelength regime because that dictates the types of sensors that can be used. Usually, high resolution, reliability, and absolute accuracy are more expensive. Most of the commercial systems have internal standards, but the reliability of the standards depends on both the optical design and the electronics associated with it. One should be careful to compare specifications in selecting a particular instrument. Drift and reliability specifications are sometimes difficult to evaluate. This is especially true for digital displays, where the display may indicate precision far superior to the absolute value of accuracy.

Example 13.1: Reference Source an LED. One instrument employs a rotating sector to expose the detector alternately to radiation from the detector and from an infrared light-emitting diode (LED) (Fig. 13.24).[20]
In Fig. 13.24 the light from the target passes through a focusing lens and interference filter, is chopped by the rotating sectored disk, and falls on the detector. Light from the LED passes through the sectored disk during the dark cycle of the target to reach the detector, producing a square-wave ac signal output from the detector. This is amplified twice by high-gain amplifiers and rectified using the demodulator signal for synchronous rectification. The rectifier output is fed back through an operational amplifier to increase or decrease the LED current to bring the ac signal to zero. The LED is calibrated so that its current is used to indicate temperature. The heat delivered by the diode depends exponentially on temperature, so the accuracy of instrument depends on the stability of the LED and on the accuracy and stability of the electronics used for display. This usually includes linearization, analog-to-digital, and digital display net-

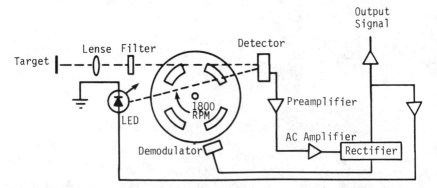

Figure 13.24. Schematic of an infrared LED reference instrument. Note that the auxiliary visual sighting optics are not shown. (Courtesy of the Williamson Corporation, Concord, Mass.)

works. Recording utilizes the analog output from the linearization circuit. Control can be either analog or digital. Because of the exponential nature of the LED output, each model spans a relative narrow range of temperatures (e.g., 25–125°C), but many models are available to cover a broad temperature range (e.g., 11 models cover the range 25–1650°C a 2.8- to 3.5-μm bandpass for the target radiation).

The family of instruments available from this source includes sensors, recorders, and displays for permanent installation. Also, hand-held instruments that have a visual display in the sighting window so that temperatures can be read as a particular target area is viewed.

Example 13.2: Reference Source a Controlled Blackbody. An interesting arrangement for accurate measurements at long wavelengths uses a rotating sector disk with mirror surfaces on both sides to expose target radiation and blackbody radiation alternately to two detectors (Fig. 13.25).[23] The temperature of the blackbody is adjusted automatically until both source and blackbody are at the same temperature. The detectors are low-mass matched thermistors. Their nonlinearity is not a problem because they only determine when target and blackbody signals are balanced. The target temperature is measured with a thermocouple and displayed. This is a low-temperature instrument capable of measuring temperature with a precision of 0.05°C near room temperature.

Example 13.3: Reference Body Temperature Compensation. In this system a rotating disk chopper alternately reflects the reference cavity image back

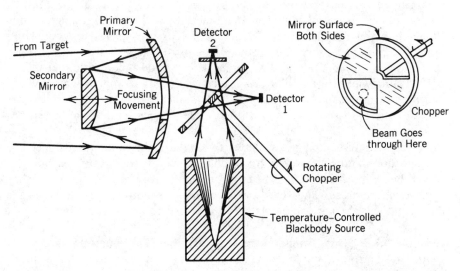

Figure 13.25. A rotating sectored disk alternately transmits as shown, or reflects when rotated 90° more. (Courtesy of Barnes Engineering Company, Stanford, Conn.)

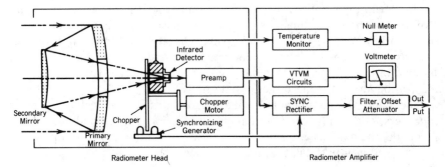

Figure 13.26. The rotating disk is silvered to reflect the blackbody cavity radiation to the detector. (Courtesy of Wahl Instruments, Inc.)

to the detector, then exposes the target to the detector (Fig. 13.26).[24] A thermistor bead senses the blackbody temperature to permit adjustment of the output signal to compensate for blackbody temperature drift.

Example 13.4: Stable Photodiode. Leeds and Northrup manufactures several Spectray temperature detectors similar to their total radiation pyrometers of Chapter 14, but using silicon diode photovoltaic sensors to measure radiant energy concentrated by mirror or lens optics.[25] The mirror system shown in Fig. 13.27 has a silvered window at the bottom of the sectional view and facing the reader in the plan view. A ring-shaped unsilvered region allows the radiant beam from the target to pass through the window and the photosensor support to reach the concave mirror and be reflected on the aperture of the sensor assembly. The silicon photo-voltaic signal is trimmed, during manufacture, with a potentiometer and supplied to the output terminals. The rear lens provides a telescope for optical alignment by focusing on the image of the target reflected from the aperture plate. Filters are used to obtain detectors designed for either 0.65– or 0.90-μm wavelengths. The Planck's law output is approximately T^{12}, and the instruments are recommended for applications where the emittance is low, where they are less sensitive to emittance than total radiation detectors.

A recorder/indicator is available to convert the nonlinear signal to temperature and display it on a linear chart. Several models are available to cover the range from 470 to 1550°C, using the 0.9-μm detector for the lower temperatures. Range cards are available for the recorders to adapt to the output signals that can be as low as 0.03 mV to as high as 129 mV.

Example 13.5: High-Temperature Optical Fiber Thermometry. Sapphire single-crystal rods have been used to provide an optical path for total radiation pyrometry for many years (see Section 14.6). Sapphire single-crystal fibers became available in the 1970s. They can be obtained com-

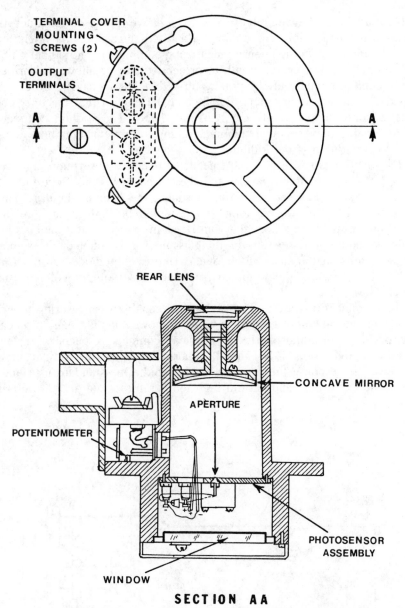

Figure 13.27. Plan and section view of the L & N spectral detector.

mercially in lengths of many meters. Recently, they have been combined with photoelectric sensors to produce high-temperature optical fiber thermometry systems with unusually rapid response and superior precison.[26] They have even been proposed to replace thermocouples in range III of IPTS.[27]

The sapphire fiber merely serves as a wave guide to conduct the radiation from the high temperature to a fused-silica fiber that then conducts the radiation to the intensity measuring equipment. The sapphire fiber can be enclosed in a metal sheath to protect it, can be allowed to project into the high temperature region as in a blackbody cavity, or can be provided with an iridium metal coating on the end so that the fiber end, itself, becomes a blackbody cavity (Fig. 13.28). The iridium coating is covered with a polycrystalline alumina coating to prevent scratching and to reduce evaporation of the iridium.

High-quality sapphire fibers are needed to reduce absorption and scattering loses to acceptable levels, and they must have smooth surfaces to reduce surface losses and scattering. Usually, the fiber is about 1.25 mm diameter and about 300 mm long. It is coupled through an air gap to a fused-silica optical glass fiber that can be as much as several meters long. After the radiation passes through the glass fiber, it is coupled across an air gap to a lens that focuses it on one or more photodiode sensors (Fig. 13.28). Temperature is determined from the sensor output voltage using Planck's law.

Commercial forms are available that are capable of measuring teperatures over the range 500–2000°C.[28,29] Note, however, that many problems may occur from atmospheric or physical contacts, especially at temperatures above 1600°C. Because the thermal mass of the iridium film is very low, the response time is remarkably rapid. One standard instrument uses an 800 ± 50-nm band path at high temperatures, and a 1000 ± 50-nm

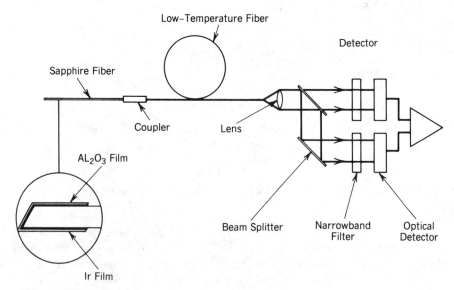

Figure 13.28. Optical fiber instrument. (Courtesy of Accufiber, Vancouver, Wash.)

band path at lower temperatures. The two can be used simultaneously to provide self-calibration.

A 1.25-mm-diameter fiber delivers about 10^{-6} W of power into a 100-nm wavelength band at 800 nm from a 1000°C blackbody, but the fiber must deliver about 10^{-13} W at 300°C and 10^{-3} W at 2000°C. The electronics should be stable and linear to 1 part in 50,000 and the detector should be stable to less than 2 parts per billion to achieve absolute accuracy of 2°C at 1000°C. Under these conditions and with the transmission losses know to within 1%, the manufacturer claims a resolution of 0.00002°C per minute, a precision of 0.2°C, and an absolute accuracy of 1°C at 1000°C and $3\frac{1}{4}$°C at 1500°C. The accuracy is limited by the accuracy of the reference standard.

13.13 SUMMARY

Both automatic optical pyrometers and infrared radiation pyrometers are extremely useful for industrial control. The rapid growth of the latter is the result of improvements in design, so that the higher intensity and unique transmittance of infrared systems can be used. Some of these instruments are truly precision instruments that are reliable and effective. Others are subject to drift or are relatively insensitive. The requirements of the application should be carefully considered and compared with performance characteristics in selecting an instrument.

REFERENCES

1. V. V. Kandyba and V. A. Kovalevskii, A photoelectric spectropyrometer of high precision. *Dokl. Akad Nauk SSSR* **108**, 633 1956.

2. R. D. Lee, The NBS photoelectric pyrometer of 1961. *Temp.: Its Meas. Control Sci. Ind.* **3** (Part 1), 507 (1961).

3. E. C. Magison, The why's of radiation pyrometry. *Instrumentation* **17**, No. 1 (1975).

4. G. Hass and A. F. Turner, Coatings for Infra red optics. In "Ergebnisse der Hochvacuumtechnik und der Physik dunner Sichten" (M. Aurwater, ed.). Wiss. Verlagsges., Stuttgart, 1957.

5. J. M. Lloyd, "Thermal Imaging Systems." Plenum Press, New York, 1975.

6. A. E. Martin, "Infra Red Instrumentation and Techniques," p. 35. Am. Elsevier, New York, 1966.

7. H. Kunz, Representation of the temperature scale above 1337.58 K with photoelectric direct current pyrometers. *Metrologia* **5**, 88–102 (1969).

8. J. Bezemer, Spectral sensitivity corrections for optical standard pyrometers. *Metrologia* **10**, 47–52 (1974).

9. H. J. Kostkowski and R. D. Lee, Theory and methods of optical pyrometry. *NBS Monogr.* (*U.S.*) **41** (1962).

10. P. B. Coates, Wavelength specification in optical and photoelectric pyrometry. *Metrologia* **13**, 1–5 (1977).

11. M. Ohtsuka, R. E. Bedford, and C. K. Ma, A note on the temperature dependence of residual films on glass. *Metrologia* **15**, 165–166 (1979).

12. W. L. Wolfe and G. J. Zissis, eds., "The Infrared Handbook," Chapter 11. Office of Naval Research, Washington, D.C., 1978.

13. A. Hadni, "Essentials of Modern Physics Applied to the Study of the Infrared." Pergamon Press, New York, 1967.

14. United Detector Technology, Inc., "Planar Diffused Silicon Photodiodes." Santa Monica, California.

15. J. F. Bobelot, J. Magill, R. W. Ohse, and M. Hoch, Microsecond and Submicrosecond multi-wavelength pyrometry for pulsed heating technique diagnostics. *Temp.: Its Meas. Control Sci. Ind.* **5** (Part 1), 439–446 (1982).

16. RCA, "Photomultiplier Tubes." RCA, Solid State Div. Electro Optics and Devices, Lancaster, Pennsylvania, 1976.

17. B. Woerner, A Photoelectric direct current spectral pyrometer with linear characteristics. *Temp.: Its Meas. Control Sci. Ind.* **5** (Part 1), 429–432 (1982).

18. P. Villers, W. Falbel, and G. Falbel, Thermopile. U.S. Pat. 3, 424, 624 (1969).

19. R. D. Baxter, Small-size high-performance radiation thermopile. U.S. Pat. 4, 111, 717 (1978).

20. Williamson, 70 Domino Drive, P.O. Box 1270, Concord, Massachusetts 01742.

21. B. Bernard, Infra-red techniques for temperature measurement of transparent objects. *Ind. Heat.*, Dec., 2366–2380 (1966).

22. Ircon, Inc., 7555 N. Linder Avenue, Skokie, Ilinois 60077.

23. Barnes Engineering Company, 30 Commerce Road, Stamford, Connecticut 06904.

24. Wahl Instruments, Inc., 5750 Hannum Avenue, Culver City, California 90230.

25. Leeds and Northrup Company, "Spectray 65, Spectray 90, and Low Temperature Spectray 90 Spectrally Selective Temperature Detectors," No. 177360, Issue 9. North Wales, Pennsylvania, 1979.

26. R. R. Dils, High temperature optical fiber thermomentry. *J. Appl. Phys.* **54**, 1198–1201 (1983).

27. H. B. Amick, Optical fiber thermometry. *Test Meas. World*, **6**, April, pp. 73–78 (1986).

28. J. R. Cooper, Fiber optic sensors—High temperature precision for industry. *Ind. Heat.* **53**, January, pp. 43–44 (1986).

29. Accufiber, Inc., 2000 E. Columbia Way, Vancouver, Washington 98660.

14

TOTAL RADIATION
PYROMETERS

14.1 INTRODUCTION

Total radiation pyrometers are instruments that measure temperature when all the thermal radiation within a small solid angle from a radiant source is focused on a temperature sensor such as a thermopile. They are described as total because all wavelengths of Planck's law radiation are, in principle, focused on sensor. Actually, they are really partial radiation pyrometers.

When the thermal radiation is focused on the sensor it increases the temperature of the sensor. If the sensor is a thermopile that has an approximately linear relationship in output voltage to measuring junction (sensor) temperature, the radiant energy of the source is converted to an output voltage that varies as approximately the source temperature to the fourth power.

The principal advantage of a total radiation pyrometer is that it provides a very stable, noncontact, output signal for process control. This is especially useful for processes operating at temperatures above 1450°C, where type B or type S thermocouples may have short life because of contamination or other changes accompanying the high temperature. They are also very useful for process control in the range 200–1450°C, where it is essential to have a noncontact system, such as controlling the temperature of a continuously moving sheet of material. Total radiation pyrometers usually must be adjusted at the time they are installed to indicate the correct temperature, but need very little attention thereafter.

Early attempts to determine the rate of heat transfer from an incandescent body were unsuccessful. For example, Newton, about 1700, assumed that heat radiating from a body was proportional to the temperature difference between the body and its surroundings,

$$\dot{Q} = K_c(t_x - t_s),$$ (14.1)

where $\dot{Q}$ is the rate of heat flow, K_c is a constant, t_x is the temperature of the body, and t_s is the temperature of the surroundings. This equation is completely unsatisfactory for radiant heat transfer but is often a good approximation for convective transfer, and will be used later in the chapter to describe sensor response. Equation (14.1) is often identified as Newton's law of cooling (or heating).

In the early 1800s, Dulong and Petit, based on experiments at about 300°C, proposed that the heat flow rate from a body was given by

$$\dot{Q} = st,$$ (14.2)

where t is the body temperature and s is a proportionality constant. This, too, is completely unsatisfactory.

The development of Planck's law in 1900 was discussed in Chapter 11 as Eq. (11.41) for hemispherical radiation,

$$W = \pi \int_0^\infty C_1 \lambda^{-5}(\exp(C_1/\lambda T) - 1)^{-1} \, d\lambda.$$ (14.3)

When integrated over all wavelengths the Stefan–Boltzmann law results,

$$W = \sigma T^4.$$ (14.4)

The Stefan–Boltzmann constant for hemispherical radiation has a value, $\sigma = 5.670 \times 10^{-8} \, \mathrm{Wm^{-2} \, K^{-4}}$. Nonblackbodies emit less radiation than a blackbody. They have a total emittance of less than 1. Then the hemispherical radiant heat flux from a body is given by

$$W = e\sigma T^4,$$ (14.5)

where e is the effective emittance for all wavelength and is less than 1. The bodies are not gray bodies because the emittance varies with wavelength. These equations gave the scientific basis for total radiation pyrometers.

The pioneering work that demonstrated that radiation pyrometry might be useful for temperature measurement was conducted by McSweeney in 1828.[1] He used a concave mirror to focus the light emitted from a furnace on the bulb of a thermometer. The temperature rise of the thermometer was an indication of the temperature of the source. At that time there was no way to convert the thermometer temperature to furnace temperature because the Stefan–Boltzmann law had not yet been discovered.

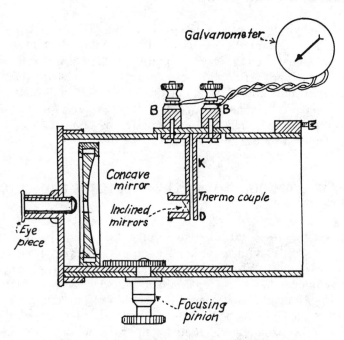

Figure 14.1. Fery mirror instrument. (From Ref. 3.)

In 1902, C. Féry proposed a total radiation pyrometer that is basically the same as the lens containing instruments now in use.[2] In 1904 he proposed a mirror instrument that is basically the same as the mirror instruments now in use.[3] It did not have lens absorptions and gave a higher output than his telescope instrument. The mirror instrument contained two inclined mirrors that could be observed through an ocular. They served as a split-image range finder, so the image of the target could be focused on the thermocouple sensor (Fig. 14.1).

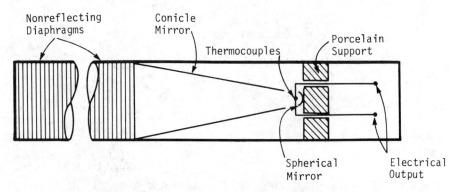

Figure 14.2. Schematic diagram of the Thwing pyrometer.

Another early instrument was designed by C. B. Thwing.[4] It employed a conical mirror to deliver the radiation, with multiple reflection, to a small concave mirror that focused it onto a thermopile. Both mirror instruments were open and did not suffer absorption losses from lenses or windows (Fig. 14.2). However, dust and oxidation of the mirror was a constant problem. Modern instruments have mirrors or lenses. It is necessary for us to consider the principles of total radiation heat transfer as applied to instruments containing lenses, windows, and mirrors.

14.2 THE STEFAN–BOLTZMANN LAW FOR TOTAL RADIATION PYROMETERS

According to Eq. (14.3), Planck's law is integrated over all wavelengths to get the Stefan–Boltzmann law for application to radiation pyrometry.[5] Real pyrometers, however, do not respond to all wavelengths. A typical total radiation pyrometer has a lens or a mirror and window to focus radiation from the target onto a sensor (Figs. 14.3 and 14.4).

The transmittance of the lens or the window is cut off at both short and long wavelengths. Common materials for the windows or lenses are Corning 7740 borosilicate glass, fused silica glass, and cyrstalline calcium fluoride (Figure 14.5).[6,7] These materials are opaque at short and long wavelengths but are transparent in the visible and near infrared. When they are used, Planck's law cannot be integrated over all wavelengths. Therefore, we need to investigate the effect of these cutoffs on the thermal radiant intensity.

From Fig. 14.5 it is apparent that the short-wavelength cutoff is at $0.300\,\mu m$ or shorter for all the materials. Because Planck's law radiant intensity is negligible in those regions of the spectrum at all practical temperatures, the short-wavelength cutoff can be neglected (Fig. 14.6). This is not true for the long-wavelength cutoff. It causes a great deal of loss

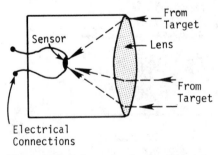

Figure 14.3. Schematic of a lens-focused pyrometer.

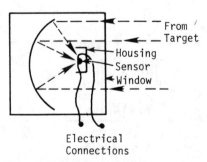

Figure 14.4. Schematic of a mirror-focused pyrometer.

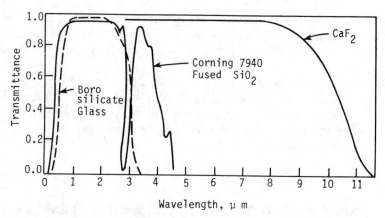

Figure 14.5. Transmittance of window and lens materials. (From T. R. Harrison, "Radiation Pyrometry and Its Underlying Principles of Heat Transfer." Copyright 1960, John Wiley & Sons, Inc., New York. Reprinted by permission.)

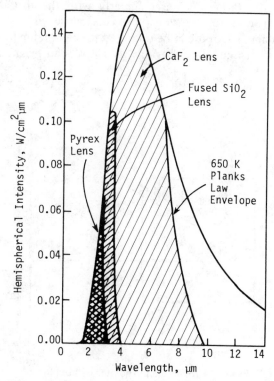

Figure 14.6. Effect of thick window or lenses on integrated Planck's law areas.

of radiation, especially for lower-temperature applications. Because of this cutoff, low-temperature instruments have low-temperature limits that depend on the sensitivity of the sensing device and on the design parameters.

Figure 14.6 is drawn to show the effect of a thick window or lens on the integrated Planck's law intensity. If the lens or window is thin, transmittance will be greater than shown. The design of the optical system and the sensitivity of the sensor determines how much output is obtained from the energy that is received.

14.3   GEOMETRY OF OPTICAL SYSTEMS WITHOUT LENSES OR MIRRORS

It is possible to expose a sensor such as a thermopile to a radiant source without using a lens or mirror for focusing. A window in front of the thermopile is needed to keep dust out. This sort of instrument has a solid angle of entrance that depends on the diameter of the opening, D_0, the distance to the sensor, L_s, and the sensor diameter, D_s (Fig. 14.7).

The size of the target increases linearly with the distance from the pyrometer L_t. Therefore, this must be very close to most targets. Its principal advantage is that it can be operated to lower temperatures without lenses or mirrors. Temperatures as low as 100°C can be measured with a lens-free device.

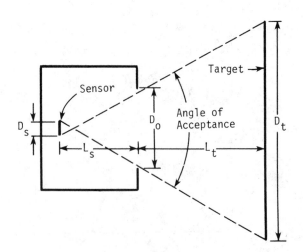

Figure 14.7. Simple pyrometer without lens.

14.4 GEOMETRY OF LENS SYSTEMS FOR TOTAL RADIATION PYROMETRY

Most practical total radiation pyrometers employ a lens or mirror to focus the radiation from the source onto the sensing elements. This makes it possible to position the pyrometer at an appropriate distance from the source. The pyrometer focus depends on its design. For a single-lens system the distance of the sensor behind the lens depends on the focal length of the lens and the distance to the target (Fig. 14.8). This is made adjustable and pyrometers must be used within the target size and distance limitations specified by the manufacturer. Using simple lens equations, at focus,

$$\frac{1}{L_s} + \frac{1}{L_t} = \frac{1}{f} \tag{14.6}$$

and

$$\frac{D_t}{D_s} = \frac{L_t}{L_s}, \tag{14.7}$$

where f is the focal length of the lens. The focal length is a function of the

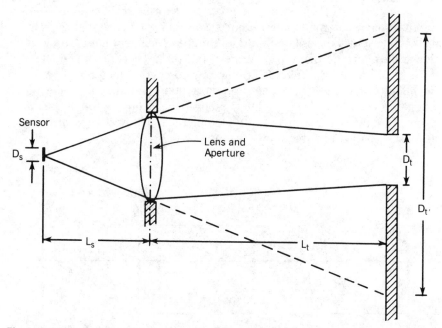

Figure 14.8. Geometry of a simple lens pyrometer with the image of the target focused on the sensor. The intensity reaching the sensor from D_t is the same as would reach it from D_t' without the lens.

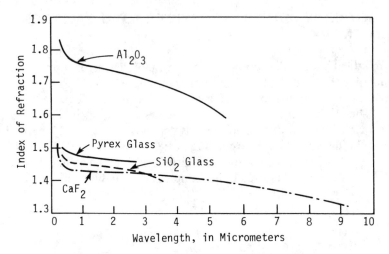

Figure 14.9. Index of refraction of lens materials. (From T. R. Harrison, "Radiation Pyrometry and Its Underlying Principles of Heat Transfer." Copyright 1960, John Wiley & Sons, Inc., New York. Reprinted by permission.)

wavelength because the index of refraction depends on wavelength (Fig. 14.9).

The higher index of refraction produces shorter focal lengths. Lenses therefore have chromatic aberration because all wavelengths do not focus at the same spot. For visible light optics multiple-element lenses are used to reduce the aberration, but for pyrometers this is not practical. The effect of chromatic aberration can be determined by comparing the optical paths for the highest and the lowest index of refraction encountered in the lens.[6] When this is done, the ray pattern shows the solid angles of radiation received by the sensor (Fig. 14.10). Commercial instruments usually give a target size shaped like the outer envelope of Fig. 14.10.

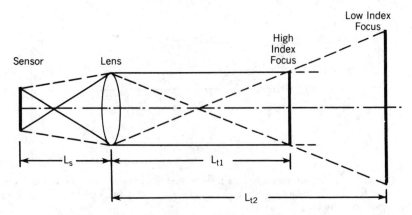

Figure 14.10. Ray pattern for a lens showing the effect of chromatic aberration.

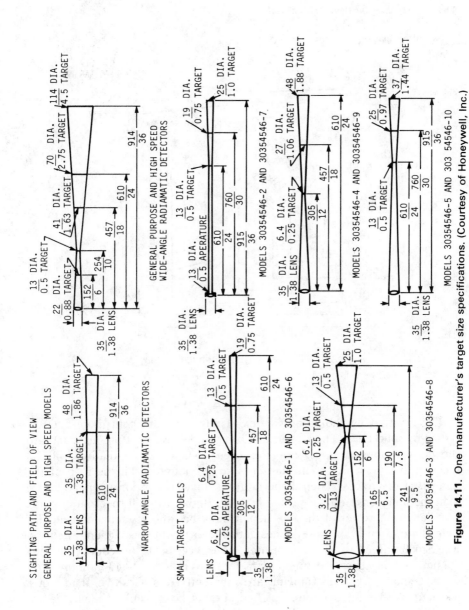

Figure 14.11. One manufacturer's target size specifications. (Courtesy of Honeywell, Inc.)

417

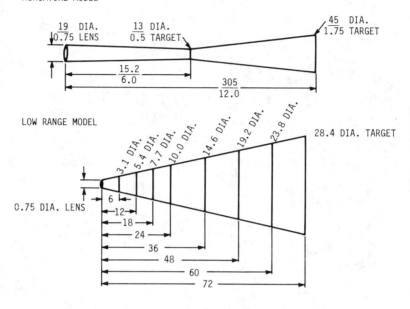

MINIATURE MODEL

LOW RANGE MODEL

Figure 14.11. (Continued)

The rays having the highest index of refraction are bent more and are shown by solid lines. Those with the lowest index are bent the least and are shown with dashed lines. For the fixed distance L_s, radiation will be received by the sensor if the target is large enough to fill the field of the outermost ray traces. Then if the target is anywhere in target space larger than the outer envelope of rays in Fig. 14.10, the rays will reach the sensor. The gathering power of the lens depends on the ratio of its diameter to L_s. Lenses are often $1\frac{1}{2}$ or more inches in diameter to obtain sufficient intensity. If a small target is to be measured, it is necessary to reduce the lens diameter or the sensing-element diameter. This reduces the radiant intensity so that small-target instruments are inherently less sensitive than other instruments. Target diameters of 6 to 12 mm ($\frac{1}{4}$ to $\frac{1}{2}$ in.) in diameter are usually considered to be small (Fig. 14.11).

14.5 GEOMETRY OF MIRROR OPTIC SYSTEMS FOR TOTAL RADIATION PYROMETERS

Mirror optic systems do not have lenses for focusing and do not have the absorption associated with them. The reflectance of one or two mirrors is

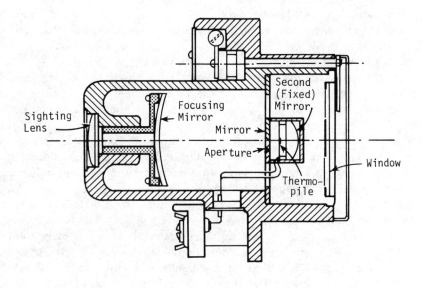

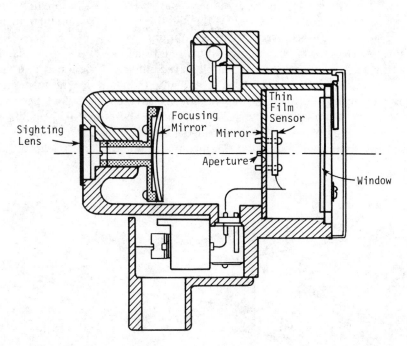

Figure 14.12. Geometry of typical mirror system pyrometers. (Courtesy of Leeds & Northrup Company.)

greater, and broader band, than the transmittance of lenses; so mirror systems should provide more nearly total radiation sensing and higher outputs. However, mirrors must be protected from dust and oxidation or their reflectance will diminish with time. Therefore, mirror systems are hermetically sealed and have a thin glass entrance window.

The entrance window is usually made from fused silica. The cutoff frequency is the infrared depends on the water content of fused silica. The OH bend and stretch modes, even at the very low concentration found in glass, are sufficient to prevent transmission in the 2.5–3.0-mu region of the spectrum. Absence of FeO absorptions is also essential. Special grades of fused silica are available with higher infrared transmittance. Making the window thin and of low absorption glass helps to improve transmittance. The window and its seals must not be subjected to excessive pressure. One manufacturer recommends that gas pressures not exceed 40 psi.

Two typical mirror systems are shown in Fig. 14.12.[8,9] The instrument at the top has two focusing mirrors to concentrate the incoming radiation on the thermopile. This gives a higher thermopile temperature at low radiation levels, so this geometry is used for low-temperature total radiation instruments. The simpler system is used for higher temperatures and for rapid response using a thin-film thermopile of low thermal mass. In both systems the small auxiliary sighting lens allows one to observe the image of the target on the mirror. Alignment of the sighting path perpendicular to the mirror, within 1°, is required for some mounting systems.

The large mirror in both the two-mirror and single-mirror systems can be moved axially to control the solid angle of receptance of the sensor. Moving the mirror closer to the aperture A reduces the radiation passing through it. The circular center area of the window is silvered to reflect rays and prevent them from heating the sensor assembly. Therefore, the ring-shaped transparent region of the window and the aperture in front of the sensor define the solid angle limits for the incoming radiation to the sensor. The

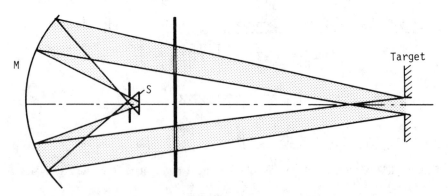

Figure 14.13. Mirror-focusing geometry for small target size.

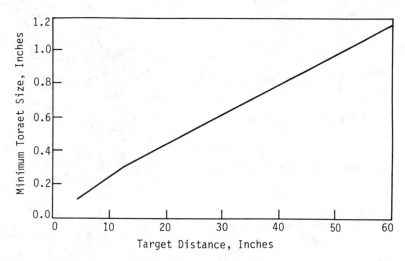

Figure 14.14. Minimum target size for a commercial 24-in. single-mirror instrument.

curvature of the large mirror is slightly converting, so that the target can be a spot (Fig. 14.13).

The two-mirror system of one manufacturer has fixed focal lengths of 13, 17, 24, or 24 in.[5] They are recommended for use at 12–15, 15–20, 20–32, and greater than 32 in. from the target, respectively. The single-mirror system is also fixed focus and must be selected at the time of purchase. Focus distances available are 4, 6, 7–12, 12–16, 16–24, and 24–∞. Required minimum target areas are shown in Fig. 14.14. Notice the small target size that is possible with mirror optics.

14.6 SAPPHIRE LIGHT GUIDE TOTAL RADIATION PYROMETERS

Miniature total radiation pyrometers are sometimes equipped with sapphire light guides to define the optical path from the target to the instrument.[5] They are especially useful for very high temperature applications, where the furnace window must be as small as possible. The sapphire can be sealed into a housing for gastight furnaces.

The light guide is made of single-crystal synthetic sapphire, which has melting point of 2050°C and is available as long, slender rods ground and polished to circular cross section. Rods are usually $\frac{1}{8}$ in. in diameter with lengths such as 6, 12, or 18 in. The index of refraction varies considerably with wavelength (Fig. 14.9). Because the index is so high, light entering the circular end at grazing incidence is refracted, for the minimum index of refraction, at 38.68° (Fig. 14.15).

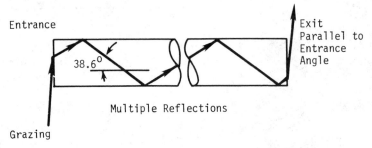

Figure 14.15. Entrance and exit diagrams of sapphire light guides.

This is true according to Snell's law,

$$n = \frac{\sin \theta_i}{\sin \theta_r}$$

where θ_i is the angle of incidence and θ_r is the angle of refraction (measured from the normal). Taking the smallest value of n for sapphire as 1.60 and with θ_i at 90° for grazing incidence,

$$\sin \theta_r = \frac{\sin 90°}{1.60} = \frac{1}{1.60}$$

$$\theta_r = 38.68°.$$

As this radiation reaches the cylindrical surface of the rod it is totally reflected at this and all smaller angles of incidence. However, the rays entering at angles to the axis are polarized on entering. Sapphire is a hexagonal crystal that is optically birefringent. Single crystals usually grow with the optic axis at an angle to the geometric axis, so that propogation through the crystal is very complex. Losses of 29–80% for 6- to 18-in. lengths are common. When the radiation reaches the exit end, it bends away from the normal in accordance with Snell's law as it emerges from the exit plane. Therefore, if hemispherical radiation enters, hemispherical radiation leaves. Then the shape and area of the sensor, and its distance from the exit plane, become very important. The solid angle of the sensor determines the portion of the hemispherical radiation reaching it. And roughly the same solid angle at the entrance plane determines the radiation transmitted to the sensor. Bringing a planar sensor as close as possible to the exit plane increases the solid angle and produces a higher output signal. Thus an axial adjustment should be very effective in adjusting the output to calibrate and to correct for emittance effects.

14.7   THE INTENSITY OF RADIATION REACHING THE SENSOR

The various losses in the focusing path between the source and the sensor reduce the intensity of radiation reaching the sensor. For an infinitesimal area of the target and an infinitesimal area of the sensor, we can construct a ray path diagram as we have done previously and examine the intensity loss as a function of wavelength. To determine the total radiation reaching the sensor, we must integrate over the target area, the sensor area, and all wavelengths from zero to infinity. This can be done in principle. In practice we cannot, because we do not know all the material parameters as a function of wavelength. Experimental measurements with spectrometers can be used to determine absorption and reflection losses in the visible and infrared regions. Most of the Planck's law intensity lies in the infrared region, where infrared spectrometry is required. By operating the spectrometer source at a certain temperature, the transmittance can be measured as a function of wavelength and specimen thickness to obtain empirical data for design purposes. Then the transmittance–wavelength curve can be divided into regions of approximately constant transmittance. The area in each region can be determined and the areas summed to get the total transmittance. This has been done by Harrison to obtain design data for CaF_2, fused SiO_2, and borosilicate glass lenses.[6] As explained in Chapter 11, when Planck's law is integrated from zero to a particular wavelength, the fraction of the total integrated intensity from zero to infinity can be determined. Then it is possible to calculate an equivalent wavelength cutoff for an experimental transmittance curve by equating the fraction transmitted, determined experimentally, to the fraction transmitted computed for Planck's law integration to the equivalent cutoff wavelength. In other words, the equivalent wavelength cutoff is determined by making the area under the experimental curve equal the area under the Planck's law integrated curve. This is useful in comparing designs and in computations. The design with the longest-wavelength cutoff has the highest transmittance, and tables are available for computations based on Planck's law integration.

Computation of the intensity reaching the sensor can, in principle, be computed at each surface reflection and each lens or window absorption. In principle, then, the intensity reaching the sensor is given by

$$W_s = A_s I_t \prod_m (1 - r_m) \prod_n e^{-K_n x_n} . \tag{14.8}$$

Here A_s is the sensor area; I_t is the intensity of the target integrated over the target area that is geometrically situated so that it can reach the sensor; r_m is the reflectance at each of the m lens, window, or mirror surfaces through which the radiation passes or from which it is reflected; and the last term is the product of the lens or window absorptions for each of the n

lenses and windows of thickness x and absorption coefficient K through which the radiation passes. Each of the terms is wavelength dependent. Since the wavelength shifts with temperature of the source, each is also temperature dependent. This is especially true for the absorption coefficient terms.

The thickness of a lens is greater at the center than at its edges. According to Harrison, for a simple planoconvex lens, the fraction transmitted through it is given by

$$\frac{I}{I_o} = \frac{1}{K(x_{max} - x_{min})}\left(e^{-Kx_{min}} - e^{-Kx_{max}}\right). \tag{14.9}$$

Then the lens loss portions of Eq. (14.8) must be replaced by Eq. (14.9). Harrison has made calculations for lens designs represented by one manufacturer.[6] These calculations are useful for designing systems.

For the user the calculations point out the effect of various factors. For example, the radiation reaching the sensor from either lens or mirror systems is less intense near the optic axis than away from it. Since the area available for transmittance varies as the diameter squared, both geometric and absorption effects cause the target to receive more radiation through the outer anular portion of the optical path. Therefore, radiation from around the target area may sometimes affect sensor output, and this may vary with temperature because absorption, reflectance, and wavelength depend on temperature. In practice, this effect must be controlled. One way of controlling it is to use a sighting tube as the target, making the end of the sighting tube large enough to keep sighting effects constant. Sighting tubes are discussed in Section 14.12.

If the sensor is assumed to have a temperature rise proportional to the intensity of the radiation it receives, if it has a linear output voltage in response to a change in sensor temperature, if the target intensity is assumed to depend on the Stefan–Boltzmann equation, and if lens thickness variations are neglected, the output of the instrument should have the form

$$V_o = MGF\Pi_m(1 - r_m)\Pi_n e^{-K_m x_n}\sigma T_t^4. \tag{14.10}$$

Here M is the proportionality constant for sensor temperature to voltage conversion, G is a factor for the target and sensor geometry, F is a factor to represent the fraction of integrated Planck's law intensity transmitted by the system, $\Pi_m(1 - r_m)$ is the product of the transmittances for the surfaces in the optical system, $\Pi_n e^{-K_n x_n}$ is the product of the absorption of the n windows or lenses, σ is the Stefan–Boltzmann constant, and T_t is the target temperature. If the various factors are considered constant, the output voltage has the general form

$$V_o = NT_t^4, \tag{14.11}$$

where N is a constant for the particular measuring instrument. When the output of a total radiation pyrometer is calibrated against target temperature, the actual output is found to very slightly from Eq. (14.11). Usually, the exponent is nearly constant. For some commercial instruments it is very close to 4. For others it is almost 5. Inherent in Eq. (14.11) is the assumption that the sensor has a temperature directly proportional to radiant intensity reaching it. We need to examine this assumption before concluding our discussion of output characteristics.

14.8 SENSOR TEMPERATURE RESPONSE

The temperature of the sensor depends on an energy balance. Radiant energy from the target heats the "front" area of the sensor. The back surface, and that portion of the front surface not in the solid angle of rays to the target, radiates from the sensor to the instrument housing (Fig. 14.16). The sensor is also cooled by conduction of heat from the sensor to the housing through its supports. Cooling by convection could also occur, but is not significant if the enclosure is small or if it is evacuated. In most commercial instruments, convection is unimportant but radiation and con-

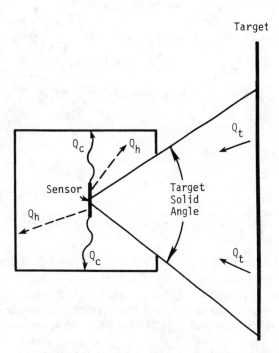

Figure 14.16. The heat from the target by radiation in a solid angle is dissipated by radiation and conduction out of the sensor to the housing.

duction are. When sighted on a target the sensor temperature increases until the thermal radiation from the target, heating the sensor, is equal to the thermal radiation and conduction to the housing, cooling the sensor. That is,

$$Q_{R_{in}} = Q_{R_{out}} + Q_{C_{out}}. \tag{14.12}$$

When a steady-state balance between heat into the sensor and heat out to the housing is reached the sensor temperature remains constant.

Substituting the Stefan–Boltzmann equation for radiant heat transfer terms, and using the one-dimensional steady-state Fourier equation for the conduction term gives

$$\alpha \sigma e_t F_t A_t T_t^4 = [\pi + (\pi - \alpha)] \sigma t_s A_s (T_s^4 - T_h^4) + A_c k (T_s - T_h). \tag{14.13}$$

Here α is the solid angle of radiation received from the target by the sensor, σ is the Stefan–Boltzmann constant, e is the emittance, F is the fraction of radiant energy reaching the target, T is the absolute temperature, A is the area, and k is the thermal conductivity. The subscripts t, s, and h refer to target, sensor, and housing, respectively. Neglected in this equation is the radiation of the sensor back to the target. The housing temperature can be assumed constant for the moment. Note that

$$\begin{aligned} T_s^4 - T_h^4 &= (T_s^2 - T_h^2)(T_s^2 + T_h^2) \\ &= (T_s - T_h)(T_s + T_h)(T_s^2 + T_h^2). \end{aligned} \tag{14.14}$$

Substituting Eq. (14.14) into (14.13) gives

$$\alpha \sigma e_t F_t A_t T_t^4 = \{[\pi + (\pi - \alpha)] \sigma e_s A_s (T_s + T_h)(T_s^2 + T_h^2) + A_c k\}(T_s - T_h). \tag{14.15}$$

Solving for T_s gives

$$T_s = T_h + \frac{\alpha \sigma e_t F_t A_t T_t^4}{[\pi + (\pi - \alpha)] \sigma e_s A_s (T_s + T_h)(T_s^2 + T_h^2) + A_c k}. \tag{14.16}$$

From Eq. (14.16) we see that for a particular design, all the terms in the numerator could be constant except T_t^4. This is also true for the first term in the denominator, except the $(T_s + T_h)$ terms. The thermal conductivity for the sensor support should depend on the average support temperature, so the second term, too, is a function of $(T_s + T_h)$. However, if T_h is constant, $T_t^4 \gg (T_s + T_h)$, and $A_t \gg A_s$, T_s will depend on T_t^4 and the output will depend on T_t^4, as indicated in Eqs. (14.10) and (14.11). For constant T_h an increase in target temperature, increasing T_s, should actually tend to have less than T_t^4 dependency because of the $(T_s + T_h)$ dependence of the denominator. It is possible to thermostat the housing to keep T_h constant. It is also possible to reduce e_s on its back side to reduce the effect of the first term in the denominator.

Equation (14.16) shows linear dependence of T_s on T_h. If not thermostated, the housing temperature is dependent on sensor temperature. Then the heat balance must consider the heat removed from the housing by external radiation, conduction, and convection. This is not easy to model. If housing temperature is allowed to increase with sensor temperature, it is clear that the sensor temperature will rise above the temperature expected for a constant housing temperature.

Thermopiles are usually used to determine sensor temperature. Each thermocouple responds, producing a voltage as approximately a linear function of the difference in temperature between its measuring and reference junction. In the usual construction the reference junction is thermally attached to the housing. Then it is $(T_s - T_h)$ that produces a voltage output. From Eq. (14.16) we see that the $(T_s + T_h)$ terms in the denominator will reduce the output as T_s and T_h rise when T_h is not controlled. Therefore, some type of temperature compensation is needed.

14.9 ELECTRICAL SHUNT COMPENSATION

This method of ambient temperature compensation of a thermocouple is discussed in detail in Chapter 10. In that method of high-resistance shunt is connected across the output terminals and the voltage drop across the shunt is the signal measured (Fig. 14.17).

The resistance R_s is a wire-wound resistor in good thermal contact with the reference junctions at T_h. It is made of fine (about 0.005 in.) copper and nickel wire. As explained in Chapter 10, the lengths of copper and nickel wire are chosen so that R_h increases in resistance as T_h increases, in just the

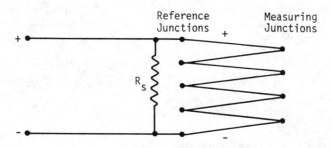

Figure 14.17. The shunt R_s is connected across the output terminals and is at temperature T_h.

right amount to compensate for the reduction in Seebeck voltage accompanying the increase in T_h. The resistance of copper increases with temperature more than does that of nickel. The right proportions are chosen empirically to achieve the desired compensation.

14.10   THERMAL SHUNT COMPENSATION

If thermal energy is allowed to flow by an auxiliary path, it will reduce T_s and increase T_h, thus reducing output somewhat. If the auxiliary path, the thermal shunt, is electrically insulated and has a negative coefficient of thermal conductivity, the second term in the denominator of Eq. (14.15) will decrease as T_s and T_h increase. Nickel has a negative thermal conductivity temperature dependence. Empirical adjustment of such a shunt is a suitable method of ambient temperature compensation (Fig. 14.18).

Note that the thermal shunt is directly comparable, in principle, with the

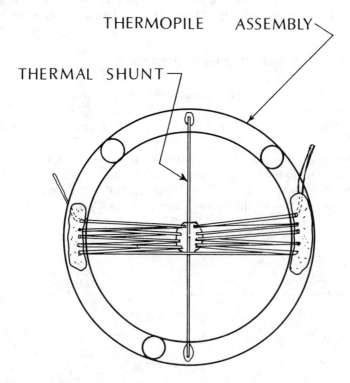

THERMOPILE   ASSEMBLY

THERMAL   SHUNT

Figure 14.18. A sensor using a thermal shunt for ambient temperature compensation. (From T. R. Harrison, "Radiation Pyrometry and Its Underlying Principles of Heat Transfer." Copyright 1960, John Wiley & Sons, Inc., New York. Reprinted by permission.)

electrical shunt. Both reduce the output, but the controlled increase in output to compensate for reduced output from increased ambient temperature is then possible.

14.11 OUTPUT CHARACTERISTICS

Total radiation pyrometers provide rapid response because the sensors have low thermal mass. Their output is approximately proportional to T_t^4.

A comparison of the output of two different designs of radiation pyrometers shows that both designs have nearly fourth-power response to target temperature (Figs 14.19 and 14.20).

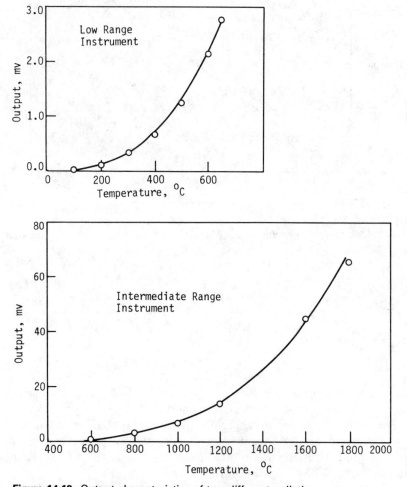

Figure 14.19. Output characteristics of two different radiation pyrometers.

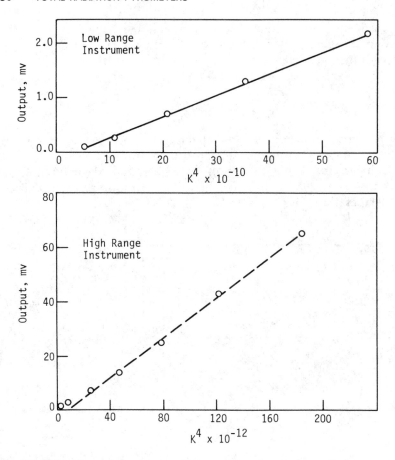

Figure 14.20. Output response plotted against T^4 for two different total radiation pyrometers.

The output equations are approximately

$$V = NT_t^4 \qquad (14.17)$$

but the exponent can be slightly greater than 4. The proportionality constant N varies from one design to another.

The output response is inherently nonlinear. Radiation pyrometers are more sensitive at high temperatures than at low temperatures. Microprocessor linearization circuits are commercially available. They are usually built into digital display and control circuits.

14.12   TARGET TUBES AND SIGHTING TUBES

Sighting into an open furnace often causes problems caused by dust, gases, leaks into the furnace, emittance variations within the furnace, and by accidental placement of intervening objects between the pyrometer and the furnace. Ambient temperature variations for a remotely positioned pyrometer may also cause problems that a target tube or sighting tube may help to reduce by its thermal contact with the furnace. Therefore, one or the other is usually employed when using a total radiation pyrometer.

Target tubes are closed end tubes that are installed with the closed end projecting into the furnace far enough for it to be heated to the desired process temperature. By employing a tube with a large enough diameter, and by properly aligning the pyrometer on the tube, the problems of radiation from nontarget areas can be eliminated. The emittance of the closed end is constant and nearly equal to 1. Heating a target tube requires time, so they reduce the response time of the pyrometer assembly, and this sometimes prevents their use.

Selection of the material for a target tube depends on the process, especially the gaseous atmosphere and the temperature. Ceramic target tubes sometimes fail from rapid heating and cooling, so thermal shock situations should be avoided. Target tubes are available in metals such as Inconel and type 347 stainless steel. They have machined ends for ease of alignment, coupling to mountings, and sealing the furnace opening. They are limited to temperatures of about 1150°C, are corroded by sulfurous atmospheres, and must be replaced periodically after oxidation or corrosion. Ceramic targets are brittle and so must not receive impacts and must be free to expand and contract. They are available in high-mullite porcelain, silicon carbide, silicon-nitride-bonded silicon carbide, and recrystallized aluminum oxide. The high thermal conductivity and thermal shock resisistance of silicon carbide makes that material an excellent choice to about 1650°C if some permeability can be permitted. The properties of the high-alumina porcelain, mullite, and aluminum oxide tubes vary from one manufacturer to another and caution is needed to be sure that the properties are suitable for the application. Properly manufactured tubes should be vacuum tight to about 1350–1500°C, depending on composition and method of manufacture. When mounted horizontally, some high-alumina porcelain tubes will soften and bend at about 1350°C. Some compositions are especially designed for molten-salt-bath applications. A typical target tube is shown in Fig. 14.21.

Sighting tubes are open-ended tubes used for mounting, to limit the field of view, and are especially useful when fast response or direct sighting on the product is needed. For example, sighting on clinker in a rotary cement kiln permits automatic combustion control that would not be possible with a target tube. Often, it is necessary to keep the sighting path free of dust or

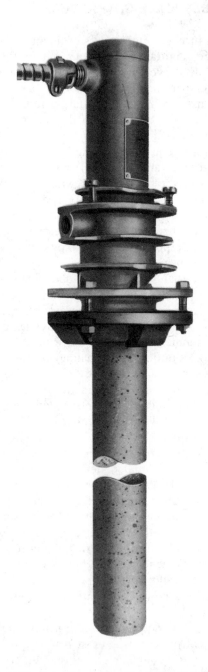

Figure 14.21. Typical sighting tube. (Courtesy of Leeds & Northrup.)

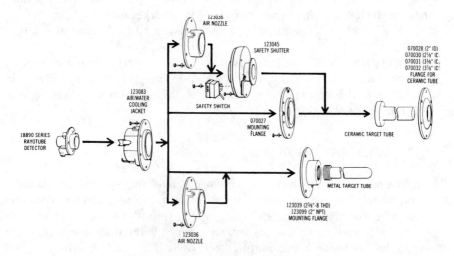

TYPICAL ASSEMBLIES WITH CLOSED-END TARGET TUBES

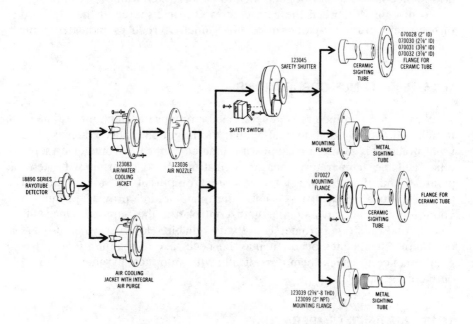

TYPICAL ASSEMBLIES WITH OPEN-END SIGHTING TUBES

Figure 14.22. Safety shutters. (From Refs. 8 and 9.)

gases by admitting compressed air at a low rate to purge the tube. Sighting tubes are available in the same compositions as target tubes. Where thermal shock is a problem, sometimes sighting tubes are more durable than target tubes.

A safety shutter is sometimes used to prevent overheating if a sighting tube ruptures or if ambient cooling fails. One design has a bimetallic strip that releases the spring-loaded shutter if the temperature exceeds the maximum safe value (Fig. 14.22).

14.13 AUXILIARY WINDOWS

Many furnaces, especially those with protective atmospheres or vacuum, must have windows to prevent exchange of furnace gases and air. Windows are often borosilicate glass, 96% SiO_2 glass, fused SiO_2, or sapphire. The windows will have two reflecting surfaces, which will reduce intensity. If the window has a shorter-wavelength cutoff, or is thick and absorbing, it will greatly reduce the sensor temperature. Evaporation of metal or accumulation of dust inside the furnace, as well as dust on the exterior surface, will also reduce intensity. It is a good idea to calibrate the window by removing it and placing it between the pyrometer and a test source at the expected furnace temperature to determine how much it reduces indicated temperature.

14.14 EMITTANCE CORRECTIONS

Total radiation pyrometers in general need emittance corrections at time of installation and any time thereafter that the target emittance changes. The usual practice is to use a disappearing filament optical pyrometer or some other suitable temperature measuring instrument to determine the temperature and then to adjust the indicated output of the total radiation pyrometer to agree with it. Usually, the manufacturer provides a built-in voltage divider, or an amplifier gain control, for the purpose. Once adjusted, total radiation pyrometers are very reliable. However, when process or equipment changes occur, it may be necessary to readjust them. It is good practice to check them periodically with an optical pyrometer sighted on the same target.

14.15 AMBIENT TEMPERATURE

If ambient temperature varies excessively, the output will be unstable. Housings are provided by the manufacturers so that air or water cooling can be used to regulate the temperature of the pyrometer. See Sections 14.9 and 14.10 for built-in temperature compensation.

14.16 LINEARIZING CIRCUITRY

Manufacturers of total radiation pyrometry provide linearizing circuitry to convert the voltage from the detector, which depends on T^4, to a linear output. This can be done with a microprocessor to convert the output voltage versus temperature curve into a series of linear segments, increasing slope with temperature. Then, for a particular segment, a linear output analog control signal can be provided. Digital-to-analog conversion, digital display, and computer interface capabilities are commonly provided. Emittance correction, alarms peak selection, and delay functions are available.

REFERENCES

1. P. H. Dike, W. T. Gray, and F. K. Schroyer, "Optical Pyrometry," Tech. Publ. A1.4000/1966. Leeds and Northrup Company, North Wales, Pennsylvania, 1966.

2. C. Féry, La mesure des température élevées et la loi de Stefan. *C. R. Habd. Seances Acad. Sci.* **134**, 977 (1902).

3. C. Féry, Télescope pyrométrique. (1904); Figure from W. P. Wood and J. M. Cork, "Pyrometry," p. 117. McGraw-Hill, New York, 1927.

4. C. B. Thwing, A New Radiation Pyrometer. *J. Franklin Inst.* **165**, 363–370 (1908); Figure after W. P. Wood and J. M. Cork, "Pyrometry," p. 122. McGraw-Hill, New York, 1927.

5. Honeywell, Inc., "Radiamatic Detectors and Accessories," Bull. S939-2C. Honeywell Inc., Process & Control Div., Fort Washington, Pennsylvania.

6. T. D. Harrison, "Radiation Pyrometry and its Underlying Principles of Radiant Heat Transfer." Wiley, New York, 1960.

7. Corning Glass Works, "Properties of Corning's Glass and Glass-Ceramic Families." Corning, New York, 1979.

8. Leeds and Northrup Company, "Rayotube Temperature Detectors and Assemblies." North Wales, Pennsylvania, 1967.

9. Leeds and Northrup Company, "18890-Series Thin-Film Temperature Detectors and Assemblies." North Wales, Pennsylvania, 1983.

15

NOVEL METHODS OF TEMPERATURE MEASUREMENT

15.1 INTRODUCTION

In this chapter we discuss a number of the less common methods of temperature measurement. Most of them are relatively new. Some may be important methods in the future, and references are included to aid in further study.

15.2 PHOSPHOR QUENCHING (THERMOGRAPHIC PHOSPHORS)

Some phosphors exhibit phosphorescence, that is, a luminescence on excitation that has temperature dependence. The basic electronic structure of a phosphor, somewhat oversimplified, is as shown in Fig. 15.1.

When illuminated with an excitation source having energy, $\Delta E_e = h\nu$, where h is Planck's constant and ν is the frequency of the source, electrons will be elevated from the valence band to the conduction band. For example, ultraviolet light has sufficient energy to excite many commercial phosphors. Fluorescence occurs when the electrons decay back to the valence band through an intermediate level contributed by an impurity deliberately chosen for the purpose. Then the frequency of the emitted fluorescence is given by $\Delta E_F = h\nu$. If that decay is independent of temperature, the material is described as fluorescent. However, if the decay is temperature dependent, it is called phosphorescence. The latter occurs if the impurity level is so chosen that thermal excitation can raise electrons from the valence band to the impurity level. Then thermal radiation will

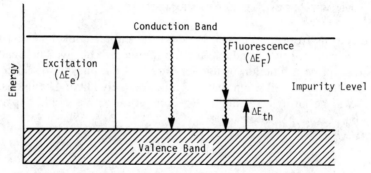

Figure 15.1. Schematic of the energy levels of a quenched phosphor.

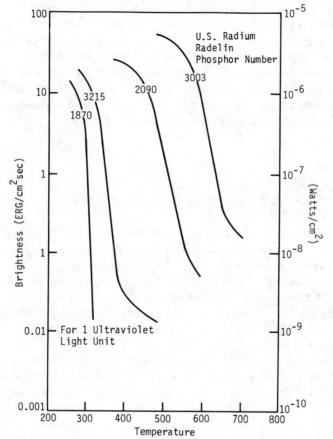

Note:
1 Ultraviolet Unit Corrosponds to a 100 Watt Mercury Arc Lamp Plus Ultraviolet Filter 53 Inches Distant, or Two 15 Watt Fluorescent Blacklite Lamps (Purple Glass) 30 Inches Distant.

Figure 15.2. Temperature dependence of phosphorescence. (From P. Czysz and P. Dixon, "Thermographic Heat Transfer Measurement." © 1968, Chilton Company, Radnor, PA 19089.)

excite electrons from the valence band to the impurity level so that there will be fewer empty sites available for the fluorescent decay. The availability of sites will be a function of temperature and the phosphorescence may be completely quenched when they are all effectively filled. The intensity of emission will be an inverse function of temperature. We would expect the form to be

$$I = I_0(1 - e^{-\Delta E_{Th}/kT}) \tag{15.1}$$

where k is Boltzmann's constant, T is absolute temperature, and I_0 is the unperturbed fluorescence.

Thermal quenching phosphors are commercially available (Fig. 15.2). They have been painted on aerodynamic models for wind-tunnel measurements. The temperature of the *Gemini* space capsule at various altitudes of simulated reentry to the earth's atmosphere was studied in this way.[1] By using ultraviolet light for excitation, photographing the thermally quenched phosphor intensity, and measuring the film with a densitometer, temperature differences of 0.2°C were measured. Response time was less than 1 ms. This method is related to the method described next.

15.3   FLUOROPTIC TEMPERATURE MEASUREMENT

This new commercial method is based on measuring the ratio of the intensities of radiation emitted by two different phosphor ions when subjected to a common excitation source.[2] The ratio of intensities is a smooth, nearly linear function of temperature (Fig. 15.3). The emitted intensities depend on the intensity of the incident radiation, but the ratio does not if both fluorescing ions receive the same excitation intensity. The rare earth phosphor $(Gd_{0.99}Eu_{0.01})_2O_2S$ has discrete $4f$ electron transitions from the rare earth ions. They can be excited by near-ultraviolet light (Fig. 15.4).

The phosphor is deposited at the end of a fused-quartz optical fiber that is covered with a black, carbon-free Teflon coating. Ultraviolet light is projected down the fiber to excite the phosphor. The return signals are filtered and the intensities are measured with photo diodes (Fig. 15.5).

The optical fiber can be 2–15 m long. It is connected to a microprocessor-controlled instrument that contains the tungsten halogen excitation source, emission measuring sensors, digital temperature display, and signal processors. Analog and digital output signals are provided. Advantages are: the detector is tiny, has small thermal mass, and is easily placed. Optical signals are not perturbed by electrical or magnetic fields, and the probe can be used whenever Teflon coatings will survive.

Accuracy is stated as ±0.5°C from 20 to 50°C and ±1° for the rest of the range −50 to 200°C. The small size of the sensor, 0.8 mm in diameter, the

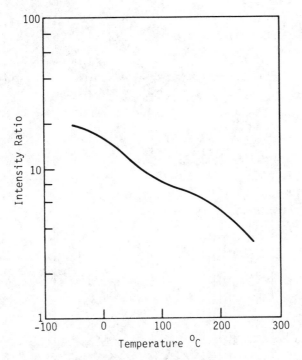

Figure 15.3. Intensity ratio of 570–670 nm fluorescence from $(Gd_{0.99}Eu_{0.01})_2O_2S$ phosphor as a function of temperature. (From K. A. Wickersheim and R. B. Alves, "Fluoroptic Temperature Measuring," brochure, Luxtron, Mountain View, Calif., 1979.)

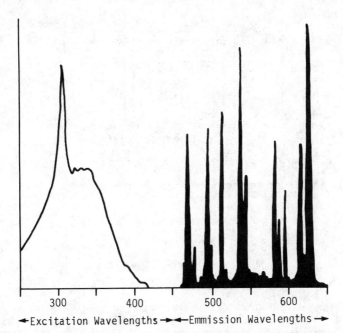

←Excitation Wavelengths→ ←Emmission Wavelengths→

Figure 15.4. Excitation and emission spectra for $(Gd_{0.99}Eu_{0.01})_2O_2S$ phosphor. (From K. A. Wickersheim and R. B. Alves, "Fluoroptic Temperature Measuring," brochure, Luxtron, Mountain View, Calif., 1979.)

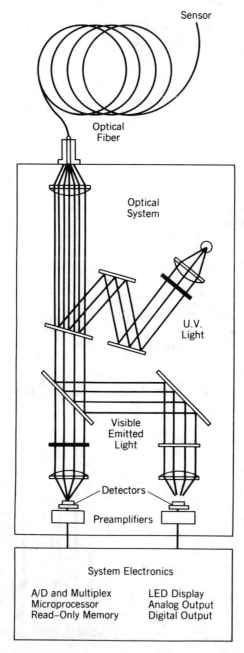

Figure 15.5. Optical ray diagram for the Luxtron fluoroptic temperature sensing system. (From K. A. Wickersheim and R. B. Alves, "Fluoroptic Temperature Measuring," brochure, Luxtron, Mountain View, Calif., 1979.)

desirable magnetic and electrical properties, and the corrosion resistance of the Teflon makes this a very interesting instrument for many applications.

15.4 PIEZOELECTRIC RESONANT FREQUENCY

Quartz crystals, cut at a particular orientation to their crystallographic axes and at a particular thickness, have been used as stable electronic frequency sources since that use was discovered by Cady in 1921. For most cuts the temperature shifted the resonant frequency, and this effect has been used to maintain the temperature of frequency source enclosures to get maximum output at the desired frequency. Most electronic applications require very stable frequency sources that do not vary with temperature. The GT orientation was found to be the most stable and could be produced with essentially zero-temperature coefficient by balancing two opposing displacements. When the width-to-length ratio was varied, a linear temperature dependence was produced.[3]

An improved cut, the LC cut (for linear temperature coefficient), was developed in 1965. It produces a linear change in resonant frequency of about 1000 Hz/°C over the temperature range −80 to 250°C. This is used commercially by Hewlett-Packard to produce a very sensitive and reliable quartz thermometer.[4] In that instrument the beat frequency between an LC-cut and a GT-cut crystal is digitized, the frequency counted, and converted to temperature. The specifications show an accuracy of ±0.04°C from −40 to 150°C and ±0.075°C from −80 to 250°C. The instrument includes a microprocessor to determine resonant frequency, convert it to temperature, and display the results on a seven-digit LED display panel. Two probes can be used and their temperature difference can be displayed. Internal calibration and failure detection are incorporated. The instrument can be provided with an interface bus for computer control and for data processing and recording. Response time is 0.1, 1, and 10 s for the low 0.01°C, medium 0.001°C, and high 0.0001°C resolution settings. An analog output of 0–10 V is included.

15.5 THERMAL NOISE

This is a method that is of great fundamental interest because it is the only method that rivals gas thermometry, in principle, as a true thermodynamic method of temperature measurement. It is included here in the chapter under novel methods because practical realization of noise thermometry is not yet really available. Because of its fundamental importance there is a great deal of current research on thermal noise methods. Thermal noise provides such a weak signal that all other sources of noise, or of systematic error, must be controlled for the method to be successful. If current research proves the method reliable, it will be a new basis for establishing the thermodynamic temperature scale.

Every body containing charged particles emits thermal noise in accordance with the energy density of the charges if it is used to convey energy from one thermodynamic state to another. The noise emitted by the electrons of an electrical resistor is the basis for thermal noise measurements. The effect was discovered by Johnson, who showed that the noise produced by an amplifier had a fundamental lower limit that originated in the thermal noise of the unbiased resistor used in the input stage.[5] He showed that the noise produced by thermal fluctuation of the electrons in the resistor produced a time-average voltage squared given by

$$\langle V^2 \rangle = 4kTR(f_2 - f_1),\tag{15.2}$$

where k is Boltzmann's constant, T is absolute temperature, R is the resistance, and f_2 and f_1 are the maximum and minimum frequencies passed through the amplifier used in measuring $\langle V^2 \rangle$.

As a result of this discovery, Nyquist made a theoretical analysis of the phenomenon using a model of two resistors connected in a resonant circuit.[6] He deduced from the second law of thermodynamics and the electrical properties of the circuit that the thermal noise is a universal function of frequency, resistance, and temperature only. Then, using a model of two resistors connected by a nondissipative transmission line, he combined statistical mechanical considerations with circuit theory and concluded that the relationship given in Eq. (15.2) is general. He drew attention to the analogy between thermal vibrations of the electrons in a resistor and the thermal vibrations of molecules in a gas. He pointed out that if the energy per degree of freedom is taken as

$$\varepsilon = h\nu/(e^{h\nu/kT} - 1),\tag{15.3}$$

then his reasoning gives

$$V^2 \, d\nu = 4R_\nu h \, d\nu/(e^{h\nu/kT} - 1).\tag{15.4}$$

Equation (15.4) is appropriate at high temperatures and reduces to Eq. (15.2) at room temperature for audio frequencies.

Callen and Welton made a fundamental quantum-mechanical analysis of the relationship and confirmed the forms of Nyquist and Johnson.[7] Their results are general. They also apply to viscous flow, Brownian movement, and other dissipative systems. Planck's radiation law is a direct result of the energy density they calculate. Thus over the entire range of temperatures of interest, the general method of forces transferred through a dissipative system is of great importance. Therefore, thermal noise has received a great deal of attention.

One of the areas where thermal noise has a particular appeal is in the region of ultra-low temperatures. Because the noise signal is directly proportional to absolute temperature, the voltage fluctuations are extremely small. However, a program at the National Bureau of Standards using an R SQUID, a Josephson junction in parallel with a resistor, coupled to a radio-frequency oscillator, has given very encouraging results. In the NBS program, Fig. 15.6, the circuit shown in the dashed box is maintained at a cryogenic temperature T, whereas the other components are at room temperature. The R-SQUID is formed by a resistor R (typically $10^{-5}\,\Omega$) connected in parallel with a point-contact Nb Josephson junction J. The circuit is biased by a dc current I_0 generated by a dry cell and a series resistor R_B. The voltage drop induced in R causes the Josephson junction to oscillate at an audio frequency (AF; typically, 1–100 kHz). A radio-frequency (RF) oscillator injects an RF current into a tank circuit coupled to the R-SQUID via a coaxial cable. The reflected power, phase or amplitude modulated by the AF Josephson signal, is amplified by an RF amplifier RFA. Subsequent demodulation yields the AF signal, which is passed through a bandpass filter. At this point the AF signal consists of the sum of the amplified signal from the Josephson junction and noise generated by the RFA. This noisy signal is presented to a frequency counter, which repeatedly measures the frequency. An on-line computer calculates several variances which are used to detect the presence of extraneous noise signals and to define the noise temperature. A provisional

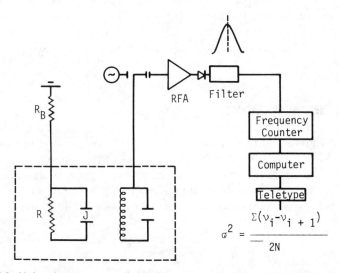

Figure 15.6. Noise thermometer circuit. [From R. J. Soulen and D. Van Vechten, Noise thermometry at NBS using a Josephine junction, in "Temperature: Its Measurement and Control in Science and Industry" (James F. Shooley, Ed.), Vol. 5, Part 1, American Institute of Physics, New York, 1982, p. 116.]

scale in the previously inaccessible region 0.01–0.52 K has been developed and is maintained as NBS-CTS-1.[8]

There is a great deal of interest in noise thermometry at high temperatures because the signal is directly proportional to absolute temperature. A recent review gave 37 pertinent references.[9]

There are three different signals from a noise thermometer that can be utilized.[9] In addition to the time-average voltage of Eq. (15.2), the short-circuited noise current is

$$\langle I^2 \rangle = 4kT(f_2 - f_1)/R . \tag{15.5}$$

By multiplying Eq. (15.2) by (15.3) and taking the square root, the noise power is obtained:

$$\langle P \rangle = 4kT(f_2 - f_1) . \tag{15.6}$$

This equation does not contain R. It is particularly appealing where drift in R may occur, as at very high temperatures.

Most of the recent measurements have been based on a comparison technique originated by Garrison and Lawson in 1949.[10] They adjusted a standard resistor at a reference temperature to produce the same noise as for the sensing resistor. The parallel capacitance of the sensing and reference resistors were adjusted so each had the same RC product. They used sequential sampling between dual channels for the standard and the reference, to minimize drift and nonlinearity of the amplifiers. Then the sensor temperature was computed from the reference temperature and the resistance ratio. Modern instrumentation has made improvements possible. Successful experiments in industrial environments may soon bring about commercial instruments.[11]

15.6 THERMAL IMAGING

Thermal imaging is a noncontact method of converting the thermal radiation pattern of an object or scene into a visible image. Because the emittance of the target areas are usually not everywhere equal, the image is not a temperature image. It must be interpreted in terms of emittance as well as temperature. Spraying an object black, when that is permissible, will produce a temperature image.

15.6.1 Plumbicon Tube

For ordinary temperatures, above and below room temperature, a special camera is needed. For very bright objects, such as incandescent filaments, an ordinary TV camera equipped with a Plumbicon tube is satisfactory.[12] It is more sensitive and more nearly linear in response than a Vidicon tube. De Bie and Klomp used such a camera to observe hot spots in tungsten lamp filaments. They displayed a pictorial image on one TV monitor and the intensity of a single line scan on another monitor. For the former they used a short focal distance, maximum aperture, and maximum transmittance. For the latter they used a blue filter to get improved apparent homogeneity of the PbO photosensitive layer of the Plumbicon tube. Filters were used at high temperature and their experiments covered 1700–3200 K. A calibrated tungsten strip lamp was placed in front of the incandescent target and adjusted in intensity to equal that of the target. They estimated the error at 0.3% at 2000 K.

15.6.2   Infrared Television with Solid State Detectors

Several companies manufacture special infrared television cameras for use to about 12 μm.[13–14] Those using long wavelengths require cryogenic cooling. The Agema Thermovision 870 uses a facet tracking system in which silicon or germanium lenses are used to focus the image into a mirror optic scanning system (Fig. 15.7). Vertical scanning is produced by an oscillating aspherical mirror just behind the objective lens assembly. The reflected ray is folded by reflection from an aspherical mirror, a flat mirror, and from three mirrors mounted on a component to have a single axis. Two of the mirrors are aspherical. Horizontal scanning is accomplished by a ten-facet polygon mirror rotated at 15,000 RPM. The high efficiency of the facet tracking system gives good response using a thermoelectrically cooled

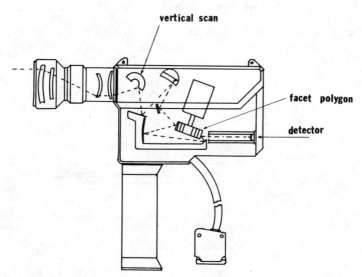

Figure 15.7. Agema video camera showing a single ray path to the detector. (Courtesy Agema Infrared Systems.)

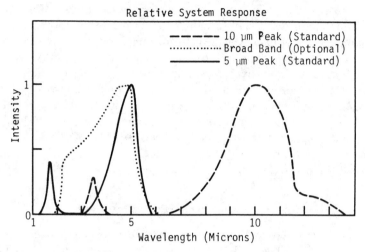

Figure 15.8. Relative system response.

3–5 μm detector. Cryogenic cooling of an $(HgTe)_x(CdTe)_{1-x}$ detector can be used to reach longer wavelengths (Fig. 15.8).

The thermal image can be presented as a black-on-white or a white-on-black print, as isotherms, as an intensity-modulated line scan (superimposed on the image, if desired), as an amplitude-modulated relief map, and as a color image. Electronic synchronization for photo recording is provided.

15.6.3 Pyroelectric Television Imaging

A pyroelectric detector is a ferroelectric crystal that produces an electric charge on crystal faces perpendicular to the polar axis in response to a

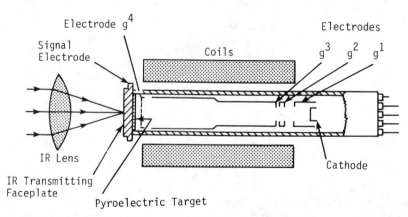

Figure 15.9. Design of the Pyricon television tube. (Courtesy of Thomas Edison Tubes and Devices, Inc., Dover, NJ.)

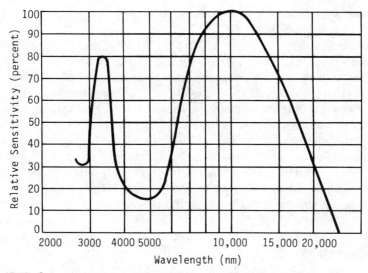

Figure 15.10. Spectral response of the Pyricon tube. (Courtesy of Thomas Edison Tubes and Devices, Inc., Dover, NJ.)

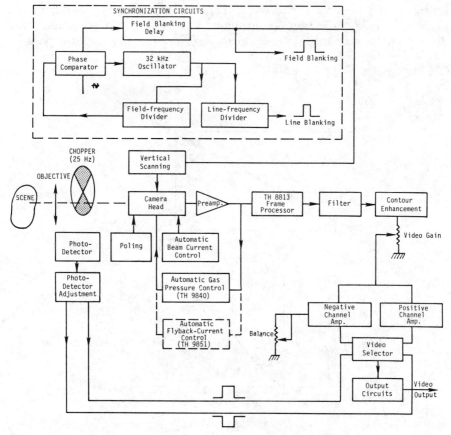

Figure 15.11. Schematic diagram of the video circuit. (Courtesy of Thomas Edison Tubes and Devices, Inc., Dover, NJ.)

change in temperature. The polycrystalline detector must be poled to align the pyroelectric axis prior to use. (Poling is the process of cooling the crystal through its ferroelectric transition temperature under a dc field to align all domains in the same direction.) A television tube based on tryglycine sulfate is manufactured by Thomson CSF as a Pyricon tube (Fig. 15.9).[15] The tube is sensitive to long wavelengths (Fig. 15.10), so very low temperatures can be detected compared to other thermal imaging methods.

The pyroelectric effect produces a signal only during the time the temperature is changing. The original tubes gave an image only when the target moved or the camera was moved. Adding a chopper in conjunction with a specially designed television circuit produced a camera that does not need to be moved (Fig. 15.11). This system is so sensitive that cancer diagnosis through detection of small differences in body temperature is one of the applications.

15.7 INFRARED PHOTOGRAPHY

Eastman manufactures infrared film that is widely used to make infrared pictures at the near-infrared wavelengths. This method can be quantitative if the target temperature at one location is known.[16] The target is photo-

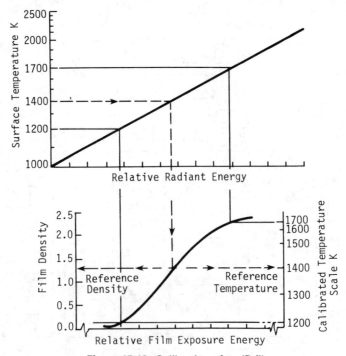

Figure 15.12. Calibration of an IR film.

graphed, and a standard stepped gray scale is photographed. The film is processed and the gray scale intensity is matched to the known target temperature. This is used as a calibration point and the target picture is analyzed with a densitometer to produce a temperature profile (Fig. 15.12).

The temperature of the known point on the surface, upper figure dashed arrow, is correlated with the film density gray scale (lower figure) to calibrate the film. The abscissa of the upper scale is calculated from Wien's law. Target emittance must be constant, albeit unknown, for this to give a thermal map of the target.

15.8   SPECTROSCOPIC TEMPERATURE MEASUREMENTS

There are many different methods of spectroscopic temperature measurement of flames and plasmas, but they are not yet commercial systems. These include:

1. Absolute emission coefficient of a neutral line[17,18]
2. Line intensity ratios[17,18]
3. Line reversal[17,18]
4. Rotational Raman spectroscopy[19] and vibrational Raman spectroscopy[20]
5. Coherent anti-Stokes' Raman spectroscopy[19,21–25]
6. Laser tomography[26]

These methods are too experimental and too complex for adequate discussion. References are provided.

15.9   INDEX OF REFRACTION METHODS

The index of refraction of a fluid is a function of temperature. Schlieren, shadowgraph, and interferometer techniques can be used to visualize the density variations in a fluid and produce temperature isotherms based on the fluid density. These methods are best applied to axially symmetric fields and are most often used to study the temperature distribution in fluids around heated or cooled objects.[27]

15.10   TEMPERATURE-SENSITIVE INDICATORS (CRAYONS, PAINTS, TRANSFERS)

Various crayons and paints are available for detecting the maximum temperature that an object has experienced. These are readily available through the large temperature measurement companies. They have special value where it is necessary to know if an object such as a furnace shell or an output transistor has ever reached a critical temperature. They are convenient to use and give an instant visual display if the object has been overheated.

These devices are made with a finely divided pigment dispersed in, or covered by, an organic vehicle with the desired temperature as its melting point. When the vehicle melts, the color of the pigment deepens because the light scattering by the pigment powder is reduced as its surface is wet. The pigments are color coded to improve visual recognition and to differentiate one indicator from another. Some are compounded in a series of 5° melting temperatures. Higher temperatures usually have about 25 or 50° intervals between one indicator and the next. They are not reusable.

15.11 LIQUID-CRYSTAL SENSORS

Certain cholesteric liquid crystals undergo changes in reflectance and color as the temperature changes. Liquid crystals are long-chain molecules with periodicity in one direction when bundled together, but with wormlike nonperiodicity in the other two directions. The alignment is very thermal sensitive, and the color changes are related to interference associated with the crystal spacing. The color and contrast occur in a temperature range of only a few degrees. By mixing different compounds, the range can be extended to about 15°.[28]

Liquid crystals are used to make inexpensive display thermometers; sometimes they are enclosed in a plastic film shaped like 68°, 70°, 72°, 74°, and so on, so that the brightest-colored number indicates the approximate ambient temperature.

15.12 ULTRASONIC THERMOMETRY

The velocity of sound in a body, for longitudinal waves, is given by

$$v = \left(\frac{E}{\rho}\right)^{1/2}, \tag{15.7}$$

where E is Young's modulus of elasticity and ρ is the density. Both are functions of temperature and decrease as temperature increases. Usually, the velocity of sound decreases when temperature increases, and the internal friction increases as temperature increases. The change in the sound velocity is relatively easy to measure, but the increased attenuation of the signal makes measurement more difficult.

Ultrasonic thermometry is attractive in situations, such as in nuclear reactors, where other methods may not be practical. A recent study at Sandia Laboratories concluded that the method may be suitable in some reactor situations.[29]

The study used a long rod of the test metal connected to a Remendur driver rod. The test rod was matched in acoustic impedance to the driver rod. The long, slender rod had two ring-shaped grooves machined at the opposite end to define a test section. An ultrasonic pulse was induced with the magnetostrictive Remendur driver. It traveled down the rod, reflected from the other end, and returned. The grooves outlining the gauge section caused reflections visible in the signal. Timing the reflection signals permitted calculation of the velocity.[30] Some metals recrystallized at the very high temperatures studied, and noise appeared in the response. Titanium was found suitable to 900°C, 302 stainless steel to 1300°C, rhodium to 1500°C, and rhenium to 2600°C. Accuracy and reliability remain to be determined.

15.13 MICROWAVE RADIOMETRY

Microwave radiometry has been used to measure sea surface temperatures from the Seasat satellite. Although still developmental, temperature measurements are believed to have a root-mean-square sensitivity of 1.2°C or better.[31]

15.14 ACOUSTIC THERMOMETRY

Equation 15.7 can be modified for gases to obtain a more useful relationship. For a monatomic gas such as ^{4}He, the velocity can be written as

$$v^2 = (C_p/C_v)(RT/M)(1 + \alpha P + \beta P^2),\qquad(15.8)$$

where C_p and C_v are the specific heats at constant pressure and constant volume respectively, R is the gas constant, M is the molecular weight, and α and β are functions of the gas virial coefficients. Plumb and Catland give a good discussion of the theory and the apparatus required.[32] This method is very important to research at temperatures near absolute zero.[32,33,34,35] There are no commercial instruments at this time.

REFERENCES

1. P. Czysz and W. P. Dixon, Thermographic heat transfer measurement. *Instrum. Control Syst.*, **41**, 71–76 (1968).

2. Luxtron Corp., "Fluoroptic Temperature Sensing Brochure." Mountain View, California, 1981.

C. Ovrén, M. Adolfsson, B. Hök, and T. Brogardh, New opportunities with fibre-optic measurement. *Sensor Rev.*, October, pp. 199–205 (1985).

3. W. P. Mason, "Piezoelectric Crystals and Their Application to Ultrasonics." Van Nostrand, New York, 1950.

4. Hewlett Packard Corporation, "Model 2804A Quartz Thermometer," Hewlett Packard Catalog, p. 639. Palo Alto, 1982.

5. J. B. Johnson, Thermal agitation of electrons in conductors, *Phys. Rev.* **32**, 97–109 (1928).

6. H. Nyquist, Thermal agitation of electrical charge in conductors. *Phys. Rev.*, **32**, 110–113 (1928).

7. H. B. Callen and T. A. Welton, Irreversibility and generalized noise. *Phys. Rev.* **83**, 34–40 (1951).

8. R. J. Soulen, Jr. and D. Van Vechten, Noise thermometry at NBS using a Josephson junction. *Temp.: Its Meas. Control Sci. Ind.* **5** (Part 1), 115–123 (1982).

9. T. V. Blalock and R. L. Shepard, A decade of progress in high temperature Johnson noise thermometry. *Temp.: Its Meas. Control Sci. Ind.* **5** (Part 2), 1219–1223 (1982).

10. J. B. Garrison and A. W. Lawson, An absolute noise thermometer for high temperatures and high pressures. *Rev. Sci. Instrum.* **20**, 785 (1949).

11. H. Brixy, R. Hecker, and K. F. Rittinghaus, Application of noise thermometry in industry under plant conditions. *Temp.: Its Meas. Control Sci. Ind.* **5** (Part 2), 1225–1237 (1982).

12. J. R. deBie W. G. Klomp, Temperature measurement using a plumbicon camera tube." *Conf. Ser.—Inst. Phys.* **26**, 306–314 (1975).

13. J. Agerskans, Thermal imaging: A technical review. *Conf. Ser.—Inst. Phys.* **26**, 375–388 (1975).

13a. Agema Infrared Systems AB, Agema Infrared Systems Brochure, Danderyd, Sweden, 1987.

14. Probeye Thermal Video Systems, Industrial Products Division, Hughes Aircraft Co., Probeye Thermal Video Systems Brochure, Carlsbad, California, 1987.

15. Thomson-CSF Components Corp., Bulletin APV 6152. "Pyricon Thermal Television Pickup." Thomson-CSF Electron Tube Div., Clifton, New Jersey, 1971.

16. F. G. Pollack and R. O. Hinckel, Surface temperature mapping with infrared photographic pyrometer. *NASA Tech. Brief* **69-10113** (1969).

17. R. H. Tourin, "Spectroscopic Gas Temperature Measurements." Am. Elsevier, New York, 1966.

18. E. Pfender, Spectroscopic temperature determination in high temperature gases. In "Measurements in Heat Transfer" (E. R. G. Eckert and R. J. Goldstein, eds.), Chapter 6, pp. 295–336, Hemisphere Publ. Corp., Washington, D.C., 1976.

19. M. C. Drake, C. Asawaroengchai, D. L. Drapcho, K. D. Viers, and G. M. Rosenblatt, The use of rotational raman scattering for measurement of gas temperature. *Temp.: Its Meas. Control Sci. Ind.* **5** (Part 1), 621–629 (1982).

20. M. C. Drake, M. Lapp, and C. M. Penney, Use of the vibrational raman effect for gas temperature measurements. *Temp.: Its Meas. Control Sci. Ind.* **5** (Part 1), 631–638 (1982).

21. J. P. Taran and M. Pealat, Practical CARS temperature measurement. *Temp.: Its Meas. Control Sci. Ind.* **5** (Part 1), 575–582 (1982).

22. G. L. Switzer and L. P. Goss, A hardened CARS system for temperature and species-concentration measurements in practical combustion environments. *Temp.: Its Meas. Control Sci. Ind.* **5** (Part 1), 583–587 (1982).

23. J. F. Verdieck, J. A. Shirley, R. J. Hall, and A. C. Eckbreth, CARS thermometry in reacting systems, *Temp.: Its Meas. Control Sci. Ind.* **5** (Part 1), 595–608 (1982).

24. L. A. Rahn, S. C. Johnston, R. L. Farrow, and P. L. Mattern, CARS thermometry in an internal combustion engine. *Temp.: Its Meas. Control Sci. Ind.* **5** (Part 1), 609–613 (1982).

25. D. Klick, K. A. Marko, and L. Rimai, Temperature measurements for combustion diagnostics from high resolution single-pulse CARS N_2 spectra. *Temp.: Its Meas. Control Sci. Ind.* **5** (Part 1), 615–627 (1982).

26. H. G. Semerjian, R. J. Santoro, P. J. Emmerman, and R. Boulard, Laser tomography for temperature measurements in flames. *Temp.: Its Meas. Control Sci. Ind.* **5** (Part 1), 649 (1982).

27. R. J. Goldstein, Optical tecnniques for temperature measurements. In "Measurements in Heat Transfer" (E. R. G. Eckert and R. J. Goldstein, eds.), Chapter 5, pp. 241–293. Hemisphere Publ. Corp., Washington, D.C., 1976.

28. T. C. Rozzell, *J. Microwave Power* **9**(3), 241–249 (1974), RAMAL, Inc., Box 275, Sandy, Utah 84070.

29. G. A. Carlson and H. G. Plein, "Refractory Metals for Ultrasonic Thermometry Applications," NUREG/CR-0368, SAND78-1382. Sandia Laboratories, Albuquerque, New Mexico, 1978.

30. G. A. Carlson, W. H. Sullivan and H. G. Plein, "Application of Ultrasonic Thermometry in LMFBR Safety Research," IEEE Cat. No. 77CH1264-ISU, p. 24, Inst. Electr. Electron. Eng., New York, 1977.

31. R. Hofer, E. J. Njoku, and J. W. Waters, Microwave radiometric measurements of sea surface temperature from the seasat satellite: First results. *Science* **212**, 1385–1387 (1981).

32. H. Plumb and G. Gatland, Acoustical thermometry and the NBS provisional temperature scale 2—20K. *Metrologia* **2**, 127–139 (1966).

33. J. S. Rodgers, R. J. Tainsh, M. S. Anderson and C. A. Swenson, Comparison between gas thermometer acoustic and platinum resistance temperature scales between 2 and 20K. *Metrologia* **4**, 47–59 (1968).

34. A. R. Colclough, Systematic errors in primary acoustic thermometry in the range 2–20K. *Metrologia* **9**, 75–98 (1973).

35. A. R. Colclough, Primary acoustic thermometry: Principles and current trends. *Temp.: Its Meas. Control Sci. Ind.* **5** (Part 1), 65–75 (1982).

16

PYROMETRIC CONES

16.1 INTRODUCTION

Pyrometric cones are little triangular ceramic prisms which, when set at a slight angle, bend over in an arc so that the tip reaches the level of the base at a particular temperature if heated at a particular rate (Fig. 16.1).

The bending of the cone at a particular temperature is caused by the formation of a viscous liquid within the cone body, so that the cone bends as the result of viscous flow. The shape of the cone, the viscosity of the liquid phase, the relative amounts of liquid and crystalline solid within the cone body, the height, and the angle from the vertical affect the rate at which the cone will soften and bend by viscous flow under the influence of gravity. The endpoint temperature when the tip of the cone touches the supporting plaque is calibrated for each cone composition when heated at a standard rate. The cones are made so that the endpoint temperatures are usually about 25–40°C apart, but the spacing is uneven and varies with size and heating rate (Table 16.1).[1]

Obviously, the viscosity of the liquid depends on temperature. Over a fairly narrow range of temperatures, for a particular composition, the viscosity η can be expected to have Arrhenius temperature dependence:

$$\eta = \eta_0 e^{\Delta H_\eta / RT}. \tag{16.1}$$

Here η_0 is a constant, ΔH_η is the activation energy for viscous flow, R is the gas constant, and T is absolute temperature. The composition of the liquid affects η_0 and ΔH_η. The composition and amount of liquid depend on

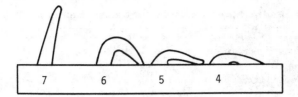

Figure 16.1. Pyrometric cone plaque supporting pyrometric cones No. 4, 5, 6, and 7, after heating to a temperature equivalent to cone 6.

the cone composition, the temperature, and the heating rate because cones are made from mixtures of ceramic mineral powders. These powders react with each other to form some liquid. But the reactions are controlled by the kinetics of the physicochemical processes involved. Equilibrium is not established, so all factors that affect reaction kinetics affect the viscous deformation process. Nevertheless, when properly used, the cones have a remarkably good endpoint temperature if heated at a standard rate.

Pyrometric cones are sensitive to heating rate. They do not deform at the same temperature when heated at different rates. The reason for this is that the kinetics of the chemical reactions causing liquid formation require time and occur more rapidly at higher temperatures. Therefore, cones are sensitive to both temperature and time. This is one of the reasons why pyrometric cones are very useful for ceramic processing. Firing a ceramic body to maturity depends also on heating time and temperature. The degree of vitrification or sintering depends on the same variables that affect pyrometric ones. ["Vitrification" is a word meaning development of a glassy bond phase to produce a strong ceramic article. Sintering is consolidation of powders into a strong compact without the presence of a liquid phase (i.e., by solid-state reactions).] Therefore, they are often a better indication of proper firing schedule than is temperature alone. It is difficult to translate heating rate, time at a constant temperature, and temperature into the integral form of completed chemical reactions necessary to indicate body maturity. Pyrometric cones serve this purpose. They are also useful for nonceramic process.

A further use of pyrometric cones is to compare their endpoint with the endpoint of cones made from other compositions. For example, a clay mined from a natural deposit can be compared in melting behavior with a standard cone by forming it into a cone, placing it in a plaque with a series of standard cones, and heating the plaque at a standard rate. The furnace is shut off when the tip of the unknown cone touches the plaque. Then the standard cones are compared to it to find out which standard cone has the equivalent deformation behavior. The standard cone touching the plaque at the same time as the unknown is used in designating the pyrometric cone equivalent of the unknown. Thus if the clay and cone 30 both touched the

TABLE 16.1 Orton Standard Pyrometric Cones Temperature Equivalents

Cone Type Heating Rate Cone number	Large Regular		Large Iron Free		Self-Supporting Regular		Self-Supporting Iron Free		Small Regular	Small PCE	Cone Type Heating Rate Cone number
	108°F/hr	270°F/hr	108°F/hr	270°F/hr	108°F/hr	270°F/hr	108°F/hr	270°F/hr	540°F/hr	270°F/hr	
022	1074	1092			1087	1094			1157		022
021	1105	1132			1112	1143			1195		021
020	1148	1173			1159	1180			1227		020
019	1240	1265			1243	1267			1314		019
018	1306	1337			1314	1341			1391		018
017	1348	1386			1353	1391			1445		017
016	1407	1443			1411	1445			1517*		016
015	1449	1485			1452	1488			1549*		015
014	1485	1528			1488	1531			1616		014
013	1539	1578			1542	1582			1638		013
012	1571	1587			1575	1591			1652		012
011	1603	1623			1607	1627			1684		011
010	1629*	1641*	1623	1656	1632	1645			1686*		010
09	1679*	1693*	1683	1720	1683	1697			1751*		09
08	1733*	1751*	1733	1773	1737	1755			1801*		08
07	1783*	1803*	1778	1816	1787	1807			1846*		07
06	1816*	1830*	1816	1843	1819	1834			1873*		06
05 1/2	1852	1873	1852	1886	1855	1877			1908		05 1/2
05	1888*	1915*	1890	1929	1891	1918			1944*		05
04	1922*	1940*	1940	1967	1926	1944			2008*		04
03	1987*	2014*	1989	2007	1990	2017			2068*		03
02	2014*	2048*	2016	2050	2017	2052			2098*		02
01	2043*	2079*	2052	2088	2046	2082			2152*		01
1	2077*	2109*	2079	2111	2080	2113			2154*		1
2	2088*	2124*	Not Mfg	Not Mfg	2091	2127			2154*		2
3	2106*	2134*	2104	2136	2109	2138			2185*		3
4	2134*	2167*			2142	2169			2208*		4
5	2151*	2185*			2165	2199			2230*		5
6	2194*	2232*			2199	2232			2291*		6
7	2219*	2264*			2228	2273			2307*		7
8	2257*	2305*			2273	2314			2372*		8
9	2300*	2336*			2300	2336			2403*		9
10	2345*	2381*			2345	2381			2426*		10
11	2361*	2399*			2361	2399			2437*		11
12	2383*	2419*			2383	2419			2471*	2439*	12
13	2410*	2455*			2428	2458			2460	2460*	13
13 1/2	Not Mfg	Not Mfg			2466	2493			Not Mfg	Not Mfg	13 1/2
14	2530*	2491*			2489	2523			2548	2548*	14
14 1/2	Not Mfg	Not Mfg			2527	2568			Not Mfg	Not Mfg	14 1/2
15	2595*	2608*			2583	2602			2606	2606*	15
15 1/2	Not Mfg	Not Mfg			2617	2633			Not Mfg	Not Mfg	15 1/2
16	2651*	2683*			2655	2687			2716	2716*	16
17	2691*	2705*			2694	2709			2754	2754*	17
18	2732*	2743*			2736	2746			2772	2772*	18
19	2768*	2782*			2772	2786			2806	2806*	19
20	2808*	2820*			2811	2824			2847	2847*	20
21	2847*	2856*_			2851	2860				2883*	21
23	2887*	2894*			2890	2898				2921*	23
26	2892*	2921*								2950*	26
27	2937*	2961*								2984*	27
28	2937*	2971*								2995*	28
29	2955*	2993*								3018*	29
30	2977*	3009*								3029*	30
31	3022*	3054*								3061*	31
31 1/2	ND	ND								3090*	31 1/2
32	3103*	3123*								3123*	32
32 1/2	3124*	3146*								3135*	32 1/2
33	3150*	3166*								3169*	33
34	3195*	3198*								3205*	34
35	3243*	3243*								3245*	35
36	3268*	3265*								3279*	36
37	ND	ND								3308*	37
38	ND	ND								3362	38
39	ND	ND								3389	39
40	ND	ND								3425	40
41	ND	ND								3578	41
42	ND	ND								3659	42

The temperature equivalent tables are designed to be a guide for the selection of cones to use during firing. The temperature listed may only have a relative value to the user. However, the values do provide a good starting point and once the proper cones are determined for a particular firing condition, excellent firing control can be maintained.

NOTES:

ND = Not determined

* Temperature equivalents as determined by the National Bureau of Standards by H.P. Beerman (See Journal of the American Ceramic Society Volume 39, 1956). Large cones at 2 inch mounting height, Small & PCE cones at 15/16 inch.

1. The temperature equivalents in this table apply only to Orton Standard Pyrometric Cones, heated at the rate indicated in air atmosphere.
2. The rates of heating shown at the head of each column of temperature equivalents were maintained during the last several hundred degrees of temperature rise.
3. The temperature equivalents are not necessarily those at which cones will deform under firing conditions different from those under which the calibration determinations were made.
4. For reproducible results, care should be taken to insure that the cones are set in a plaque with the bending face at the correct angle of 8° from the vertical with the cone tips at a uniform height above the plaque.

plaque at the same time, the clay would be reported to have a pyrometric cone equivalent (PCE) of cone 30.

16.2 HISTORICAL PERSPECTIVE

Pyrometric cones were first used by Luath and Vogt in the manufacture of Sèvres porcelain before 1882. In 1886, Hermann Seger, director of the royal pottery works in Berlin, published a memoir describing a series of cones having endpoint temperatures between 600 and 1800°C.[2] The cones were made from mixtures of powdered materials made from quartz sand, Norwegian feldspar, calcium carbonate, and Zettlitz kaolin. The latter is a pure clay having the following chemical analysis:

SiO_2	46.9%
Al_2O_3	38.6%
Fe_2O_3	0.8%
Alkali oxides	1.1%

Some of the cones also contained Al_2O_3, Fe_2O_3, Na_2CO_3, and/or H_3BO_3. Seger's work was completed before it was possible to measure temperature at the high temperatures necessary for firing porcelain. Seger based his cone series on the degree of maturity in vitrifying a ceramic body and specifically excluded temperature as a method of determining firing conditions. He experimentally varied composition in a systematic way to produce the desired variations in firing maturity. From the composition and quantity of the raw materials, he computed the chemical analysis of his standard cones. Seger produced the cones for his own and for public use, and sold them internationally to the ceramic industry. He did not specify endpoint temperatures but Le Châtelier gives endpoint temperatures of Seger cones in his book of 1907[3] (Table 16.2).

Edward E. Orton, Jr. founded the Department of Ceramics at Ohio State University in 1894. He developed a complete cone series in the ceramic department in 1896 for use in the United States. He later formed the Standard Pyrometric Cone Company, which manufactured cones until his death in 1932. His will decreed establishment of a nonprofit foundation to continue manufacture for the benefit of the ceramic industry. The National Bureau of Standards made an analysis of temperature equivalents in 1926 and 1954.[4,5]

The numbering of the pyrometric cones is not uniform for historical reasons. The original Seger series had cones numbered from 1 to 36, with the larger number being the higher temperature. This was extended downward to lower temperatues, counting down from 01 to 022. Still later the higher-temperature cones up to 42 were added.

Pyrometric cones are influenced by kiln atmosphere. The iron and lead

TABLE 16.2 Le Châtelier's Temperatures and Composition of Seger Cones

No.	Temp.	Composition
38	1890	$1\ Al_2O_2 + 1\ SiO_2$
36	1850	$1\ Al_2O_2 + 1.5\ SiO_2$
35	1830	$1\ Al_2O_2 + 2\ SiO_2$
34	1810	$1\ Al_2O_2 + 2.5\ SiO_2$
33	1790	$1\ Al_2O_2 + 3\ SiO_2$
32	1770	$1\ Al_2O_2 + 4\ SiO_2$
31	1750	$1\ Al_2O_2 + 5\ SiO_2$
30	1730	$1\ Al_2O_2 + 6\ SiO_2$
29	1710	$1\ Al_2O_2 + 8\ SiO_2$
28	1690	$1\ Al_2O_2 + 10\ SiO_2$
27	1670	$1 \begin{Bmatrix} 0.3\ K_2O \\ 0.7\ CaO \end{Bmatrix} + 20(Al_2O_3 + 10\ SiO_2)$
26	1650	$1 \begin{Bmatrix} 0.3\ K_2O \\ 0.7\ CaO \end{Bmatrix} + 7.2(Al_2O_3 + 10\ SiO_2)$
25	1630	$1 \begin{Bmatrix} 0.3\ K_2O \\ 0.7\ CaO \end{Bmatrix} + 6.6(Al_2O_3 + 10\ SiO_2)$
24	1610	$1 \begin{Bmatrix} 0.3\ K_2O \\ 0.7\ CaO \end{Bmatrix} + 6.0(Al_2O_3 + 10\ SiO_2)$
23	1590	$1 \begin{Bmatrix} 0.3\ K_2O \\ 0.7\ CaO \end{Bmatrix} + 5.4(Al_2O_3 + 10\ SiO_2)$
22	1570	$1 \begin{Bmatrix} 0.3\ K_2O \\ 0.7\ CaO \end{Bmatrix} + 4.9(Al_2O_3 + 10\ SiO_2)$
21	1550	$1 \begin{Bmatrix} 0.3\ K_2O \\ 0.7\ CaO \end{Bmatrix} + 4.4(Al_2O_3 + 10\ SiO_2)$
20	1530	$1 \begin{Bmatrix} 0.3\ K_2O \\ 0.7\ CaO \end{Bmatrix} + 3.9(Al_2O_3 + 10\ SiO_2)$
19	1510	$1 \begin{Bmatrix} 0.3\ K_2O \\ 0.7\ CaO \end{Bmatrix} + 3.5(Al_2O_3 + 10\ SiO_2)$
18	1490	$1 \begin{Bmatrix} 0.3\ K_2O \\ 0.7\ CaO \end{Bmatrix} + 3.1(Al_2O_3 + 10\ SiO_2)$
17	1470	$1 \begin{Bmatrix} 0.3\ K_2O \\ 0.7\ CaO \end{Bmatrix} + 2.7(Al_2O_3 + 10\ SiO_2)$
16	1450	$1 \begin{Bmatrix} 0.3\ K_2O \\ 0.7\ CaO \end{Bmatrix} + 2.4(Al_2O_3 + 10\ SiO_2)$
15	1430	$1 \begin{Bmatrix} 0.3\ K_2O \\ 0.7\ CaO \end{Bmatrix} + 2.1(Al_2O_3 + 10\ SiO_2)$
14	1410	$1 \begin{Bmatrix} 0.3\ K_2O \\ 0.7\ CaO \end{Bmatrix} + 1.8(Al_2O_3 + 10\ SiO_2)$
13	1390	$1 \begin{Bmatrix} 0.3\ K_2O \\ 0.7\ CaO \end{Bmatrix} + 1.6(Al_2O_3 + 10\ SiO_2)$
12	1370	$1 \begin{Bmatrix} 0.3\ K_2O \\ 0.7\ CaO \end{Bmatrix} + 1.4(Al_2O_3 + 10\ SiO_2)$
11	1350	$1 \begin{Bmatrix} 0.3\ K_2O \\ 0.7\ CaO \end{Bmatrix} + 1.2(Al_2O_3 + 10\ SiO_2)$

TABLE 16.2 *Continued*

No.	Temp.	Composition
10	1330	$1\left\{\begin{array}{l}0.3\,K_2O\\0.7\,CaO\end{array}\right\} + 1.0(Al_2O_3 + 10\,SiO_2)$
9	1310	$1\left\{\begin{array}{l}0.3\,K_2O\\0.7\,CaO\end{array}\right\} + 0.9(Al_2O_3 + 10\,SiO_2)$
8	1290	$1\left\{\begin{array}{l}0.3\,K_2O\\0.7\,CaO\end{array}\right\} + 0.8(Al_2O_3 + 10\,SiO_2)$
7	1270	$1\left\{\begin{array}{l}0.3\,K_2O\\0.7\,CaO\end{array}\right\} + 0.7(Al_2O_3 + 10\,SiO_2)$
6	1250	$1\left\{\begin{array}{l}0.3\,K_2O\\0.7\,CaO\end{array}\right\} + 0.6(Al_2O_3 + 10SiO_2)$
5	1230	$1\left\{\begin{array}{l}0.3\,K_2O\\0.7\,CaO\end{array}\right\} + 0.5(Al_2O_3 + 10\,SiO_2)$
4	1210	$1\left\{\begin{array}{l}0.3\,K_2O\\0.7\,CaO\end{array}\right\} + 0.5\,Al_2O_3 + 4\,SiO_2$
3	1190	$1\left\{\begin{array}{l}0.3\,K_2O\\0.7\,CaO\end{array}\right\} + \left\{\begin{array}{l}0.45\,Al_2O_3\\0.05\,Fe_2O_3\end{array}\right\} + 4\,SiO_2$
2	1170	$1\left\{\begin{array}{l}0.3\,K_2O\\0.7\,CaO\end{array}\right\} + \left\{\begin{array}{l}0.4\,Al_2O_3\\0.1\,Fe_2O_3\end{array}\right\} + 4\,SiO_2$
1	1150	$1\left\{\begin{array}{l}0.3\,K_2O\\0.7\,CaO\end{array}\right\} + \left\{\begin{array}{l}0.3\,Al_2O_3\\0.2\,Fe_2O_3\end{array}\right\} + 4\,SiO_2$
01	1130	$1\left\{\begin{array}{l}0.3\,K_2O\\0.7\,CaO\end{array}\right\} + \left\{\begin{array}{l}0.3\,Al_2O_3\\0.2\,Fe_2O_3\end{array}\right\} + \left\{\begin{array}{l}3.95\,SiO_2\\0.05\,B_2O_3\end{array}\right.$
02	1110	$1\left\{\begin{array}{l}0.3\,K_2O\\0.7\,CaO\end{array}\right\} + \left\{\begin{array}{l}0.3\,Al_2O_3\\0.2\,Fe_2O_3\end{array}\right\} + \left\{\begin{array}{l}3.90\,SiO_2\\0.10\,B_2O_3\end{array}\right.$
03	1090	$1\left\{\begin{array}{l}0.3\,K_2O\\0.7\,CaO\end{array}\right\} + \left\{\begin{array}{l}0.3\,Al_2O_3\\0.2\,Fe_2O_3\end{array}\right\} + \left\{\begin{array}{l}3.85\,SiO_2\\0.15\,B_2O_3\end{array}\right.$
04	1070	$1\left\{\begin{array}{l}0.3\,K_2O\\0.7\,CaO\end{array}\right\} + \left\{\begin{array}{l}0.3\,Al_2O_3\\0.2\,Fe_2O_3\end{array}\right\} + \left\{\begin{array}{l}3.80\,SiO_2\\0.20\,B_2O_3\end{array}\right.$
05	1050	$1\left\{\begin{array}{l}0.3\,K_2O\\0.7\,CaO\end{array}\right\} + 1\left\{\begin{array}{l}0.3\,Al_2O_3\\0.2\,Fe_2O_3\end{array}\right\} + \left\{\begin{array}{l}3.75\,SiO_2\\0.25\,B_2O_3\end{array}\right.$
06	1030	$1\left\{\begin{array}{l}0.3\,K_2O\\0.7\,CaO\end{array}\right\} + 1\left\{\begin{array}{l}0.3\,Al_2O_3\\0.2\,Fe_2O_3\end{array}\right\} + \left\{\begin{array}{l}3.70\,SiO_2\\0.30\,B_2O_3\end{array}\right.$
07	1010	$1\left\{\begin{array}{l}0.3\,K_2O\\0.7\,CaO\end{array}\right\} + 1\left\{\begin{array}{l}0.3\,Al_2O_3\\0.2\,Fe_2O_3\end{array}\right\} + \left\{\begin{array}{l}3.65\,SiO_2\\0.35\,B_2O_3\end{array}\right.$
08	990	$1\left\{\begin{array}{l}0.3\,K_2O\\0.7\,CaO\end{array}\right\} + 1\left\{\begin{array}{l}0.3\,Al_2O_3\\0.2\,Fe_2O_3\end{array}\right\} + \left\{\begin{array}{l}3.60\,SiO_2\\0.40\,B_2O_3\end{array}\right.$
09	970	$1\left\{\begin{array}{l}0.3\,K_2O\\0.7\,CaO\end{array}\right\} + 1\left\{\begin{array}{l}0.3\,Al_2O_3\\0.2\,Fe_2O_3\end{array}\right\} + \left\{\begin{array}{l}3.55\,SiO_2\\0.45\,B_2O_3\end{array}\right.$
010	950	$1\left\{\begin{array}{l}0.3\,K_2O\\0.7\,CaO\end{array}\right\} + 1\left\{\begin{array}{l}0.3\,Al_2O_3\\0.2\,Fe_2O_3\end{array}\right\} + \left\{\begin{array}{l}3.5\,SiO_2\\0.5\,B_2O_3\end{array}\right.$
011	920	$1\left\{\begin{array}{l}0.5\,Na_2O\\0.5\,PbO\end{array}\right\} + 0.80\,Al_2O_3 + \left\{\begin{array}{l}3.6\,SiO_2\\1.0\,B_2O_3\end{array}\right.$
012	890	$1\left\{\begin{array}{l}0.5\,Na_2O\\0.5\,PbO\end{array}\right\} + 0.75\,Al_2O_3 + \left\{\begin{array}{l}3.5\,SiO_2\\1.0\,B_2O_3\end{array}\right.$

TABLE 16.2 *Continued*

No.	Temp.	Composition
013	800	$1\begin{Bmatrix}0.5\,Na_2O\\0.5\,PbO\end{Bmatrix} + 0.70\,Al_2O_3 + \begin{Bmatrix}3.4\,SiO_2\\1.0\,B_2O_3\end{Bmatrix}$
014	830	$1\begin{Bmatrix}0.5\,Na_2O\\0.5\,PbO\end{Bmatrix} + 0.65\,Al_2O_3 + \begin{Bmatrix}3.3\,SiO_2\\1.0\,B_2O_3\end{Bmatrix}$
015	800	$1\begin{Bmatrix}0.5\,Na_2O\\0.5\,PbO\end{Bmatrix} + 0.60\,Al_2O_3 + \begin{Bmatrix}3.2\,SiO_2\\1.0\,B_2O_3\end{Bmatrix}$
016	770	$1\begin{Bmatrix}0.5\,Na_2O\\0.5\,PbO\end{Bmatrix} + 0.55\,Al_2O_3 + \begin{Bmatrix}3.1\,SiO_2\\1.0\,B_2O_3\end{Bmatrix}$
017	740	$1\begin{Bmatrix}0.5\,Na_2O\\0.5\,PbO\end{Bmatrix} + 0.50\,Al_2O_3 + \begin{Bmatrix}3.0\,SiO_2\\1.0\,B_2O_3\end{Bmatrix}$
018	710	$1\begin{Bmatrix}0.5\,Na_2O\\0.5\,PbO\end{Bmatrix} + 0.40\,Al_2O_3 + \begin{Bmatrix}2.8\,SiO_2\\1.0\,B_2O_3\end{Bmatrix}$
019	680	$1\begin{Bmatrix}0.5\,Na_2O\\0.5\,PbO\end{Bmatrix} + 0.30\,Al_2O_3 + \begin{Bmatrix}2.6\,SiO_2\\1.0\,B_2O_3\end{Bmatrix}$
020	650	$1\begin{Bmatrix}0.5\,Na_2O\\0.5\,PbO\end{Bmatrix} + 0.20\,Al_2O_3 + \begin{Bmatrix}2.4\,SiO_2\\1.0\,B_2O_3\end{Bmatrix}$
021	620	$1\begin{Bmatrix}0.5\,Na_2O\\0.5\,PbO\end{Bmatrix} + 0.10\,Al_2O_3 + \begin{Bmatrix}2.2\,SiO_2\\1.0\,B_2O_3\end{Bmatrix}$
022	590	$1\begin{Bmatrix}0.5\,Na_2O\\0.5\,PbO\end{Bmatrix} + 0.00\,Al_2O_3 + \begin{Bmatrix}2.0\,SiO_2\\1.0\,B_2O_3\end{Bmatrix}$

Source: Ref. 3.

Figure 16.2. Self-supporting pyrometric cones. [From D. A. Fronk and M. Vukovich, Jr., Deformation behavior of pyrometric cones and the testing of self-supporting cones, *B. Am. Cer. Soc.* [2] **53**, 156–158 (1974).]

oxide bearing cones are sensitive to reducing conditions. In 1932 an iron-free series of cones 010 through cone 3 was introduced by the Orton Foundation. This was improved in 1976. The Orton Foundation introduced self-supporting cones in 1973 (Fig. 16.2).[6]

16.3 ORTON PYROMETRIC CONES

The Orton Foundation manufactures pyrometric cones in three sizes: small, $1\frac{1}{8}$ in. high; regular, $2\frac{1}{2}$ in. high; and self-supporting, $2\frac{1}{2}$ in. high. The small cones that are intended for PCE determinations have had organic binders removed by calcination. All others contain organic binders to improve their dry strength. The cones and their endpoint temperatures for different heating rates are shown in Table 16.1.

16.4 PYROMETRIC CONE PERFORMANCE

The end point temperature depends on the kinetics of the dissolution and viscous flow processes which control the amount, composition, and viscosity of the liquid phase. Therefore, the deformation temperature depends both on heating rate and on time at temperature (Figs. 16.3 and 16.4).[6] If the temperature is high enough to produce some liquid, holding at that temperature will usually "soak down" cones, which would not deform until a higher temperature were reached if the standard heating rate were applied. As shown in Fig. 16.3, this can be two cones higher. Sometimes it is as much as five cones higher.

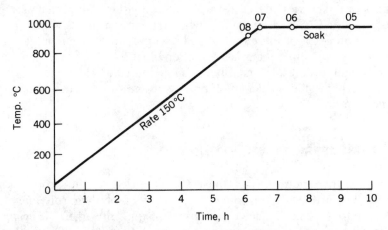

Figure 16.3. Effect of holding time. [From D. A. Fronk and M. Vukovich, Jr., Deformation behavior of pyrometric cones and the testing of self-supporting cones, *B. Am. Cer. Soc.* [2] **53**, 156–158 (1974).]

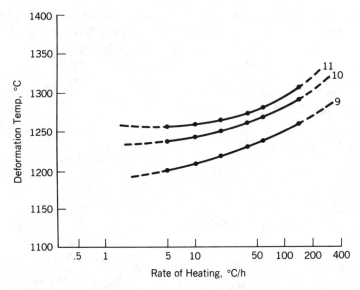

Figure 16.4. Effect of heating rate. [From D. A. Fronk and M. Vukovich, Jr., Deformation behavior of pyrometric cones and the testing of self-supporting cones, *B. Am. Cer. Soc.* [2] **53**, 156–158 (1974).]

The furnace atmosphere is also important. All except the PCE cones contain an organic binder which must be oxidized before sintering occurs. Oxidation is complete by the time the cones reach 500°C if an oxidizing atmosphere is available. If not, sintering will close the pores and prevent oxidation, and the retained carbon will alter the deformation behavior. If carbon is retained, either from the organic binder or from carbonscious materials in unknown materials, it will cause bloating at high temperatures because CO or CO_2 will be emitted when the carbon reacts with either an internal or an external source of oxygen. Even in an oxidizing atmosphere, cones of clays and other materials will often bloat because of gas generating reactions. Sulfates are particularily common causes of bloating. Therefore, the PCE test specifies a calcination process to remove volatiles and oxidize organic material.

16.5 PYROMETRIC CONE-SETTING PROCEDURES

The regular pyrometric cones must be supported at the proper angle. Usually, this is done by inserting them in order of their softening temperature into a cone plaque. Commercial plaques are available from the cone manufacturers. Plaques can also be made by mixing 80% −20 mesh tabular alumina with 20% plastic kaolin clay, tempering with water to a stiff mud consistency, and forming into a bar shape by hand. The height and angle that the cone projects above the plaque must be controlled accurately. Large cones should project 2 in. above the plaque, and small cones should project $\frac{15}{16}$ in. All cones should be set with the face stamped with the cone number inclined 8° from the vertical. A jig cut 82° to the horizontal is convenient for mounting cones. Cones should be put into plaques in the correct order. The low-numbered cones deform first and should not be set at an angle which will cause them to interfere with the higher-temperature cones nearby.

When controlling firing processes, four cones are often used (Fig. 16.1). This usually includes two cones lower than the desired endpoint cone, the endpoint cone, and one higher than the endpoint cone. Then, when used, the time and temperature of the first two cones will help establish the conditions necessary to reach the desired endpoint cone. The higher cone is helpful in evaluating the degree of overfiring if the endpoint cone conditions are exceeded slightly. It is sometimes called a "guard" cone. It is not unusual for cone plaques set at different places in a large kiln to have different time and temperature conditions—so that the four-cone series is often useful.

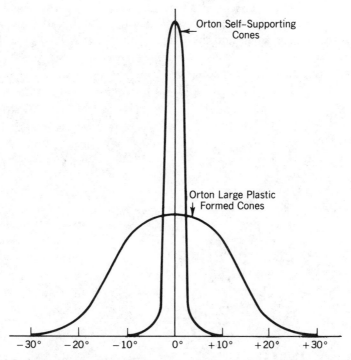

Figure 16.5. Approximate endpoint precision for self-supporting and plastic-formed cones. [From D. A. Fronk and M. Vukovich, Jr., Deformation behavior of pyrometric cones and the testing of self-supporting cones, *B. Am. Cer. Soc.* [2] **53**, 156–158 (1974).]

The endpoint of the cones is remarkably reproducible when heated at the standard rate. This depends in part on the skill with which the cones are placed so that they are at the correct angle and height. The self-supporting large cones need only be set on a flat horizontal surface to be used. Then the endpoint temperature variation may be only about ±2° (Fig. 16.5).

16.6 DETERMINATION OF PYROMETRIC CONE EQUIVALENCE

The PCE test has been standardized because it is useful in evaluating the refractoriness of materials used in the production of ceramic products, and as a method for testing the quality of ceramic products such as refractories. A standard method of test has been adopted by the American Society of Testing and Materials ASTM designation C24-79, and that publication should be consulted by anyone making PCE tests.[7]

The procedure for PCE determinations starts with preparation of the raw material. It is necessary to crush and grind a small representative sample of the material to pass a U.S. 70 standard sieve, calcine the powder at least 30 minutes at 925–980°C, pass the calcined powder through the sieve again, mix it with a small amount of water and an organic binder, form it into cones, dry the cones, and sand them to the required dimensions. The unknown cones and the standard cones are then mounted in a disk-shaped plaque at the proper angle and the plaque is dried. It is then placed into a

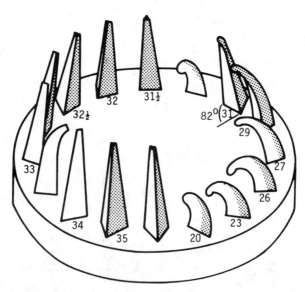

Figure 16.6. Cone plaque after heating. (From "1984 Annual Book of ASTM Standards," 1984. Reprinted with permission from the American Society for Testing and Materials, 1916 Race Street, Philadelphia, PA 19103.)

furnace and fires at a specified rate, usually using natural gas and air to reach about cone 28, and natural gas and oxygen to reach the higher cones. Observations of the bending behavior are made during heating. When the unknown cone has bent to touch the plaque, the furnace is shut off, the plaque is removed, and the unknown cones are compared to the standard cones (Fig. 16.6). the ASTM procedure gives details for making molds and plaques, setting jigs, and for furnace operation.

REFERENCES

1. The Edward Orton, Jr. Ceramic Foundation, "The Properties and Uses of Orton Standard Pyrometric Cones." Columbus, Ohio, 1978.
2. H. Seger, *Ton. Ztg.*, pp. 135, 229 (1886).
3. H. Le Châtelier and O. Boudouard, "High Temperature Measurements" (G. K. Burgess, transl.), 2nd ed. Wiley, New York, 1907.
4. C. O. Fairchild and M. F. Peters, Characteristics of pyrometric cones. *J. Am. Ceram. Soc.* **9**, 200 (1926).
5. H. P. Beerman, Calibration of pyrometric cones. *J. Am. Ceram. Soc.* **39**, 47–53 (1956).
6. D. A. Fronk and M. Vukovich, Jr., Deformation behavior of pyrometric cones and the testing of self-supporting cones. *Bull. Am. Ceram. Soc.* **53**, 156–158 (1974).
7. American Society for Testing and Materials, Standard method for pyrometric cone equivalent (PCE) of refractory materials. Part 17. "1984 Annual Book of ASTM Standards," ANSI/ASTM C24-79, Philadelphia, Pennsylvania, 1984.

17

CALIBRATION METHODS

17.1 INTRODUCTION

The purpose of this chapter is to explain methods of calibration that are useful for the student, scientist, engineer, or industrial practitioner so that he or she can have confidence in temperature measurements within the limits necessary for the application. Other chapters include material on installation effects, interpolation methods, temperature scales, and particular methods of measuring temperatures that will also need to be consulted. Some of the experimental methods will be explained in some detail after a basis for decision making has been prepared.

17.2 CLASSIFICATION OF STANDARDIZATION METHODS

Standardization can be classified as primary, secondary, tertiary, and quaterary. Primary standardization is calibration of a temperature measuring instrument in which the fundamental interpolation equation is determined for the instrument. This requires measurements to calibrate the sensing device as well as suitable connections and instrumentation to convert the sensing output signal to temperature. The minimum number of measurement temperatures depends on the number of coefficients in the interpolating equation. For example, if a type S thermocouple is to be fit by an equation,

$$V = a + bt + ct^2,$$ (17.1)

three measurements of V at three different well-known temperatures will be required to determine the constants a, b, and c. The temperatures must be precisely known and thermodynamic invariant points, such as the primary and secondary points of IPTS, can be used. If the equation is correct for the sensor being calibrated, and if all errors in measurement have been properly controlled, primary calibration should produce a primary standard. This is the basic approach of standards laboratories. The primary standardization goal is to produce an instrument that fits, to the best possible degree, the best estimate of the true thermodynamic temperature scale. Primary standardization is so important that tremendous scientific effort has been devoted to it during this century. Much of the research on primary standardization has been published in the international journal *Metrologia*. It is not the purpose of this book to provide a background for those interested in primary standardization. However, some of the principles are of benefit to the general user and will be presented.

Secondary standardization is calibration of a temperature measuring instrument against another instrument that has previously been calibrated as a primary standard. This can be done over the range of temperature of interest for the two instruments. Since this method is not limited to thermodynamic invarient points, measurements can be made at many temperatures and statistical techniques used to correlate the secondary standard with the primary standard. This helps reduce the error accompanying the comparison. Obviously, the errors in the primary standard are incorporated in the secondary standard. National standards laboratories maintain many secondary standard devices having precisely known limits of accuracy for use in the calibration services that they provide.

Tertiary standards result from comparisons of devices with reliable secondary standards. For example, the National Bureau of Standards will calibrate any suitable temperature measuring sensor against its secondary standards when submitted in accordance with their requirements and on payment of a fee. This is done routinely as a service to instrument companies and other users to provide accurate temperature measurements standards for scientific and industrial use. After calibration the device is returned with the calibration data, appropriate equations, and a statement of the limits of accuracy.

Tertiary standards are often the basis for quarterary standards. A sensor is calibrated against a Bureau of Standards calibrated device. Quaterary standards are often provided by instrument companies with the statement "calibration traceable to the National Bureau of Standards." Because of the cost of tertiary calibration by a standards laboratory, quaterary standardization is the most common classification used in scientific and industrial work. The choice of level of standardization depends on the accuracy required of the classification. The reproducibility of primary standardization over the approximate range 0.8–692.73 K (419.58°C) is believed to be better than 1 mK (Fig. 17.1).[1] This is far better than most applications

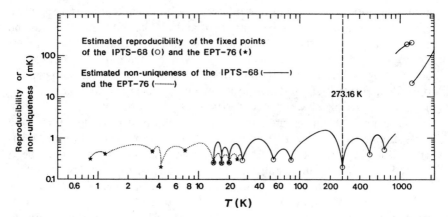

Figure 17.1. Estimated uncertainties from reproducibility or nonuniqueness of the IPTS. (From Ref. 1.)

require. Quaterary standardization is perfectly satisfactory for most applications. Note that in most cases it is the sensor only that is calibrated. (An exception is the thermometer.) One must be very careful that the instrumentation is suitable to obtain the accuracy available in the sensor.

Calibration requires special furnaces and instrumentation. Often, temperature sensors in industrial applications deteriorate during use. Therefore, many companies maintain a comparison furnace for checking the working sensors against their quarterary standards. This practice is often economically feasible and avoids risking their quaterary standards in the industrial environment. Again, one should be careful not to ignore instrumentation problems that may degrade the performance of a calibrated sensor.

17.3 THERMODYNAMIC BASIS FOR PRIMARY STANDARDS

17.3.1 Phase Rule Basis

According to the phase rule the degrees of freedom F available to a system is given by

$$F = C - P + 2, \tag{17.2}$$

where C is the number of chemical components of the system, P the number of phases present, and the 2 represents the independent variables pressure and temperature. If the pressure is held constant externally, Eq. (17.2) becomes

$$F = C - P + 1. \tag{17.3}$$

These equations are the basis for the thermodynamic invariant points used for primary calibration. For example, if the pressure is held constant, the ice-point degrees of freedom is given by Eq. (17.3) as

$$F = 1 - 2 + 1 = 0 . \tag{17.4}$$

Here we have one component, H_2O, and two phases, ice and water. As long as thermal equilibrium exists and the pressure is constant, a mixture of water and ice has zero degrees of freedom, so the temperature is invariant. This is a thermodynamic invariant point. A similar example is the steam point, where the pressure is held constant at 1 atm. Many of the primary calibration points and most of the secondary calibration points of IPTS are the thermodynamic invariant points represented by Eq. (17.3). However, that method requires external control of pressure at a precise value.

Equation (17.2) applies at the triple point, where solid liquid and vapor are in equilibrium. When all three phases are in equilibrium, for a single-component system, Eq. (17.2) gives

$$F = 1 - 3 + 2 = 0 .$$

This is better, in principle, than Eq. (17.3) because no external variable must be controlled. It is only necessary to provide a pure single-component system and heat or cool it until all three phases appear in equilibrium. This is the basis for adopting the triple point of water as the single defining point for the IPTS. It is fortunate that the triple point of water at 273.16 K is only 0.01° higher than the ice point at 273.15 K. Many primary and secondary standards are now based on triple points, and there is much interest in adopting more triple-point calibration points as the necessary technology is developed.

17.3.2 Effect of Impurities

The phase rule applies because the chemical potential of every chemical species present must be equal in all phases if chemical equilibrium exists. Otherwise, transport of material by diffusion to change the composition of the phases would be possible and there would not be a chemical equilibrium. For a single-component system, then, at the triple point, the chemical potential of that component is the same in all three phases—solid, liquid, and gas—and zero degrees of freedom result. The phase rule as we have used it in Eq. (17.2) and (17.3) is for a single-component system. Actually, it is not possible to produce a completely pure, single-component system. Some level of impurities is always present. Therefore, we need to understand the effects of impurities on the phase relationships.

Every high school physics student learns that adding salt to water

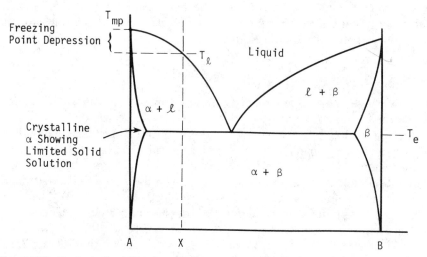

Figure 17.2. Typical phase diagram showing freezing-point depression when impurity B is added to A. Freezing does not begin for the composition X until cooled to T_l, ΔT below T_{mp}. Freezing is not complete until the temperature reaches T_e, well below T_{mp} of pure A.

reduces the melting point, allowing us to freeze ice cream with a brine–water mixture. In general, impurities lower the freezing point (Fig. 17.2).

In a few systems where the solvent (the major constituent) can dissolve more of the solute (the impurity) in the solid phase than it can in the liquid phase, the freezing point is raised by the presence of impurities. Antimony in tin is an example (Fig. 17.3).

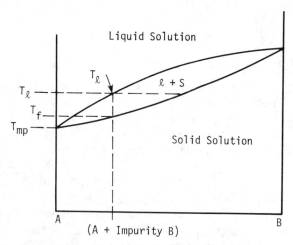

Figure 17.3. Unusual solid solution where the freezing point is raised by the presence of the impurity at ($A +$ impurity B) composition. Freezing begins at T_l and is complete at T_f.

Diffusion is many orders of magnitude faster in liquids than in solids. The equilibrium diagrams in Figs. 17.2 and 17.3 are drawn with an exaggerated amount of impurity to show the effect of the impurity. In real freezing-point determinations, the composition of the solid deviates from that shown because there is not time for impurities to diffuse in the solid. However, because of the extremely low concentration of impurities in the freezing-point materials used for temperature standards, equilibrium diagrams are suitable.

17.3.3 Cooling Curves

When a furnace cools, heat is transferred through its walls to the surroundings. The driving force for cooling is the temperature difference between the furnace and its surroundings. When most furnaces are shut off, their cooling temperature–time curves are exponential because the driving force reduces with temperature (Fig. 17.4). The furnace cools very much like a sensor subjected to a step function (see Chapter 18).

Freezing-point determinations are conducted in furnace with very small differences in temperatures so that the cooling curve appears to be approximately linear over the time period necessary for freezing. Usually, a freezing-point experiment is conducted by heating the metal only a few degrees above the melting point and then reducing the furnace power so that the metal cools very slowly. The heat removed is then nearly at a constant rate. Above the freezing point, the liquid cools at a rate dependent on the heat removal rate and its heat capacity. Equating the heat lost from a unit mass of the metal to the heat flowing from the furnace, per unit mass, gives

$$C_p \frac{dt}{d\theta} = \frac{dQ}{d\theta}. \tag{17.5}$$

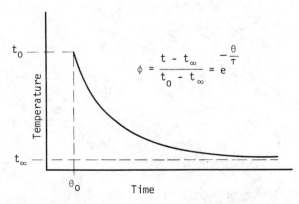

$$\phi = \frac{t - t_\infty}{t_0 - t_\infty} = e^{-\frac{\theta}{\tau}}$$

Figure 17.4. Idealized cooling curve for a furnace cooling from t_0 to t_∞.

Solving for the time rate of temperature change,

$$\frac{dt}{d\theta} = \frac{1}{C_p}\frac{dQ}{d\theta}.$$ (17.6)

If $dQ/d\theta$ is constant and the mass is constant, the slope of the cooling curve is the reciprocal of the specific heat (Fig. 17.5).

Many high-purity liquids can be supercooled. Classical homogeneous nucleation theory explains this as the result of the necessity for surface formation. The liquid must be supercooled enough to provide the energy necessary to form new surface. Then a classical cooling curve will look like Fig. 17.5.

The amount of supercooling, and the temperature recovery accompanying nucleation (temperature recalescence), depend on the solid–liquid surface energy and the free energies of the solid and the liquid as a function of temperature. High surface energy and low free-energy difference favors supercooling. If a liquid is impure as shown in phase diagrams in Figs. 17.2 and 17.3, the temperature will not be constant after nucleation (Fig. 17.6).

In both cases, the freezing plateau does not have a constant temperature. It is very important to obtain freezing-point standards that have high enough purity to give suitable freezing plateaus. Zone-refined solids and purified gases for cryogenic temperatures are available that are suitable for standards. These usually have fewer than 10 parts per million total impurities. However, the freezing plateau is never truly constant. Some slope

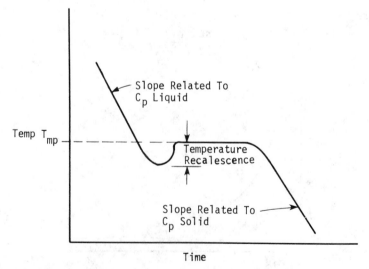

Figure 17.5. Typical cooling curve for a high-purity element or compound. Supercooling and subsequent recalescence exaggerated.

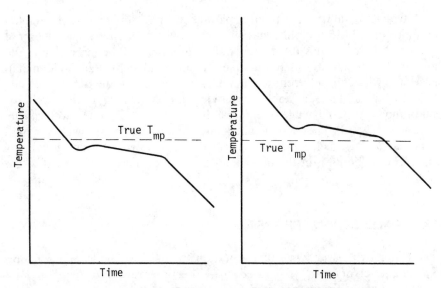

Figure 17.6. Cooling curves for impure materials. Left: impurity lowers the freezing point in Fig. 17.2. Right: impurity raises the freezing point as in Fig. 17.3.

always exists with the most precise measurements. It is only necessary that the plateau be constant on the scale of temperature measurement required for the application. Only sophisticated standards laboratories and similar precise applications need be concerned with the millikelvin deviations from constant temperature. A rule of thumb that has statistical support is that the standard need only be measured to have an error less than about one-tenth the error expected in applying the temperature measuring instrument.

Melting plateaus are better than freezing plateaus because recalescence is not observed. If the metal slowly is cooled below freezing, segregation of impurities will occur during freezing. Remelting with the furnace temperature constant only slightly above the melting point (so the metal will be at uniform temperature) will melt the segregated impurities first (assuming freezing-point depression by segregation of impurities). Remelting after rapid cooling, where there is not time for impurity segregation, should show a flatter melt plateau. If the cooling rate affects the melting plateau, it is evidence of impurities.[2]

In any melting- or freezing-point determination it is essential to have the true equilibrium interface adjacent to the sensor being calibrated. The geometry of triple-point and melting- or freezing-point apparatus is chosen to provide a long cylindrical liquid–solid interface surrounding the sensor. Operating procedures are adapted to ensure that interface is produced and maintained.

A typical metal freezing-point apparatus will have the metal held in a tall graphite crucible. (Graphite has high thermal conductivity and does not react with many metals.) A central well of graphite will be inserted into the top so that the well will be completely surrounded by the liquid metal. Freezing will occur adjacent to the well and adjacent to the outer wall. The objective is to have a stable, unmoving interface next to the well for calibration purposes while the metal interface at the outer well slowly advances inward. The frozen metal adjacent to the outer wall has high conductivity and helps maintain temperature uniformity. Impurity segregation in the liquid is at the outer interface, not at the interface next to the well.

17.4 CALIBRATION ERRORS

There are two basic types of calibration errors, systematic and random. Systematic errors can be caused by either theoretical or experimental errors. Random errors are statistical in nature and can be analyzed by statistical methods. Good statistical analysis makes it possible to calculate the probability of making a certain amount of random error. Systematic errors can be completely unsuspected, so great care is needed to avoid systematic errors.

Temperature measurements are always taken for a specific purpose. The limits of accuracy of any temperature measurement depend on that purpose. It is very wasteful to attempt $\frac{1}{10}$-degree accuracy if $\pm 5°$ is adequate for the purpose. Often, temperature is a state variable used to determine the value of a particular property of interest. We may ask ourselves, "What accuracy do I need in measuring temperature?" The answer to this question can be determined if the limits for the property are known, and if its functional dependence on temperature is known, by taking the total derivative. As an example, let us suppose that a property $V(T, I, R)$ is to be measured and has the relationship

$$V = I^2 R^{-1} e^{Q/kT} . \qquad (17.7)$$

Then

$$dV = \frac{\partial V}{\partial I} \, dI + \frac{\partial V}{\partial R} \, dR + \frac{\partial V}{\partial T} \, dT \qquad (17.8)$$

and

$$\frac{dV}{V} = \frac{1}{V} \left(\frac{\partial V}{\partial I} \, dI + \frac{\partial V}{\partial R} \, dR + \frac{\partial V}{\partial T} \, dT \right). \qquad (17.9)$$

Taking the partial derivatives of Eq. (17.7) and substituting into Eq. (17.9) gives

$$\frac{dV}{V} = \frac{1}{V}\left(2IR^{-1}e^{Q/kT}\,dI - I^2R^{-2}e^{Q/kT}\,dR - I^2R^{-1}\,e^{Q/kT}\,\frac{Q}{kT^2}\,dT\right). \quad (17.10)$$

Substituting for V on the right side gives

$$\frac{dV}{V} = \frac{2\,dI}{I} - \frac{dR}{R} - \frac{Q}{kT^2}\,dT\,. \quad (17.11)$$

[Note that because dR can be either positive or negative it is the absolute value of dR/R that is important. The signs of terms in Eq. (17.7) are not significant.] If we replace the differentials by increments, the result is

$$\frac{\Delta V}{V} = \frac{2\Delta I}{I} - \frac{\Delta R}{R} - \frac{Q}{kT^2}\,\Delta T\,. \quad (17.12)$$

From this we see that errors in measurement of I, R, and T do not have equal effect on the final error in $\Delta V/V$. To have the same fractional contribution to the total error in $\Delta V/V$, we need to measure I twice as accurately as R, and the contribution from errors in ΔT depends on Q/kT^2. The usual goal is to reduce the error in the property of interest to some specific value (e.g., 1% for V). This is best accomplished by reducing the error in measuring each of the parameters on which it depends sufficiently, but so that each contributes about equally to the total error. In our example there is no advantage in measuring R to $\pm 10^{-6}$ if I is known to $\pm 10\%$.

17.5 CUMULATION OF UNCERTAINTY

Every variable affecting a temperature measurement contributes to the uncertainty of the measurement. If these variables are independent and the measurement errors random, the total variance should be the root mean square of the individuals. For our example from Eq. (17.7),

$$S_V^2 = S_I^2 + S_R^2 + S_T^2 \quad (17.13)$$

or

$$S_V = (S_I^2 + S_R^2 + S_T^2)^{1/2}\,. \quad (17.14)$$

17.6 CONSTANT-TEMPERATURE ZONES

For a calibration to be accurate, the entire sensor must be in a uniform temperature region that must be constant in temperature during the calibration. Large sensors require large regions of constant temperature. This situation requires care in the geometric aspects of the design as well as other factors. Cryostats often require carefully planned heat flow paths. High-temperature furnaces often require multiple windings so that temperature gradients can be controlled by controlling the electrical power to various zones of the furnace. A high-thermal-conductivity block, such as a copper block drilled to take the sensor, is often useful in obtaining a uniform temperature zone.

Constant temperatures are often achieved by providing a large thermal mass surrounding the sensors (e.g., metal freezing-point apparatuses). For rapid changes in temperature, as in tertiary calibrations at high temperatures, a low thermal mass is desirable. A thin metal tube heated electrically, with the end openings thermally insulated and the sensors at midlength, can be suitable if the power supplies are suitably controlled.

Sometimes temperature varies with time because the power input controller cycles on and off sinusoidally. If the controller is properly adjusted, this may be caused by too much insulation between the control sensor and the heating elements. Moving the control sensor close to the heating elements, with a large thermal mass in the constant-temperature zone to damp out temperature oscillations, often is a satisfactory solution.

Auxiliary sensors should be used to determine the uniformity of temperature when possible. Methods for doing this are discussed at various places later in this chapter.

17.7 IMMERSION ERRORS

Any new calibration facility, or any new sensor design in an existing facility, should be tested for immersion error. This error results because the sensor is heated or cooled by conduction of heat along its length extending outside the constant-temperature zone. Immersion error depends on the sensor design. A heavy sensor with a massive high-conductivity stem will need to be immersed to a great depth in the constant-temperature region, and temperature changes in that region can be induced by insertion of such a sensor. To test for immersion errors, the sensor should be withdrawn small measured increments until temperature changes are detected. If the sensor temperature drops on the first few increments of withdrawal, the immersion length was too short and the sensor temperature may not be homogeneous. It is the sensor design, not the difference in temperature, that most affects immersion errors. A high-thermal-conductivity leak path will require very deep immersion regardless of the temperature difference between ambient and testing temperatures.

Sometimes, when exploring a zone for temperature uniformity or testing for immersion error, it is difficult to determine if the variations are caused by sensor variability. This is especially true for thermocouples, where homogeneity variations for the portion of the wires in the thermal gradient may cause fluctuations. Substituting different sensors often will be helpful.

17.8 CALIBRATION IN A TEST FACILITY VERSUS CALIBRATION IN SITU

Sensors are best calibrated in a test facility where temperature uniformity and immersion error can be controlled. When this is done, there is always the question of whether the sensor responds the same in an application as it did in the calibration apparatus. This is particularly true when the sensor must be modified to fit it into the apparatus. The same conditions that are needed for calibration are really needed for calibration in situ. However, if it is not possible to control temperature uniformity, the size of the uniform temperature region, or the immersion error, corrections can sometimes be determined, or at least estimated, by calibration in situ.

The instrumentation used in calibration is often superior to that used in an application, especially an industrial application, where older equipment, or less expensive equipment, may be the only equipment available. Then, testing the equipment response, and the control functions associated with it, may also require measurements in situ. This will often include using a standard signal source to test the instrumentation response—to simulate the same signal that would arrive from the sensor. For example, to test thermocouples, a standard millivolt potentiometer can be used to provide voltages representing the range of signals expected from the thermocouple to test the instrument response. A run-up box is commercially available for use when a large number of instruments need to be tested at a particular voltage. The run-up box is a battery source and a voltage divider that can be set to produce a voltage equal to that from a potentiometer. The output voltage of the run-up box is adjusted to the desired millivolt output and then used to test the instruments.

As a necessary condition for reliable instrumentation, standards laboratories must have good ambient temperature control, often constant to ±1°C or better. Testing in situ without ambient temperature control may be misleading for both instrument and sensor evaluation. If the sensors and instruments are not evaluated during ambient temperature extremes, the performance may be far inferior to that indicated under comfortable ambient conditions.

The entire system in an application needs to be tested. One of the best ways to do this is to conduct a primary calibration in situ, using a thermodynamic invariant point in the range of temperature needed for the application. This will often help in estimating the uncertainty of the apparatus if the measurements receive enough replication for statistical analysis. Various methods of primary calibrations will be discussed next.

17.9 THE TRIPLE POINT OF WATER

This is the defining point of IPTS and can be achieved with a reproducibility of 0.01 mK. It is considered to be exactly 273.16 K, which makes it exactly 0.01°C (Fig. 17.7).

Water has its greatest density at 4°C, and errors can result if separation from the water–ice interface occurs. Concentration of impurities on melting a large volume of ice and strain in crystalline ice can also introduce errors. The triple-point apparatus has evolved to a very great degree, so that the requirements for water purification, cell construction and filling, and procedures for use are well established. Commercial cells are available (Fig. 17.8)[3] and the operating procedures well documented.[4] Briefly, the procedure requires cooling the cell in shaved wet ice for 2 hours, freezing a mantle of ice about 10 mm thick around the central well by introducing solid CO_2 or its equivalent, melting a layer of ice to produce a liquid film around the central well in equilibrium with the ice, immersing the apparatus in an ice water bath (which is only 0.01°C cooler than the triple-point apparatus), and waiting a day before beginning measurements. The cell operates better if a plastic container is used to support it and separate it

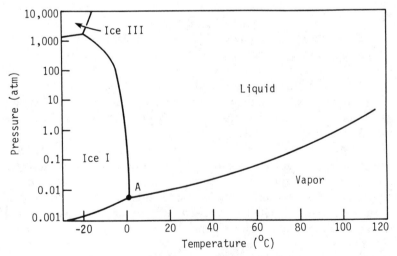

Figure 17.7. Phase diagram for water showing the triple point *A*, a thermodynamic invarient point.

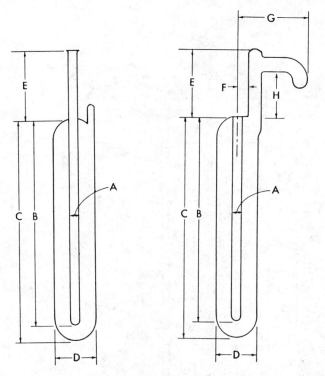

Figure 17.8. Two commercial triple-point-of-water cells. (Courtesy of Jarrett Instrument Company.)

from actual contact with the ice bath (Figure 17.9).[5] If the mantle of ice is carefully maintained to prevent cracking or melting off at the bottom, such a triple point can be used for months.

17.10 CRYOGENIC TRIPLE POINTS

Cells for cryogenic triple points are, of necessity, small, so special care is needed to avoid impurities, including those adsorbed on the apparatus or absorbed from other fluids. They are constructed with internal heat shields to slow heat transfer and sealed in an evacuated enclosure that is immersed in a cryogenic fluid. The cells themselves can be sealed and must be strong enough to resist the internal pressure produced by warming to room temperature. As such, they are readily transportable and very useful. Sealed cells have been made for CO_2, Hg, Xe, Kr, CH_4, Ar, O_2, N_2, Ne, and H_2.[6]

In use the cells are frozen, then melted with successive tiny bursts of electric power while monitoring temperature with time very precisely. The equilibrium temperature after each burst of power is used to determine the rate of rise in temperature. This is rapid at first, nearly flat during melting,

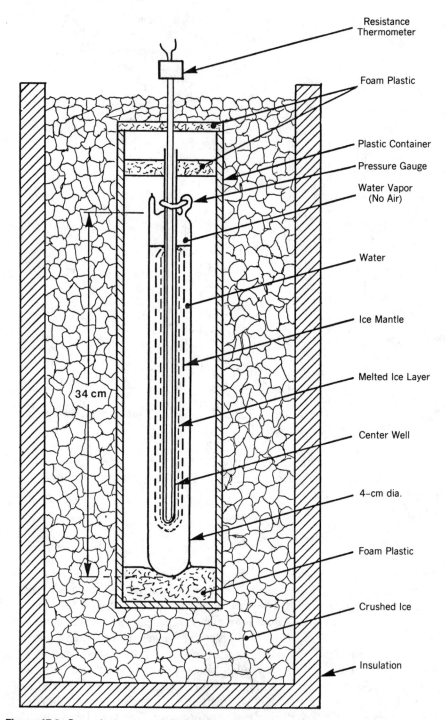

Figure 17.9. Properly supported triple point of water cell. (From J. V. McAllan, "The Effect of Pressure on the Water Triple Point Temperature," CSIRO, Division of Applied Physics, National Measurement Laboratory, Sydney, Australia.)

and rapid when entirely liquid. The most reliable portion of the plateau (usually about 15% at each end is rejected) is extrapolated forward to meet the line after melting, and the inflection point obtained. This can be taken as the triple point. A better procedure is to determine fraction melted, F, from the plateau–time relation. Large values of $1/F$ from about 1000 to 50 usually vary linearly with temperature. Extrapolating to $1/F = 1$ gives the triple point.

A recent round-robin evaluation of sealed triple-point cells gave very good results.[7] Differences in the triple points of Ar, O_2, H_2, Ne, CH_4, and D_2 were found to be the result of instrument variations rather than from the cells. Excellent agreement was obtained.

This leads to a new philosophy of calibration, especially for cryogenic systems. The availability of easily transportable sealed durable cells allows any laboratory to perform the type of calibration that was previously limited to standards laboratories, or similar laboratories with sophisticated equipment. It will only be necessary to buy the cell and follow the appropriate recipe supplied with it to obtain and use a reliable primary standard.

17.11 THE ICE POINT

The ice point is perfectly suitable as a calibrating point for many applications requiring accuracy to about 1 mK. Because ice floats, it is essential that the ice be finely divided, wet on the surface, but drained to prevent warm water settling out. If the ice floats on top of water in a Dewar flask, errors of a degree or two can occur if the sensor projects into the water. Because of poor heat transfer between coarse ice and a sensor, the ice should be shaved or finely crushed and the sensor should project into the mixture far enough to prevent excessive immersion errors. Often about 10 cm is needed to reach 0.01°C and about 30 cm is needed to reach 0.001°C. Ordinary distilled or dionized water is suitable. For 0.1°C error, most city tap water is suitable. Thermocouples and similar devices can be calibrated by immersing a high-conductivity-metal block in the ice bath and inserting them into holes drilled in the block.

17.12 THE TRIPLE POINT OF GALLIUM

Because of the difficulties in measuring the steam point, the triple point of gallium is very useful as a calibration point just above room temperature, at 29.774°C. It is an alternative to the steam point in calibrating a standard platinum resistance thermometer in range II of IPTS. The triple-point apparatus is made from polymers such as nylon and Teflon because the expansion would break glass (Fig. 17.10).[8]

In use the gallium is frozen near the ice point, heated with an oil bath to

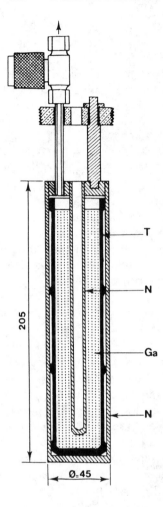

Figure 17.10. Gallium triple-point cell. T, Teflon; N, nylon; G, gallium. [From J. Bonhoure and R. Pello, Température du Point Triple du Gallium, *Metrologia* **19**, 15–20 (1983). By permission of Springer-Verlag.]

about 40°C to melt the gallium adjacent to the well and to the outer wall of the cell, and then held in an oil bath about 0.01°C above the triple-point temperature. Sensors should be preheated to test temperature to avoid excessive melting of the inner sheath.

17.13 THE MELTING POINT OF GALLIUM

Commercial melting-point apparatus for the melting point of gallium at 29.7714°C are available. The cell can be removed for freezing. When returned to the apparatus it will automatically make the initial melt and hold it at the melting point plateau for about 12 hours. An uncertainty of ±0.002°C is claimed (Fig. 17.11).[9]

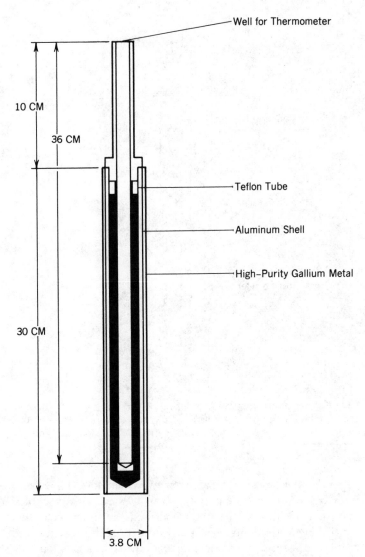

Figure 17.11. Gallium melting-point apparatus. (Courtesy of Yellow Springs Instrument Co., Yellow Springs, OH.)

17.14 FREEZING POINTS OF TIN, LEAD, ZINC, ANTIMONY, ALUMINUM, SILVER, AND GOLD

Apparatus for freezing-point determinations are basically as described previously for triple-point apparatus, but are made from refractory materials such as graphite and fused silica (Fig. 17.12).[10] The fused silica tube for the central well is replaced with graphite in some designs.

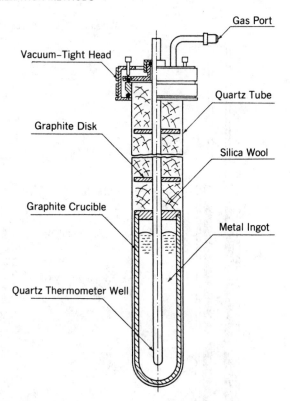

Figure 17.12. Typical high-temperature freezing-point cell. [From G. Bongiovanni, L. Crovini, and P. Marcarino, Effect of dissolved oxygen and freezing techniques on the silver point, *Metrologia* **11**, 125–132 (1975). By permission of Springer-Verlag.]

The general procedure is the same for all the metals. They are heated a few degrees above the freezing point until melted, slowly cooled to the freezing point, nucleated at the outer wall and the inner well, and held at the plateau for measurements. The temperature of the furnace is held just below the plateau temperature, so that cooling is so slow that the plateau temperature is maintained for hours or days. Because these cells must have a long well to reduce immersion error, they require a large amount of high-purity metal. The volume is approximately 200 ml.

Tin freezes at 231.9681°C, but it must be nucleated after it supercools about 4°C below the freezing point. This can be done by removing the crucible assembly from the furnace until temperature recalescence occurs, then returning it to the furnace. This nucleates the freezing at the outer wall. Withdrawing the sensor until it cools, then reinserting it, nucleates the solid around the inner well. A nitrogen atmosphere, not carbon black, should be used to prevent oxidation.

Zinc freezes at 419.58°C. It nucleates easily. Immediately after nucleation removing the sensor, cooling it, and then returning it produces the desired stable frozen metal next to the well. Nitrogen may be used as a protective atmosphere.

Antimony freezes at 630.755°C. High-purity antimony supercools about 10–40°C. The same procedure as described for tin should be used. Argon is necessary as a protective atmosphere because oxidation lowers the freezing point. This enhances nucleation, so oxidation can be detected by its nucleation behavior.

The freezing point of aluminum is 660.46°C. It nucleates similar to zinc. Aluminum reacts with silica, so graphite fiber insulation and a graphite well should be used. Argon gas should be used as a protective atmosphere.

Silver freezes at 961.93°C. Graphite or alumina are needed to replace the fused silica, with graphite for metal contact. The nucleation procedure is the same as for zinc. An inert gas, argon, is needed to prevent oxidation. Internal oxidation depresses the freezing point about 4 mK.

Gold melts at 1064.43 and is not susceptible to oxidation. The nucleation procedure is the same as for zinc. The protective atmosphere is needed to prevent oxidation of the graphite crucible.

17.15 CONSTANT-TEMPERATURE SOURCES

Commercial constant-temperature baths, blocks, and fluidized beds are available for calibrations at or near room temperature. Water baths at low temperatures and oil baths at temperatures to about 350°C can be rapidly stirred, well insulated, and regulated (Fig. 17.13). They are widely used for calibration purposes. Constant-temperature blocks are also used to prevent oil spills or fire hazards (Fig. 17.14).[11]

Constant-temperature fluidized beds are also becoming popular because they can operate at higher temperatures, do not cause a fire hazard, and transfer heat more rapidly than air would at the same temperature. One such commercial unit has air-suspended aluminum oxide particles as the fluidized bed, an internal heat exchanger to preheat air used to produce fluidizing, an air requirement of 30 ℓ/m at 10 psi for a 4-in.-diameter by 18-in.-deep working volume.[12] Temperature stability is given as ±0.1°C at 100°C and ±0.2°C at 600°C.

17.16 COMPARISON FURNACES

There are almost as many designs of calibration furnaces as there are calibraters. Many fine designs exist. Some have low thermal mass, excellent regulation, and rapid response (Fig. 17.15).[13] Others have large thermal mass and slow response, so that secondary calibration can be conducted

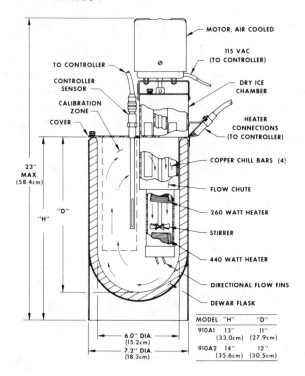

Figure 17.13. A stirred liquid calibration facility. (Courtesy of Rosemount, Inc.)

(Fig. 17.16).[14] The latter is available commercially. Many commercial tube furnaces are well adapted to this application.

17.17 HEAT PIPES

A heat pipe is a sealed container with walls sufficiently porous or roughened to provide capillarity so that a small amount of liquid will be distributed by capillary action to cover the walls completely. This wet wall container is a marvelous distributor of heat by vaporization where locally hot and condensation where locally cold. The temperature everywhere is determined by the pressure of the liquid vapor within the container. That pressure can be controlled by external pressure through an inert gas connection to the container, where the vapor–gas contact zone is a stable diffusion gradient.[15] In recent years this has been very helpful in providing isothermal chambers for blackbody sources (e.g., Fig. 17.16) and other calibration chambers. Limitations are the availability of a working liquid of suitable vapor pressure and of containers to contain them. Neur and Brost recommend the liquids and containers listed in Table 17.1.[16] One should note, however,

Figure 17.14. Constant-temperature calibration block. (Courtesy of Kaye Instruments.)

TABLE 17.1 Working Fluids and Compatible Wall Materials for Heat-Pipe Blackbodies

Working Fluid	Wall Material	Temperature Range (K)[a]
Liquid nitrogen LN_2	Stainless steel	70–110
Methane, CH_4	Copper, aluminium	100–150
Carbon tetrafluoride, CF_4	Copper, aluminium	100–200
Freon, $CCIF_3$	Copper, aluminium	120–300
Ammonia, NH_3	Stainless steel, nickel, aluminium	230–330
Acetone, C_3H_6O	Copper, stainless steel	230–420
Methanol, CH_4O	Copper, nickel	240–420
Water, H_2O	Copper, nickel, titanium	300–550
Organic fluids	Stainless steel, super alloys, carbon steel	400–600
Mercury + additives, Hg+	Stainless steel	450–800
Potassium, K	Stainless steel, nickel	700–1000
Sodium, Na	Stainless steel, nickel, Inconel	800–1350

[a]Recommended temperature range limited by fluid properties and strength of wall materials.

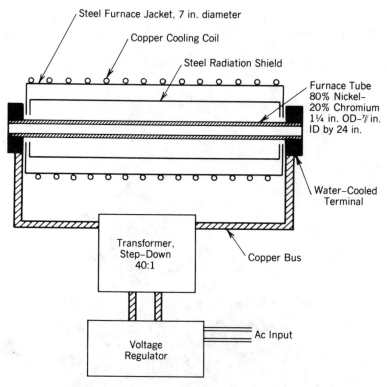

Steel Furnace Jacket, 7 in. diameter

Copper Cooling Coil

Steel Radiation Shield

Furnace Tube
80% Nickel–
20% Chromium
1¼ in. OD–⁷⁄ in.
ID by 24 in.

Water–Cooled
Terminal

Transformer,
Step–Down
40:1

Copper Bus

Voltage
Regulator

Ac Input

Figure 17.15. Low thermal mass furnace. (From "1984 Annual Book of ASTM Standards," 1984. Reprinted with permission from the American Society for Testing and Materials, 1916 Race Street, Philadelphia, PA 19103.)

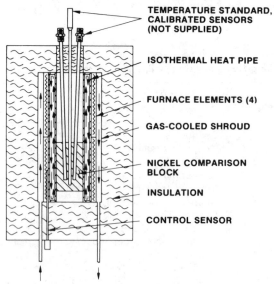

TEMPERATURE STANDARD,
CALIBRATED SENSORS
(NOT SUPPLIED)

ISOTHERMAL HEAT PIPE

FURNACE ELEMENTS (4)

GAS-COOLED SHROUD

NICKEL COMPARISON
BLOCK

INSULATION

CONTROL SENSOR

Figure 17.16. A calibration furnace using a liquid sodium heat pipe for the range 250 to 1100°C. (Courtesy of Rosemount, Inc.)

that even a small amount of hot sodium metal explodes violently on contact with air should the container rupture or air be emitted.

17.18 LOW-TEMPERATURE BLACKBODY SOURCES

Low-temperature blackbody sources are commercially available for calibration of infrared pyrometers. They usually have a copper cavity heated electrically to produce a uniform temperature and a window aperature designed to emit with an emittance greater than 0.98. They are painted flat black inside. The most important features are the temperature regulation and uniformity and the design of the aperature and exit cone.

17.19 HIGH-TEMPERATURE BLACKBODY SOURCES

High-temperature blackbody sources are needed for silver, gold, copper, paladium, platinum, rhodium, and similar melting points. One of the most sophisticated designs includes a sodium heat pipe to ensure temperature uniformity (Fig. 17.17).[17]

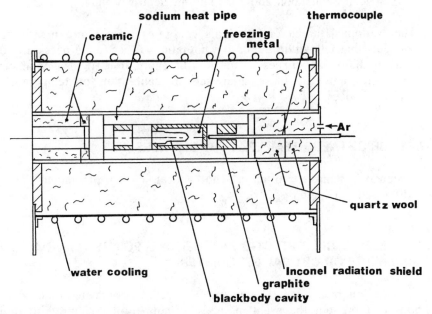

Figure 17.17. Design of a high-temperature metal freezing-point furnace. [From M. Ohtsuka and R. E. Bedford, "Measurement of the thermodynamic temperature interval between the freezing points of silver and copper, in "Temperature: Its Measurement and Control in Science and Industry" (James F. Tooley, Ed.), Vol. 5, Part 1, American Institute of Physics, New York, 1982, p. 176.]

17.20 NATIONAL BUREAU OF STANDARDS CALIBRATION SERVICES AND STANDARD REFERENCE MATERIALS

High-purity materials for primary standards are available from the U.S. National Bureau of Standards. NBS periodically issues Special Publication 260, which describes the standards, and publishes an annual price list. The properties of the platinum standard for resistance and thermocouple applications are reviewed in PB-277 172 as Pt-67 or SRM-1967.

The NBS, as a part of its Standard Reference Materials Program, publishes information about their application as standards, including temperature measurement. Important recent publications include:

NBS Special Publication 260-77, "Application of Some Metal SRM's as Thermodynamic Fixed Points," 1982.

NBS Special Publication 260-62, "Temperature Reference Standard for Use Below 0.5 K," 1979.

NBS Special Publication 260-56, "Standard Reference Materials: Standard Thermocouple Materials, Pt67: SRM 1967," 1978.

NBS Special Publication 260-44, "Preparation and Use of Superconductive Fixed Point Devices," 1972.

NBS Special Publication 260-34, "Thermoelectric Voltage," 1972.

The National Bureau of Standards calibrates temperature measuring instruments that fall within the design limitations it has found necessary to ensure reliability, and issues certificates of calibration. It also conducts training programs for those interested in calibration of temperature measuring devices.

17.21 BRIGHTNESS TEMPERATURE SOURCES

Tungsten strip lamps are often used to calibrate optical pyrometers. They are discussed in Chapter 12.

17.22 ELECTRICAL DETECTION OF MELTING POINTS USING A WIRE BRIDGE OR OPTICAL PYROMETER

Sometimes laboratory experiments are, of necessity, conducted under conditions that prevent the usual methods of temperature calibration from being used. For example, a small specimen within a severe temperature gradient may be subjected to some other appropriate measurement, but the temperature sensor is subjected to immersion errors so that calibration is not suitable. Sometimes this can be solved with a thermocouple with the sensing junction bridged by a tiny length of a wire. The wire is selected to have a melting point near the desired temperature range. If a platinum type S thermocouple has a few millimeters of pure gold wire welded between the measuring junction ends, it will act satisfactorily as a thermocouple until the gold melts. A brief thermal arrest will occur just at the gold point. If the gold is supported by surface tension, electrical continuity will continue.[18] If the wire is too long and unsupported, the circuit will open and the EMF will be lost when melting occurs. Accuracy of 0.2°C is possible.[19]

A ribbon of pure metal can be heated slowly in a furnace and its temperature measured with an optical pyrometer. When the wire melts and opens the circuit, the melting point can be determined by extending a plot of apparent temperature versus time to the time of melting.

17.23 LIQUID-IN-GLASS THERMOMETER CALIBRATION

A lucid description of the laboratory equipment required and the procedures to be used for liquid-in-glass thermometer calibration is given by Wise and Soulen.[20]

REFERENCES

1. An excellent source of standards information and calibration methods is: "Supplementary Information for the IPTS-68 and the EPT 76," a monograph from Bureau International des Poids et Mésures, Pavillon de Breleuil, F92310, Sevres, France (1983). (Figure 17.1 is from p. 22 of the monograph).

2. E. H. McClaren and E. G. Murdock, The freezing points of high purity metals as precision standards. *Can. J. Phys.* **46** (Part 8b), 401–444 (1968).

3. Jarrett Instrument Company, Inc., 2910 Lindell Court, Wheaton, Maryland 20902.

4. J. L. Riddle, G. T. Furokawa, and H. H. Plumb, Platinum resistance thermometry. *NBS Monogr.* (*U.S.*) **126** (1973); 4b. G. T. Furokawa and W. R. Biggs, Reproducibility of some triple point of water cells. *Temp.: Its Meas. Control Sci. Ind.* **5** (Part 1), 291–297 (1982).

5. J. V. McAllen, The effect of pressure on the water triple point temperature. (*Temp.: Its Meas. Control Sci. Ind.* **5** (Part 1), 285–290 (1982).

6. F. Pavese and D. Ferri, Ten years of research on sealed cells for phase transition studies at IMGC. (*Temp.: Its Meas. Control Sci. Ind.* **5** (Part 1), 217–227 (1982).

7. F. Pavese, On the use of first-generation sealed cells in an international comparison of triple-point temperatures of gases. *Temp.: Its Meas. Control Sci. Ind.* **5** (Part 1), 209–215 (1982).

8. J. Bonhoure and R. Pello, Température du point triple du gallium. *Metrologia* **19**, 15–20 (1983).

9. Yellow Springs Instrument Company, Inc., Yellow Springs, Ohio 45387.

10. G. Bongiovanni, L. Crovini, and P. Marcarino, Effects of dissolved oxygen and freezing techniques on the silver point. *Metrologia* **11**, 125–132 (1975).

11. Kaye Instruments, 15 DeAngelo Drive, Bedford, Massachusetts 01730.

12. Isotech Technology Ltd., Pine Grove, Southport, Merseyside PR9 9AG, England.

13. American Society for Testing and Materials, "Standard Method for Calibration of Thermocouples by Comparison Techniques," ASTM E220-86. Philadelphia, Pennsylvania, 1986.

14. Rosemount, Inc., 1256 Trapp Road, Eagan, MN 55121.

15. C. A. Brusse, J. P. Labrande, and C. Bassini, The gas-controlled heat pipe: A temperature-pressure transducer. *Conf. Ser.—Phys.* **26**, 428–438 (1973).

16. G. Neuer and O. Brost, Heat pipes for the realization of isothermal conditions at temperature reference sources. *Conf. Ser.—Inst. Phys.* **26**, 446–452 (1973).

17. M. Ohtsuka and R. E. Bedford, Measurement of the thermodynamic temperature interval between the freezing points of silver and copper. *Temp.: Its Meas. Control Sci. Ind.* **5** (Part 1), 175–181 (1982).

18. C. K. Ma, M. Ohtsuka, and R. E. Bedford, Novel method for measuring high temperature furnace gradients. *Rev. Sci. Instrum.* **51**, 52–54 (1980).

19. P. I. Roberts, The importance of thermocouple calibrations. *Meas. Control* **13**, 213–217 (1980).

20. J. A. Wise and R. A. Soulen, Jr., Thermometer calibration: A model for state calibration laboratories. *NBS Monogr*, (*U.S.*) **74** (1986).

18

INSTALLATION EFFECTS

18.1 INTRODUCTION

Every temperature measurement situation is also a heat transfer problem. The presence of the sensor may change the temperature. There must be a temperature difference if a sensing element is to be heated, so dynamic considerations (Chapter 19) also require understanding of heat transfer principles. The measurement itself extracts heat. This could be through an infrared window, a thermoelectric current, or conduction up the stem of a thermometer, but must occur in order to produce a temperature signal. It is beyond the scope of this book to consider heat transfer in detail, and the reader is referred to standard texts for that purpose. But some of the errors resulting from heat transfer considerations and some of the simple methods of analysis will be discussed to illustrate the importance of sensor design on a temperature measurement.

18.2 THE PURPOSE OF A TEMPERATURE MEASUREMENT

Selection of a sensor must conform to the requirements imposed by the purpose of a temperature measurement. For example, a sensor hanging in front of a window on a cold winter day, heated by convection from a radiator below the window, will indicate a temperature modified by radiation out through the window, cold air descending next to the window and hot air rising from the radiator. If the purpose is to measure air temperature, it should be shielded from radiation out of the window, and its location in the convection currents should be selected for the type of air temperature desired. If personal comfort is to be measured, the position should be changed to a location people might occupy, but radiation out of the window should not be excluded. Always keep in mind what the purpose of the measurement is in selecting a sensor and installing it.

18.3 THERMAL MASS

When a cold thermometer is inserted into hot water, it cools the water in proportion to its mass m, specific heat C_p, and the temperature difference between t_1 before insertion and its final temperature t_2. The product mC_p is called the thermal mass. For our thermometer, the enthalpy required to heat it is given by

$$\Delta H = \int_{t_1}^{t_2} mC_p \, dt \, . \qquad (18.1)$$

If specific heat is constant, this is simply

$$\Delta H = mC_p(t_2 - t_1) \, . \qquad (18.2)$$

Whenever the system is tiny, the sensor must be chosen so that its thermal mass is also tiny, just to prevent altering the system. The mass and specific heat of the measuring system can be estimated and Eq. (18.2) can be used to estimate the effect of introducing a sensor on the temperature of the system. In some systems, such as living systems, the temperature change accompanying insertion of the sensor can be critical. In other systems a long time will be required for a system to return to equilibrium if a massive sensor has been inserted, altering the temperature. Therefore, any sensor should have a thermal mass compatible with the application.

18.4 IMMERSION ERRORS (INSERTION ERRORS)

If a sensor, such as our thermometer stem, emerges from the system being measured, the sensor will provide a heat conduction path. Heat transfer will occur also by radiation and convection from the stem. Radiation may also occur from the sensing element because its emittance may be high compared to the emittance from the system without the sensing element. Every such situation is complex and each one must be considered individually. All these real situations are too complex for accurate mathematical description. However, certain models have been developed which allow the heat transfer effects to be calculated. Engineering judgment is needed to evaluate the simplified models for a particular application. Sometimes a model can be

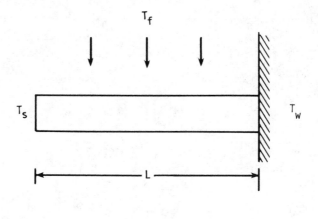

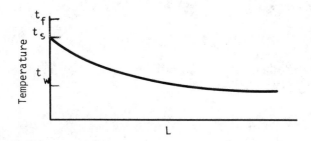

Figure 18.1. Immersion error model; a long, thin rod.

used and the error expected evaluated experimentally before it is accepted as a reasonable approximation.

The simplest model is the one for immersion error (Fig. 18.1).[1] Let us assume that the sensor is a small slender rod projecting through the wall of a vessel to measure the temperature of the fluid in the vessel. If the fluid is hot, heat is conducted to the vessel wall along the length of the rod, cooling the sensor. Heat is delivered to the sensor by convection from the fluid. In this model radiation is neglected. Then for any location along the length of the rod the hot fluid will deliver heat to the rod which will be conducted to the cooler wall. The general Fourier heat transfer equation in one dimension (assuming that there is no radial gradient in the rod) becomes

$$\frac{d^2t}{dx^2} = \frac{Q}{k},$$

(18.3)

where Q is the heat transferred by convection per unit volume and k is the thermal conductivity of the rod. Using Newton's law of cooling to represent the heat transfer, this becomes

$$\frac{d^2t}{dx^2} = \frac{2\pi rh(t_f - t)\,dx}{\pi r^2\,dx\,k} \tag{18.4}$$

$$= \frac{2h}{rk}(t_f - t). \tag{18.5}$$

Here h is the convective heat transfer coefficient.
It is convenient to substitute $\phi = (t_f - t)$. Then

$$\frac{d^2\phi}{dx^2} = \frac{2h\phi}{rk}. \tag{18.6}$$

This has the solution

$$\phi = C_1 e^{\sqrt{(2h/rk)}x} + C_2 e^{-\sqrt{(2h/rk)}x}. \tag{18.7}$$

If the rod is very thin, so that the end loss can be neglected, the slope dt/dx is zero at the very end. Then

$$\frac{d\phi}{dx} = 0, \qquad x = L. \tag{18.8}$$

Taking the derivative of Eq. (18.7) and applying 18.8 gives

$$\frac{d\phi}{dx} = C_1 \frac{2h}{rk} e^{\sqrt{(2h/rk)}L} - C_2 \frac{2h}{rk} e^{-\sqrt{(2h/rk)}L} = 0. \tag{18.9}$$

At the wall, where $x = 0$, the value of ϕ is $(t_f - t_w)$. Substituting this boundary condition in Eq. (3.17) gives

$$\phi_w = (t_f - t_w) = C_1 + C_2. \tag{18.10}$$

Dividing Eq. (18.7) by (18.10) and substituting the values of C_1 and C_2 from Eq. (18.9) gives

$$\frac{\phi}{\phi_w} = \frac{e^{\sqrt{(2h/rk)}(L-x)} + e^{-\sqrt{(2h/rk)}(L-x)}}{e^{(2h/rk)L} + e^{(2h/rk)L}}$$

$$= \frac{\cosh[\sqrt{(2h/rk)}(L-x)]}{\cosh[\sqrt{(2h/rk)}L]}. \tag{18.11}$$

For the extreme end of the rod, where $x = L$, the error in temperature measurement, from Eq. (18.11), is

$$\phi_{\text{error}} = t_f - t_s = \frac{\phi_{\text{wall}}}{\cosh[\sqrt{(2h/rk)}L]} \tag{18.12}$$

$$= \frac{t_f - t_w}{\cosh[\sqrt{(2h/rk)}L]}, \tag{18.13}$$

where t_s is the sensor temperature.

From this result we see that the error is directly proportional to the difference between the fluid temperature and the wall temperature, so reducing the thermal gradient between them should help. It is often feasible to insulate the wall at the sensor and raise its temperature to reduce the error. Also, we see that reducing the radius and the thermal conductivity of the rod is desirable, and increasing the heat transfer coefficient and the rod length is desirable. Often, the heat transfer coefficient can be increased by increasing the velocity of the fluid flowing over the rod.

Notice that this derivation does not represent many real factors that usually exist. Radiation has been neglected. The long, thin rod does not represent the true geometry of many sensors. For example, a thermocouple is usually two insulated wires. The sensor is often in a massive well not satisfactorily represented by the long, thin model. Heat transfer to a well must then be accompanied by heat transfer to the sensor. The actual shape and location of the sensor, its thermal properties, and many other factors then become important. With all its limitations, however, this simple model is useful.

Insertion errors are of a great deal of concern. If they are known, they can be compensated for in many applications. Immersion errors in calibration devices are tested by varying the immersion depth, increasing the depth into an isothermal enclosure until no change is detected with additional immersion. When the long, thin rod model is appropriate, the temperature change with immersion depth should conform to Eq. (18.11).

18.5 INSULATED WIRE MODEL

Sparrow has analyzed the effect of insulating a single wire, or a double wire, on the heat transfer from a sensor such as a thermocouple.[2] For the single wire he assumed that the axial heat transfer in the insulation was zero, and that the central wire had no radial thermal gradient. The radial heat flow through the insulation couples the convective heat flow into the insulation with the axial heat flow out through the wire. His result is that the radial heat flow is given by

$$\frac{dQ_r}{dx} = \frac{t_f - t}{(1/2\pi r_i h) + \ln(r_i/r_w)/2\pi k_i}, \tag{18.14}$$

where the i subscript indicates insulation and the w subscript indicates wire. Clearly, the presence of the insulation decreases the rate of heat transfer and increases the error. For the two-wire model, Eckert assumed that the wires were at the same temperature at a given x, used the sum of the thermal conductivities of the two wires with the appropriate area, and used the appropriate weight average values for the external and internal insulation.

$$r_i' = \frac{L_1 + L_2}{4} \tag{18.15}$$

$$r_w' = \sqrt{2}\,r_w\,, \tag{18.16}$$

where L_1 and L_2 are the major and minor outside diameters of the insulation. These he substituted in Eq. (18.14) to show the effect of the insulation on the rate of heat transfer.

18.6 ERRORS FROM RADIANT HEAT TRANSFER

Radiant exchange with a sensor may change the indicated temperature. Typically, a sensor is placed in a convective environment with cooler walls. Then radiation from the sensor to the walls is possible. Since these are not blackbody conditions, the emittance of the sensor and the wall must be considered. The heat is transferred in accordance with the Stefan–Boltzmann equation, per unit area,

$$\dot{Q} = \sigma(e_h T_h^4 - e_c T_c^4)\,, \tag{18.17}$$

where σ is the Stefan–Boltzmann constant, e is emittance, and T is absolute temperature. The h subscript indicates hot and the c indicates cold.

The literature abounds with special cases, such as the flow of steam in pipes, in which the radiation heat transfer has been represented by a linear approximation parallel to that for convection,

$$\dot{Q} = h_r A(T_r - T_c)\,, \tag{18.18}$$

where h_r is a radiation coefficient and A is the external area. This is obviously a poor approximation but is used where radiation is not very great, where the temperature range to be represented is small, and where a linear relationship allows other mathematical models to be utilized. Care should be exercised when using the linear approximation that the error introduced is acceptable.

A model often used to evaluate error attributed to radiation losses is that of a sensor heated by convection in an enclosure (Fig. 18.2). The con-

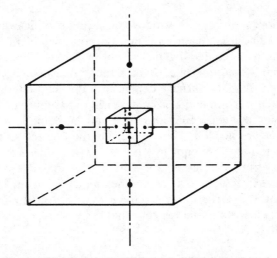

Figure 18.2. Model of radiation loss from a hot sensor.

vective heat transfer to the surface can be equated to the radiant loss at steady state if the conduction from the sensor to the shell is negligible and the gas is transparent. Then

$$\dot{Q}_c = \dot{Q}_R$$
$$h_c A(T_f - T_s) = \sigma A(e_s T_s^4 - e_w T_w^4). \tag{18.19}$$

Solving for the difference between the fluid and the sensor temperature gives

$$T_f - T_s = \frac{\sigma}{h_c}(e_s T_s^4 - e_w T_w^4). \tag{18.20}$$

The error depends primarily on the convective heat transfer coefficient, the sensor temperature to the fourth power, and the sensor emittance. Increasing the heat transfer coefficient and decreasing the sensor emittance will reduce the error.

18.7 RADIATION METHODS OF MEASUREMENT

Radiant methods of temperature measurement require a window or opening for the sensing device. If there is an opening, or if the window leaks, a hot furnace can be cooled locally by the inrush of air if the furnace is under negative pressure. Most fossil-fuel-fired furnaces are under negative pressure, so this effect is frequently encountered. Use of a target tube is often a suitable cure.

Radiant leaks through a window can also significantly reduce target temperature. The loss depends on the transmittance of the window, not on the wavelengths accepted by the sensor. Usually, the Stefan–Boltzmann law, or Planck's law integrated over a range of wavelengths, applies. The heat radiated out of the furnace depends on the solid angle available to the target. Reducing the opening size, increasing the length of a target tube, or moving the target deeper into a furnace may be helpful. When intermittant measurements are needed, a shutter, inside the furnace so that it is heated, will prevent radiant loss until it is opened. For a massive target inside the furnace measuring the temperature as a function of time the shutter is opened, and extrapolating to zero time, should give an indication of the magnitude of the error. The response time of the radiant temperature measuring device is a limitation that should be recognized.

18.8   TEMPERATURES OF RAPIDLY MOVING FLUIDS

A temperature sensor in a rapidly moving fluid will read a different value if moving along with the fluid than it will if it is standing still because the kinetic energy of the fluid will raise the temperature of the fluid if it is stopped. We call the temperature of the fluid, unarrested, the fluid temperature t_f. If the fluid is stopped completely under adiabatic conditions, it will reach an adiabatic temperature t_a. This is often called the stagnation temperature or the recovery temperature. The actual temperature measured is usually between t_f and t_a and the latter two terms properly apply to the actual temperature under the actual stagnation conditions. If all the kinetic energy is converted to thermal energy,

$$t_a \doteq t_f + \frac{v^2}{2JgC_p}, \tag{18.21}$$

where v is velocity, g is gravitation constant, C_p is specific heat at constant pressure, and J is mechanical equivalent of heat. This equation applies only for fluids that behave like ideal gases. Real fluids, even when completely stopped under adiabatic conditions, do not deliver the temperature rise predicted by Eq. (18.21). In fact, both positive and negative deviations are possible. Therefore, a stagnation factor is used to describe the temperature rise. The stagnation factor times the dynamic recovery term in Eq. (18.21) gives the actual rise at complete stagnation. Actual devices that do not completely stagnate the fluid will give smaller temperature rise. The reader is advised to consult the literature for the conditions appropriate for a particular situation.

To reduce immersion errors in moving fluids it is often feasible to put the sensing element into the fluid stream at an elbow, or to insert the sensor so that it has a projection downstream in the flow. Then the sensor supports are heated (or cooled) by the fluid raising (or lowering) their temperature to reduce the temperature gradient from the sensor to the supports, thus reducing sensor error.

18.9 SURFACE TEMPERATURE MEASUREMENT

Surface temperatures are often difficult to measure because heat is extracted or delivered if the surface is touched by a sensor. Noncontact radiation techniques are often practical, although some heat exchange occurs through the opening for the detector.

Spring contacts and rolling contacts are available from the thermocouple and thermistor manufacturers. Bare sensors and very low thermal mass sensors help reduce the heat transfer. Sometimes a thermocouple of extremely fine wire can be welded to a surface to detect very small temperature changes. For example, Sandor and Jordan[3] welded type E thermocouples of 0.13-mm wire diameter to a metal airfoil skin using a 2500-μF capacitor charged to 20 V. They were able to measure transient temperature pulses, induced in the metal by plastic deformation, of only 0.1°C at 10-ms intervals.

It is a common technique to embed the sensor wires in grooves in the surface, or in holes drilled almost to the surface from the back side, to attempt to reduce immersion error. Thin-film resistance elements are supplied with sticky tape to attach them to a surface. Taping wires to a surface can reduce conduction losses. Sensors are also provided with the sensing element mounted on a plate that can be bolted to a surface. The plates are often copper, aluminum, or stainless steel.

A very fine way to measure constant surface temperatures is to use a flat metal plate in contact with the surface. If a sensor is placed on each side and a heater or cooler is used on the side away from the surface to be measured, the true temperature of the surface will be obtained when both sensors indicate the same temperature. Then no heat is conducted to or from the surface.

18.10 CONSUMABLE THERMOCOUPLES

Sometimes it is necessary to measure the temperature of an object that is being consumed or destroyed during the measurement time. For example, a furnace wall may be slowly abraded or dissolved away in an industrial process. This will usually destroy an embedded sensing element. For the *Apollo* vessel the heat shield ablated during reentry into the earth's atmosphere. Special thermocouples were made to embed in the heat shield. The wires were fine enough and close enough together so that they were continuously rewelded as they burned away.[4]

The temperature of molten metal is often measured with small, throwaway thermocouples. The manufacturer supplies them in a quick-change assembly to expedite using them for molten metal testing.[5]

18.11   SHIELDING FROM THERMAL RADIATION

When a blackbody wall is interposed between two other blackbody walls that are radiating toward each other, the presence of the wall between them reduces the radiation exchange between the hot and the cold outer walls by 50%. If the emittance is less than 1, the exchange rate is reduced further. For this reason thin metal radiation shields are often used to reduce the rate of radiant heat transfer. For fluid flow past a sensor, the radiation shields are often one or more concentric tubes aligned in the direction of fluid flow and extending far enough forward and backward in the flow direction to reduce the solid angle of open radiation to a small value.

18.12   HIGH-TEMPERATURE GASES

The temperature of high-temperature gases is measured with standard devices with suitable protection tubes if the furnace atmosphere and the temperature are compatible with a protection tube system. Radiation methods can sometimes be used for higher temperatures. If a flame is not highly radiant, very long optical paths are required. Then special methods, such as Schlieren, laser Raman spectroscopy, or calorimeter probes may be suitable.

A calorimeter probe is a water-jacketed cylindrical tube through which the hot gas can be pumped. The gas is cooled as it passes through the tube until it reaches a temperature low enough to permit measuring its temperature with a standard sensor. At steady-state conditions the water jacket inlet and outlet temperatures are also measured. Then the gas temperature at the entrance can be calculated. The water jacket serves as a calorimeter. The heat extracted is calculated from the water flow rate and the inlet and outlet temperature difference. The temperature of the gas at the inlet is calculated from the sensor temperature. The tare technique, where measurements are made with and without gas flowing through the central tube, makes it possible to reduce the errors caused by external heating of the water jacket. Then the enthalpy change of the gas can be calculated and the gas temperature calculated from it, the thermodynamic properties of the gas, the gas flow rate through the probe, and the sensor temperature.[6]

18.13   SUMMARY

The method selected to measure temperature and the way the sensor is installed must be carefully considered if the temperature measurement is of significance. In choosing the sensor and the installation method, some of the factors that must be considered are:

1. Purpose of the measurement
2. Types of sensors that are compatible with the sensor environment during the measurement
3. Response time required
4. Heat capacity of the system to be measured and the heat capacity of the sensing system
5. Mechanisms of sensor error through conduction convection of radiation, and their relative and absolute importance
6. System size and geometry, and sensor size and geometry
7. Suitability of the sensor signal for control or other functions
8. Cost

REFERENCES

1. E. R. G. Eckert and R. M. Drake, Jr., "Heat and Mass Transfer," p. 39 McGraw-Hill, New York, 1959.
2. E. M. Sparrow, Error estimates in temperature measurement. *In* "Measurements in Heat Transfer" (E. R. G. Eckert and R. J. Goldstein, eds.), 2nd ed., Chapter 1. Hemisphere Publ. Corp., Washington, D.C., 1976.
3. B. I. Sandor and E. H. Jordan, "Temperature Based Stress Analysis of Notched Members," Final Report, AFFDL-TR-78-176. Wright Patterson Air Force Base, Ohio, 1979.
4. Pyrodyne Division, William Wahl Corporation, 1001 Colorado Avenue, Santa Monica, California, 90404.
5. Leeds and Northrup Company North Wales, Pennsylvania 19454.
6. J. Grey, Probe measurements in high temperature gases and dense plasmas. *In* "Measurements in Heat Transfer" (E. R. G. Eckert and R. J. Goldstein, eds.), 2nd ed., Chapter 7. Hemisphere Publ. Corp., Washington, D.C., 1976.

19

DYNAMIC RESPONSE OF SENSORS

The selection and design of sensors must consider the dynamic response characteristics needed for the application. If a sensor must respond in microseconds (as·for the temperature rise of a shock tube) the dynamic response requirement precludes using most methods of temperature measurement. Slow response requirements are not usually a problem. Rapid response requirements often are. Sensors with rapid response are often fragile and it is frequently difficult to satisfy both the need for rapid response and the need for longevity. Only after we understand the dynamics of sensor response can we consider the other factors in sensor selection. We start with a simple analysis and proceed to more complex situations, but only give an introduction to the subject. A great deal of information is available in heat transfer and instrumentation literature and should be consulted where a more thorough treatment is needed.[1-3]

19.2 CONVECTION COOLING A RODLIKE SENSOR

The simplest model of a sensor for analysis of dynamic effects is the same as used in Chapter 18 for heat transfer modeling. It is a long, slender rod heated or cooled by convection under conditions that allow radiation to be neglected (Fig. 19.1).

If we assume that the rod has a high thermal conductivity, so that its internal temperature is radially uniform, and if it is suddenly plunged into

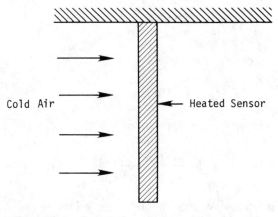

Figure 19.1. Convention cooling a long, thin rod.

the cooling airstream, the heat extracted by convection can be equated to the heat delivered as the rod is cooled. For a unit length of rod, for convection,

$$\dot{Q}_c = h_c A(t_\theta - t_\infty),$$

and for heat delivery,

$$\dot{Q}_a = -V\rho C_p \frac{dt}{d\theta}.$$

Equating gives

$$h_c A(t_\theta - t_\infty) = -V\rho C_p \frac{dt}{d\theta},$$

where A is area and V is volume. In terms of radius,

$$2\pi r h_c(t_\theta - t_\infty) = \pi r^2 \rho C_p \frac{dt}{d\theta}, \tag{19.1}$$

where r is the sensor radius, h_c is the convective heat transfer coefficient (a function of sensor shape and gas velocity), t_θ is the temperature at time θ, t_∞ is the temperature of the gas (assumed constant), ρ is the sensor density, and C_p is its specific heat at constant pressure. Separating variables and integrating gives

$$\frac{dt}{t_\theta - t_\infty} = \frac{-2h_c}{r\rho C_p} d\theta \tag{19.2}$$

$$\ln(t_\theta - t_\infty) = \frac{-2h_c}{r\rho C_p}\theta + C.$$ (19.3)

For the boundary condition $\theta = 0$, $t_\theta = t_0$, so

$$C = \ln(t_0 - t_\infty).$$ (19.4)

Substituting for C gives

$$\ln\frac{t_\theta - t_\infty}{t_0 - t_\infty} = \frac{-2h_c}{r\rho C_p}\theta.$$ (19.5)

It is often convenient to use the fraction change in temperature difference, ϕ, as a reduced variable. Then

$$\ln\phi = -\left(\frac{2h_c}{r\rho C_p}\right)\theta.$$ (19.6)

The term in parentheses on the right has the units of reciprocal time. The time constant for the change, τ, is given by

$$\frac{1}{\tau} = \frac{2h_c}{r\rho C_p}$$

or

$$\tau = \frac{r\rho C_p}{2h_c}.$$ (19.7)

Then the general equation for the cooling of the sensor is

$$\phi = e^{-t/\tau}.$$ (19.8)

A plot of the cooling curve is shown in Fig. 19.2. Plotting $\ln\phi$ as a function of time allows us to obtain the time constant, τ (Fig. 19.3).

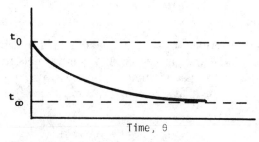

Figure 19.2.   Cooling curve for a small, thin rod.

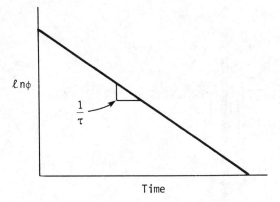

Figure 19.3. In ϕ as a function of time.

Many temperature measuring devices can be reasonably well represented by this simple model. A bare wire thermocouple, a thermistor bead, and other sensors having small size, high thermal conductivity, and being used at temperatures where convection dominates fit this model well. In instrumentation literature this is considered a first-order response.

19.3   SENSORS IN ENCLOSURES

When a sensor is placed inside an enclosure, such as a thermocouple in a ceramic tube or a resistor in a glass envelope, the kinetics become more complex. We will use a second simple model of a thermometer resting on the bottom of a thermometer well (Fig. 19.4).

For this model heat must be removed from the well by convection, and the thermometer bulb must be cooled by conduction. Then, assuming that the well temperature at any time is uniform throughout, heat is transferred from the well to the air by convection as in the previous model.

$$\dot{Q}_1 = h_c A_w (t_w - t_a) = V_w \rho_w C_{p_w} \frac{dt_w}{d\theta}. \qquad (19.9)$$

Heat is transferred from the thermometer bulb to the well at a different rate, by convection, according to

$$\dot{Q}_2 = M(t_w - t_s), \qquad (19.10)$$

where M is the contact heat conduction coefficient between the well and the thermometer bulb. Again, assuming that the thermometer bulb is homogeneous in temperature at any time, the heat delivered cools the

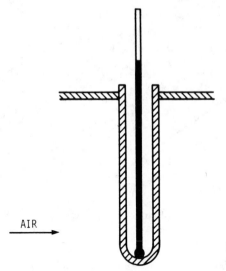

Figure 19.4. Thermometer in a thermometer well suddenly cooled by the air.

AIR

thermometer according to

$$\dot{Q}_2 = V_s \rho_s C_{p_s} \frac{dt_s}{d\theta}. \tag{19.11}$$

Solving for T_a from Eq. (19.9) gives

$$t_a = t_w - \frac{V_w \rho_w C_{p_w}}{h_c A_w} \frac{dt_w}{d\theta} \tag{19.12}$$

$$= t_w - \tau_w \frac{dt_w}{d\theta}, \tag{19.13}$$

where τ_w has been substituted as in Eq. (19.2). Similarly, for Eqs. (19.10) and (19.11),

$$t_w = \frac{V_s \rho_s C_{p_s}}{M} \frac{dt_s}{d\theta} + t_s \tag{19.14}$$

$$= \tau_s \frac{dt_s}{d\theta} + t_s. \tag{19.15}$$

Substituting the value of t_w from Eq. (19.15) into (19.13) eliminates the wall temperature to give us the overall response,

$$t_a = \tau_s \frac{dt_s}{d\theta} + t_s + \tau_w \tau_s \frac{d^2 t_s}{d\theta^2} + \tau_w \frac{dt_s}{d\theta} \qquad (19.16)$$

$$t_a = \tau_w \tau_s \frac{d^2 t_s}{d\theta^2} + (\tau_s + \tau_w) \frac{dt_s}{d\theta} + t_s . \qquad (19.17)$$

Equation (19.17) can be solved by various methods, including the Laplace transform,[4] to get

$$\phi = \frac{t_s - t_\infty}{t_0 - t_\infty} = 1 - \frac{1}{\tau_w - \tau_s} (\tau_w e^{-\theta/\tau_w} - \tau_s e^{-\theta/\tau_s}) . \qquad (19.18)$$

In Eq. (19.18) we have substituted t_∞ for t_a to make it parallel to Eq. (19.8). Note that if τ_w is zero, Eq. (19.18) reduces to Eq. (19.8). We have two relaxation times in Eq. (19.18) because of the thermal resistance between the thermometer well and the thermometer. If the thermometer did not touch bottom, natural convection within the well would need to begin (neglecting radiation) before heat could be transferred to the thermometer. This would cause a delay. Then after convection is established it would have a time constant τ_s just like τ_w above, but much larger because h_c in the well would be very small. Then a delay time would need to be added to Eq. (19.18). The net result would be as shown in Figs. 19.5 and 19.6.

As shown in Fig. 19.6, if the relaxation time for the well is much shorter than that for the sensor, so that the well cools first and then the sensor cools, it may be possible to determine the relaxation times from a logarith-

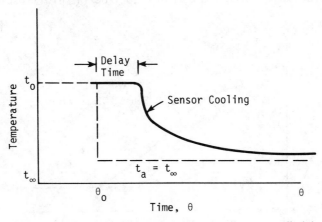

Figure 19.5. Sensor cooling with delay time plus cooling controlled by two time constants for a step change in air temperature.

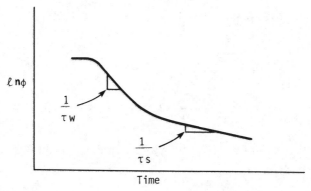

Figure 19.6. Schematic plot of ln ϕ versus θ for two widely different relaxation times.

mic plot. Usually, however, the plot gives an irregular curve without linear regions. In instrumentation literature this model is described as a second-order response with time delay.

19.4 RESPONSE TO A RAMP CHANGE

Temperature sensors are often used with a controller to provide a constant heating rate. A constant heating rate is often called a ramp function. A single sensor response as modeled in Section 19.2 will have a single time constant to describe its temperature response. For a ramp change for one temperature to another, where the temperature during the change is given by $t_a = t_0 - k\theta$, the sensor response will be as shown in Fig. 19.7.

The temperature of the sensor lags behind the air temperature because of the thermal capacitance of the sensor and the thermal resistance of the

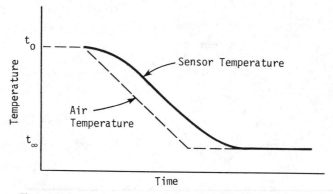

Figure 19.7. Response of a bare sensor to a ramp change.

convective heat transfer represented by the time constant τ. The response of the sensor is given by

$$t_s = t_0 + \frac{1}{\tau} e^{-\theta/\tau} \int_0^\theta (t_0 - k\theta) \, e^{\theta/\tau} \, d\theta \, . \qquad (19.19)$$

At any time after the transient of $e^{-\theta/\tau}$ approaches zero, the sensor lags behind the air temperature by

$$t_s - t_a = k\tau \, . \qquad (19.20)$$

As the ramp approaches the new value of temperature t_∞, a second curved transient response occurs. This effect is often seen in real temperature measuring situations.

If a thermometer well is used, the shape of the two transient sections will depend on the relative values of the two time constants. When the transient affects are gone the sensor will lag behind the air temperature by

$$t_s - t_a = k(\tau_w + \tau_s) \, . \qquad (19.21)$$

19.5 REDUCING THERMAL LAG TIME

Reduction of thermal time constants is necessary whenever the thermal response is slower than needed for the purpose at hand. It is very important to recognize that the temperature response for a particular application may be limited by other components, not by the sensor. If the instrumentation response, the power response, or anything else in the response cycle has a longer time constant than the sensor, the longest response time will control and nothing done to short response components will bring an improvement. In a control cycle the longest time constant dominates. This is discussed further in Chapter 20.

To reduce response time of a sensor, changes that reduce its thermal mass (its thermal capacity) will be helpful. Thermocouples, flake thermistors, bead thermistors, and thin-film resistance detectors are commercially available with extremely short time constants. In comparing response times it is important to recognize that some companies rate the response time as the characteristic "relaxation" time we have been using. Some use the time to achieve 50% of the change for a step function, and some use the time to achieve 90, 95, or 99% of the change for a step function. When the characteristic time constant, or relaxation time, in Eq. (19.8) is used, the value of ϕ when the time equals the time constant is given by Eq. (19.22). When $\theta = \tau$,

$$\phi = e^{-\tau/\tau} = e^{-1} = \frac{1}{2.718} = 0.368 .$$

(19.22)

When $\phi = 0.90$, the time is given by

$$\ln \phi = \ln 0.9 = -\frac{\theta}{\tau}$$

$$\theta = -\tau(\ln 0.9) .$$

(19.23)

Radiation instruments with photodetectors or thin-film detectors often have rapid rise time.

When used in a thermal well the response time depends on the rapidity of heat transfer. Replacing a low conductivity well by a high thermal conductivity well, or using a grease or liquid to improve thermal contact, or machining the sensor or the well to improve thermal contact can be helpful. Reducing the thermal mass of the well will also help. For convective heat transfer, increasing the gas velocity will tremendously improve the heat transfer rate.

In many processes the response time of the sensor is adequate but the time required for heat to be transferred from the heating elements to the ware to the sensor is too long. If the heating elements must heat ware or furnace parts of high thermal capacity before the sensor can be heated, the control response may be too long. Then moving the sensor closer to the heating elements will shorten the response time.

19.6 DETERMINING SENSOR TIME CONSTANTS

The time constant for a sensor can be determined experimentally or calculated from Eq. (19.7). As an example, let us determine the time constant for a bare type K thermocouple cooling in air, and then calculate the convective heat transfer coefficient, h_c. Our experiment is set up as follows. We prepare a beaker of boiling water, insert the thermocouple into a thermometer well in the beaker, and allow it to reach equilibrium temperature. Then we quickly remove it to a stand so that it is suspended in the air, where it cools by convection. The EMF from the thermocouple is amplified with a small dc amplifier with a gain of 100 and is then recorded at a speed of 25 mm/s during the cooling period (Fig. 19.8). Values of ϕ and $\ln \phi$ are determined from the cooling curve and plotted (Table 19.1 and Figs. 19.8 and 19.9).

From the slope of the line in Fig. 19.9 the time constant was found to be

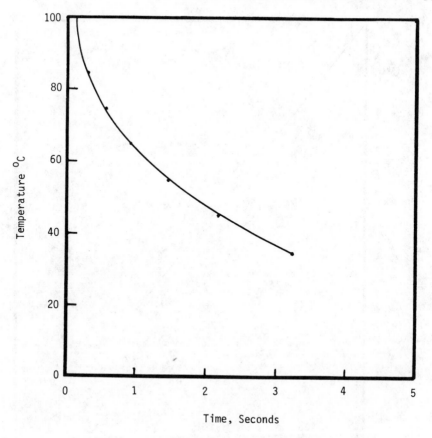

Figure 19.8. Cooling curve, bare type K thermocouple in air.

20.3 s, and it can be equated to its theoretical value from Eq. (19.7),

$$\frac{r\rho C_p}{2h_c} = 20.3 \text{ s} .$$

(19.24)

Our bare thermocouple is composed of two wires, somewhat apart,

TABLE 19.1

$t(°C)$	ϕ	$\ln \phi$	$\theta(s)$
85	0.800	−0.223	0.35
75	0.6667	−0.405	0.600
65	0.5333	−0.629	0.937
55	0.600	−0.916	1.438
45	0.2667	−1.322	2.175
35	0.1333	−2.015	3.250

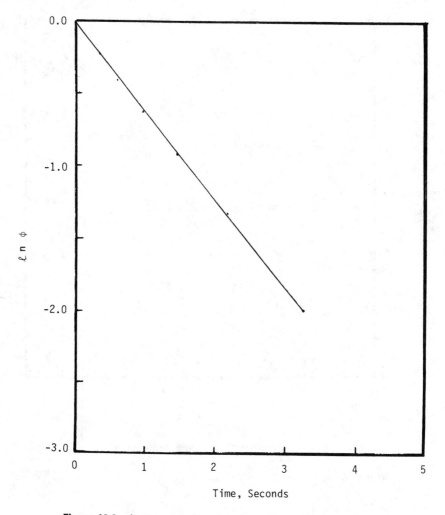

Figure 19.9. In ϕ versus time, bare type K thermocouple in air.

terminated at a weld bead. If we assume that the convective heat transfer is controlled by the wires adjacent to the bead, we can calculate h_c from τ and properties of the wire. Taking, for a 24-gauge thermocouple,

$$r = 0.0078 \text{ cm}$$

$$\rho = 8.9 \text{ g/cm}$$

$$C_p = 0.106 \text{ cal/g-}°\text{C}$$

and substituting in Eq. (19.24) gives

$$h_c = \frac{r\rho C_p}{2(20.3)}$$

$$= 0.00018 \text{ cal/cm}^2\text{-}°C\text{-s} . \tag{19.25}$$

This is a reasonable value when compared with values tabulated in the literature.

In performing such an experiment the temperature range tested should be the same as the temperature range for the application because the properties of materials change with temperature. Note, too, that this example applies at a temperature where radiation is relatively unimportant. Experiments at higher temperatures should be conducted under conditions where the emittances of the experiment are good approximations for the emittances of the process of interest.

19.7 CALCULATION OF THERMOCOUPLE TIME CONSTANTS

An experimental study of bare loop thermocouples, by R. J. Moffat, produced a general equation for the relaxation time of thermocouples of that configuration (Figs. 19.10 and 19.11).[5] He found

$$\tau = 3.5 \times 10^3 \rho C_p d^{5/4} G^{15.8/\sqrt{\tau}} \tag{19.26}$$

if the following conditions are met:

1. Bare loop configuration
2. Temperature 160–1600°F
3. Diameter 0.016–0.051 in.
4. Pressure = 1 atm.
5. Wire long enough to eliminate conduction effects

In his equation ρ is the average density of the two wires, C_p is the average specific heat, d is the diameter of the wires, and G is the mass flow rate. Condition 5 is met, when the flow rate G is greater than 5 lb/ft^2-s, if the wire length is at least five wire diameters. Lower flow rates require longer wires to reduce conduction error. If the wires are twisted, τ is increased. An effective diameter of $1.5d$ gives good representation. A large bead increases response time and can be compensated for by the equation

$$\frac{\tau}{\tau_0} = \left(\frac{D}{d}\right)^{3/8},$$

where D is the bead diameter.

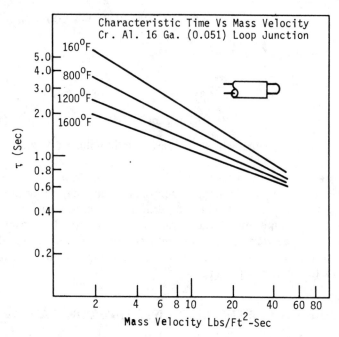

Figure 19.10. Log characteristic time versus log mass velocity for different fluid temperatures. [From R. W. Moffatt, How to specify thermocouple response, *'SA J.* **4**(6), 219–223 (1957).]

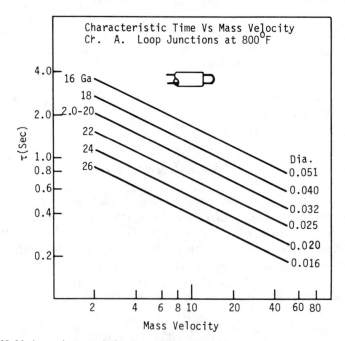

Figure 19.11. Log characteristic time versus log mass velocity for 800°F fluid temperature and different thermocouple diameters. [From R. W. Moffat, How to specify thermocouple response, *ISA J.* **4**(6) 219–223 (1957).]

Because there are so many different combinations of sensor and environment, it is difficult to make many generalizations about response times. Manufacturers usually provide some response-time information about sensors, and Moffat's findings allow one to make estimates of probable effects.

REFERENCES

1. L. C. Burmeister, "Convective Heat Transfer." Wiley, New York, 1983.
2. E. M. Sparrow and R. D. Cess, "Radiation Heat Transfer." McGraw-Hill, New York, 1978.
3. E. O. Doeblin, "Measurement Systems." McGraw-Hill, New York, 1983.
4. I. Lefkowitz, Methods of dynamic analysis. *ISA J*. **2**(6), 203–205 (1955).
5. R. J. Moffat, How to specify thermocouple response. *ISA J*. **4**(6), 219–223 (1957).

20

TEMPERATURE INSTRUMENTATION AND CONTROL

20.1 INTRODUCTION

The subjects of instrumentation and control are so complex that many books have been written on each subject. This chapter presents only a few of the basic ideas so that you will recognize when a temperature measurement system has instrumentation or control considerations in addition to simple temperature measurement considerations.

The need for temperature control is obvious, and the need for instruments to bring about that control is also obvious. The limits of control, the acceptable variations in temperature for a process, are less obvious. The types of instrumentation involve so many combinations of precision, cost, reliability, and utility that it is impossible to present a suitable analysis without analyzing a particular system. We will, rather, describe a typical control situation and present some of the concepts involved in the instrumentation.

20.2 THE FEEDBACK LOOP

Controls can be classified as feedforward and feedback devices. Feedforward control measures the process variables going into a system and controls them to bring about a desired result. Feedback control measures the output from a system and adjusts the process variables to maintain the desired result within acceptable limits. Feedback control is much more powerful than feedforward control because it provides corrective action to maintain the result at the desired level. The feedback of information to initiate corrective action, if necessary, is called the feedback loop. It appears not only in control of a process, but as an integral part of almost all the components used to control a process. We explain the feedback control loop in the case of a potentiometric recorder and then apply it to other components.

A potentiometric recorder for a thermocouple has an electric schematic as shown in Fig. 20.1. The voltage from the thermocouple TC has opposite polarity to the voltage produced by the IR drop across the calibrated potentiometer slide wire SW. Manual adjustment of the potentiometer by moving contact C so that the IR drop on the slide wire is equal and opposite in polarity to the voltage from the thermocouple would be possible if a galvanometer were provided to allow us to determine if current was flowing in the thermocouple circuit, as described in Chapter 10. In that manual process we would have moved the contact on the side wire left or right, depending on whether the IR drop was too large or too small until the galvanometer balanced. In other words, we would have detected an error signal, the difference between the thermocouple voltage and the potentiometer IR drop, and adjusted the potentiometer to bring that error signal to zero. We would be a part of the control loop because we read the galvanometer and took corrective action. That is, we served as comparitor and corrector.

The electrical schematic (Fig. 20.1) includes the same elements. The error signal appears at the input of the operational amplifier, which amplifies the error and drives a phase-sensitive motor to move the slide wire contact C just as we would have done manually. The slide wire is calibrated. Therefore, the setting can be displayed visually or electronically as needed. If the temperature of the thermocouple changes, an error signal appears which is amplified and used to correct the potentiometer setting.

The loop nature of this typical feedback device is shown in Fig. 20.2.

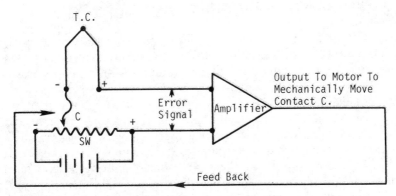

Figure 20.1. Electrical schematic diagram of a potentiometer feedback loop.

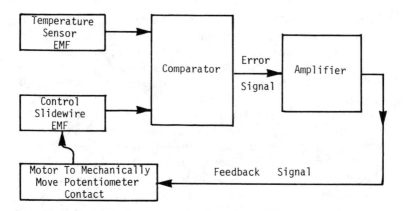

Figure 20.2. Block diagram of the functions in Fig. 20.1.

The feedback principle is very important. Note that this is negative feedback action. If the potentiometer setting is too high, the feedback reduces the potentiometer voltage.

Many components of proportional control must move a value, move an indicator setting, or make an adjustment where an electrical signal is sent, and the device responds to comply with that signal. Each of these components has built into it this negative feedback loop to assure that the setting will respond to the electrical signal. Note that the device will give a linear response to an electrical signal if the potentiometer is a linear resistance. Most of these components are driven by a low-voltage current. Many control signals are 4–20 mA. If the potentiometer has a resistance of 1000 Ω, this represents a signal of 4–20 V. Low voltages like this are not dangerous and require inexpensive wiring, so they are almost universally used in process instruments.

20.3 TYPICAL CONTROL SITUATION

Every control situation includes a number of components, for example, a temperature sensor, a measuring instrument, controller, a valve actuated by the controller, heat controlled by the valve, and a furnace with ware that responds to the heat. Each component will have a time constant associated with it. Usually, the temperature sensor and controller have short-time constants, and the load (furnace and heater) has a long time constant. If so, the response of the process is controlled by the longest time constant. If several parts have similar time constants, the response is very complex and unstable oscillations may result. Since each component of a process has its own time constant that relates to the capacitance and resistance of the component, it will be necessary to understand, and sometimes to modify,

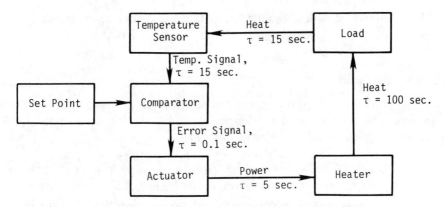

Figure 20.3. Typical temperature control loop.

the parameters for each component in order to produce a satisfactory system. Obviously, control of the process depends on all parts of the system, and a change to shorten the response time of an electronic component will be ineffective if some other component has a longer relaxation time.

20.3.1 A Typical Temperature Control Loop

Almost every temperature control situation can be described by a temperature control loop (Fig. 20.3). Such loops have positive gain with negative feedback and can oscillate. It is necessary to design the system to prevent such oscillations. This is usually done by choosing very different time constants for the various components, so that oscillations in one are not magnified by the next. It is also necessary to adjust the gain of loop components to prevent excess gain and subsequent oscillation.

The response of each component can be complex, and zero-, first-, and second-order responses are common. The transfer function from input to output can be described mathematically and represent a major portion of texts on control theory. The combined transfer function for the entire control loop is even more complex. Some modern computer control systems have over 100 different control algorithms stored and available for process control situations. A temperature controller might be one small part of such a process control system.

20.4 TEMPERATURE CONTROLLERS

Every controller must:

1. Receive a signal representing temperature.

2. Compare it with a signal representing the desired temperature.
3. Initiate corrective action.

20.4.1 Set Point

The signal received from the sensor can be any of those discussed in earlier chapters. The set point is the comparative signal which must be provided. The set-point signal can be achieved by:

1. Manual setting (simple and inexpensive)
2. A constant-speed motor-driven signal (ramp)
3. A combination of motor drives and clocks (ramp up, hold for time period, ramp down)
4. A cam-operated set point (replace cam—new temperature program)
5. An electronic set point following a graph of time–temperature

Commercial forms of all the above are available. Instruments that provide for an adjustable time–temperature program are usually called "program controllers."

The trend in recent years has been to convert analog signals from sensors to digital signals and to use direct digital control (DDC). Then ramps, holding periods, and nonlinear time–temperature programs can be controlled from a computer terminal.

20.4.2 The Comparator

The comparator is a device to compare the temperature measuring signal to the set-point signal. Mechanical and pneumatic devices are still in use for

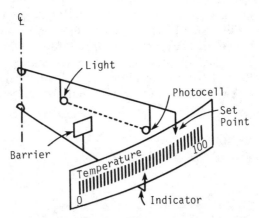

Figure 20.4. Typical millivoltmeter instrument.

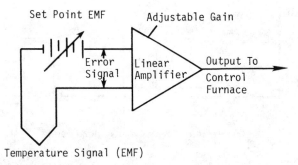

Figure 20.5. Electronic control instrument. Note the similarity between the controller and the potentiometer in Fig. 20.2.

this purpose, but most equipment now has a millivoltmeter or an electronic potentiometer as the comparator. The discussion here will be for analog instruments. Digital instrumentation are discussed later.

1. The Millivoltmeter. A millivoltmeter calibrated to read temperature directly is the simplest form of comparator. All that is needed is a set-point indicator so arranged that when the temperature exceeds the set point, the power is switched off. These instruments are most commonly used with thermocouples. One common type has a photocell illuminated by a light beam which controls an external relay (Fig. 20.4). When the flag carried by the temperature indicator shuts off the light to the photocell, the relay actuates to turn off the furnace. Sometimes a degree of proportional control is provided where the photocell signal reduces the furnace power in increments.

2. The Electronic Controller. Electronic controllers may be very complex. Functionally, however, they consist of a high-gain operational amplifier which amplifies an error signal sufficiently to adjust the power to a heater or cooler to maintain the desired temperature (Fig. 20.5). This example is for a furnace where a thermocouple provides an EMF opposite in polarity to a potential provided by another source (shown here as a variable-voltage battery). The difference between the sensor signal and the set-point signal is usually amplified with a linear amplifier and that amplified signal is used to initiate control action.

20.5 CONTROL ACTION

20.5.1 On–Off Control

The simplest control action is on–off control. The power is either completely on or completely off. With a steady-state thermal loss, off–on control

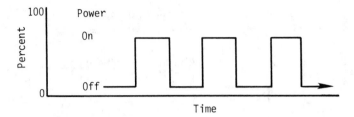

Figure 20.6. Power cycle at steady state.

consists of equally spaced intervals of on and off cycles (Fig. 20.6). Often thermal lags result in a periodic temperature response (Fig. 20.7).

Power is turned off when the temperature reaches the set point, but the temperature drifts on upward until heat already delivered is dissipated enough to start the temperature back down. Power remains off until the set-point temperature is reached again, but the temperature continues to drop until the heat delivered with the power on equals the heat being removed by cooling. Then the temperature increases, the set point is reached, and another cycle begins. The period of the cycle depends on the time constant of the heating and cooling process. Although shown here as somewhat sinusoidal, the slopes of the heating and cooling rates may be very different. When heating is much faster than cooling, the rise will be steep and the cooling long, and vice versa.

20.5.2 Proportional Control

In proportional control the amount of control action (correction) depends on the magnitude of the error signal. Proportional controllers often have adjustable proportional bandwidth, adjustable integral correction (also called offset, droop, or automatic reset), and adjustable derivative correction (often called rate control).

20.5.3 Proportional Bandwidth

The proportional bandwidth (PBW) is that fraction (in percent) of the total range of the instrument over which the full range of power control will act.

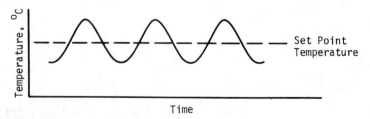

Figure 20.7. Typical temperature response to on–off control.

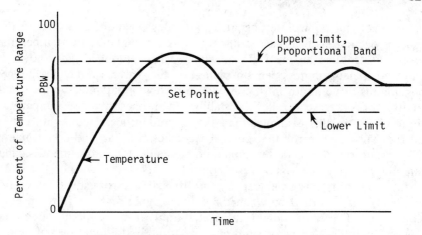

Figure 20.8. Example of overshoot and oscillation.

The smaller PBW is, the more sensitive the control function will be. For example, a 10% PBW means that the full power range will be applied between −5 and +5% of the range from the set point. Five percent below the set point will produce full power; 5% above will produce minimum power. If a controller is set for 10% PBW and the furnace is started from a cold condition, the temperature might change with time as shown in Fig. 20.8. The severe overshoot is the result of the long time constant for heat transfer in the furnace, and a narrow PBW. Increasing PBW will decrease the tendency to overshoot but decrease the temperature control sensitivity once the process reaches its normal control level.

The proportional bandwidth is really the sensitivity that is controlled by the gain G of the linear amplifier used to amplify the error signal (Fig. 20.9). Then

$$E_o = GE_i .\tag{20.1}$$

20.5.4 Integral Control

When an error signal exists and is amplified to call for power, the usual method of control is to use the amplified error signal to hold a control valve

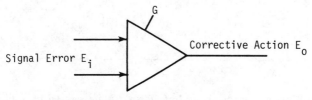

Figure 20.9. Variable-gain amplifier.

or an electrical control setting at a particular position. If a change in a process occurs, such as increasing the ware throughput, the temperature drop will produce an error signal which, when amplified, will move the valve to a new, more open position. For the valve to stay in the new position, its drive motor must continue to receive an amplified error signal. In other words, the controller will stabilize the process at a new temperature slightly less, in this example, than the desired process temperature (Fig. 20.10). This is an error that could be corrected by an additional manual adjustment, but that is better done automatically. This adjustment is called integral, droop, or automatic reset control.

Integral control is produced by an adjustable error signal integration amplifier that is turned on when the control signal first enters the proportional band. The error signal is integrated until, at the set point, integration stops. The integrated error signal is used to add additional output control signal (on heating) in order to hold the process at the desired temperature.

The integrating operational amplifier can be set for gain G' and for integrating time interval $(\theta_2 - \theta_1)$.

$$E' = G' \int_{\theta_1}^{\theta_2} E_i \, d\theta . \tag{20.2}$$

The integrating time is often expressed as the number of resets per minute.

Many process controllers have proportional–integral (PI) control. This is often sufficient for the requirements if long time constants are not present in the system. For temperature control of continuous processes with large thermal masses, so that time constants are large, derivative control is also needed.

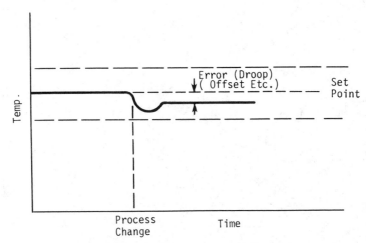

Figure 20.10. Uncorrected deviation from the set point caused by a process change requiring increased power input.

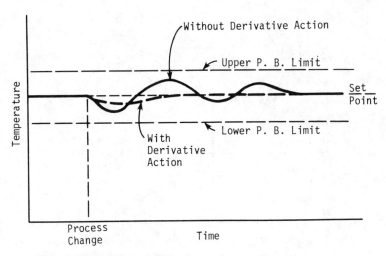

Figure 20.11. Effect of derivative action on process response.

20.5.5 Derivative Control

Derivative control is used to compensate for long time constants, opposing the corrective action of proportional and integral control. When properly adjusted it helps damp out the oscillations accompanying a narrow proportional bandwidth (high proportional amplifier gain Fig. 20.11). The derivative operational amplifier has an output given by

$$E'' = G'' \frac{dE_i}{d\theta}.$$ (20.3)

20.5.6 Combined Functions

When all three functions are combined, the equivalent circuit can be shown as in Fig. 20.12. Obviously, the gain of all three amplifiers must be adjusted to obtain adequate process control. This is not simple, and for new processes, requires experimentation. Selection of the control functions is based on the time constants and the sensitivity requirements of the process. Process components must be compatible with the control function required of the process. It is usually necessary to experiment, starting with proportional control, then integral, and then derivative. Sudden changes in set point give the same type of oscillations as sudden changes in process demand. So it is often convenient to work with set-point changes. Increasing the proportional gain until oscillations appear, adding integral and increasing it until oscillations appear, then adding derivative if necessary to reduce those oscillations is the usual method. Selecting the integrating time allows the

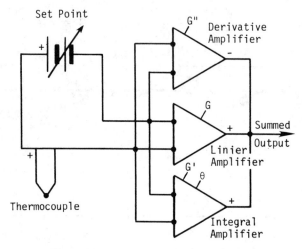

Figure 20.12. Combined control functions.

integration interval to be less than the natural frequency of the process with proportional gain only. For systems where the time constants can be estimated, computer simulation of the process will reduce the required experimentation.

20.5.7 Approach Control

Approach control is a special control for periodic systems. It is designed to prevent overshoot when using a narrow PBW on first heat-up. In effect, it increases the proportional bandwidth during initial heat-up and returns to

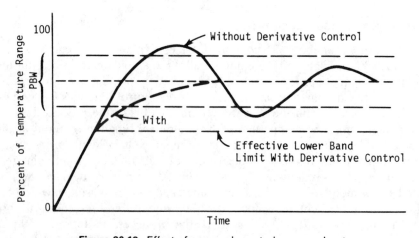

Figure 20.13. Effect of approach control on overshoot.

the narrow bandwidth after reaching the control band. It is an adjustable corrective action which depends on the distance from the temperature of the set point. When properly adjusted, it corrects for the time lags in the system—in effect lowering the lower proportional band limit. For the heating situation described in Fig. 20.8, the effect of approach control might be as shown in Fig. 20.13. It is especially useful for periodic operations.

20.6 ANALOG VERSUS DIGITAL CONCEPTS

Most temperature measuring devices produce an analog signal. An analog signal is one that varies continuously with temperature. Digital signals are signals that have discrete steps. Analog signals can be converted to digital signals, and vice versa. Conversion of an analog signal to a digital signal introduces an error. For example, a signal that changes continuously between 4 and 5 will be represented as 4 from about 3.5 to 4.5 and as 5 from about 4.5 to 5.5. The conversion error is a maximum halfway between digits, in addition to other, usually smaller errors. In principle, however, the error in conversion can be reduced to a negligible value by digitizing in small-enough increments, such as $1 \mu V$ for a thermocouple signal. Computers operate as digital systems. The growth of computer control will undoubtedly continue. Digital analysis of data and digital control of processes is changing rapidly, but certainly digital concepts are necessary in modern control systems.

20.7 LEVELS OF COMPUTER CONTROL

There are three levels of computer control. Each individual process controller, such as a feedback system to adjust a value to maintain a temperature, represents the lowest level. Often, this is microprocessor controlled with response specified in read-only memory to perform routinely a particular function.

The next higher level is microcomputer or minicomputer controlled, where the decision making, sequencing, and adjusting actions are possible. This allows one to adjust the process response by adjusting the set points of various microprocessor-controlled units of the process, by specifying sequences, and by appropriate software to control the overall response of a larger system. At this level an entire processing plant could be controlled from a central control room.

The highest level integrates process control with management-level control. Order, inventory, cost accounting, and other important management considerations are integrated with plant operation and process control in a single high-level computer and data highway system. Information from plant processes becomes available to management, and information from management becomes available to plant operators. This newest, highest level has yet to be widely adopted. Because different types of computers and different computer languages have been used in the past in process operation than in management systems, technological change may proceed slowly. However, these systems are expected to be extremely powerful and will be adopted as rapidly as cost and technical considerations permit.

20.8 CONTROL ELEMENT AND COMPUTER ENVIRONMENTS

Many industrial processes have severe environments around them. A sensor in a refinery may experience ambient temperatures of −15 to +40°C. A sensor in a steel mill may have ambient temperatures of 100°C or more, and may be exposed to corrosive fumes leaking from a furnace. The environment for industrial sensors, then, is often extreme. It is essential that they be sealed, protected, and reliable. The instrumentation purchased for an application should be compatible with the conditions.

Computers require moderate temperatures and a clean, noncorrosive atmosphere. They are housed in air-conditioned space that is adapted to human comfort. The critical systems for control should be placed in the computer area whenever possible.

20.9 DATA ACQUISITION AND CONVERSION

A typical process has a number of temperatures controlled from a central point. The sequence of control usually has the following elements:

1. Temperature sensors as appropriate for the process being controlled
2. Signal processing to prepare the analog signal for interpretation
3. Filtering to reduce noise and prevent erroneous signals
4. Analog multiplexing so that any particular sensor can be questioned
5. Sample-and-hold circuitry so that the signal from the particular sensor will be reliably represented
6. Analog-to-digital conversion to make the signal available to the computer
7. A programming sequencer under control of the computer to trigger the analog multiplexer, sample-and-hold, and analog-to-digital conversion circuits
8. A computer or data processor for analysis of data, display, and

decision making to provide control signals to the programming sequencer

9. Power amplification of the analog signal
10. Control valve action in response to the analog signal

We will discuss some of these steps individually.

20.10 SIGNAL PROCESSING

Temperature sensors, in general, have nonlinear response, although resistance, diode and thermocouple instruments are sometimes treated as linear in some inexpensive commercial instruments. The conversion of the nonlinear output to a correct indicated temperature can be accomplished by using a nonlinear scale on a linear recorder. This is often done for analog indicating and display instruments for thermocouples and resistance thermometers that are designed to read temperature directly. For a type S thermocouple the Seebeck coefficient gets smaller at higher temperatures. Then the scale diversions are closer together at the upper end of the scale. A better method would be to convert the nonlinear sensor response to a linear output for display and control purposes. This is usually done with a separate instrument called a signal processor.

Signal processors can be used for many purposes suitable to an application. Standard signal processor functions include (1) summing, (2) averaging, (3) amplifying (or reducing), (4) selecting the lowest (or highest) input, (5) limiting the output to a set value, (6) multiplying signals, (7) taking the square root, and (8) adding or subtracting a fixed value (a bias).

It is usually less expensive to adjust the analog signal coming from a group of sensors before analog-to-digital (A/D) conversion because good A/D converters are expensive and a single A/D converter can handle many multiplexed analog signals.

For temperature measurement the nonlinear signals, especially from thermistor and radiation sensors, require good signal processing. For this purpose piecewise linearization is often used. This method divides the curve of signal versus temperature from the sensor into a series of linear segments. Many inexpensive instruments have only a few segments. Some instruments have 30 or more segments. Then the signal is sent to an amplifier network that selects the appropriate linear segment, subtracts a bias for the lower segments, amplifies the signal with the gain appropriate to the linear segment, and adds a bias for the linear equivalent of the lower segments.

Any such linearization will have an error that will depend on the equation for the nonlinear portion, the method of fitting the linear curve, and the length of the section. If we fit a curve with a line by least squares the line will intersect the curve at two points. It will have maximum error at the middle and both ends of the line. Selected instrumentation should specify the number of linear segments for the temperature range and the maximum error resulting from linearization.

20.11   FILTERING

Filtering is important to remove electrical noise. Filters are classified as passive or active. A passive filter is made of resistors, capacitors, and sometimes, inductors. Active filters include resistors, capacitors, and operational amplifiers, rarely inductors. High-pass, low-pass, bandpass, and notch (blocking) filters are discussed in standard texts. They are important to temperature devices because electrical noise can interfere with satisfactory operation. An electrical noise signal of only a few microvolts can be misinterpreted as a meaningful signal, and these signals must be eliminated (for practical purposes) before amplification.

20.12   SIGNAL TRANSMITTING

Many temperature sensors provide a signal that is very weak and which would be subject to electrical noise. Therefore, the initial filtering and amplification is often performed at the sensor location, and a stronger signal is produced by a signal transmitter to be sent to a computer room. Often that signal is 0–24 V dc. It the filtering, linearization, and amplification in the signal transmitter are reliable in the hostile environment of many industrial situations, the amplified signal will usually not have problems of signal noise.

20.13   SOURCES OF NOISE

Temperature sensors are often used around rotating machinery and around furnace power situations where large electrical currents are switched. The radio-frequency field from switching and the 60-Hz field from transformers and rotating machinery cause induced voltages in electrical conductors. The high-frequency noise can be reduced with shielded cable and a low-pass filter, and the 60-Hz noise can be reduced with a bandpass filter, shorting it to ground or an active notch filter. Connectors should be as short as possible. They should be shielded. For resistance devices and other devices requiring an excitation signal, the excitation connectors (the current connecting wires) should be twisted in one shielded pair and the voltage connecting wires should be twisted in another shielded pair. Twisting reduces inductively coupled low-frequency noise. Shielding reduces capacitatively coupled high-frequency noise.

20.14 AMPLIFIERS

Modern instrument amplifiers usually incorporate differential operational amplifiers because they suppress common-mode noise. Common-mode noise comes from induced voltages from unwanted electromagnetic fields that affect, and therefore appear in, both wires of a two-wire signal transmitter as well as biasing voltages common to both wires. If the noise is identical in each wire, a differential amplifier should, in principle, not respond to it, but respond only to the signal coming from the sensor. In practice, the signals are not actually identical, and the amplifiers do not reject all response to common-mode signals. Therefore, when buying instrumentation it is important to select equipment with common-mode rejection ratios suitable for the noise that may be present.

20.15 MICROPROCESSOR-CONTROLLED INSTRUMENTS

Microprocessor-controlled temperature measuring instruments are growing very rapidly. They can control data acquisition filtering and linearization, and have programming and PID control function. Their efficiency, improved accuracy, convenience, and relatively low cost make them increasingly popular. Many new special-purpose instruments have ROMs to prevent programming error and pushbutton controls for simple and versatile program selection. These are available as complete packages ready for installation and use. Many have facilities for multiplexing input and output for many control points. They are especially useful for batch-type systems where temperature ramp and hold times are frequently changed.

Microprocessors must be interfaced to the signal sensors. Well-designed instruments have interface hardware that allow one merely to connect the sensor to the proper terminals. And they have alarm signals, program selection prompting, malfunction or programming error flags, display and/or recording capabilities, and so on.

20.16 MICROCOMPUTER-CONTROLLED INSTRUMENTS

Microcomputers are more powerful than microprocessor-controlled instruments, and are also growing in use very rapidly. The problem with microcomputer use is the need to interface with the data acquisition system and the output control system. Many new graduating engineers have the necessary skills to set up a microcomputer system. The time and effort required for a particular application may not justify setting up a system if commercial systems are available already designed and programmed. However, commercial instrumentation using the IEEE 488 bus is available from many manufacturers. Interfacing with such instruments and the auxiliary computer functions, extended memory, printers, and plotters is not usually a problem. Using transistor-transistor logic (TTL) to multiplex and providing power for controlled processes may be. The TTL circuitry must be protected from power surges and excess demand. Outputs require optical coupling or other suitable methods to prevent overloading the TTL circuitry. Usually, a digital controller accepting TTL output signals is available from the companies that manufacture microcomputer-controlled interfaces.

20.17 SUMMARY

The instrumentation and control functions are vital to successful process control. Temperature control usually requires PID control capabilities and often requires set-point programming capabilities. The instrument manufacturers make a complete spectrum of control systems, from the simple analog manual set-point millivoltmeter to the companywide master computer system with fibre optics communications highway. In selecting a system reliability, process requirements and cost must be considered. Manufacturers' literature is helpful. However, the requirements for good temperature control are so important that examination of existing operating systems is very helpful. Warming an instrument with a hair dryer can sometimes detect drift from ambient temperature changes that you would never suspect from a manufacturer's literature.

RECOMMENDED READING

J. W. Brewer, "Control Systems," Prentice-Hall, Englewood Cliffs, New Jersey, 1974.

E. O. Doeblin, "Measurement Systems." McGraw-Hill, New York, 1983.

APPENDIX: THERMOCOUPLE REFERENCE TABLES

Tables for Type B, E, J, K, NICROSIL-NISIL, R, S, and T provided by Leeds and Northrup.

Tables for Tungsten/Rhenium thermocouples provided by ASTM.

TYPE B THERMOCOUPLES

TEMPERATURES IN DEGREES C (IPTS 1968).   REFERENCE JUNCTION AT 0 C

DEG C	0	1	2	3	4	5	6	7	8	9	10
	THERMOELECTRIC VOLTAGE IN ABSOLUTE MILLIVOLTS										
0	0.000	-0.000	-0.000	-0.001	-0.001	-0.001	-0.001	-0.001	-0.002	-0.002	-0.002
10	-0.002	-0.002	-0.002	-0.002	-0.002	-0.002	-0.002	-0.002	-0.003	-0.003	-0.003
20	-0.003	-0.003	-0.003	-0.003	-0.003	-0.002	-0.002	-0.002	-0.002	-0.002	-0.002
30	-0.002	-0.002	-0.002	-0.002	-0.002	-0.001	-0.001	-0.001	-0.001	-0.001	-0.000
40	-0.000	-0.000	-0.000	0.000	0.000	0.001	0.001	0.001	0.002	0.002	0.002
50	0.002	0.003	0.003	0.003	0.004	0.004	0.004	0.005	0.005	0.006	0.006
60	0.006	0.007	0.007	0.008	0.008	0.009	0.009	0.010	0.010	0.011	0.011
70	0.011	0.012	0.012	0.013	0.014	0.014	0.015	0.015	0.016	0.017	0.017
80	0.017	0.018	0.019	0.020	0.020	0.021	0.022	0.022	0.023	0.024	0.025
90	0.025	0.026	0.026	0.027	0.028	0.029	0.030	0.031	0.031	0.032	0.033
100	0.033	0.034	0.035	0.036	0.037	0.038	0.039	0.040	0.041	0.042	0.043
110	0.043	0.044	0.045	0.046	0.047	0.048	0.049	0.050	0.051	0.052	0.053
120	0.053	0.055	0.056	0.057	0.058	0.059	0.060	0.062	0.063	0.064	0.065
130	0.065	0.066	0.068	0.069	0.070	0.071	0.073	0.074	0.075	0.077	0.078
140	0.078	0.079	0.081	0.082	0.083	0.085	0.086	0.088	0.089	0.091	0.092
150	0.092	0.093	0.095	0.096	0.098	0.099	0.101	0.102	0.104	0.106	0.107
160	0.107	0.109	0.110	0.112	0.113	0.115	0.117	0.118	0.120	0.122	0.123
170	0.123	0.125	0.127	0.128	0.130	0.132	0.133	0.135	0.137	0.139	0.140
180	0.140	0.142	0.144	0.146	0.148	0.149	0.151	0.153	0.155	0.157	0.159
190	0.159	0.161	0.163	0.164	0.166	0.168	0.170	0.172	0.174	0.176	0.178
200	0.178	0.180	0.182	0.184	0.186	0.188	0.190	0.192	0.194	0.197	0.199
210	0.199	0.201	0.203	0.205	0.207	0.209	0.211	0.214	0.216	0.218	0.220
220	0.220	0.222	0.225	0.227	0.229	0.231	0.234	0.236	0.238	0.240	0.243
230	0.243	0.245	0.247	0.250	0.252	0.254	0.257	0.259	0.262	0.264	0.266
240	0.266	0.269	0.271	0.274	0.276	0.279	0.281	0.284	0.286	0.289	0.291
250	0.291	0.294	0.296	0.299	0.301	0.304	0.307	0.309	0.312	0.314	0.317
260	0.317	0.320	0.322	0.325	0.328	0.330	0.333	0.336	0.338	0.341	0.344
260	0.317	0.320	0.322	0.325	0.328	0.330	0.333	0.336	0.338	0.341	0.344
270	0.344	0.347	0.349	0.352	0.355	0.358	0.360	0.363	0.366	0.369	0.372
280	0.372	0.375	0.377	0.380	0.383	0.386	0.389	0.392	0.395	0.398	0.401
290	0.401	0.404	0.406	0.409	0.412	0.415	0.418	0.421	0.424	0.427	0.431
300	0.431	0.434	0.437	0.440	0.443	0.446	0.449	0.452	0.455	0.458	0.462
310	0.462	0.465	0.468	0.471	0.474	0.477	0.481	0.484	0.487	0.490	0.494
320	0.494	0.497	0.500	0.503	0.507	0.510	0.513	0.517	0.520	0.523	0.527
330	0.527	0.530	0.533	0.537	0.540	0.544	0.547	0.550	0.554	0.557	0.561
340	0.561	0.564	0.568	0.571	0.575	0.578	0.582	0.585	0.589	0.592	0.596
350	0.596	0.599	0.603	0.606	0.610	0.614	0.617	0.621	0.625	0.628	0.632
360	0.632	0.636	0.639	0.643	0.647	0.650	0.654	0.658	0.661	0.665	0.669
370	0.669	0.673	0.677	0.680	0.684	0.688	0.692	0.696	0.699	0.703	0.707
380	0.707	0.711	0.715	0.719	0.723	0.727	0.730	0.734	0.738	0.742	0.746
390	0.746	0.750	0.754	0.758	0.762	0.766	0.770	0.774	0.778	0.782	0.786
400	0.786	0.790	0.794	0.799	0.803	0.807	0.811	0.815	0.819	0.823	0.827
410	0.827	0.832	0.836	0.840	0.844	0.848	0.853	0.857	0.861	0.865	0.870
420	0.870	0.874	0.878	0.882	0.887	0.891	0.895	0.900	0.904	0.908	0.913
430	0.913	0.917	0.921	0.926	0.930	0.935	0.939	0.943	0.948	0.952	0.957
440	0.957	0.961	0.966	0.970	0.975	0.979	0.984	0.988	0.993	0.997	1.002
450	1.002	1.006	1.011	1.015	1.020	1.025	1.029	1.034	1.039	1.043	1.048
460	1.048	1.052	1.057	1.062	1.066	1.071	1.076	1.081	1.085	1.090	1.095
470	1.095	1.100	1.104	1.109	1.114	1.119	1.123	1.128	1.133	1.138	1.143
480	1.143	1.148	1.152	1.157	1.162	1.167	1.172	1.177	1.182	1.187	1.192
490	1.192	1.197	1.202	1.206	1.211	1.216	1.221	1.226	1.231	1.236	1.241
500	1.241	1.246	1.252	1.257	1.262	1.267	1.272	1.277	1.282	1.287	1.292
510	1.292	1.297	1.303	1.308	1.313	1.318	1.323	1.328	1.334	1.339	1.344
520	1.344	1.349	1.354	1.360	1.365	1.370	1.375	1.381	1.386	1.391	1.397
530	1.397	1.402	1.407	1.413	1.418	1.423	1.429	1.434	1.439	1.445	1.450
540	1.450	1.456	1.461	1.467	1.472	1.477	1.483	1.488	1.494	1.499	1.505
550	1.505	1.510	1.516	1.521	1.527	1.532	1.538	1.544	1.549	1.555	1.560
560	1.560	1.566	1.571	1.577	1.583	1.588	1.594	1.600	1.605	1.611	1.617
570	1.617	1.622	1.628	1.634	1.639	1.645	1.651	1.657	1.662	1.668	1.674
580	1.674	1.680	1.685	1.691	1.697	1.703	1.709	1.715	1.720	1.726	1.732
590	1.732	1.738	1.744	1.750	1.756	1.762	1.767	1.773	1.779	1.785	1.791
600	1.791	1.797	1.803	1.809	1.815	1.821	1.827	1.833	1.839	1.845	1.851

TYPE B THERMOCOUPLES

TEMPERATURES IN DEGREES C (IPTS 1968).   REFERENCE JUNCTION AT 0 C

DEG C	0	1	2	3	4	5	6	7	8	9	10
	THERMOELECTRIC VOLTAGE IN ABSOLUTE MILLIVOLTS										
600	1.791	1.797	1.803	1.809	1.815	1.821	1.827	1.833	1.839	1.845	1.851
610	1.851	1.857	1.863	1.869	1.875	1.882	1.888	1.894	1.900	1.906	1.912
620	1.912	1.918	1.924	1.931	1.937	1.943	1.949	1.955	1.961	1.968	1.974
630	1.974	1.980	1.986	1.993	1.999	2.005	2.011	2.018	2.024	2.030	2.036
640	2.036	2.043	2.049	2.055	2.062	2.068	2.074	2.081	2.087	2.094	2.100
650	2.100	2.106	2.113	2.119	2.126	2.132	2.139	2.145	2.151	2.158	2.164
660	2.164	2.171	2.177	2.184	2.190	2.197	2.203	2.210	2.216	2.223	2.230
670	2.230	2.236	2.243	2.249	2.256	2.263	2.269	2.276	2.282	2.289	2.296
680	2.296	2.302	2.309	2.316	2.322	2.329	2.336	2.343	2.349	2.356	2.363
690	2.363	2.369	2.376	2.383	2.390	2.396	2.403	2.410	2.417	2.424	2.430
700	2.430	2.437	2.444	2.451	2.458	2.465	2.472	2.478	2.485	2.492	2.499
710	2.499	2.506	2.513	2.520	2.527	2.534	2.541	2.548	2.555	2.562	2.569
720	2.569	2.576	2.583	2.590	2.597	2.604	2.611	2.618	2.625	2.632	2.639
730	2.639	2.646	2.653	2.660	2.667	2.674	2.682	2.689	2.696	2.703	2.710
740	2.710	2.717	2.724	2.732	2.739	2.746	2.753	2.760	2.768	2.775	2.782
750	2.782	2.789	2.797	2.804	2.811	2.818	2.826	2.833	2.840	2.848	2.855
760	2.855	2.862	2.869	2.877	2.884	2.892	2.899	2.906	2.914	2.921	2.928
770	2.928	2.936	2.943	2.951	2.958	2.966	2.973	2.980	2.988	2.995	3.003
780	3.003	3.010	3.018	3.025	3.033	3.040	3.048	3.055	3.063	3.070	3.078
790	3.078	3.086	3.093	3.101	3.108	3.116	3.124	3.131	3.139	3.146	3.154
800	3.154	3.162	3.169	3.177	3.185	3.192	3.200	3.208	3.215	3.223	3.231
810	3.231	3.239	3.246	3.254	3.262	3.269	3.277	3.285	3.293	3.301	3.308
820	3.308	3.316	3.324	3.332	3.340	3.347	3.355	3.363	3.371	3.379	3.387
830	3.387	3.395	3.402	3.410	3.418	3.426	3.434	3.442	3.450	3.458	3.466
840	3.466	3.474	3.482	3.490	3.498	3.506	3.514	3.522	3.530	3.538	3.546
850	3.546	3.554	3.562	3.570	3.578	3.586	3.594	3.602	3.610	3.618	3.626
860	3.626	3.634	3.643	3.651	3.659	3.667	3.675	3.683	3.691	3.700	3.708
870	3.708	3.716	3.724	3.732	3.741	3.749	3.757	3.765	3.773	3.782	3.790
880	3.790	3.798	3.806	3.815	3.823	3.831	3.840	3.848	3.856	3.865	3.873
890	3.873	3.881	3.890	3.898	3.906	3.915	3.923	3.931	3.940	3.948	3.957
900	3.957	3.965	3.973	3.982	3.990	3.999	4.007	4.016	4.024	4.032	4.041
910	4.041	4.049	4.058	4.066	4.075	4.083	4.092	4.100	4.109	4.117	4.126
920	4.126	4.135	4.143	4.152	4.160	4.169	4.177	4.186	4.195	4.203	4.212
930	4.212	4.220	4.229	4.238	4.246	4.255	4.264	4.272	4.281	4.290	4.298
940	4.298	4.307	4.316	4.325	4.333	4.342	4.351	4.359	4.368	4.377	4.386
950	4.386	4.394	4.403	4.412	4.421	4.430	4.438	4.447	4.456	4.465	4.474
960	4.474	4.483	4.491	4.500	4.509	4.518	4.527	4.536	4.545	4.553	4.562
970	4.562	4.571	4.580	4.589	4.598	4.607	4.616	4.625	4.634	4.643	4.652
980	4.652	4.661	4.670	4.679	4.688	4.697	4.706	4.715	4.724	4.733	4.742
990	4.742	4.751	4.760	4.769	4.778	4.787	4.796	4.805	4.814	4.824	4.833
1,000	4.833	4.842	4.851	4.860	4.869	4.878	4.887	4.897	4.906	4.915	4.924
1,010	4.924	4.933	4.942	4.952	4.961	4.970	4.979	4.989	4.998	5.007	5.016
1,020	5.016	5.025	5.035	5.044	5.053	5.063	5.072	5.081	5.090	5.100	5.109
1,030	5.109	5.118	5.128	5.137	5.146	5.156	5.165	5.174	5.184	5.193	5.202
1,040	5.202	5.212	5.221	5.231	5.240	5.249	5.259	5.268	5.278	5.287	5.297
1,050	5.297	5.306	5.316	5.325	5.334	5.344	5.353	5.363	5.372	5.382	5.391
1,060	5.391	5.401	5.410	5.420	5.429	5.439	5.449	5.458	5.468	5.477	5.487
1,070	5.487	5.496	5.506	5.516	5.525	5.535	5.544	5.554	5.564	5.573	5.583
1,080	5.583	5.593	5.602	5.612	5.621	5.631	5.641	5.651	5.660	5.670	5.680
1,090	5.680	5.689	5.699	5.709	5.718	5.728	5.738	5.748	5.757	5.767	5.777
1,100	5.777	5.787	5.796	5.806	5.816	5.826	5.836	5.845	5.855	5.865	5.875
1,110	5.875	5.885	5.895	5.904	5.914	5.924	5.934	5.944	5.954	5.964	5.973
1,120	5.973	5.983	5.993	6.003	6.013	6.023	6.033	6.043	6.053	6.063	6.073
1,130	6.073	6.083	6.093	6.102	6.112	6.122	6.132	6.142	6.152	6.162	6.172
1,140	6.172	6.182	6.192	6.202	6.212	6.223	6.233	6.243	6.253	6.263	6.273
1,150	6.273	6.283	6.293	6.303	6.313	6.323	6.333	6.343	6.353	6.364	6.374
1,160	6.374	6.384	6.394	6.404	6.414	6.424	6.435	6.445	6.455	6.465	6.475
1,170	6.475	6.485	6.496	6.506	6.516	6.526	6.536	6.547	6.557	6.567	6.577
1,180	6.577	6.588	6.598	6.608	6.618	6.629	6.639	6.649	6.659	6.670	6.680
1,190	6.680	6.690	6.701	6.711	6.721	6.732	6.742	6.752	6.763	6.773	6.783
1,200	6.783	6.794	6.804	6.814	6.825	6.835	6.846	6.856	6.866	6.877	6.887

TYPE B THERMOCOUPLES

DEG C	0	1	2	3	4	5	6	7	8	9	10
				THERMOELECTRIC VOLTAGE IN ABSOLUTE MILLIVOLTS							
1,200	6.783	6.794	6.804	6.814	6.825	6.835	6.846	6.856	6.866	6.877	6.887
1,210	6.887	6.898	6.908	6.918	6.929	6.939	6.950	6.960	6.971	6.981	6.991
1,220	6.991	7.002	7.012	7.023	7.033	7.044	7.054	7.065	7.075	7.086	7.096
1,230	7.096	7.107	7.117	7.128	7.138	7.149	7.159	7.170	7.181	7.191	7.202
1,240	7.202	7.212	7.223	7.233	7.244	7.255	7.265	7.276	7.286	7.297	7.308
1,250	7.308	7.318	7.329	7.339	7.350	7.361	7.371	7.382	7.393	7.403	7.414
1,260	7.414	7.425	7.435	7.446	7.457	7.467	7.478	7.489	7.500	7.510	7.521
1,270	7.521	7.532	7.542	7.553	7.564	7.575	7.585	7.596	7.607	7.618	7.628
1,280	7.628	7.639	7.650	7.661	7.671	7.682	7.693	7.704	7.715	7.725	7.736
1,290	7.736	7.747	7.758	7.769	7.780	7.790	7.801	7.812	7.823	7.834	7.845
1,300	7.845	7.855	7.866	7.877	7.888	7.899	7.910	7.921	7.932	7.943	7.953
1,310	7.953	7.964	7.975	7.986	7.997	8.008	8.019	8.030	8.041	8.052	8.063
1,320	8.063	8.074	8.085	8.096	8.107	8.118	8.128	8.139	8.150	8.161	8.172
1,330	8.172	8.183	8.194	8.205	8.216	8.227	8.238	8.249	8.261	8.272	8.283
1,340	8.283	8.294	8.305	8.316	8.327	8.338	8.349	8.360	8.371	8.382	8.393
1,350	8.393	8.404	8.415	8.426	8.437	8.449	8.460	8.471	8.482	8.493	8.504
1,360	8.504	8.515	8.526	8.538	8.549	8.560	8.571	8.582	8.593	8.604	8.616
1,370	8.616	8.627	8.638	8.649	8.660	8.671	8.683	8.694	8.705	8.716	8.727
1,380	8.727	8.738	8.750	8.761	8.772	8.783	8.795	8.806	8.817	8.828	8.839
1,390	8.839	8.851	8.862	8.873	8.884	8.896	8.907	8.918	8.929	8.941	8.952
1,400	8.952	8.963	8.974	8.986	8.997	9.008	9.020	9.031	9.042	9.053	9.065
1,410	9.065	9.076	9.087	9.099	9.110	9.121	9.133	9.144	9.155	9.167	9.178
1,420	9.178	9.189	9.201	9.212	9.223	9.235	9.246	9.257	9.269	9.280	9.291
1,430	9.291	9.303	9.314	9.326	9.337	9.348	9.360	9.371	9.382	9.394	9.405
1,440	9.405	9.417	9.428	9.439	9.451	9.462	9.474	9.485	9.497	9.508	9.519
1,450	9.519	9.531	9.542	9.554	9.565	9.577	9.588	9.599	9.611	9.622	9.634
1,460	9.634	9.645	9.657	9.668	9.680	9.691	9.703	9.714	9.726	9.737	9.748
1,470	9.748	9.760	9.771	9.783	9.794	9.806	9.817	9.829	9.840	9.852	9.863
1,480	9.863	9.875	9.886	9.898	9.909	9.921	9.933	9.944	9.956	9.967	9.979
1,490	9.979	9.990	10.002	10.013	10.025	10.036	10.048	10.059	10.071	10.082	10.094
1,500	10.094	10.106	10.117	10.129	10.140	10.152	10.163	10.175	10.187	10.198	10.210
1,510	10.210	10.221	10.233	10.244	10.256	10.268	10.279	10.291	10.302	10.314	10.325
1,520	10.325	10.337	10.349	10.360	10.372	10.383	10.395	10.407	10.418	10.430	10.441
1,530	10.441	10.453	10.465	10.476	10.488	10.500	10.511	10.523	10.534	10.546	10.558
1,540	10.558	10.569	10.581	10.593	10.604	10.616	10.627	10.639	10.651	10.662	10.674
1,550	10.674	10.686	10.697	10.709	10.721	10.732	10.744	10.756	10.767	10.779	10.790
1,560	10.790	10.802	10.814	10.825	10.837	10.849	10.860	10.872	10.884	10.895	10.907
1,570	10.907	10.919	10.930	10.942	10.954	10.965	10.977	10.989	11.000	11.012	11.024
1,580	11.024	11.035	11.047	11.059	11.070	11.082	11.094	11.105	11.117	11.129	11.141
1,590	11.141	11.152	11.164	11.176	11.187	11.199	11.211	11.222	11.234	11.246	11.257
1,600	11.257	11.269	11.281	11.292	11.304	11.316	11.328	11.339	11.351	11.363	11.374
1,610	11.374	11.386	11.398	11.409	11.421	11.433	11.444	11.456	11.468	11.480	11.491
1,620	11.491	11.503	11.515	11.526	11.538	11.550	11.561	11.573	11.585	11.597	11.608
1,630	11.608	11.620	11.632	11.643	11.655	11.667	11.678	11.690	11.702	11.714	11.725
1,640	11.725	11.737	11.749	11.760	11.772	11.784	11.795	11.807	11.819	11.830	11.842
1,650	11.842	11.854	11.866	11.877	11.889	11.901	11.912	11.924	11.936	11.947	11.959
1,660	11.959	11.971	11.983	11.994	12.006	12.018	12.029	12.041	12.053	12.064	12.076
1,670	12.076	12.088	12.099	12.111	12.123	12.134	12.146	12.158	12.170	12.181	12.193
1,680	12.193	12.205	12.216	12.228	12.240	12.251	12.263	12.275	12.286	12.298	12.310
1,690	12.310	12.321	12.333	12.345	12.356	12.368	12.380	12.391	12.403	12.415	12.426
1,700	12.426	12.438	12.450	12.461	12.473	12.485	12.496	12.508	12.520	12.531	12.543
1,710	12.543	12.555	12.566	12.578	12.590	12.601	12.613	12.624	12.636	12.648	12.659
1,720	12.659	12.671	12.683	12.694	12.706	12.718	12.729	12.741	12.752	12.764	12.776
1,730	12.776	12.787	12.799	12.811	12.822	12.834	12.845	12.857	12.869	12.880	12.892
1,740	12.892	12.903	12.915	12.927	12.938	12.950	12.961	12.973	12.985	12.996	13.008
1,750	13.008	13.019	13.031	13.043	13.054	13.066	13.077	13.089	13.100	13.112	13.124
1,760	13.124	13.135	13.147	13.158	13.170	13.181	13.193	13.204	13.216	13.228	13.239
1,770	13.239	13.251	13.262	13.274	13.285	13.297	13.308	13.320	13.331	13.343	13.354
1,780	13.354	13.366	13.378	13.389	13.401	13.412	13.424	13.435	13.447	13.458	13.470
1,790	13.470	13.481	13.493	13.504	13.516	13.527	13.539	13.550	13.562	13.573	13.585
1,800	13.585	13.596	13.607	13.619	13.630	13.642	13.653	13.665	13.676	13.688	13.699
1,810	13.699	13.711	13.722	13.733	13.745	13.756	13.768	13.779	13.791	13.802	13.814
1,820	13.814										

TYPE E THERMOCOUPLES

DEG C	0	1	2	3	4	5	6	7	8	9	10
					THERMOELECTRIC VOLTAGE IN ABSOLUTE MILLIVOLTS						
-190	-8.561	-8.588	-8.615	-8.642	-8.669	-8.696	-8.722	-8.748	-8.774	-8.799	-8.824
-180	-8.273	-8.303	-8.333	-8.362	-8.391	-8.420	-8.449	-8.477	-8.505	-8.533	-8.561
-170	-7.963	-7.995	-8.027	-8.058	-8.090	-8.121	-8.152	-8.183	-8.213	-8.243	-8.273
-160	-7.631	-7.665	-7.699	-7.733	-7.767	-7.800	-7.833	-7.866	-7.898	-7.931	-7.963
-150	-7.279	-7.315	-7.351	-7.387	-7.422	-7.458	-7.493	-7.528	-7.562	-7.597	-7.631
-140	-6.907	-6.945	-6.983	-7.020	-7.058	-7.095	-7.132	-7.169	-7.206	-7.243	-7.279
-130	-6.516	-6.556	-6.596	-6.635	-6.675	-6.714	-6.753	-6.792	-6.830	-6.869	-6.907
-120	-6.107	-6.149	-6.190	-6.231	-6.273	-6.314	-6.354	-6.395	-6.436	-6.476	-6.516
-110	-5.680	-5.724	-5.767	-5.810	-5.853	-5.896	-5.938	-5.981	-6.023	-6.065	-6.107
-100	-5.237	-5.282	-5.327	-5.371	-5.416	-5.460	-5.505	-5.549	-5.593	-5.637	-5.680
-90	-4.777	-4.824	-4.870	-4.916	-4.963	-5.009	-5.055	-5.100	-5.146	-5.191	-5.237
-80	-4.301	-4.350	-4.398	-4.446	-4.493	-4.541	-4.588	-4.636	-4.683	-4.730	-4.777
-70	-3.811	-3.860	-3.910	-3.959	-4.009	-4.058	-4.107	-4.156	-4.204	-4.253	-4.301
-60	-3.306	-3.357	-3.408	-3.459	-3.509	-3.560	-3.610	-3.661	-3.711	-3.761	-3.811
-50	-2.787	-2.839	-2.892	-2.944	-2.996	-3.048	-3.100	-3.152	-3.203	-3.254	-3.306
-40	-2.254	-2.308	-2.362	-2.416	-2.469	-2.522	-2.575	-2.628	-2.681	-2.734	-2.787
-30	-1.709	-1.764	-1.819	-1.874	-1.929	-1.983	-2.038	-2.092	-2.146	-2.200	-2.254
-20	-1.151	-1.208	-1.264	-1.320	-1.376	-1.432	-1.487	-1.543	-1.599	-1.654	-1.709
-10	-0.581	-0.639	-0.696	-0.754	-0.811	-0.868	-0.925	-0.982	-1.038	-1.095	-1.151
0	0.000	-0.059	-0.117	-0.176	-0.234	-0.292	-0.350	-0.408	-0.466	-0.524	-0.581
0	0.000	0.059	0.118	0.176	0.235	0.295	0.354	0.413	0.472	0.532	0.591
10	0.591	0.651	0.711	0.770	0.830	0.890	0.950	1.011	1.071	1.131	1.192
20	1.192	1.252	1.313	1.373	1.434	1.495	1.556	1.617	1.678	1.739	1.801
30	1.801	1.862	1.924	1.985	2.047	2.109	2.171	2.233	2.295	2.357	2.419
40	2.419	2.482	2.544	2.607	2.669	2.732	2.795	2.858	2.921	2.984	3.047
50	3.047	3.110	3.173	3.237	3.300	3.364	3.428	3.491	3.555	3.619	3.683
60	3.683	3.748	3.812	3.876	3.941	4.005	4.070	4.134	4.199	4.264	4.329
70	4.329	4.394	4.459	4.524	4.590	4.655	4.720	4.786	4.852	4.917	4.983
80	4.983	5.049	5.115	5.181	5.247	5.314	5.380	5.446	5.513	5.579	5.646
90	5.646	5.713	5.780	5.846	5.913	5.981	6.048	6.115	6.182	6.250	6.317
100	6.317	6.385	6.452	6.520	6.588	6.656	6.724	6.792	6.860	6.928	6.996
110	6.996	7.064	7.133	7.201	7.270	7.339	7.407	7.476	7.545	7.614	7.683
120	7.683	7.752	7.821	7.890	7.960	8.029	8.099	8.168	8.238	8.307	8.377
130	8.377	8.447	8.517	8.587	8.657	8.727	8.797	8.867	8.938	9.008	9.078
140	9.078	9.149	9.220	9.290	9.361	9.432	9.503	9.573	9.644	9.715	9.787
150	9.787	9.858	9.929	10.000	10.072	10.143	10.215	10.286	10.358	10.429	10.501
160	10.501	10.573	10.645	10.717	10.789	10.861	10.933	11.005	11.077	11.150	11.222
170	11.222	11.294	11.367	11.439	11.512	11.585	11.657	11.730	11.803	11.876	11.949
180	11.949	12.022	12.095	12.168	12.241	12.314	12.387	12.461	12.534	12.608	12.681
190	12.681	12.755	12.828	12.902	12.975	13.049	13.123	13.197	13.271	13.345	13.419
200	13.419	13.493	13.567	13.641	13.715	13.789	13.864	13.938	14.012	14.087	14.161
210	14.161	14.236	14.310	14.385	14.460	14.534	14.609	14.684	14.759	14.834	14.909
220	14.909	14.984	15.059	15.134	15.209	15.284	15.359	15.435	15.510	15.585	15.661
230	15.661	15.736	15.812	15.887	15.963	16.038	16.114	16.190	16.266	16.341	16.417
240	16.417	16.493	16.569	16.645	16.721	16.797	16.873	16.949	17.025	17.101	17.178
250	17.178	17.254	17.330	17.406	17.483	17.559	17.636	17.712	17.789	17.865	17.942
260	17.942	18.018	18.095	18.172	18.248	18.325	18.402	18.479	18.556	18.633	18.710
270	18.710	18.787	18.864	18.941	19.018	19.095	19.172	19.249	19.326	19.404	19.481
280	19.481	19.558	19.636	19.713	19.790	19.868	19.945	20.023	20.100	20.178	20.256
290	20.256	20.333	20.411	20.488	20.566	20.644	20.722	20.800	20.877	20.955	21.033
300	21.033	21.111	21.189	21.267	21.345	21.423	21.501	21.579	21.657	21.735	21.814
310	21.814	21.892	21.970	22.048	22.127	22.205	22.283	22.362	22.440	22.518	22.597
320	22.597	22.675	22.754	22.832	22.911	22.989	23.068	23.147	23.225	23.304	23.383
330	23.383	23.461	23.540	23.619	23.698	23.777	23.855	23.934	24.013	24.092	24.171
340	24.171	24.250	24.329	24.408	24.487	24.566	24.645	24.724	24.803	24.882	24.961
350	24.961	25.041	25.120	25.199	25.278	25.357	25.437	25.516	25.595	25.675	25.754
360	25.754	25.833	25.913	25.992	26.072	26.151	26.230	26.310	26.389	26.469	26.549
370	26.549	26.628	26.708	26.787	26.867	26.947	27.026	27.106	27.186	27.265	27.345
380	27.345	27.425	27.504	27.584	27.664	27.744	27.824	27.903	27.983	28.063	28.143
390	28.143	28.223	28.303	28.383	28.463	28.543	28.623	28.703	28.783	28.863	28.943

TYPE E THERMOCOUPLES

TEMPERATURES IN DEGREES C (IPTS 1968).   REFERENCE JUNCTION AT 0 C

DEG C	0	1	2	3	4	5	6	7	8	9	10
				THERMOELECTRIC VOLTAGE IN ABSOLUTE MILLIVOLTS							
400	28.943	29.023	29.103	29.183	29.263	29.343	29.423	29.503	29.584	29.664	29.744
410	29.744	29.824	29.904	29.984	30.065	30.145	30.225	30.305	30.386	30.466	30.546
420	30.546	30.627	30.707	30.787	30.868	30.948	31.028	31.109	31.189	31.270	31.350
430	31.350	31.430	31.511	31.591	31.672	31.752	31.833	31.913	31.994	32.074	32.155
440	32.155	32.235	32.316	32.396	32.477	32.557	32.638	32.719	32.799	32.880	32.960
450	32.960	33.041	33.122	33.202	33.283	33.364	33.444	33.525	33.605	33.686	33.767
460	33.767	33.848	33.928	34.009	34.090	34.170	34.251	34.332	34.413	34.493	34.574
470	34.574	34.655	34.736	34.816	34.897	34.978	35.059	35.140	35.220	35.301	35.382
480	35.382	35.463	35.544	35.624	35.705	35.786	35.867	35.948	36.029	36.109	36.190
490	36.190	36.271	36.352	36.433	36.514	36.595	36.675	36.756	36.837	36.918	36.999
500	36.999	37.080	37.161	37.242	37.323	37.403	37.484	37.565	37.646	37.727	37.808
510	37.808	37.889	37.970	38.051	38.132	38.213	38.293	38.374	38.455	38.536	38.617
520	38.617	38.698	38.779	38.860	38.941	39.022	39.103	39.184	39.264	39.345	39.426
530	39.426	39.507	39.588	39.669	39.750	39.831	39.912	39.993	40.074	40.155	40.236
540	40.236	40.316	40.397	40.478	40.559	40.640	40.721	40.802	40.883	40.964	41.045
550	41.045	41.125	41.206	41.287	41.368	41.449	41.530	41.611	41.692	41.773	41.853
560	41.853	41.934	42.015	42.096	42.177	42.258	42.339	42.419	42.500	42.581	42.662
570	42.662	42.743	42.824	42.904	42.985	43.066	43.147	43.228	43.308	43.389	43.470
580	43.470	43.551	43.632	43.712	43.793	43.874	43.955	44.035	44.116	44.197	44.278
590	44.278	44.358	44.439	44.520	44.601	44.681	44.762	44.843	44.923	45.004	45.085
600	45.085	45.165	45.246	45.327	45.407	45.488	45.569	45.649	45.730	45.811	45.891
610	45.891	45.972	46.052	46.133	46.213	46.294	46.375	46.455	46.536	46.616	46.697
620	46.697	46.777	46.858	46.938	47.019	47.099	47.180	47.260	47.341	47.421	47.502
630	47.502	47.582	47.663	47.743	47.824	47.904	47.984	48.065	48.145	48.226	48.306
640	48.306	48.386	48.467	48.547	48.627	48.708	48.788	48.868	48.949	49.029	49.109
650	49.109	49.189	49.270	49.350	49.430	49.510	49.591	49.671	49.751	49.831	49.911
660	49.911	49.992	50.072	50.152	50.232	50.312	50.392	50.472	50.553	50.633	50.713
670	50.713	50.793	50.873	50.953	51.033	51.113	51.193	51.273	51.353	51.433	51.513
680	51.513	51.593	51.673	51.753	51.833	51.913	51.993	52.073	52.152	52.232	52.312
690	52.312	52.392	52.472	52.552	52.632	52.711	52.791	52.871	52.951	53.031	53.110
700	53.110	53.190	53.270	53.350	53.429	53.509	53.589	53.668	53.748	53.828	53.907
710	53.907	53.987	54.066	54.146	54.226	54.305	54.385	54.464	54.544	54.623	54.703
720	54.703	54.782	54.862	54.941	55.021	55.100	55.180	55.259	55.339	55.418	55.498
730	55.498	55.577	55.656	55.736	55.815	55.894	55.974	56.053	56.132	56.212	56.291
740	56.291	56.370	56.449	56.529	56.608	56.687	56.766	56.845	56.924	57.004	57.083
750	57.083	57.162	57.241	57.320	57.399	57.478	57.557	57.636	57.715	57.794	57.873
760	57.873	57.952	58.031	58.110	58.189	58.268	58.347	58.426	58.505	58.584	58.663
770	58.663	58.742	58.820	58.899	58.978	59.057	59.136	59.214	59.293	59.372	59.451
780	59.451	59.529	59.608	59.687	59.765	59.844	59.923	60.001	60.080	60.159	60.237
790	60.237	60.316	60.394	60.473	60.551	60.630	60.708	60.787	60.865	60.944	61.022
800	61.022	61.101	61.179	61.258	61.336	61.414	61.493	61.571	61.649	61.728	61.806
810	61.806	61.884	61.962	62.041	62.119	62.197	62.275	62.353	62.432	62.510	62.588
820	62.588	62.666	62.744	62.822	62.900	62.978	63.056	63.134	63.212	63.290	63.368
830	63.368	63.446	63.524	63.602	63.680	63.758	63.836	63.914	63.992	64.069	64.147
840	64.147	64.225	64.303	64.380	64.458	64.536	64.614	64.691	64.769	64.847	64.924
850	64.924	65.002	65.080	65.157	65.235	65.312	65.390	65.467	65.545	65.622	65.700
860	65.700	65.777	65.855	65.932	66.009	66.087	66.164	66.241	66.319	66.396	66.473
870	66.473	66.551	66.628	66.705	66.782	66.859	66.937	67.014	67.091	67.168	67.245
880	67.245	67.322	67.399	67.476	67.553	67.630	67.707	67.784	67.861	67.938	68.015
890	68.015	68.092	68.169	68.246	68.323	68.399	68.476	68.553	68.630	68.706	68.783
900	68.783	68.860	68.936	69.013	69.090	69.166	69.243	69.320	69.396	69.473	69.549
910	69.549	69.626	69.702	69.779	69.855	69.931	70.008	70.084	70.161	70.237	70.313
920	70.313	70.390	70.466	70.542	70.618	70.694	70.771	70.847	70.923	70.999	71.075
930	71.075	71.151	71.227	71.304	71.380	71.456	71.532	71.608	71.683	71.759	71.835
940	71.835	71.911	71.987	72.063	72.139	72.215	72.290	72.366	72.442	72.518	72.593
950	72.593	72.669	72.745	72.820	72.896	72.972	73.047	73.123	73.199	73.274	73.350
960	73.350	73.425	73.501	73.576	73.652	73.727	73.802	73.878	73.953	74.029	74.104
970	74.104	74.179	74.255	74.330	74.405	74.480	74.556	74.631	74.706	74.781	74.857
980	74.857	74.932	75.007	75.082	75.157	75.232	75.307	75.382	75.458	75.533	75.608
990	75.608	75.683	75.758	75.833	75.908	75.983	76.058	76.133	76.208	76.283	76.358
1,000	76.358										

TYPE J THERMOCOUPLES

TEMPERATURES IN DEGREES C (IPTS 1968).   REFERENCE JUNCTION AT 0 C

DEG C	0	1	2	3	4	5	6	7	8	9	10
					THERMOELECTRIC VOLTAGE IN ABSOLUTE MILLIVOLTS						
-190	-7.659	-7.683	-7.707	-7.731	-7.755	-7.778	-7.801	-7.824	-7.846	-7.868	-7.890
-180	-7.402	-7.429	-7.455	-7.482	-7.508	-7.533	-7.559	-7.584	-7.609	-7.634	-7.659
-170	-7.122	-7.151	-7.180	-7.209	-7.237	-7.265	-7.293	-7.321	-7.348	-7.375	-7.402
-160	-6.821	-6.852	-6.883	-6.914	-6.944	-6.974	-7.004	-7.034	-7.064	-7.093	-7.122
-150	-6.499	-6.532	-6.565	-6.598	-6.630	-6.663	-6.695	-6.727	-6.758	-6.790	-6.821
-140	-6.159	-6.194	-6.228	-6.263	-6.297	-6.331	-6.365	-6.399	-6.433	-6.466	-6.499
-130	-5.801	-5.837	-5.874	-5.910	-5.946	-5.982	-6.018	-6.053	-6.089	-6.124	-6.159
-120	-5.426	-5.464	-5.502	-5.540	-5.578	-5.615	-5.653	-5.690	-5.727	-5.764	-5.801
-110	-5.036	-5.076	-5.115	-5.155	-5.194	-5.233	-5.272	-5.311	-5.349	-5.388	-5.426
-100	-4.632	-4.673	-4.714	-4.755	-4.795	-4.836	-4.876	-4.916	-4.956	-4.996	-5.036
-90	-4.215	-4.257	-4.299	-4.341	-4.383	-4.425	-4.467	-4.508	-4.550	-4.591	-4.632
-80	-3.785	-3.829	-3.872	-3.915	-3.958	-4.001	-4.044	-4.087	-4.130	-4.172	-4.215
-70	-3.344	-3.389	-3.433	-3.478	-3.522	-3.566	-3.610	-3.654	-3.698	-3.742	-3.785
-60	-2.892	-2.938	-2.984	-3.029	-3.074	-3.120	-3.165	-3.210	-3.255	-3.299	-3.344
-50	-2.431	-2.478	-2.524	-2.570	-2.617	-2.663	-2.709	-2.755	-2.801	-2.847	-2.892
-40	-1.960	-2.008	-2.055	-2.102	-2.150	-2.197	-2.244	-2.291	-2.338	-2.384	-2.431
-30	-1.481	-1.530	-1.578	-1.626	-1.674	-1.722	-1.770	-1.818	-1.865	-1.913	-1.960
-20	-0.995	-1.044	-1.093	-1.141	-1.190	-1.239	-1.288	-1.336	-1.385	-1.433	-1.481
-10	-0.501	-0.550	-0.600	-0.650	-0.699	-0.748	-0.798	-0.847	-0.896	-0.945	-0.995
0	0.000	-0.050	-0.101	-0.151	-0.201	-0.251	-0.301	-0.351	-0.401	-0.451	-0.501
0	0.000	0.050	0.101	0.151	0.202	0.253	0.303	0.354	0.405	0.456	0.507
10	0.507	0.558	0.609	0.660	0.711	0.762	0.813	0.865	0.916	0.967	1.019
20	1.019	1.070	1.122	1.174	1.225	1.277	1.329	1.381	1.432	1.484	1.536
30	1.536	1.588	1.640	1.693	1.745	1.797	1.849	1.901	1.954	2.006	2.058
40	2.058	2.111	2.163	2.216	2.268	2.321	2.374	2.426	2.479	2.532	2.585
50	2.585	2.638	2.691	2.743	2.796	2.849	2.902	2.956	3.009	3.062	3.115
60	3.115	3.168	3.221	3.275	3.328	3.381	3.435	3.488	3.542	3.595	3.649
70	3.649	3.702	3.756	3.809	3.863	3.917	3.971	4.024	4.078	4.132	4.186
80	4.186	4.239	4.293	4.347	4.401	4.455	4.509	4.563	4.617	4.671	4.725
90	4.725	4.780	4.834	4.888	4.942	4.996	5.050	5.105	5.159	5.213	5.268
100	5.268	5.322	5.376	5.431	5.485	5.540	5.594	5.649	5.703	5.758	5.812
110	5.812	5.867	5.921	5.976	6.031	6.085	6.140	6.195	6.249	6.304	6.359
120	6.359	6.414	6.468	6.523	6.578	6.633	6.688	6.742	6.797	6.852	6.907
130	6.907	6.962	7.017	7.072	7.127	7.182	7.237	7.292	7.347	7.402	7.457
140	7.457	7.512	7.567	7.622	7.677	7.732	7.787	7.843	7.898	7.953	8.008
150	8.008	8.063	8.118	8.174	8.229	8.284	8.339	8.394	8.450	8.505	8.560
160	8.560	8.616	8.671	8.726	8.781	8.837	8.892	8.947	9.003	9.058	9.113
170	9.113	9.169	9.224	9.279	9.335	9.390	9.446	9.501	9.556	9.612	9.667
180	9.667	9.723	9.778	9.834	9.889	9.944	10.000	10.055	10.111	10.166	10.222
190	10.222	10.277	10.333	10.388	10.444	10.499	10.555	10.610	10.666	10.721	10.777
200	10.777	10.832	10.888	10.943	10.999	11.054	11.110	11.165	11.221	11.276	11.332
210	11.332	11.387	11.443	11.498	11.554	11.609	11.665	11.720	11.776	11.831	11.887
220	11.887	11.943	11.998	12.054	12.109	12.165	12.220	12.276	12.331	12.387	12.442
230	12.442	12.498	12.553	12.609	12.664	12.720	12.776	12.831	12.887	12.942	12.998
240	12.998	13.053	13.109	13.164	13.220	13.275	13.331	13.386	13.442	13.497	13.553
250	13.553	13.608	13.664	13.719	13.775	13.830	13.886	13.941	13.997	14.052	14.108
260	14.108	14.163	14.219	14.274	14.330	14.385	14.441	14.496	14.552	14.607	14.663
270	14.663	14.718	14.774	14.829	14.885	14.940	14.995	15.051	15.106	15.162	15.217
280	15.217	15.273	15.328	15.383	15.439	15.494	15.550	15.605	15.661	15.716	15.771
290	15.771	15.827	15.882	15.938	15.993	16.048	16.104	16.159	16.214	16.270	16.325
300	16.325	16.380	16.436	16.491	16.547	16.602	16.657	16.713	16.768	16.823	16.879
310	16.879	16.934	16.989	17.044	17.100	17.155	17.210	17.266	17.321	17.376	17.432
320	17.432	17.487	17.542	17.597	17.653	17.708	17.763	17.818	17.874	17.929	17.984
330	17.984	18.039	18.095	18.150	18.205	18.260	18.316	18.371	18.426	18.481	18.537
340	18.537	18.592	18.647	18.702	18.757	18.813	18.868	18.923	18.978	19.033	19.089
350	19.089	19.144	19.199	19.254	19.309	19.364	19.420	19.475	19.530	19.585	19.640
360	19.640	19.695	19.751	19.806	19.861	19.916	19.971	20.026	20.081	20.137	20.192
370	20.192	20.247	20.302	20.357	20.412	20.467	20.523	20.578	20.633	20.688	20.743
380	20.743	20.798	20.853	20.909	20.964	21.019	21.074	21.129	21.184	21.239	21.295
390	21.295	21.350	21.405	21.460	21.515	21.570	21.625	21.680	21.736	21.791	21.846

TYPE J THERMOCOUPLES

DEG C	0	1	2	3	4	5	6	7	8	9	10
				THERMOELECTRIC VOLTAGE IN ABSOLUTE MILLIVOLTS							
400	21.846	21.901	21.956	22.011	22.066	22.122	22.177	22.232	22.287	22.342	22.397
410	22.397	22.453	22.508	22.563	22.618	22.673	22.728	22.784	22.839	22.894	22.949
420	22.949	23.004	23.060	23.115	23.170	23.225	23.280	23.336	23.391	23.446	23.501
430	23.501	23.556	23.612	23.667	23.722	23.777	23.833	23.888	23.943	23.999	24.054
440	24.054	24.109	24.164	24.220	24.275	24.330	24.386	24.441	24.496	24.552	24.607
450	24.607	24.662	24.718	24.773	24.829	24.884	24.939	24.995	25.050	25.106	25.161
460	25.161	25.217	25.272	25.327	25.383	25.438	25.494	25.549	25.605	25.661	25.716
470	25.716	25.772	25.827	25.883	25.938	25.994	26.050	26.105	26.161	26.216	26.272
480	26.272	26.328	26.383	26.439	26.495	26.551	26.606	26.662	26.718	26.774	26.829
490	26.829	26.885	26.941	26.997	27.053	27.109	27.165	27.220	27.276	27.332	27.388
500	27.388	27.444	27.500	27.556	27.612	27.668	27.724	27.780	27.836	27.893	27.949
510	27.949	28.005	28.061	28.117	28.173	28.230	28.286	28.342	28.398	28.455	28.511
520	28.511	28.567	28.624	28.680	28.736	28.793	28.849	28.906	28.962	29.019	29.075
530	29.075	29.132	29.188	29.245	29.301	29.358	29.415	29.471	29.528	29.585	29.642
540	29.642	29.698	29.755	29.812	29.869	29.926	29.983	30.039	30.096	30.153	30.210
550	30.210	30.267	30.324	30.381	30.439	30.496	30.553	30.610	30.667	30.724	30.782
560	30.782	30.839	30.896	30.954	31.011	31.068	31.126	31.183	31.241	31.298	31.356
570	31.356	31.413	31.471	31.528	31.586	31.644	31.702	31.759	31.817	31.875	31.933
580	31.933	31.991	32.048	32.106	32.164	32.222	32.280	32.338	32.396	32.455	32.513
590	32.513	32.571	32.629	32.687	32.746	32.804	32.862	32.921	32.979	33.038	33.096
600	33.096	33.155	33.213	33.272	33.330	33.389	33.448	33.506	33.565	33.624	33.683
610	33.683	33.742	33.800	33.859	33.918	33.977	34.036	34.095	34.155	34.214	34.273
620	34.273	34.332	34.391	34.451	34.510	34.569	34.629	34.688	34.748	34.807	34.867
630	34.867	34.926	34.986	35.046	35.105	35.165	35.225	35.285	35.344	35.404	35.464
640	35.464	35.524	35.584	35.644	35.704	35.764	35.825	35.885	35.945	36.005	36.066
650	36.066	36.126	36.186	36.247	36.307	36.368	36.428	36.489	36.549	36.610	36.671
660	36.671	36.732	36.792	36.853	36.914	36.975	37.036	37.097	37.158	37.219	37.280
670	37.280	37.341	37.402	37.463	37.525	37.586	37.647	37.709	37.770	37.831	37.893
680	37.893	37.954	38.016	38.078	38.139	38.201	38.262	38.324	38.386	38.448	38.510
690	38.510	38.572	38.633	38.695	38.757	38.819	38.882	38.944	39.006	39.068	39.130
700	39.130	39.192	39.255	39.317	39.379	39.442	39.504	39.567	39.629	39.692	39.754
710	39.754	39.817	39.880	39.942	40.005	40.068	40.131	40.193	40.256	40.319	40.382
720	40.382	40.445	40.508	40.571	40.634	40.697	40.760	40.823	40.886	40.950	41.013
730	41.013	41.076	41.139	41.203	41.266	41.329	41.393	41.456	41.520	41.583	41.647
740	41.647	41.710	41.774	41.837	41.901	41.965	42.028	42.092	42.156	42.219	42.283
750	42.283	42.347	42.411	42.475	42.538	42.602	42.666	42.730	42.794	42.858	42.922
760	42.922	42.986	43.050	43.114	43.178	43.242	43.306	43.370	43.435	43.499	43.563
770	43.563	43.627	43.692	43.756	43.820	43.885	43.949	44.014	44.078	44.142	44.207
780	44.207	44.271	44.336	44.400	44.465	44.529	44.594	44.658	44.723	44.788	44.852
790	44.852	44.917	44.981	45.046	45.111	45.175	45.240	45.304	45.369	45.434	45.498
800	45.498	45.563	45.627	45.692	45.757	45.821	45.886	45.950	46.015	46.080	46.144
810	46.144	46.209	46.273	46.338	46.403	46.467	46.532	46.596	46.661	46.725	46.790
820	46.790	46.854	46.919	46.983	47.047	47.112	47.176	47.241	47.305	47.369	47.434
830	47.434	47.498	47.562	47.627	47.691	47.755	47.819	47.884	47.948	48.012	48.076
840	48.076	48.140	48.204	48.269	48.333	48.397	48.461	48.525	48.589	48.653	48.716
850	48.716	48.780	48.844	48.908	48.972	49.036	49.099	49.163	49.227	49.291	49.354
860	49.354	49.418	49.481	49.545	49.608	49.672	49.735	49.799	49.862	49.926	49.989

Because of the known instability of Type J thermocouples above 870C this temperature has been used as the upper limit in the above tables. We do not normally recommend the use of Type J couples above 760C except in special circumstances.

For those occasional applications requiring operation at higher temperatures the values listed opposite are provided. This extension is a mathematical extrapolation based on limited calibration data and caution should be exercised in its use. The basis for the extended data is discussed in NBS Monograph 125.

900 C	51.88 mV
1000	57.94
1100	63.78

TYPE K THERMOCOUPLES

DEG C	0	1	2	3	4	5	6	7	8	9	10
					THERMOELECTRIC VOLTAGE IN ABSOLUTE MILLIVOLTS						
0	0.000	0.039	0.079	0.119	0.158	0.198	0.238	0.277	0.317	0.357	0.397
10	0.397	0.437	0.477	0.517	0.557	0.597	0.637	0.677	0.718	0.758	0.798
20	0.798	0.838	0.879	0.919	0.960	1.000	1.041	1.081	1.122	1.162	1.203
30	1.203	1.244	1.285	1.325	1.366	1.407	1.448	1.489	1.529	1.570	1.611
40	1.611	1.652	1.693	1.734	1.776	1.817	1.858	1.899	1.940	1.981	2.022
50	2.022	2.064	2.105	2.146	2.188	2.229	2.270	2.312	2.353	2.394	2.436
60	2.436	2.477	2.519	2.560	2.601	2.643	2.684	2.726	2.767	2.809	2.850
70	2.850	2.892	2.933	2.975	3.016	3.058	3.100	3.141	3.183	3.224	3.266
80	3.266	3.307	3.349	3.390	3.432	3.473	3.515	3.556	3.598	3.639	3.681
90	3.681	3.722	3.764	3.805	3.847	3.888	3.930	3.971	4.012	4.054	4.095
100	4.095	4.137	4.178	4.219	4.261	4.302	4.343	4.384	4.426	4.467	4.508
110	4.508	4.549	4.590	4.632	4.673	4.714	4.755	4.796	4.837	4.878	4.919
120	4.919	4.960	5.001	5.042	5.083	5.124	5.164	5.205	5.246	5.287	5.327
130	5.327	5.368	5.409	5.450	5.490	5.531	5.571	5.612	5.652	5.693	5.733
140	5.733	5.774	5.814	5.855	5.895	5.936	5.976	6.016	6.057	6.097	6.137
150	6.137	6.177	6.218	6.258	6.298	6.338	6.378	6.419	6.459	6.499	6.539
160	6.539	6.579	6.619	6.659	6.699	6.739	6.779	6.819	6.859	6.899	6.939
170	6.939	6.979	7.019	7.059	7.099	7.139	7.179	7.219	7.259	7.299	7.338
180	7.338	7.378	7.418	7.458	7.498	7.538	7.578	7.618	7.658	7.697	7.737
190	7.737	7.777	7.817	7.857	7.897	7.937	7.977	8.017	8.057	8.097	8.137
200	8.137	8.177	8.216	8.256	8.296	8.336	8.376	8.416	8.456	8.497	8.537
210	8.537	8.577	8.617	8.657	8.697	8.737	8.777	8.817	8.857	8.898	8.938
220	8.938	8.978	9.018	9.058	9.099	9.139	9.179	9.220	9.260	9.300	9.341
230	9.341	9.381	9.421	9.462	9.502	9.543	9.583	9.624	9.664	9.705	9.745
240	9.745	9.786	9.826	9.867	9.907	9.948	9.989	10.029	10.070	10.111	10.151
250	10.151	10.192	10.233	10.274	10.315	10.355	10.396	10.437	10.478	10.519	10.560
260	10.560	10.600	10.641	10.682	10.723	10.764	10.805	10.846	10.887	10.928	10.969
270	10.969	11.010	11.051	11.093	11.134	11.175	11.216	11.257	11.298	11.339	11.381
280	11.381	11.422	11.463	11.504	11.546	11.587	11.628	11.669	11.711	11.752	11.793
290	11.793	11.835	11.876	11.918	11.959	12.000	12.042	12.083	12.125	12.166	12.207
300	12.207	12.249	12.290	12.332	12.373	12.415	12.456	12.498	12.539	12.581	12.623
310	12.623	12.664	12.706	12.747	12.789	12.831	12.872	12.914	12.955	12.997	13.039
320	13.039	13.080	13.122	13.164	13.205	13.247	13.289	13.331	13.372	13.414	13.456
330	13.456	13.497	13.539	13.581	13.623	13.665	13.706	13.748	13.790	13.832	13.874
340	13.874	13.915	13.957	13.999	14.041	14.083	14.125	14.167	14.208	14.250	14.292
350	14.292	14.334	14.376	14.418	14.460	14.502	14.544	14.586	14.628	14.670	14.712
360	14.712	14.754	14.796	14.838	14.880	14.922	14.964	15.006	15.048	15.090	15.132
370	15.132	15.174	15.216	15.258	15.300	15.342	15.384	15.426	15.468	15.510	15.552
380	15.552	15.594	15.636	15.679	15.721	15.763	15.805	15.847	15.889	15.931	15.974
390	15.974	16.016	16.058	16.100	16.142	16.184	16.227	16.269	16.311	16.353	16.395
400	16.395	16.438	16.480	16.522	16.564	16.607	16.649	16.691	16.733	16.776	16.818
410	16.818	16.860	16.902	16.945	16.987	17.029	17.072	17.114	17.156	17.199	17.241
420	17.241	17.283	17.326	17.368	17.410	17.453	17.495	17.537	17.580	17.622	17.664
430	17.664	17.707	17.749	17.792	17.834	17.876	17.919	17.961	18.004	18.046	18.088
440	18.088	18.131	18.173	18.216	18.258	18.301	18.343	18.385	18.428	18.470	18.513
450	18.513	18.555	18.598	18.640	18.683	18.725	18.768	18.810	18.853	18.895	18.938
460	18.938	18.980	19.023	19.065	19.108	19.150	19.193	19.235	19.278	19.320	19.363
470	19.363	19.405	19.448	19.490	19.533	19.576	19.618	19.661	19.703	19.746	19.788
480	19.788	19.831	19.873	19.916	19.959	20.001	20.044	20.086	20.129	20.172	20.214
490	20.214	20.257	20.299	20.342	20.385	20.427	20.470	20.512	20.555	20.598	20.640
500	20.640	20.683	20.725	20.768	20.811	20.853	20.896	20.938	20.981	21.024	21.066
510	21.066	21.109	21.152	21.194	21.237	21.280	21.322	21.365	21.407	21.450	21.493
520	21.493	21.535	21.578	21.621	21.663	21.706	21.749	21.791	21.834	21.876	21.919
530	21.919	21.962	22.004	22.047	22.090	22.132	22.175	22.218	22.260	22.303	22.346
540	22.346	22.388	22.431	22.473	22.516	22.559	22.601	22.644	22.687	22.729	22.772
550	22.772	22.815	22.857	22.900	22.942	22.985	23.028	23.070	23.113	23.156	23.198
560	23.198	23.241	23.284	23.326	23.369	23.411	23.454	23.497	23.539	23.582	23.624
570	23.624	23.667	23.710	23.752	23.795	23.837	23.880	23.923	23.965	24.008	24.050
580	24.050	24.093	24.136	24.178	24.221	24.263	24.306	24.348	24.391	24.434	24.476
590	24.476	24.519	24.561	24.604	24.646	24.689	24.731	24.774	24.817	24.859	24.902
600	24.902	24.944	24.987	25.029	25.072	25.114	25.157	25.199	25.242	25.284	25.327

TYPE K THERMOCOUPLES

TEMPERATURES IN DEGREES C (IPTS 1968).   REFERENCE JUNCTION AT 0 C

DEG C	0	1	2	3	4	5	6	7	8	9	10

THERMOELECTRIC VOLTAGE IN ABSOLUTE MILLIVOLTS

DEG C	0	1	2	3	4	5	6	7	8	9	10
600	24.902	24.944	24.987	25.029	25.072	25.114	25.157	25.199	25.242	25.284	25.327
610	25.327	25.369	25.412	25.454	25.497	25.539	25.582	25.624	25.666	25.709	25.751
620	25.751	25.794	25.836	25.879	25.921	25.964	26.006	26.048	26.091	26.133	26.176
630	26.176	26.218	26.260	26.303	26.345	26.387	26.430	26.472	26.515	26.557	26.599
640	26.599	26.642	26.684	26.726	26.769	26.811	26.853	26.896	26.938	26.980	27.022
650	27.022	27.065	27.107	27.149	27.192	27.234	27.276	27.318	27.361	27.403	27.445
660	27.445	27.487	27.529	27.572	27.614	27.656	27.698	27.740	27.783	27.825	27.867
670	27.867	27.909	27.951	27.993	28.035	28.078	28.120	28.162	28.204	28.246	28.288
680	28.288	28.330	28.372	28.414	28.456	28.498	28.540	28.583	28.625	28.667	28.709
690	28.709	28.751	28.793	28.835	28.877	28.919	28.961	29.002	29.044	29.086	29.128
700	29.128	29.170	29.212	29.254	29.296	29.338	29.380	29.422	29.464	29.505	29.547
710	29.547	29.589	29.631	29.673	29.715	29.756	29.798	29.840	29.882	29.924	29.965
720	29.965	30.007	30.049	30.091	30.132	30.174	30.216	30.257	30.299	30.341	30.383
730	30.383	30.424	30.466	30.508	30.549	30.591	30.632	30.674	30.716	30.757	30.799
740	30.799	30.840	30.882	30.924	30.965	31.007	31.048	31.090	31.131	31.173	31.214
750	31.214	31.256	31.297	31.339	31.380	31.422	31.463	31.504	31.546	31.587	31.629
760	31.629	31.670	31.712	31.753	31.794	31.836	31.877	31.918	31.960	32.001	32.042
770	32.042	32.084	32.125	32.166	32.207	32.249	32.290	32.331	32.372	32.414	32.455
780	32.455	32.496	32.537	32.578	32.619	32.661	32.702	32.743	32.784	32.825	32.866
790	32.866	32.907	32.948	32.990	33.031	33.072	33.113	33.154	33.195	33.236	33.277
800	33.277	33.318	33.359	33.400	33.441	33.482	33.523	33.564	33.604	33.645	33.686
810	33.686	33.727	33.768	33.809	33.850	33.891	33.931	33.972	34.013	34.054	34.095
820	34.095	34.136	34.176	34.217	34.258	34.299	34.339	34.380	34.421	34.461	34.502
830	34.502	34.543	34.583	34.624	34.665	34.705	34.746	34.787	34.827	34.868	34.909
840	34.909	34.949	34.990	35.030	35.071	35.111	35.152	35.192	35.233	35.273	35.314
850	35.314	35.354	35.395	35.435	35.476	35.516	35.557	35.597	35.637	35.678	35.718
860	35.718	35.758	35.799	35.839	35.880	35.920	35.960	36.000	36.041	36.081	36.121
870	36.121	36.162	36.202	36.242	36.282	36.323	36.363	36.403	36.443	36.483	36.524
880	36.524	36.564	36.604	36.644	36.684	36.724	36.764	36.804	36.844	36.885	36.925
890	36.925	36.965	37.005	37.045	37.085	37.125	37.165	37.205	37.245	37.285	37.325
900	37.325	37.365	37.405	37.445	37.484	37.524	37.564	37.604	37.644	37.684	37.724
910	37.724	37.764	37.803	37.843	37.883	37.923	37.963	38.002	38.042	38.082	38.122
920	38.122	38.162	38.201	38.241	38.281	38.320	38.360	38.400	38.439	38.479	38.519
930	38.519	38.558	38.598	38.638	38.677	38.717	38.756	38.796	38.836	38.875	38.915
940	38.915	38.954	38.994	39.033	39.073	39.112	39.152	39.191	39.231	39.270	39.310
950	39.310	39.349	39.388	39.428	39.467	39.507	39.546	39.585	39.625	39.664	39.703
960	39.703	39.743	39.782	39.821	39.861	39.900	39.939	39.979	40.018	40.057	40.096
970	40.096	40.136	40.175	40.214	40.253	40.292	40.332	40.371	40.410	40.449	40.488
980	40.488	40.527	40.566	40.605	40.645	40.684	40.723	40.762	40.801	40.840	40.879
990	40.879	40.918	40.957	40.996	41.035	41.074	41.113	41.152	41.191	41.230	41.269
1,000	41.269	41.308	41.347	41.385	41.424	41.463	41.502	41.541	41.580	41.619	41.657
1,010	41.657	41.696	41.735	41.774	41.813	41.851	41.890	41.929	41.968	42.006	42.045
1,020	42.045	42.084	42.123	42.161	42.200	42.239	42.277	42.316	42.355	42.393	42.432
1,030	42.432	42.470	42.509	42.548	42.586	42.625	42.663	42.702	42.740	42.779	42.817
1,040	42.817	42.856	42.894	42.933	42.971	43.010	43.048	43.087	43.125	43.164	43.202
1,050	43.202	43.240	43.279	43.317	43.356	43.394	43.432	43.471	43.509	43.547	43.585
1,060	43.585	43.624	43.662	43.700	43.739	43.777	43.815	43.853	43.891	43.930	43.968
1,070	43.968	44.006	44.044	44.082	44.121	44.159	44.197	44.235	44.273	44.311	44.349
1,080	44.349	44.387	44.425	44.463	44.501	44.539	44.577	44.615	44.653	44.691	44.729
1,090	44.729	44.767	44.805	44.843	44.881	44.919	44.957	44.995	45.033	45.070	45.108
1,100	45.108	45.146	45.184	45.222	45.260	45.297	45.335	45.373	45.411	45.448	45.486
1,110	45.486	45.524	45.561	45.599	45.637	45.675	45.712	45.750	45.787	45.825	45.863
1,120	45.863	45.900	45.938	45.975	46.013	46.051	46.088	46.126	46.163	46.201	46.238
1,130	46.238	46.275	46.313	46.350	46.388	46.425	46.463	46.500	46.537	46.575	46.612
1,140	46.612	46.649	46.687	46.724	46.761	46.799	46.836	46.873	46.910	46.948	46.985
1,150	46.985	47.022	47.059	47.096	47.134	47.171	47.208	47.245	47.282	47.319	47.356
1,160	47.356	47.393	47.430	47.468	47.505	47.542	47.579	47.616	47.653	47.689	47.726
1,170	47.726	47.763	47.800	47.837	47.874	47.911	47.948	47.985	48.021	48.058	48.095
1,180	48.095	48.132	48.169	48.205	48.242	48.279	48.316	48.352	48.389	48.426	48.462
1,190	48.462	48.499	48.536	48.572	48.609	48.645	48.682	48.718	48.755	48.792	48.828
1,200	48.828	48.865	48.901	48.937	48.974	49.010	49.047	49.083	49.120	49.156	49.192

TYPE K THERMOCOUPLES

DEG C	0	1	2	3	4	5	6	7	8	9	10
				THERMOELECTRIC VOLTAGE IN ABSOLUTE MILLIVOLTS							
1,200	48.828	48.865	48.901	48.937	48.974	49.010	49.047	49.083	49.120	49.156	49.192
1,210	49.192	49.229	49.265	49.301	49.338	49.374	49.410	49.446	49.483	49.519	49.555
1,220	49.555	49.591	49.627	49.663	49.700	49.736	49.772	49.808	49.844	49.880	49.916
1,230	49.916	49.952	49.988	50.024	50.060	50.096	50.132	50.168	50.204	50.240	50.276
1,240	50.276	50.311	50.347	50.383	50.419	50.455	50.491	50.526	50.562	50.598	50.633
1,250	50.633	50.669	50.705	50.741	50.776	50.812	50.847	50.883	50.919	50.954	50.990
1,260	50.990	51.025	51.061	51.096	51.132	51.167	51.203	51.238	51.274	51.309	51.344
1,270	51.344	51.380	51.415	51.450	51.486	51.521	51.556	51.592	51.627	51.662	51.697
1,280	51.697	51.733	51.768	51.803	51.838	51.873	51.908	51.943	51.979	52.014	52.049
1,290	52.049	52.084	52.119	52.154	52.189	52.224	52.259	52.294	52.329	52.364	52.398
1,300	52.398	52.433	52.468	52.503	52.538	52.573	52.608	52.642	52.677	52.712	52.747
1,310	52.747	52.781	52.816	52.851	52.886	52.920	52.955	52.989	53.024	53.059	53.093
1,320	53.093	53.128	53.162	53.197	53.232	53.266	53.301	53.335	53.370	53.404	53.439
1,330	53.439	53.473	53.507	53.542	53.576	53.611	53.645	53.679	53.714	53.748	53.782
1,340	53.782	53.817	53.851	53.885	53.920	53.954	53.988	54.022	54.057	54.091	54.125
1,350	54.125	54.159	54.193	54.228	54.262	54.296	54.330	54.364	54.398	54.432	54.466
1,360	54.466	54.501	54.535	54.569	54.603	54.637	54.671	54.705	54.739	54.773	54.807
1,370	54.807	54.841	54.875								

NICROSIL-NISIL THERMOCOUPLES

°C	0	1	2	3	4	5	6	7	8	9	10

THERMOELECTRIC VOLTAGE IN ABSOLUTE MILLIVOLTS

°C	0	1	2	3	4	5	6	7	8	9	10
0	0.000	0.026	0.052	0.078	0.104	0.130	0.156	0.182	0.208	0.235	0.261
10	0.261	0.287	0.313	0.340	0.366	0.392	0.419	0.445	0.472	0.498	0.525
20	0.525	0.552	0.578	0.605	0.632	0.658	0.685	0.712	0.739	0.766	0.793
30	0.793	0.820	0.847	0.874	0.901	0.928	0.955	0.982	1.010	1.037	1.064
40	1.064	1.092	1.119	1.146	1.174	1.201	1.229	1.257	1.284	1.312	1.340
50	1.340	1.367	1.395	1.423	1.451	1.479	1.507	1.535	1.563	1.591	1.619
60	1.619	1.647	1.675	1.703	1.731	1.760	1.788	1.816	1.845	1.873	1.902
70	1.902	1.930	1.959	1.987	2.016	2.045	2.073	2.102	2.131	2.160	2.188
80	2.188	2.217	2.246	2.275	2.304	2.333	2.362	2.392	2.421	2.450	2.479
90	2.479	2.508	2.538	2.567	2.596	2.626	2.655	2.685	2.714	2.744	2.774
100	2.774	2.803	2.833	2.863	2.892	2.922	2.952	2.982	3.012	3.042	3.072
110	3.072	3.102	3.132	3.162	3.192	3.222	3.253	3.283	3.313	3.343	3.374
120	3.374	3.404	3.435	3.465	3.495	3.526	3.557	3.587	3.618	3.649	3.679
130	3.679	3.710	3.741	3.772	3.802	3.833	3.864	3.895	3.926	3.957	3.988
140	3.988	4.019	4.050	4.082	4.113	4.144	4.175	4.207	4.238	4.269	4.301
150	4.301	4.332	4.364	4.395	4.427	4.458	4.490	4.522	4.553	4.585	4.617
160	4.617	4.648	4.680	4.712	4.744	4.776	4.808	4.840	4.872	4.904	4.936
170	4.936	4.968	5.000	5.032	5.064	5.097	5.129	5.161	5.193	5.226	5.258
180	5.258	5.291	5.323	5.355	5.388	5.420	5.453	5.486	5.518	5.551	5.584
190	5.584	5.616	5.649	5.682	5.715	5.747	5.780	5.813	5.846	5.879	5.912
200	5.912	5.945	5.978	6.011	6.044	6.077	6.110	6.144	6.177	6.210	6.243
210	6.243	6.277	6.310	6.343	6.377	6.410	6.443	6.477	6.510	6.544	6.577
220	6.577	6.611	6.645	6.678	6.712	6.745	6.779	6.813	6.847	6.880	6.914
230	6.914	6.948	6.982	7.016	7.050	7.084	7.118	7.152	7.186	7.220	7.254
240	7.254	7.288	7.322	7.356	7.390	7.424	7.459	7.493	7.527	7.561	7.596
250	7.596	7.630	7.664	7.699	7.733	7.767	7.802	7.836	7.871	7.905	7.940
260	7.940	7.975	8.009	8.044	8.078	8.113	8.148	8.182	8.217	8.252	8.287
270	8.287	8.322	8.356	8.391	8.426	8.461	8.496	8.531	8.566	8.601	8.636
280	8.636	8.671	8.706	8.741	8.776	8.811	8.846	8.881	8.916	8.952	8.987
290	8.987	9.022	9.057	9.093	9.128	9.163	9.199	9.234	9.269	9.305	9.340
300	9.340	9.376	9.411	9.446	9.482	9.517	9.553	9.589	9.624	9.660	9.695
310	9.695	9.731	9.767	9.802	9.838	9.874	9.909	9.945	9.981	10.017	10.053
320	10.053	10.088	10.124	10.160	10.196	10.232	10.268	10.304	10.340	10.376	10.412
330	10.412	10.448	10.484	10.520	10.556	10.592	10.628	10.664	10.700	10.736	10.773
340	10.773	10.809	10.845	10.881	10.917	10.954	10.990	11.026	11.063	11.099	11.135
350	11.135	11.172	11.208	11.244	11.281	11.317	11.354	11.390	11.426	11.463	11.499
360	11.499	11.536	11.573	11.609	11.646	11.682	11.719	11.755	11.792	11.829	11.865
370	11.865	11.902	11.939	11.975	12.012	12.049	12.086	12.122	12.159	12.196	12.233
380	12.233	12.270	12.307	12.343	12.380	12.417	12.454	12.491	12.528	12.565	12.602
390	12.602	12.639	12.676	12.713	12.750	12.787	12.824	12.861	12.898	12.935	12.972
400	12.972	13.009	13.046	13.084	13.121	13.158	13.195	13.232	13.270	13.307	13.344
410	13.344	13.381	13.419	13.456	13.493	13.530	13.568	13.605	13.642	13.680	13.717
420	13.717	13.755	13.792	13.829	13.867	13.904	13.942	13.979	14.017	14.054	14.092
430	14.092	14.129	14.167	14.204	14.242	14.279	14.317	14.354	14.392	14.430	14.467
440	14.467	14.505	14.542	14.580	14.618	14.655	14.693	14.731	14.769	14.806	14.844
450	14.844	14.882	14.920	14.957	14.995	15.033	15.071	15.108	15.146	15.184	15.222
460	15.222	15.260	15.298	15.336	15.373	15.411	15.449	15.487	15.525	15.563	15.601
470	15.601	15.639	15.677	15.715	15.753	15.791	15.829	15.867	15.905	15.943	15.981
480	15.981	16.019	16.057	16.095	16.134	16.172	16.210	16.248	16.286	16.324	16.362
490	16.362	16.401	16.439	16.477	16.515	16.553	16.591	16.630	16.668	16.706	16.744
500	16.744	16.783	16.821	16.859	16.898	16.936	16.974	17.012	17.051	17.089	17.127
510	17.127	17.166	17.204	17.243	17.281	17.319	17.358	17.396	17.435	17.473	17.511
520	17.511	17.550	17.588	17.627	17.665	17.704	17.742	17.781	17.819	17.858	17.896
530	17.896	17.935	17.973	18.012	18.050	18.089	18.127	18.166	18.205	18.243	18.282
540	18.282	18.320	18.359	18.397	18.436	18.475	18.513	18.552	18.591	18.629	18.668
550	18.668	18.707	18.745	18.784	18.823	18.861	18.900	18.939	18.978	19.016	19.055
560	19.055	19.094	19.132	19.171	19.210	19.249	19.288	19.326	19.365	19.404	19.443
570	19.443	19.481	19.520	19.559	19.598	19.637	19.676	19.714	19.753	19.792	19.831
580	19.831	19.870	19.909	19.948	19.987	20.025	20.064	20.103	20.141	20.181	20.220
590	20.220	20.259	20.298	20.337	20.376	20.415	20.454	20.492	20.531	20.570	20.609
600	20.609	20.648	20.687	20.726	20.765	20.804	20.843	20.882	20.921	20.960	20.999

NICROSIL-NISIL THERMOCOUPLES

THERMOELECTRIC VOLTAGE IN ABSOLUTE MILLIVOLTS

°C	0	1	2	3	4	5	6	7	8	9	10
600	20.609	20.648	20.687	20.726	20.765	20.804	20.843	20.882	20.921	20.960	20.999
610	20.999	21.038	21.077	21.116	21.156	21.195	21.234	21.273	21.312	21.351	21.390
620	21.390	21.429	21.468	21.507	21.546	21.585	21.624	21.663	21.703	21.742	21.781
630	21.781	21.820	21.859	21.898	21.937	21.976	22.015	22.055	22.094	22.133	22.172
640	22.172	22.211	22.250	22.289	22.329	22.368	22.407	22.446	22.485	22.524	22.564
650	22.564	22.603	22.642	22.681	22.720	22.760	22.799	22.838	22.877	22.916	22.956
660	22.956	22.995	23.034	23.073	23.112	23.152	23.191	23.230	23.269	23.309	23.348
670	23.348	23.387	23.426	23.466	23.505	23.544	23.583	23.623	23.662	23.701	23.740
680	23.740	23.780	23.819	23.858	23.897	23.937	23.976	24.015	24.054	24.094	24.133
690	24.133	24.172	24.212	24.251	24.290	24.329	24.369	24.408	24.447	24.487	24.526
700	24.526	24.565	24.604	24.644	24.683	24.722	24.762	24.801	24.840	24.880	24.919
710	24.919	24.958	24.997	25.037	25.076	25.115	25.155	25.194	25.233	25.273	24.312
720	25.312	25.351	25.391	25.430	25.469	25.509	25.548	25.587	25.626	25.666	25.705
730	25.705	25.744	25.784	25.823	25.862	25.902	25.941	25.980	26.020	26.059	26.098
740	26.098	26.138	26.177	26.216	26.256	26.295	26.334	26.373	26.413	26.452	26.491
750	26.491	26.531	26.570	26.609	26.649	26.688	26.727	26.767	26.806	26.845	26.885
760	26.885	26.924	26.963	27.002	27.042	27.081	27.120	27.160	27.199	27.238	27.278
770	27.278	27.317	27.356	27.396	27.435	27.474	27.513	27.553	27.592	27.631	27.671
780	27.671	27.710	27.749	27.789	27.828	27.867	27.906	27.946	27.985	28.024	28.063
790	28.063	28.103	28.142	28.181	28.221	28.260	28.299	28.338	28.378	28.417	28.456
800	28.456	28.495	28.535	28.574	28.613	28.653	28.692	28.731	28.770	28.810	28.849
810	28.849	28.888	28.927	28.966	29.006	29.045	29.084	29.123	29.163	29.202	29.241
820	29.241	29.280	29.320	29.359	29.398	29.437	29.476	29.516	29.555	29.594	29.633
830	29.633	29.672	29.712	29.751	29.790	29.829	29.868	29.908	29.947	29.986	30.025
840	30.025	30.064	30.103	30.143	30.182	30.221	30.260	30.299	30.338	30.378	30.417
850	30.417	30.456	30.495	30.534	30.573	30.612	30.652	30.691	30.730	30.769	30.808
860	30.808	30.847	30.886	30.925	30.965	31.004	31.043	31.082	31.121	31.160	31.199
870	31.199	31.238	31.277	31.316	31.355	31.394	31.434	31.473	31.512	31.551	31.590
880	31.590	31.629	31.668	31.707	31.746	31.785	31.824	31.863	31.902	31.941	31.980
890	31.980	32.019	32.058	32.097	32.136	32.175	32.214	32.253	32.292	32.331	32.370
900	32.370	32.409	32.448	32.487	32.526	32.565	32.604	32.643	32.682	32.721	32.760
910	32.760	32.799	32.838	32.877	32.916	32.955	32.993	33.032	33.071	33.110	33.149
920	33.149	33.188	33.227	33.266	33.305	33.344	33.383	33.421	33.460	33.499	33.538
930	33.538	33.577	33.616	33.655	33.693	33.732	33.771	33.810	33.849	33.888	33.927
940	33.927	33.965	34.004	34.043	34.082	34.121	34.159	34.198	34.237	34.276	34.315
950	34.315	34.353	34.392	34.431	34.470	34.508	34.547	34.586	34.625	34.663	34.702
960	34.702	34.741	34.780	34.818	34.857	34.896	34.935	34.973	35.012	35.051	35.089
970	35.089	35.128	35.167	35.205	35.244	35.283	35.321	35.360	35.399	35.437	35.476
980	35.476	35.515	35.553	35.592	35.631	35.669	35.708	35.747	35.785	35.824	35.862
990	35.862	35.901	35.940	35.978	36.017	36.055	36.094	36.133	36.171	36.210	36.248
1,000	36.248	36.287	36.325	36.364	36.402	36.441	36.479	36.518	36.556	36.595	36.633
1,010	36.633	36.672	36.710	36.749	36.787	36.826	36.864	36.903	36.941	36.980	37.018
1,020	37.018	37.057	37.095	37.134	37.172	37.210	37.249	37.287	37.326	37.364	37.403
1,030	37.403	37.441	37.479	37.518	37.556	37.594	37.633	37.671	37.710	37.748	37.786
1,040	37.786	37.825	37.863	37.901	37.940	37.978	38.016	38.055	38.093	38.131	38.169
1,050	38.169	38.208	38.246	38.284	38.323	38.361	38.399	38.437	38.476	38.514	38.552
1,060	38.552	38.590	38.629	38.667	38.705	38.743	38.781	38.820	38.858	38.896	38.934
1,070	38.934	38.972	39.010	39.049	39.087	39.125	39.163	39.201	39.239	39.277	39.316
1,080	39.316	39.354	39.392	39.430	39.468	39.506	39.544	39.582	39.620	39.658	39.696
1,090	39.696	39.734	39.772	39.810	39.848	39.886	39.924	39.962	40.000	40.038	40.076
1,100	40.076	40.114	40.152	40.190	40.228	40.266	40.304	40.342	40.380	40.418	40.456
1,110	40.456	40.494	40.532	40.570	40.607	40.645	40.683	40.721	40.759	40.797	40.835
1,120	40.835	40.872	40.910	40.948	40.986	41.024	41.062	41.099	41.137	41.175	41.213
1,130	41.213	41.250	41.288	41.326	41.364	41.401	41.439	41.477	41.515	41.552	41.590
1,140	41.590	41.628	41.665	41.703	41.741	41.778	41.816	41.854	41.891	41.929	41.966
1,150	41.966	42.004	42.042	42.079	42.117	42.154	42.192	42.230	42.267	42.305	42.342
1,160	42.342	42.380	42.417	42.455	42.492	42.530	42.567	42.605	42.642	42.680	42.717
1,170	42.717	42.754	42.792	42.829	42.867	42.904	42.941	42.979	43.016	43.054	43.091
1,180	43.091	43.128	43.166	43.203	43.240	43.278	43.315	43.352	43.390	43.427	43.464
1,190	43.464	43.501	43.539	43.576	43.613	43.650	43.687	43.725	43.762	43.799	43.836
1,200	43.836	43.873	43.910	43.948	43.985	44.022	44.059	44.096	44.133	44.170	44.207

NICROSIL-NISIL THERMOCOUPLES

TEMPERATURES IN DEGREES C (IPTS 1968) ●   REFERENCE JUNCTION AT 0 C

THERMOELECTRIC VOLTAGE IN ABSOLUTE MILLIVOLTS

°C	0	1	2	3	4	5	6	7	8	9	10
1,200	43.836	43.873	43.910	43.948	43.985	44.022	44.059	44.096	44.133	44.170	44.207
1,210	44.207	44.244	44.281	44.319	44.356	44.393	44.430	44.467	44.504	44.541	44.578
1,220	44.578	44.614	44.651	44.688	44.725	44.762	44.799	44.836	44.873	44.910	44.947
1,230	44.947	44.984	45.020	45.057	45.094	45.131	45.168	45.205	45.241	45.278	45.315
1,240	45.315	45.352	45.388	45.425	45.462	45.499	45.535	45.572	45.609	45.645	45.682
1,250	45.682	45.719	45.755	45.792	45.829	45.865	45.902	45.938	45.975	46.011	46.048
1,260	46.048	46.085	46.121	46.158	46.194	46.231	46.267	46.304	46.340	46.377	46.413
1,270	46.413	46.449	46.486	46.522	46.559	46.595	46.632	46.668	46.704	46.741	46.777
1,280	46.777	46.813	46.850	46.886	46.922	46.959	46.995	47.031	47.067	47.104	47.140
1,290	47.140	47.176	47.212	47.249	47,285	47.321	47.357	47.393	47.430	47.466	47.502
1,300	47.502										

TYPE R THERMOCOUPLES

TEMPERATURES IN DEGREES C (IPTS 1968). REFERENCE JUNCTION AT 0 C

DEG C	0	1	2	3	4	5	6	7	8	9	10
	THERMOELECTRIC VOLTAGE IN ABSOLUTE MILLIVOLTS										
-50	-0.226										
-40	-0.188	-0.192	-0.196	-0.200	-0.204	-0.207	-0.211	-0.215	-0.219	-0.223	-0.226
-30	-0.145	-0.150	-0.154	-0.158	-0.163	-0.167	-0.171	-0.175	-0.180	-0.184	-0.188
-20	-0.100	-0.105	-0.109	-0.114	-0.119	-0.123	-0.128	-0.132	-0.137	-0.141	-0.145
-10	-0.051	-0.056	-0.061	-0.066	-0.071	-0.076	-0.081	-0.086	-0.091	-0.095	-0.100
0	0.000	-0.005	-0.011	-0.016	-0.021	-0.026	-0.031	-0.036	-0.041	-0.046	-0.051
0	0.000	0.005	0.011	0.016	0.021	0.027	0.032	0.038	0.043	0.049	0.054
10	0.054	0.060	0.065	0.071	0.077	0.082	0.088	0.094	0.100	0.105	0.111
20	0.111	0.117	0.123	0.129	0.135	0.141	0.147	0.152	0.158	0.165	0.171
30	0.171	0.177	0.183	0.189	0.195	0.201	0.207	0.214	0.220	0.226	0.232
40	0.232	0.239	0.245	0.251	0.258	0.264	0.271	0.277	0.283	0.290	0.296
50	0.296	0.303	0.310	0.316	0.323	0.329	0.336	0.343	0.349	0.356	0.363
60	0.363	0.369	0.376	0.383	0.390	0.397	0.403	0.410	0.417	0.424	0.431
70	0.431	0.438	0.445	0.452	0.459	0.466	0.473	0.480	0.487	0.494	0.501
80	0.501	0.508	0.515	0.523	0.530	0.537	0.544	0.552	0.559	0.566	0.573
90	0.573	0.581	0.588	0.595	0.603	0.610	0.617	0.625	0.632	0.640	0.647
100	0.647	0.655	0.662	0.670	0.677	0.685	0.692	0.700	0.708	0.715	0.723
110	0.723	0.730	0.738	0.746	0.754	0.761	0.769	0.777	0.784	0.792	0.800
120	0.800	0.808	0.816	0.824	0.831	0.839	0.847	0.855	0.863	0.871	0.879
130	0.879	0.887	0.895	0.903	0.911	0.919	0.927	0.935	0.943	0.951	0.959
140	0.959	0.967	0.975	0.983	0.992	1.000	1.008	1.016	1.024	1.032	1.041
150	1.041	1.049	1.057	1.065	1.074	1.082	1.090	1.099	1.107	1.115	1.124
160	1.124	1.132	1.140	1.149	1.157	1.166	1.174	1.183	1.191	1.200	1.208
170	1.208	1.217	1.225	1.234	1.242	1.251	1.259	1.268	1.276	1.285	1.294
180	1.294	1.302	1.311	1.319	1.328	1.337	1.345	1.354	1.363	1.372	1.380
190	1.380	1.389	1.398	1.407	1.415	1.424	1.433	1.442	1.450	1.459	1.468
200	1.468	1.477	1.486	1.495	1.504	1.512	1.521	1.530	1.539	1.548	1.557
210	1.557	1.566	1.575	1.584	1.593	1.602	1.611	1.620	1.629	1.638	1.647
220	1.647	1.656	1.665	1.674	1.683	1.692	1.702	1.711	1.720	1.729	1.738
230	1.738	1.747	1.756	1.766	1.775	1.784	1.793	1.802	1.812	1.821	1.830
240	1.830	1.839	1.849	1.858	1.867	1.876	1.886	1.895	1.904	1.914	1.923
250	1.923	1.932	1.942	1.951	1.960	1.970	1.979	1.988	1.998	2.007	2.017
260	2.017	2.026	2.036	2.045	2.054	2.064	2.073	2.083	2.092	2.102	2.111
270	2.111	2.121	2.130	2.140	2.149	2.159	2.169	2.178	2.188	2.197	2.207
280	2.207	2.216	2.226	2.236	2.245	2.255	2.264	2.274	2.284	2.293	2.303
290	2.303	2.313	2.322	2.332	2.342	2.351	2.361	2.371	2.381	2.390	2.400
300	2.400	2.410	2.420	2.429	2.439	2.449	2.459	2.468	2.478	2.488	2.498
310	2.498	2.508	2.517	2.527	2.537	2.547	2.557	2.567	2.577	2.586	2.596
320	2.596	2.606	2.616	2.626	2.636	2.646	2.656	2.666	2.676	2.685	2.695
330	2.695	2.705	2.715	2.725	2.735	2.745	2.755	2.765	2.775	2.785	2.795
340	2.795	2.805	2.815	2.825	2.835	2.845	2.855	2.866	2.876	2.886	2.896
350	2.896	2.906	2.916	2.926	2.936	2.946	2.956	2.966	2.977	2.987	2.997
360	2.997	3.007	3.017	3.027	3.037	3.048	3.058	3.068	3.078	3.088	3.099
370	3.099	3.109	3.119	3.129	3.139	3.150	3.160	3.170	3.180	3.191	3.201
380	3.201	3.211	3.221	3.232	3.242	3.252	3.263	3.273	3.283	3.293	3.304
390	3.304	3.314	3.324	3.335	3.345	3.355	3.366	3.376	3.386	3.397	3.407
400	3.407	3.418	3.428	3.438	3.449	3.459	3.470	3.480	3.490	3.501	3.511
410	3.511	3.522	3.532	3.543	3.553	3.563	3.574	3.584	3.595	3.605	3.616
420	3.616	3.626	3.637	3.647	3.658	3.668	3.679	3.689	3.700	3.710	3.721
430	3.721	3.731	3.742	3.752	3.763	3.774	3.784	3.795	3.805	3.816	3.826
440	3.826	3.837	3.848	3.858	3.869	3.879	3.890	3.901	3.911	3.922	3.933
450	3.933	3.943	3.954	3.964	3.975	3.986	3.996	4.007	4.018	4.028	4.039
460	4.039	4.050	4.061	4.071	4.082	4.093	4.103	4.114	4.125	4.136	4.146
470	4.146	4.157	4.168	4.178	4.189	4.200	4.211	4.222	4.232	4.243	4.254
480	4.254	4.265	4.275	4.286	4.297	4.308	4.319	4.329	4.340	4.351	4.362
490	4.362	4.373	4.384	4.394	4.405	4.416	4.427	4.438	4.449	4.460	4.471
500	4.471	4.481	4.492	4.503	4.514	4.525	4.536	4.547	4.558	4.569	4.580
510	4.580	4.591	4.601	4.612	4.623	4.634	4.645	4.656	4.667	4.678	4.689
520	4.689	4.700	4.711	4.722	4.733	4.744	4.755	4.766	4.777	4.788	4.799
530	4.799	4.810	4.821	4.832	4.843	4.854	4.865	4.876	4.888	4.899	4.910
540	4.910	4.921	4.932	4.943	4.954	4.965	4.976	4.987	4.998	5.009	5.021

TYPE R THERMOCOUPLES

DEG C	0	1	2	3	4	5	6	7	8	9	10
					THERMOELECTRIC VOLTAGE IN ABSOLUTE MILLIVOLTS						
550	5.021	5.032	5.043	5.054	5.065	5.076	5.087	5.099	5.110	5.121	5.132
560	5.132	5.143	5.154	5.166	5.177	5.188	5.199	5.210	5.221	5.233	5.244
570	5.244	5.255	5.266	5.278	5.289	5.300	5.311	5.322	5.334	5.345	5.356
580	5.356	5.368	5.379	5.390	5.401	5.413	5.424	5.435	5.446	5.458	5.469
590	5.469	5.480	5.492	5.503	5.514	5.526	5.537	5.548	5.560	5.571	5.582
600	5.582	5.594	5.605	5.616	5.628	5.639	5.650	5.662	5.673	5.685	5.696
610	5.696	5.707	5.719	5.730	5.742	5.753	5.764	5.776	5.787	5.799	5.810
620	5.810	5.821	5.833	5.844	5.856	5.867	5.879	5.890	5.902	5.913	5.925
630	5.925	5.936	5.948	5.959	5.971	5.982	5.994	6.005	6.017	6.028	6.040
640	6.040	6.051	6.063	6.074	6.086	6.098	6.109	6.121	6.132	6.144	6.155
650	6.155	6.167	6.179	6.190	6.202	6.213	6.225	6.237	6.248	6.260	6.272
660	6.272	6.283	6.295	6.307	6.318	6.330	6.342	6.353	6.365	6.377	6.388
670	6.388	6.400	6.412	6.423	6.435	6.447	6.458	6.470	6.482	6.494	6.505
680	6.505	6.517	6.529	6.541	6.552	6.564	6.576	6.588	6.599	6.611	6.623
690	6.623	6.635	6.647	6.658	6.670	6.682	6.694	6.706	6.718	6.729	6.741
700	6.741	6.753	6.765	6.777	6.789	6.800	6.812	6.824	6.836	6.848	6.860
710	6.860	6.872	6.884	6.895	6.907	6.919	6.931	6.943	6.955	6.967	6.979
720	6.979	6.991	7.003	7.015	7.027	7.039	7.051	7.063	7.074	7.086	7.098
730	7.098	7.110	7.122	7.134	7.146	7.158	7.170	7.182	7.194	7.206	7.218
740	7.218	7.231	7.243	7.255	7.267	7.279	7.291	7.303	7.315	7.327	7.339
750	7.339	7.351	7.363	7.375	7.387	7.399	7.412	7.424	7.436	7.448	7.460
760	7.460	7.472	7.484	7.496	7.509	7.521	7.533	7.545	7.557	7.569	7.582
770	7.582	7.594	7.606	7.618	7.630	7.642	7.655	7.667	7.679	7.691	7.703
780	7.703	7.716	7.728	7.740	7.752	7.765	7.777	7.789	7.801	7.814	7.826
790	7.826	7.838	7.850	7.863	7.875	7.887	7.900	7.912	7.924	7.937	7.949
800	7.949	7.961	7.973	7.986	7.998	8.010	8.023	8.035	8.047	8.060	8.072
810	8.072	8.085	8.097	8.109	8.122	8.134	8.146	8.159	8.171	8.184	8.196
820	8.196	8.208	8.221	8.233	8.246	8.258	8.271	8.283	8.295	8.308	8.320
830	8.320	8.333	8.345	8.358	8.370	8.383	8.395	8.408	8.420	8.433	8.445
840	8.445	8.458	8.470	8.483	8.495	8.508	8.520	8.533	8.545	8.558	8.570
850	8.570	8.583	8.595	8.608	8.621	8.633	8.646	8.658	8.671	8.683	8.696
860	8.696	8.709	8.721	8.734	8.746	8.759	8.772	8.784	8.797	8.810	8.822
870	8.822	8.835	8.847	8.860	8.873	8.885	8.898	8.911	8.923	8.936	8.949
880	8.949	8.961	8.974	8.987	9.000	9.012	9.025	9.038	9.050	9.063	9.076
890	9.076	9.089	9.101	9.114	9.127	9.140	9.152	9.165	9.178	9.191	9.203
900	9.203	9.216	9.229	9.242	9.254	9.267	9.280	9.293	9.306	9.319	9.331
910	9.331	9.344	9.357	9.370	9.383	9.395	9.408	9.421	9.434	9.447	9.460
920	9.460	9.473	9.485	9.498	9.511	9.524	9.537	9.550	9.563	9.576	9.589
930	9.589	9.602	9.614	9.627	9.640	9.653	9.666	9.679	9.692	9.705	9.718
940	9.718	9.731	9.744	9.757	9.770	9.783	9.796	9.809	9.822	9.835	9.848
950	9.848	9.861	9.874	9.887	9.900	9.913	9.926	9.939	9.952	9.965	9.978
960	9.978	9.991	10.004	10.017	10.030	10.043	10.056	10.069	10.082	10.095	10.109
970	10.109	10.122	10.135	10.148	10.161	10.174	10.187	10.200	10.213	10.227	10.240
980	10.240	10.253	10.266	10.279	10.292	10.305	10.319	10.332	10.345	10.358	10.371
990	10.371	10.384	10.398	10.411	10.424	10.437	10.450	10.464	10.477	10.490	10.503
1,000	10.503	10.516	10.530	10.543	10.556	10.569	10.583	10.596	10.609	10.622	10.636
1,010	10.636	10.649	10.662	10.675	10.689	10.702	10.715	10.729	10.742	10.755	10.768
1,020	10.768	10.782	10.795	10.808	10.822	10.835	10.848	10.862	10.875	10.888	10.902
1,030	10.902	10.915	10.928	10.942	10.955	10.968	10.982	10.995	11.009	11.022	11.035
1,040	11.035	11.049	11.062	11.076	11.089	11.102	11.116	11.129	11.143	11.156	11.170
1,050	11.170	11.183	11.196	11.210	11.223	11.237	11.250	11.264	11.277	11.291	11.304
1,060	11.304	11.318	11.331	11.345	11.358	11.372	11.385	11.399	11.412	11.426	11.439
1,070	11.439	11.453	11.466	11.480	11.493	11.507	11.520	11.534	11.547	11.561	11.574
1,080	11.574	11.588	11.602	11.615	11.629	11.642	11.656	11.669	11.683	11.697	11.710
1,090	11.710	11.724	11.737	11.751	11.765	11.778	11.792	11.805	11.819	11.833	11.846
1,100	11.846	11.860	11.874	11.887	11.901	11.914	11.928	11.942	11.955	11.969	11.983
1,110	11.983	11.996	12.010	12.024	12.037	12.051	12.065	12.078	12.092	12.106	12.119
1,120	12.119	12.133	12.147	12.161	12.174	12.188	12.202	12.215	12.229	12.243	12.257
1,130	12.257	12.270	12.284	12.298	12.311	12.325	12.339	12.353	12.366	12.380	12.394
1,140	12.394	12.408	12.421	12.435	12.449	12.463	12.476	12.490	12.504	12.518	12.532

TYPE R THERMOCOUPLES

THERMOELECTRIC VOLTAGE IN ABSOLUTE MILLIVOLTS

DEG C	0	1	2	3	4	5	6	7	8	9	10
1,150	12.532	12.545	12.559	12.573	12.587	12.600	12.614	12.628	12.642	12.656	12.669
1,160	12.669	12.683	12.697	12.711	12.725	12.739	12.752	12.766	12.780	12.794	12.808
1,170	12.808	12.822	12.835	12.849	12.863	12.877	12.891	12.905	12.918	12.932	12.946
1,180	12.946	12.960	12.974	12.988	13.002	13.016	13.029	13.043	13.057	13.071	13.085
1,190	13.085	13.099	13.113	13.127	13.140	13.154	13.168	13.182	13.196	13.210	13.224
1,200	13.224	13.238	13.252	13.266	13.280	13.293	13.307	13.321	13.335	13.349	13.363
1,210	13.363	13.377	13.391	13.405	13.419	13.433	13.447	13.461	13.475	13.489	13.502
1,220	13.502	13.516	13.530	13.544	13.558	13.572	13.586	13.600	13.614	13.628	13.642
1,230	13.642	13.656	13.670	13.684	13.698	13.712	13.726	13.740	13.754	13.768	13.782
1,240	13.782	13.796	13.810	13.824	13.838	13.852	13.866	13.880	13.894	13.908	13.922
1,250	13.922	13.936	13.950	13.964	13.978	13.992	14.006	14.020	14.034	14.048	14.062
1,260	14.062	14.076	14.090	14.104	14.118	14.132	14.146	14.160	14.174	14.188	14.202
1,270	14.202	14.216	14.230	14.244	14.258	14.272	14.286	14.301	14.315	14.329	14.343
1,280	14.343	14.357	14.371	14.385	14.399	14.413	14.427	14.441	14.455	14.469	14.483
1,290	14.483	14.497	14.511	14.525	14.539	14.554	14.568	14.582	14.596	14.610	14.624
1,300	14.624	14.638	14.652	14.666	14.680	14.694	14.708	14.722	14.737	14.751	14.765
1,310	14.765	14.779	14.793	14.807	14.821	14.835	14.849	14.863	14.877	14.891	14.906
1,320	14.906	14.920	14.934	14.948	14.962	14.976	14.990	15.004	15.018	15.032	15.047
1,330	15.047	15.061	15.075	15.089	15.103	15.117	15.131	15.145	15.159	15.173	15.188
1,340	15.188	15.202	15.216	15.230	15.244	15.258	15.272	15.286	15.300	15.315	15.329
1,350	15.329	15.343	15.357	15.371	15.385	15.399	15.413	15.427	15.442	15.456	15.470
1,360	15.470	15.484	15.498	15.512	15.526	15.540	15.555	15.569	15.583	15.597	15.611
1,370	15.611	15.625	15.639	15.653	15.667	15.682	15.696	15.710	15.724	15.738	15.752
1,380	15.752	15.766	15.780	15.795	15.809	15.823	15.837	15.851	15.865	15.879	15.893
1,390	15.893	15.908	15.922	15.936	15.950	15.964	15.978	15.992	16.006	16.021	16.035
1,400	16.035	16.049	16.063	16.077	16.091	16.105	16.119	16.134	16.148	16.162	16.176
1,410	16.176	16.190	16.204	16.218	16.232	16.247	16.261	16.275	16.289	16.303	16.317
1,420	16.317	16.331	16.345	16.360	16.374	16.388	16.402	16.416	16.430	16.444	16.458
1,430	16.458	16.472	16.487	16.501	16.515	16.529	16.543	16.557	16.571	16.585	16.599
1,440	16.599	16.614	16.628	16.642	16.656	16.670	16.684	16.698	16.712	16.726	16.741
1,450	16.741	16.755	16.769	16.783	16.797	16.811	16.825	16.839	16.853	16.867	16.882
1,460	16.882	16.896	16.910	16.924	16.938	16.952	16.966	16.980	16.994	17.008	17.022
1,470	17.022	17.037	17.051	17.065	17.079	17.093	17.107	17.121	17.135	17.149	17.163
1,480	17.163	17.177	17.192	17.206	17.220	17.234	17.248	17.262	17.276	17.290	17.304
1,490	17.304	17.318	17.332	17.346	17.360	17.374	17.388	17.403	17.417	17.431	17.445
1,500	17.445	17.459	17.473	17.487	17.501	17.515	17.529	17.543	17.557	17.571	17.585
1,510	17.585	17.599	17.613	17.627	17.641	17.655	17.669	17.684	17.698	17.712	17.726
1,520	17.726	17.740	17.754	17.768	17.782	17.796	17.810	17.824	17.838	17.852	17.866
1,530	17.866	17.880	17.894	17.908	17.922	17.936	17.950	17.964	17.978	17.992	18.006
1,540	18.006	18.020	18.034	18.048	18.062	18.076	18.090	18.104	18.118	18.132	18.146
1,550	18.146	18.160	18.174	18.188	18.202	18.216	18.230	18.244	18.258	18.272	18.286
1,560	18.286	18.299	18.313	18.327	18.341	18.355	18.369	18.383	18.397	18.411	18.425
1,570	18.425	18.439	18.453	18.467	18.481	18.495	18.509	18.523	18.537	18.550	18.564
1,580	18.564	18.578	18.592	18.606	18.620	18.634	18.648	18.662	18.676	18.690	18.703
1,590	18.703	18.717	18.731	18.745	18.759	18.773	18.787	18.801	18.815	18.828	18.842
1,600	18.842	18.856	18.870	18.884	18.898	18.912	18.926	18.939	18.953	18.967	18.981
1,610	18.981	18.995	19.009	19.023	19.036	19.050	19.064	19.078	19.092	19.106	19.119
1,620	19.119	19.133	19.147	19.161	19.175	19.188	19.202	19.216	19.230	19.244	19.257
1,630	19.257	19.271	19.285	19.299	19.313	19.326	19.340	19.354	19.368	19.382	19.395
1,640	19.395	19.409	19.423	19.437	19.450	19.464	19.478	19.492	19.505	19.519	19.533
1,650	19.533	19.547	19.560	19.574	19.588	19.602	19.615	19.629	19.643	19.656	19.670
1,660	19.670	19.684	19.698	19.711	19.725	19.739	19.752	19.766	19.780	19.793	19.807
1,670	19.807	19.821	19.834	19.848	19.862	19.875	19.889	19.903	19.916	19.930	19.944
1,680	19.944	19.957	19.971	19.985	19.998	20.012	20.025	20.039	20.053	20.066	20.080
1,690	20.080	20.093	20.107	20.120	20.134	20.148	20.161	20.175	20.188	20.202	20.215
1,700	20.215	20.229	20.242	20.256	20.269	20.283	20.296	20.309	20.323	20.336	20.350
1,710	20.350	20.363	20.377	20.390	20.403	20.417	20.430	20.443	20.457	20.470	20.483
1,720	20.483	20.497	20.510	20.523	20.537	20.550	20.563	20.576	20.590	20.603	20.616
1,730	20.616	20.629	20.642	20.656	20.669	20.682	20.695	20.708	20.721	20.734	20.748
1,740	20.748	20.761	20.774	20.787	20.800	20.813	20.826	20.839	20.852	20.865	20.878
1,750	20.878	20.891	20.904	20.916	20.929	20.942	20.955	20.968	20.981	20.994	21.006
1,760	21.006	21.019	21.032	21.045	21.057	21.070	21.083	21.096	21.108		

TYPE S THERMOCOUPLES

DEG C	0	1	2	3	4	5	6	7	8	9	10
				THERMOELECTRIC VOLTAGE IN ABSOLUTE MILLIVOLTS							
-50	-0.236										
-40	-0.194	-0.199	-0.203	-0.207	-0.211	-0.215	-0.220	-0.224	-0.228	-0.232	-0.236
-30	-0.150	-0.155	-0.159	-0.164	-0.168	-0.173	-0.177	-0.181	-0.186	-0.190	-0.194
-20	-0.103	-0.108	-0.112	-0.117	-0.122	-0.127	-0.132	-0.136	-0.141	-0.145	-0.150
-10	-0.053	-0.058	-0.063	-0.068	-0.073	-0.078	-0.083	-0.088	-0.093	-0.098	-0.103
0	0.000	-0.005	-0.011	-0.016	-0.021	-0.027	-0.032	-0.037	-0.042	-0.048	-0.053
0	0.000	0.005	0.011	0.016	0.022	0.027	0.033	0.038	0.044	0.050	0.055
10	0.055	0.061	0.067	0.072	0.078	0.084	0.090	0.095	0.101	0.107	0.113
20	0.113	0.119	0.125	0.131	0.137	0.142	0.148	0.154	0.161	0.167	0.173
30	0.173	0.179	0.185	0.191	0.197	0.203	0.210	0.216	0.222	0.228	0.235
40	0.235	0.241	0.247	0.254	0.260	0.266	0.273	0.279	0.286	0.292	0.299
50	0.299	0.305	0.312	0.318	0.325	0.331	0.338	0.345	0.351	0.358	0.365
60	0.365	0.371	0.378	0.385	0.391	0.398	0.405	0.412	0.419	0.425	0.432
70	0.432	0.439	0.446	0.453	0.460	0.467	0.474	0.481	0.488	0.495	0.502
80	0.502	0.509	0.516	0.523	0.530	0.537	0.544	0.551	0.558	0.566	0.573
90	0.573	0.580	0.587	0.594	0.602	0.609	0.616	0.623	0.631	0.638	0.645
100	0.645	0.653	0.660	0.667	0.675	0.682	0.690	0.697	0.704	0.712	0.719
110	0.719	0.727	0.734	0.742	0.749	0.757	0.764	0.772	0.780	0.787	0.795
120	0.795	0.802	0.810	0.818	0.825	0.833	0.841	0.848	0.856	0.864	0.872
130	0.872	0.879	0.887	0.895	0.903	0.910	0.918	0.926	0.934	0.942	0.950
140	0.950	0.957	0.965	0.973	0.981	0.989	0.997	1.005	1.013	1.021	1.029
150	1.029	1.037	1.045	1.053	1.061	1.069	1.077	1.085	1.093	1.101	1.109
160	1.109	1.117	1.125	1.133	1.141	1.149	1.158	1.166	1.174	1.182	1.190
170	1.190	1.198	1.207	1.215	1.223	1.231	1.240	1.248	1.256	1.264	1.273
180	1.273	1.281	1.289	1.297	1.306	1.314	1.322	1.331	1.339	1.347	1.356
190	1.356	1.364	1.373	1.381	1.389	1.398	1.406	1.415	1.423	1.432	1.440
200	1.440	1.448	1.457	1.465	1.474	1.482	1.491	1.499	1.508	1.516	1.525
210	1.525	1.534	1.542	1.551	1.559	1.568	1.576	1.585	1.594	1.602	1.611
220	1.611	1.620	1.628	1.637	1.645	1.654	1.663	1.671	1.680	1.689	1.698
230	1.698	1.706	1.715	1.724	1.732	1.741	1.750	1.759	1.767	1.776	1.785
240	1.785	1.794	1.802	1.811	1.820	1.829	1.838	1.846	1.855	1.864	1.873
250	1.873	1.882	1.891	1.899	1.908	1.917	1.926	1.935	1.944	1.953	1.962
260	1.962	1.971	1.979	1.988	1.997	2.006	2.015	2.024	2.033	2.042	2.051
270	2.051	2.060	2.069	2.078	2.087	2.096	2.105	2.114	2.123	2.132	2.141
280	2.141	2.150	2.159	2.168	2.177	2.186	2.195	2.204	2.213	2.222	2.232
290	2.232	2.241	2.250	2.259	2.268	2.277	2.286	2.295	2.304	2.314	2.323
300	2.323	2.332	2.341	2.350	2.359	2.368	2.378	2.387	2.396	2.405	2.414
310	2.414	2.424	2.433	2.442	2.451	2.460	2.470	2.479	2.488	2.497	2.506
320	2.506	2.516	2.525	2.534	2.543	2.553	2.562	2.571	2.581	2.590	2.599
330	2.599	2.608	2.618	2.627	2.636	2.646	2.655	2.664	2.674	2.683	2.692
340	2.692	2.702	2.711	2.720	2.730	2.739	2.748	2.758	2.767	2.776	2.786
350	2.786	2.795	2.805	2.814	2.823	2.833	2.842	2.852	2.861	2.870	2.880
360	2.880	2.889	2.899	2.908	2.917	2.927	2.936	2.946	2.955	2.965	2.974
370	2.974	2.984	2.993	3.003	3.012	3.022	3.031	3.041	3.050	3.059	3.069
380	3.069	3.078	3.088	3.097	3.107	3.117	3.126	3.136	3.145	3.155	3.164
390	3.164	3.174	3.183	3.193	3.202	3.212	3.221	3.231	3.241	3.250	3.260
400	3.260	3.269	3.279	3.288	3.298	3.308	3.317	3.327	3.336	3.346	3.356
410	3.356	3.365	3.375	3.384	3.394	3.404	3.413	3.423	3.433	3.442	3.452
420	3.452	3.462	3.471	3.481	3.491	3.500	3.510	3.520	3.529	3.539	3.549
430	3.549	3.558	3.568	3.578	3.587	3.597	3.607	3.616	3.626	3.636	3.645
440	3.645	3.655	3.665	3.675	3.684	3.694	3.704	3.714	3.723	3.733	3.743
450	3.743	3.752	3.762	3.772	3.782	3.791	3.801	3.811	3.821	3.831	3.840
460	3.840	3.850	3.860	3.870	3.879	3.889	3.899	3.909	3.919	3.928	3.938
470	3.938	3.948	3.958	3.968	3.977	3.987	3.997	4.007	4.017	4.027	4.036
480	4.036	4.046	4.056	4.066	4.076	4.086	4.095	4.105	4.115	4.125	4.135
490	4.135	4.145	4.155	4.164	4.174	4.184	4.194	4.204	4.214	4.224	4.234
500	4.234	4.243	4.253	4.263	4.273	4.283	4.293	4.303	4.313	4.323	4.333
510	4.333	4.343	4.352	4.362	4.372	4.382	4.392	4.402	4.412	4.422	4.432
520	4.432	4.442	4.452	4.462	4.472	4.482	4.492	4.502	4.512	4.522	4.532
530	4.532	4.542	4.552	4.562	4.572	4.582	4.592	4.602	4.612	4.622	4.632
540	4.632	4.642	4.652	4.662	4.672	4.682	4.692	4.702	4.712	4.722	4.732

TYPE S THERMOCOUPLES

DEG C	0	1	2	3	4	5	6	7	8	9	10

THERMOELECTRIC VOLTAGE IN ABSOLUTE MILLIVOLTS

DEG C	0	1	2	3	4	5	6	7	8	9	10
550	4.732	4.742	4.752	4.762	4.772	4.782	4.792	4.802	4.812	4.822	4.832
560	4.832	4.842	4.852	4.862	4.873	4.883	4.893	4.903	4.913	4.923	4.933
570	4.933	4.943	4.953	4.963	4.973	4.984	4.994	5.004	5.014	5.024	5.034
580	5.034	5.044	5.054	5.065	5.075	5.085	5.095	5.105	5.115	5.125	5.136
590	5.136	5.146	5.156	5.166	5.176	5.186	5.197	5.207	5.217	5.227	5.237
600	5.237	5.247	5.258	5.268	5.278	5.288	5.298	5.309	5.319	5.329	5.339
610	5.339	5.350	5.360	5.370	5.380	5.391	5.401	5.411	5.421	5.431	5.442
620	5.442	5.452	5.462	5.473	5.483	5.493	5.503	5.514	5.524	5.534	5.544
630	5.544	5.555	5.565	5.575	5.586	5.596	5.606	5.617	5.627	5.637	5.648
640	5.648	5.658	5.668	5.679	5.689	5.700	5.710	5.720	5.731	5.741	5.751
650	5.751	5.762	5.772	5.782	5.793	5.803	5.814	5.824	5.834	5.845	5.855
660	5.855	5.866	5.876	5.887	5.897	5.907	5.918	5.928	5.939	5.949	5.960
670	5.960	5.970	5.980	5.991	6.001	6.012	6.022	6.033	6.043	6.054	6.064
680	6.064	6.075	6.085	6.096	6.106	6.117	6.127	6.138	6.148	6.159	6.169
690	6.169	6.180	6.190	6.201	6.211	6.222	6.232	6.243	6.253	6.264	6.274
700	6.274	6.285	6.295	6.306	6.316	6.327	6.338	6.348	6.359	6.369	6.380
710	6.380	6.390	6.401	6.412	6.422	6.433	6.443	6.454	6.465	6.475	6.486
720	6.486	6.496	6.507	6.518	6.528	6.539	6.549	6.560	6.571	6.581	6.592
730	6.592	6.603	6.613	6.624	6.635	6.645	6.656	6.667	6.677	6.688	6.699
740	6.699	6.709	6.720	6.731	6.741	6.752	6.763	6.773	6.784	6.795	6.805
750	6.805	6.816	6.827	6.838	6.848	6.859	6.870	6.880	6.891	6.902	6.913
760	6.913	6.923	6.934	6.945	6.956	6.966	6.977	6.988	6.999	7.009	7.020
770	7.020	7.031	7.042	7.053	7.063	7.074	7.085	7.096	7.107	7.117	7.128
780	7.128	7.139	7.150	7.161	7.171	7.182	7.193	7.204	7.215	7.225	7.236
790	7.236	7.247	7.258	7.269	7.280	7.291	7.301	7.312	7.323	7.334	7.345
800	7.345	7.356	7.367	7.377	7.388	7.399	7.410	7.421	7.432	7.443	7.454
810	7.454	7.465	7.476	7.486	7.497	7.508	7.519	7.530	7.541	7.552	7.563
820	7.563	7.574	7.585	7.596	7.607	7.618	7.629	7.640	7.651	7.661	7.672
830	7.672	7.683	7.694	7.705	7.716	7.727	7.738	7.749	7.760	7.771	7.782
840	7.782	7.793	7.804	7.815	7.826	7.837	7.848	7.859	7.870	7.881	7.892
850	7.892	7.903	7.915	7.926	7.937	7.948	7.959	7.970	7.981	7.992	8.003
860	8.003	8.014	8.025	8.036	8.047	8.058	8.069	8.081	8.092	8.103	8.114
870	8.114	8.125	8.136	8.147	8.158	8.169	8.180	8.192	8.203	8.214	8.225
880	8.225	8.236	8.247	8.258	8.270	8.281	8.292	8.303	8.314	8.325	8.336
890	8.336	8.348	8.359	8.370	8.381	8.392	8.404	8.415	8.426	8.437	8.448
900	8.448	8.460	8.471	8.482	8.493	8.504	8.516	8.527	8.538	8.549	8.560
910	8.560	8.572	8.583	8.594	8.605	8.617	8.628	8.639	8.650	8.662	8.673
920	8.673	8.684	8.695	8.707	8.718	8.729	8.741	8.752	8.763	8.774	8.786
930	8.786	8.797	8.808	8.820	8.831	8.842	8.854	8.865	8.876	8.888	8.899
940	8.899	8.910	8.922	8.933	8.944	8.956	8.967	8.978	8.990	9.001	9.012
950	9.012	9.024	9.035	9.047	9.058	9.069	9.081	9.092	9.103	9.115	9.126
960	9.126	9.138	9.149	9.160	9.172	9.183	9.195	9.206	9.217	9.229	9.240
970	9.240	9.252	9.263	9.275	9.286	9.298	9.309	9.320	9.332	9.343	9.355
980	9.355	9.366	9.378	9.389	9.401	9.412	9.424	9.435	9.447	9.458	9.470
990	9.470	9.481	9.493	9.504	9.516	9.527	9.539	9.550	9.562	9.573	9.585
1,000	9.585	9.596	9.608	9.619	9.631	9.642	9.654	9.665	9.677	9.689	9.700
1,010	9.700	9.712	9.723	9.735	9.746	9.758	9.770	9.781	9.793	9.804	9.816
1,020	9.816	9.828	9.839	9.851	9.862	9.874	9.886	9.897	9.909	9.920	9.932
1,030	9.932	9.944	9.955	9.967	9.979	9.990	10.002	10.013	10.025	10.037	10.048
1,040	10.048	10.060	10.072	10.083	10.095	10.107	10.118	10.130	10.142	10.154	10.165
1,050	10.165	10.177	10.189	10.200	10.212	10.224	10.235	10.247	10.259	10.271	10.282
1,060	10.282	10.294	10.306	10.318	10.329	10.341	10.353	10.364	10.376	10.388	10.400
1,070	10.400	10.411	10.423	10.435	10.447	10.459	10.470	10.482	10.494	10.506	10.517
1,080	10.517	10.529	10.541	10.553	10.565	10.576	10.588	10.600	10.612	10.624	10.635
1,090	10.635	10.647	10.659	10.671	10.683	10.694	10.706	10.718	10.730	10.742	10.754
1,100	10.754	10.765	10.777	10.789	10.801	10.813	10.825	10.836	10.848	10.860	10.872
1,110	10.872	10.884	10.896	10.908	10.919	10.931	10.943	10.955	10.967	10.979	10.991
1,120	10.991	11.003	11.014	11.026	11.038	11.050	11.062	11.074	11.086	11.098	11.110
1,130	11.110	11.121	11.133	11.145	11.157	11.169	11.181	11.193	11.205	11.217	11.229
1,140	11.229	11.241	11.252	11.264	11.276	11.288	11.300	11.312	11.324	11.336	11.348

TYPE S THERMOCOUPLES

TEMPERATURES IN DEGREES C (IPTS 1968).   REFERENCE JUNCTION AT 0 C

DEG C	0	1	2	3	4	5	6	7	8	9	10
				THERMOELECTRIC VOLTAGE IN ABSOLUTE MILLIVOLTS							
1,150	11.348	11.360	11.372	11.384	11.396	11.408	11.420	11.432	11.443	11.455	11.467
1,160	11.467	11.479	11.491	11.503	11.515	11.527	11.539	11.551	11.563	11.575	11.587
1,170	11.587	11.599	11.611	11.623	11.635	11.647	11.659	11.671	11.683	11.695	11.707
1,180	11.707	11.719	11.731	11.743	11.755	11.767	11.779	11.791	11.803	11.815	11.827
1,190	11.827	11.839	11.851	11.863	11.875	11.887	11.899	11.911	11.923	11.935	11.947
1,200	11.947	11.959	11.971	11.983	11.995	12.007	12.019	12.031	12.043	12.055	12.067
1,210	12.067	12.079	12.091	12.103	12.116	12.128	12.140	12.152	12.164	12.176	12.188
1,220	12.188	12.200	12.212	12.224	12.236	12.248	12.260	12.272	12.284	12.296	12.308
1,230	12.308	12.320	12.332	12.345	12.357	12.369	12.381	12.393	12.405	12.417	12.429
1,240	12.429	12.441	12.453	12.465	12.477	12.489	12.501	12.514	12.526	12.538	12.550
1,250	12.550	12.562	12.574	12.586	12.598	12.610	12.622	12.634	12.647	12.659	12.671
1,260	12.671	12.683	12.695	12.707	12.719	12.731	12.743	12.755	12.767	12.780	12.792
1,270	12.792	12.804	12.816	12.828	12.840	12.852	12.864	12.876	12.888	12.901	12.913
1,280	12.913	12.925	12.937	12.949	12.961	12.973	12.985	12.997	13.010	13.022	13.034
1,290	13.034	13.046	13.058	13.070	13.082	13.094	13.107	13.119	13.131	13.143	13.155
1,300	13.155	13.167	13.179	13.191	13.203	13.216	13.228	13.240	13.252	13.264	13.276
1,310	13.276	13.288	13.300	13.313	13.325	13.337	13.349	13.361	13.373	13.385	13.397
1,320	13.397	13.410	13.422	13.434	13.446	13.458	13.470	13.482	13.495	13.507	13.519
1,330	13.519	13.531	13.543	13.555	13.567	13.579	13.592	13.604	13.616	13.628	13.640
1,340	13.640	13.652	13.664	13.677	13.689	13.701	13.713	13.725	13.737	13.749	13.761
1,350	13.761	13.774	13.786	13.798	13.810	13.822	13.834	13.846	13.859	13.871	13.883
1,360	13.883	13.895	13.907	13.919	13.931	13.943	13.956	13.968	13.980	13.992	14.004
1,370	14.004	14.016	14.028	14.040	14.053	14.065	14.077	14.089	14.101	14.113	14.125
1,380	14.125	14.138	14.150	14.162	14.174	14.186	14.198	14.210	14.222	14.235	14.247
1,390	14.247	14.259	14.271	14.283	14.295	14.307	14.319	14.332	14.344	14.356	14.368
1,400	14.368	14.380	14.392	14.404	14.416	14.429	14.441	14.453	14.465	14.477	14.489
1,410	14.489	14.501	14.513	14.526	14.538	14.550	14.562	14.574	14.586	14.598	14.610
1,420	14.610	14.622	14.635	14.647	14.659	14.671	14.683	14.695	14.707	14.719	14.731
1,430	14.731	14.744	14.756	14.768	14.780	14.792	14.804	14.816	14.828	14.840	14.852
1,440	14.852	14.865	14.877	14.889	14.901	14.913	14.925	14.937	14.949	14.961	14.973
1,450	14.973	14.985	14.998	15.010	15.022	15.034	15.046	15.058	15.070	15.082	15.094
1,460	15.094	15.106	15.118	15.130	15.143	15.155	15.167	15.179	15.191	15.203	15.215
1,470	15.215	15.227	15.239	15.251	15.263	15.275	15.287	15.299	15.311	15.324	15.336
1,480	15.336	15.348	15.360	15.372	15.384	15.396	15.408	15.420	15.432	15.444	15.456
1,490	15.456	15.468	15.480	15.492	15.504	15.516	15.528	15.540	15.552	15.564	15.576
1,500	15.576	15.589	15.601	15.613	15.625	15.637	15.649	15.661	15.673	15.685	15.697
1,510	15.697	15.709	15.721	15.733	15.745	15.757	15.769	15.781	15.793	15.805	15.817
1,520	15.817	15.829	15.841	15.853	15.865	15.877	15.889	15.901	15.913	15.925	15.937
1,530	15.937	15.949	15.961	15.973	15.985	15.997	16.009	16.021	16.033	16.045	16.057
1,540	16.057	16.069	16.080	16.092	16.104	16.116	16.128	16.140	16.152	16.164	16.176
1,550	16.176	16.188	16.200	16.212	16.224	16.236	16.248	16.260	16.272	16.284	16.296
1,560	16.296	16.308	16.319	16.331	16.343	16.355	16.367	16.379	16.391	16.403	16.415
1,570	16.415	16.427	16.439	16.451	16.462	16.474	16.486	16.498	16.510	16.522	16.534
1,580	16.534	16.546	16.558	16.569	16.581	16.593	16.605	16.617	16.629	16.641	16.653
1,590	16.653	16.664	16.676	16.688	16.700	16.712	16.724	16.736	16.747	16.759	16.771
1,600	16.771	16.783	16.795	16.807	16.819	16.830	16.842	16.854	16.866	16.878	16.890
1,610	16.890	16.901	16.913	16.925	16.937	16.949	16.960	16.972	16.984	16.996	17.008
1,620	17.008	17.019	17.031	17.043	17.055	17.067	17.078	17.090	17.102	17.114	17.125
1,630	17.125	17.137	17.149	17.161	17.173	17.184	17.196	17.208	17.220	17.231	17.243
1,640	17.243	17.255	17.267	17.278	17.290	17.302	17.313	17.325	17.337	17.349	17.360
1,650	17.360	17.372	17.384	17.396	17.407	17.419	17.431	17.442	17.454	17.466	17.477
1,660	17.477	17.489	17.501	17.512	17.524	17.536	17.548	17.559	17.571	17.583	17.594
1,670	17.594	17.606	17.617	17.629	17.641	17.652	17.664	17.676	17.687	17.699	17.711
1,680	17.711	17.722	17.734	17.745	17.757	17.769	17.780	17.792	17.803	17.815	17.826
1,690	17.826	17.838	17.850	17.861	17.873	17.884	17.896	17.907	17.919	17.930	17.942
1,700	17.942	17.953	17.965	17.976	17.988	17.999	18.010	18.022	18.033	18.045	18.056
1,710	18.056	18.068	18.079	18.090	18.102	18.113	18.124	18.136	18.147	18.158	18.170
1,720	18.170	18.181	18.192	18.204	18.215	18.226	18.237	18.249	18.260	18.271	18.282
1,730	18.282	18.293	18.305	18.316	18.327	18.338	18.349	18.360	18.372	18.383	18.394
1,740	18.394	18.405	18.416	18.427	18.438	18.449	18.460	18.471	18.482	18.493	18.504
1,750	18.504	18.515	18.526	18.536	18.547	18.558	18.569	18.580	18.591	18.602	18.612
1,760	18.612	18.623	18.634	18.645	18.655	18.666	18.677	18.687	18.698		

TYPE T THERMOCOUPLES

DEG C	0	1	2	3	4	5	6	7	8	9	10
				THERMOELECTRIC VOLTAGE IN ABSOLUTE MILLIVOLTS							
-190	-5.439	-5.456	-5.473	-5.489	-5.506	-5.522	-5.539	-5.555	-5.571	-5.587	-5.603
-180	-5.261	-5.279	-5.297	-5.315	-5.333	-5.351	-5.369	-5.387	-5.404	-5.421	-5.439
-170	-5.069	-5.089	-5.109	-5.128	-5.147	-5.167	-5.186	-5.205	-5.223	-5.242	-5.261
-160	-4.865	-4.886	-4.907	-4.928	-4.948	-4.969	-4.989	-5.010	-5.030	-5.050	-5.069
-150	-4.648	-4.670	-4.693	-4.715	-4.737	-4.758	-4.780	-4.801	-4.823	-4.844	-4.865
-140	-4.419	-4.442	-4.466	-4.489	-4.512	-4.535	-4.558	-4.581	-4.603	-4.626	-4.648
-130	-4.177	-4.202	-4.226	-4.251	-4.275	-4.299	-4.323	-4.347	-4.371	-4.395	-4.419
-120	-3.923	-3.949	-3.974	-4.000	-4.026	-4.051	-4.077	-4.102	-4.127	-4.152	-4.177
-110	-3.656	-3.684	-3.711	-3.737	-3.764	-3.791	-3.818	-3.844	-3.870	-3.897	-3.923
-100	-3.378	-3.407	-3.435	-3.463	-3.491	-3.519	-3.547	-3.574	-3.602	-3.629	-3.656
-90	-3.089	-3.118	-3.147	-3.177	-3.206	-3.235	-3.264	-3.293	-3.321	-3.350	-3.378
-80	-2.788	-2.818	-2.849	-2.879	-2.909	-2.939	-2.970	-2.999	-3.029	-3.059	-3.089
-70	-2.475	-2.507	-2.539	-2.570	-2.602	-2.633	-2.664	-2.695	-2.726	-2.757	-2.788
-60	-2.152	-2.185	-2.218	-2.250	-2.283	-2.315	-2.348	-2.380	-2.412	-2.444	-2.475
-50	-1.819	-1.853	-1.886	-1.920	-1.953	-1.987	-2.020	-2.053	-2.087	-2.120	-2.152
-40	-1.475	-1.510	-1.544	-1.579	-1.614	-1.648	-1.682	-1.717	-1.751	-1.785	-1.819
-30	-1.121	-1.157	-1.192	-1.228	-1.263	-1.299	-1.334	-1.370	-1.405	-1.440	-1.475
-20	-0.757	-0.794	-0.830	-0.867	-0.903	-0.940	-0.976	-1.013	-1.049	-1.085	-1.121
-10	-0.383	-0.421	-0.458	-0.496	-0.534	-0.571	-0.608	-0.646	-0.683	-0.720	-0.757
0	0.000	-0.039	-0.077	-0.116	-0.154	-0.193	-0.231	-0.269	-0.307	-0.345	-0.383
0	0.000	0.039	0.078	0.117	0.156	0.195	0.234	0.273	0.312	0.351	0.391
10	0.391	0.430	0.470	0.510	0.549	0.589	0.629	0.669	0.709	0.749	0.789
20	0.789	0.830	0.870	0.911	0.951	0.992	1.032	1.073	1.114	1.155	1.196
30	1.196	1.237	1.279	1.320	1.361	1.403	1.444	1.486	1.528	1.569	1.611
40	1.611	1.653	1.695	1.738	1.780	1.822	1.865	1.907	1.950	1.992	2.035
50	2.035	2.078	2.121	2.164	2.207	2.250	2.294	2.337	2.380	2.424	2.467
60	2.467	2.511	2.555	2.599	2.643	2.687	2.731	2.775	2.819	2.864	2.908
70	2.908	2.953	2.997	3.042	3.087	3.131	3.176	3.221	3.266	3.312	3.357
80	3.357	3.402	3.447	3.493	3.538	3.584	3.630	3.676	3.721	3.767	3.813
90	3.813	3.859	3.906	3.952	3.998	4.044	4.091	4.137	4.184	4.231	4.277
100	4.277	4.324	4.371	4.418	4.465	4.512	4.559	4.607	4.654	4.701	4.749
110	4.749	4.796	4.844	4.891	4.939	4.987	5.035	5.083	5.131	5.179	5.227
120	5.227	5.275	5.324	5.372	5.420	5.469	5.517	5.566	5.615	5.663	5.712
130	5.712	5.761	5.810	5.859	5.908	5.957	6.007	6.056	6.105	6.155	6.204
140	6.204	6.254	6.303	6.353	6.403	6.452	6.502	6.552	6.602	6.652	6.702
150	6.702	6.753	6.803	6.853	6.903	6.954	7.004	7.055	7.106	7.156	7.207
160	7.207	7.258	7.309	7.360	7.411	7.462	7.513	7.564	7.615	7.666	7.718
170	7.718	7.769	7.821	7.872	7.924	7.975	8.027	8.079	8.131	8.183	8.235
180	8.235	8.287	8.339	8.391	8.443	8.495	8.548	8.600	8.652	8.705	8.757
190	8.757	8.810	8.863	8.915	8.968	9.021	9.074	9.127	9.180	9.233	9.286
200	9.286	9.339	9.392	9.446	9.499	9.553	9.606	9.659	9.713	9.767	9.820
210	9.820	9.874	9.928	9.982	10.036	10.090	10.144	10.198	10.252	10.306	10.360
220	10.360	10.414	10.469	10.523	10.578	10.632	10.687	10.741	10.796	10.851	10.905
230	10.905	10.960	11.015	11.070	11.125	11.180	11.235	11.290	11.345	11.401	11.456
240	11.456	11.511	11.566	11.622	11.677	11.733	11.788	11.844	11.900	11.956	12.011
250	12.011	12.067	12.123	12.179	12.235	12.291	12.347	12.403	12.459	12.515	12.572
260	12.572	12.628	12.684	12.741	12.797	12.854	12.910	12.967	13.024	13.080	13.137
270	13.137	13.194	13.251	13.307	13.364	13.421	13.478	13.535	13.592	13.650	13.707
280	13.707	13.764	13.821	13.879	13.936	13.993	14.051	14.108	14.166	14.223	14.281
290	14.281	14.339	14.396	14.454	14.512	14.570	14.628	14.686	14.744	14.802	14.860
300	14.860	14.918	14.976	15.034	15.092	15.151	15.209	15.267	15.326	15.384	15.443
310	15.443	15.501	15.560	15.619	15.677	15.736	15.795	15.853	15.912	15.971	16.030
320	16.030	16.089	16.148	16.207	16.266	16.325	16.384	16.444	16.503	16.562	16.621
330	16.621	16.681	16.740	16.800	16.859	16.919	16.978	17.038	17.097	17.157	17.217
340	17.217	17.277	17.336	17.396	17.456	17.516	17.576	17.636	17.696	17.756	17.816
350	17.816	17.877	17.937	17.997	18.057	18.118	18.178	18.238	18.299	18.359	18.420
360	18.420	18.480	18.541	18.602	18.662	18.723	18.784	18.845	18.905	18.966	19.027
370	19.027	19.088	19.149	19.210	19.271	19.332	19.393	19.455	19.516	19.577	19.638
380	19.638	19.699	19.761	19.822	19.883	19.945	20.006	20.068	20.129	20.191	20.252
390	20.252	20.314	20.376	20.437	20.499	20.560	20.622	20.684	20.746	20.807	20.869
400	20.869										

W3 Re/W25 Re Seebeck Voltages.

EMF in Millivolts

Reference Junctions at 0°C

Thermoelectric Voltage in Millivolts

DEG C	0	1	2	3	4	5	6	7	8	9	10	DEG C
0	0.000	0.010	0.019	0.029	0.039	0.048	0.058	0.068	0.078	0.088	0.098	0
10	0.098	0.108	0.118	0.128	0.138	0.148	0.158	0.169	0.179	0.189	0.199	10
20	0.199	0.210	0.220	0.231	0.241	0.252	0.262	0.273	0.284	0.294	0.305	20
30	0.305	0.316	0.327	0.338	0.348	0.359	0.370	0.381	0.392	0.403	0.415	30
40	0.415	0.426	0.437	0.448	0.459	0.471	0.482	0.493	0.505	0.516	0.528	40
50	0.528	0.539	0.551	0.562	0.574	0.586	0.597	0.609	0.621	0.633	0.644	50
60	0.644	0.656	0.668	0.680	0.692	0.704	0.716	0.728	0.740	0.752	0.765	60
70	0.765	0.777	0.789	0.801	0.814	0.826	0.838	0.851	0.863	0.876	0.888	70
80	0.888	0.901	0.913	0.926	0.939	0.951	0.964	0.977	0.989	1.002	1.015	80
90	1.015	1.028	1.041	1.054	1.067	1.080	1.093	1.106	1.119	1.132	1.145	90
100	1.145	1.158	1.171	1.185	1.198	1.211	1.225	1.238	1.251	1.265	1.278	100
110	1.278	1.292	1.305	1.319	1.332	1.346	1.360	1.373	1.387	1.401	1.414	110
120	1.414	1.428	1.442	1.456	1.470	1.484	1.498	1.512	1.526	1.540	1.554	120
130	1.554	1.568	1.582	1.596	1.610	1.624	1.638	1.653	1.667	1.681	1.696	130
140	1.696	1.710	1.724	1.739	1.753	1.768	1.782	1.797	1.811	1.826	1.840	140
150	1.840	1.855	1.869	1.884	1.899	1.914	1.928	1.943	1.958	1.973	1.988	150
160	1.988	2.002	2.017	2.032	2.047	2.062	2.077	2.092	2.107	2.122	2.137	160
170	2.137	2.153	2.168	2.183	2.198	2.213	2.229	2.244	2.259	2.275	2.290	170
180	2.290	2.305	2.321	2.336	2.352	2.367	2.383	2.398	2.414	2.429	2.445	180
190	2.445	2.460	2.476	2.492	2.507	2.523	2.539	2.555	2.570	2.586	2.602	190
200	2.602	2.618	2.634	2.650	2.665	2.681	2.697	2.713	2.729	2.745	2.761	200
210	2.761	2.777	2.794	2.810	2.826	2.842	2.858	2.874	2.891	2.907	2.923	210
220	2.923	2.939	2.956	2.972	2.988	3.005	3.021	3.037	3.054	3.070	3.087	220
230	3.087	3.103	3.120	3.136	3.153	3.169	3.186	3.203	3.219	3.236	3.253	230
240	3.253	3.269	3.286	3.303	3.319	3.336	3.353	3.370	3.387	3.403	3.420	240
250	3.420	3.437	3.454	3.471	3.488	3.505	3.522	3.539	3.556	3.573	3.590	250
260	3.590	3.607	3.624	3.641	3.658	3.676	3.693	3.710	3.727	3.744	3.761	260
270	3.761	3.779	3.796	3.813	3.831	3.848	3.865	3.883	3.900	3.917	3.935	270
280	3.935	3.952	3.970	3.987	4.005	4.022	4.040	4.057	4.075	4.092	4.110	280
290	4.110	4.127	4.145	4.162	4.180	4.198	4.215	4.233	4.251	4.269	4.286	290
300	4.286	4.304	4.322	4.340	4.357	4.375	4.393	4.411	4.429	4.447	4.464	300
310	4.464	4.482	4.500	4.518	4.536	4.554	4.572	4.590	4.608	4.626	4.644	310
320	4.644	4.662	4.680	4.698	4.716	4.735	4.753	4.771	4.789	4.807	4.825	320
330	4.825	4.843	4.862	4.880	4.898	4.916	4.935	4.953	4.971	4.990	5.008	330
340	5.008	5.026	5.045	5.063	5.081	5.100	5.118	5.136	5.155	5.173	5.192	340

W3 Re/W25 Re Seebeck Voltages.

DEG C	0	1	2	3	4	5	6	7	8	9	10	DEG C
Thermoelectric Voltage in Millivolts												
350	5.192	5.210	5.229	5.247	5.266	5.284	5.303	5.321	5.340	5.358	5.377	350
360	5.377	5.395	5.414	5.433	5.451	5.470	5.489	5.507	5.526	5.545	5.563	360
370	5.563	5.582	5.601	5.619	5.638	5.657	5.676	5.695	5.713	5.732	5.751	370
380	5.751	5.770	5.789	5.807	5.826	5.845	5.864	5.883	5.902	5.921	5.940	380
390	5.940	5.959	5.978	5.996	6.015	6.034	6.053	6.072	6.091	6.110	6.129	390
400	6.129	6.148	6.168	6.187	6.206	6.225	6.244	6.263	6.282	6.301	6.320	400
410	6.320	6.339	6.359	6.378	6.397	6.416	6.435	6.454	6.474	6.493	6.512	410
420	6.512	6.531	6.551	6.570	6.589	6.608	6.628	6.647	6.666	6.686	6.705	420
430	6.705	6.724	6.743	6.763	6.782	6.802	6.821	6.840	6.860	6.879	6.898	430
440	6.898	6.918	6.937	6.957	6.976	6.996	7.015	7.035	7.054	7.073	7.093	440
450	7.093	7.112	7.132	7.151	7.171	7.191	7.210	7.230	7.249	7.269	7.288	450
460	7.288	7.308	7.327	7.347	7.367	7.386	7.406	7.425	7.445	7.465	7.484	460
470	7.484	7.504	7.524	7.543	7.563	7.583	7.602	7.622	7.642	7.661	7.681	470
480	7.681	7.701	7.721	7.740	7.760	7.780	7.800	7.819	7.839	7.859	7.879	480
490	7.879	7.898	7.918	7.938	7.958	7.978	7.997	8.017	8.037	8.057	8.077	490
500	8.077	8.097	8.116	8.136	8.156	8.176	8.196	8.216	8.236	8.256	8.275	500
510	8.275	8.295	8.315	8.335	8.355	8.375	8.395	8.415	8.435	8.455	8.475	510
520	8.475	8.495	8.515	8.535	8.555	8.575	8.595	8.615	8.635	8.655	8.675	520
530	8.675	8.695	8.715	8.735	8.755	8.775	8.795	8.815	8.835	8.855	8.875	530
540	8.875	8.895	8.915	8.935	8.955	8.975	8.995	9.015	9.036	9.056	9.076	540
550	9.076	9.096	9.116	9.136	9.156	9.176	9.196	9.217	9.237	9.257	9.277	550
560	9.277	9.297	9.317	9.337	9.358	9.378	9.398	9.418	9.438	9.458	9.479	560
570	9.479	9.499	9.519	9.539	9.559	9.580	9.600	9.620	9.640	9.660	9.681	570
580	9.681	9.701	9.721	9.741	9.762	9.782	9.802	9.822	9.842	9.863	9.883	580
590	9.883	9.903	9.923	9.944	9.964	9.984	10.005	10.025	10.045	10.065	10.086	590
600	10.086	10.106	10.126	10.147	10.167	10.187	10.207	10.228	10.248	10.268	10.289	600
610	10.289	10.309	10.329	10.350	10.370	10.390	10.411	10.431	10.451	10.472	10.492	610
620	10.492	10.512	10.533	10.553	10.573	10.594	10.614	10.634	10.655	10.675	10.695	620
630	10.695	10.716	10.736	10.756	10.777	10.797	10.818	10.838	10.858	10.879	10.899	630
640	10.899	10.919	10.940	10.960	10.981	11.001	11.021	11.042	11.062	11.083	11.103	640

W3 Re/W25 Re Seebeck Voltages.

Thermoelectric Voltage in Millivolts

DEG C	0	1	2	3	4	5	6	7	8	9	10	DEG C
650	11.103	11.123	11.144	11.164	11.185	11.205	11.225	11.246	11.266	11.287	11.307	650
660	11.307	11.327	11.348	11.368	11.389	11.409	11.429	11.450	11.470	11.491	11.511	660
670	11.511	11.532	11.552	11.572	11.593	11.613	11.634	11.654	11.675	11.695	11.715	670
680	11.715	11.736	11.756	11.777	11.797	11.818	11.838	11.859	11.879	11.899	11.920	680
690	11.920	11.940	11.961	11.981	12.002	12.022	12.043	12.063	12.084	12.104	12.124	690
700	12.124	12.145	12.165	12.186	12.206	12.227	12.247	12.268	12.288	12.309	12.329	700
710	12.329	12.349	12.370	12.390	12.411	12.431	12.452	12.472	12.493	12.513	12.534	710
720	12.534	12.554	12.574	12.595	12.615	12.636	12.656	12.677	12.697	12.718	12.738	720
730	12.738	12.759	12.779	12.800	12.820	12.840	12.861	12.881	12.902	12.922	12.943	730
740	12.943	12.963	12.984	13.004	13.025	13.045	13.066	13.086	13.107	13.127	13.147	740
750	13.147	13.168	13.188	13.209	13.229	13.250	13.270	13.291	13.311	13.332	13.352	750
760	13.352	13.372	13.393	13.413	13.434	13.454	13.475	13.495	13.516	13.536	13.557	760
770	13.557	13.577	13.597	13.618	13.638	13.659	13.679	13.700	13.720	13.741	13.761	770
780	13.761	13.782	13.802	13.822	13.843	13.863	13.884	13.904	13.925	13.945	13.966	780
790	13.966	13.986	14.007	14.027	14.048	14.068	14.089	14.109	14.130	14.150	14.171	790
800	14.171	14.191	14.212	14.232	14.253	14.273	14.294	14.314	14.335	14.355	14.376	800
810	14.376	14.396	14.417	14.437	14.458	14.478	14.499	14.519	14.539	14.560	14.580	810
820	14.580	14.601	14.621	14.642	14.662	14.683	14.703	14.724	14.744	14.764	14.785	820
830	14.785	14.805	14.826	14.846	14.867	14.887	14.908	14.928	14.948	14.969	14.989	830
840	14.989	15.010	15.030	15.051	15.071	15.091	15.112	15.132	15.153	15.173	15.194	840
850	15.194	15.214	15.234	15.255	15.275	15.296	15.316	15.336	15.357	15.377	15.398	850
860	15.398	15.418	15.438	15.459	15.479	15.499	15.520	15.540	15.561	15.581	15.601	860
870	15.601	15.622	15.642	15.662	15.683	15.703	15.724	15.744	15.764	15.785	15.805	870
880	15.805	15.825	15.846	15.866	15.886	15.907	15.927	15.947	15.968	15.988	16.008	880
890	16.008	16.029	16.049	16.069	16.090	16.110	16.130	16.150	16.171	16.191	16.211	890
900	16.211	16.232	16.252	16.272	16.293	16.313	16.333	16.353	16.374	16.394	16.414	900
910	16.414	16.434	16.455	16.475	16.495	16.516	16.536	16.556	16.576	16.597	16.617	910
920	16.617	16.637	16.657	16.678	16.698	16.718	16.738	16.758	16.779	16.799	16.819	920
930	16.819	16.839	16.859	16.880	16.900	16.920	16.940	16.961	16.981	17.001	17.021	930
940	17.021	17.041	17.061	17.082	17.102	17.122	17.142	17.162	17.182	17.203	17.223	940
950	17.223	17.243	17.263	17.283	17.303	17.323	17.344	17.364	17.384	17.404	17.424	950
960	17.424	17.444	17.464	17.484	17.505	17.525	17.545	17.565	17.585	17.605	17.625	960
970	17.625	17.645	17.665	17.685	17.706	17.726	17.746	17.766	17.786	17.806	17.826	970
980	17.826	17.846	17.866	17.886	17.906	17.926	17.946	17.966	17.986	18.006	18.026	980
990	18.026	18.046	18.066	18.086	18.106	18.126	18.146	18.166	18.186	18.206	18.226	990

W3 Re/W25 Re Seebeck Voltages.

Thermoelectric Voltage in Millivolts

DEG C	0	1	2	3	4	5	6	7	8	9	10	DEG C
1000	18.226	18.246	18.266	18.286	18.306	18.326	18.346	18.366	18.386	18.406	18.426	1000
1010	18.426	18.446	18.466	18.486	18.506	18.526	18.546	18.565	18.585	18.605	18.625	1010
1020	18.625	18.645	18.665	18.685	18.705	18.725	18.745	18.765	18.784	18.804	18.824	1020
1030	18.824	18.844	18.864	18.884	18.904	18.923	18.943	18.963	18.983	19.003	19.023	1030
1040	19.023	19.043	19.062	19.082	19.102	19.122	19.142	19.161	19.181	19.201	19.221	1040
1050	19.221	19.241	19.260	19.280	19.300	19.320	19.340	19.359	19.379	19.399	19.419	1050
1060	19.419	19.438	19.458	19.478	19.498	19.517	19.537	19.557	19.577	19.596	19.616	1060
1070	19.616	19.636	19.655	19.675	19.695	19.714	19.734	19.754	19.774	19.793	19.813	1070
1080	19.813	19.833	19.852	19.872	19.892	19.911	19.931	19.951	19.970	19.990	20.009	1080
1090	20.009	20.029	20.049	20.068	20.088	20.108	20.127	20.147	20.166	20.186	20.206	1090
1100	20.206	20.225	20.245	20.264	20.284	20.303	20.323	20.343	20.362	20.382	20.401	1100
1110	20.401	20.421	20.440	20.460	20.479	20.499	20.518	20.538	20.557	20.577	20.597	1110
1120	20.597	20.616	20.636	20.655	20.674	20.694	20.713	20.733	20.752	20.772	20.791	1120
1130	20.791	20.811	20.830	20.850	20.869	20.889	20.908	20.927	20.947	20.966	20.986	1130
1140	20.986	21.005	21.025	21.044	21.063	21.083	21.102	21.122	21.141	21.160	21.180	1140
1150	21.180	21.199	21.218	21.238	21.257	21.276	21.296	21.315	21.334	21.354	21.373	1150
1160	21.373	21.392	21.412	21.431	21.450	21.470	21.489	21.508	21.528	21.547	21.566	1160
1170	21.566	21.585	21.605	21.624	21.643	21.663	21.682	21.701	21.720	21.740	21.759	1170
1180	21.759	21.778	21.797	21.816	21.836	21.855	21.874	21.893	21.913	21.932	21.951	1180
1190	21.951	21.970	21.989	22.008	22.028	22.047	22.066	22.085	22.104	22.123	22.143	1190
1200	22.143	22.162	22.181	22.200	22.219	22.238	22.257	22.277	22.296	22.315	22.334	1200
1210	22.334	22.353	22.372	22.391	22.410	22.429	22.448	22.467	22.486	22.506	22.525	1210
1220	22.525	22.544	22.563	22.582	22.601	22.620	22.639	22.658	22.677	22.696	22.715	1220
1230	22.715	22.734	22.753	22.772	22.791	22.810	22.829	22.848	22.867	22.886	22.905	1230
1240	22.905	22.924	22.943	22.961	22.980	22.999	23.018	23.037	23.056	23.075	23.094	1240
1250	23.094	23.113	23.132	23.151	23.170	23.188	23.207	23.226	23.245	23.264	23.283	1250
1260	23.283	23.302	23.321	23.339	23.358	23.377	23.396	23.415	23.434	23.452	23.471	1260
1270	23.471	23.490	23.509	23.528	23.546	23.565	23.584	23.603	23.622	23.640	23.659	1270
1280	23.659	23.678	23.696	23.715	23.734	23.753	23.772	23.790	23.809	23.828	23.846	1280
1290	23.846	23.865	23.884	23.903	23.921	23.940	23.959	23.977	23.996	24.015	24.033	1290

W3 Re/W25 Re Seebeck Voltages.

Thermoelectric Voltage in Millivolts

DEG C	0	1	2	3	4	5	6	7	8	9	10	DEG C
1300	24.033	24.052	24.071	24.089	24.108	24.127	24.145	24.164	24.183	24.201	24.220	1300
1310	24.220	24.238	24.257	24.276	24.294	24.313	24.331	24.350	24.369	24.387	24.406	1310
1320	24.406	24.424	24.443	24.461	24.480	24.499	24.517	24.536	24.554	24.573	24.591	1320
1330	24.591	24.610	24.628	24.647	24.665	24.684	24.702	24.721	24.739	24.758	24.776	1330
1340	24.776	24.795	24.813	24.831	24.850	24.868	24.887	24.905	24.924	24.942	24.961	1340
1350	24.961	24.979	24.997	25.016	25.034	25.053	25.071	25.089	25.108	25.126	25.145	1350
1360	25.145	25.163	25.181	25.200	25.218	25.236	25.255	25.273	25.291	25.310	25.328	1360
1370	25.328	25.346	25.365	25.383	25.401	25.419	25.438	25.456	25.474	25.493	25.511	1370
1380	25.511	25.529	25.547	25.566	25.584	25.602	25.620	25.639	25.657	25.675	25.693	1380
1390	25.693	25.712	25.730	25.748	25.766	25.784	25.803	25.821	25.839	25.857	25.875	1390
1400	25.875	25.893	25.912	25.930	25.948	25.966	25.984	26.002	26.020	26.039	26.057	1400
1410	26.057	26.075	26.093	26.111	26.129	26.147	26.165	26.183	26.201	26.220	26.238	1410
1420	26.238	26.256	26.274	26.292	26.310	26.328	26.346	26.364	26.382	26.400	26.418	1420
1430	26.418	26.436	26.454	26.472	26.490	26.508	26.526	26.544	26.562	26.580	26.598	1430
1440	26.598	26.616	26.634	26.652	26.670	26.688	26.706	26.723	26.741	26.759	26.777	1440
1450	26.777	26.795	26.813	26.831	26.849	26.867	26.885	26.902	26.920	26.938	26.956	1450
1460	26.956	26.974	26.992	27.010	27.027	27.045	27.063	27.081	27.099	27.117	27.134	1460
1470	27.134	27.152	27.170	27.188	27.206	27.223	27.241	27.259	27.277	27.294	27.312	1470
1480	27.312	27.330	27.348	27.365	27.383	27.401	27.419	27.436	27.454	27.472	27.489	1480
1490	27.489	27.507	27.525	27.543	27.560	27.578	27.596	27.613	27.631	27.649	27.666	1490
1500	27.666	27.684	27.701	27.719	27.737	27.754	27.772	27.790	27.807	27.825	27.842	1500
1510	27.842	27.860	27.878	27.895	27.913	27.930	27.948	27.965	27.983	28.001	28.018	1510
1520	28.018	28.036	28.053	28.071	28.088	28.106	28.123	28.141	28.158	28.176	28.193	1520
1530	28.193	28.211	28.228	28.246	28.263	28.281	28.298	28.316	28.333	28.350	28.368	1530
1540	28.368	28.385	28.403	28.420	28.438	28.455	28.472	28.490	28.507	28.525	28.542	1540
1550	28.542	28.559	28.577	28.594	28.611	28.629	28.646	28.663	28.681	28.698	28.715	1550
1560	28.715	28.733	28.750	28.767	28.785	28.802	28.819	28.837	28.854	28.871	28.888	1560
1570	28.888	28.906	28.923	28.940	28.957	28.975	28.992	29.009	29.026	29.044	29.061	1570
1580	29.061	29.078	29.095	29.112	29.130	29.147	29.164	29.181	29.198	29.215	29.233	1580
1590	29.233	29.250	29.267	29.284	29.301	29.318	29.335	29.353	29.370	29.387	29.404	1590
1600	29.404	29.421	29.438	29.455	29.472	29.489	29.506	29.523	29.540	29.558	29.575	1600
1610	29.575	29.592	29.609	29.626	29.643	29.660	29.677	29.694	29.711	29.728	29.745	1610
1620	29.745	29.762	29.779	29.796	29.813	29.830	29.847	29.863	29.880	29.897	29.914	1620
1630	29.914	29.931	29.948	29.965	29.982	29.999	30.016	30.033	30.049	30.066	30.083	1630
1640	30.083	30.100	30.117	30.134	30.151	30.168	30.184	30.201	30.218	30.235	30.252	1640

W5 Re/W25 Re Seebeck Voltages.

Thermoelectric Voltage in Millivolts

DEG C	0	1	2	3	4	5	6	7	8	9	10	DEG C
1650	30.252	30.268	30.285	30.302	30.319	30.336	30.352	30.369	30.386	30.403	30.419	1650
1660	30.419	30.436	30.453	30.470	30.486	30.503	30.520	30.537	30.553	30.570	30.587	1660
1670	30.587	30.603	30.620	30.637	30.653	30.670	30.687	30.703	30.720	30.737	30.753	1670
1680	30.753	30.770	30.786	30.803	30.820	30.836	30.853	30.869	30.886	30.903	30.919	1680
1690	30.919	30.936	30.952	30.969	30.985	31.002	31.018	31.035	31.051	31.068	31.085	1690
1700	31.085	31.101	31.118	31.134	31.150	31.167	31.183	31.200	31.216	31.233	31.249	1700
1710	31.249	31.266	31.282	31.299	31.315	31.331	31.348	31.364	31.381	31.397	31.413	1710
1720	31.413	31.430	31.446	31.462	31.479	31.495	31.511	31.528	31.544	31.560	31.577	1720
1730	31.577	31.593	31.609	31.626	31.642	31.658	31.675	31.691	31.707	31.723	31.740	1730
1740	31.740	31.756	31.772	31.788	31.804	31.821	31.837	31.853	31.869	31.886	31.902	1740
1750	31.902	31.918	31.934	31.950	31.966	31.983	31.999	32.015	32.031	32.047	32.063	1750
1760	32.063	32.079	32.095	32.111	32.128	32.144	32.160	32.176	32.192	32.208	32.224	1760
1770	32.224	32.240	32.256	32.272	32.288	32.304	32.320	32.336	32.352	32.368	32.384	1770
1780	32.384	32.400	32.416	32.432	32.448	32.464	32.480	32.496	32.512	32.528	32.543	1780
1790	32.543	32.559	32.575	32.591	32.607	32.623	32.639	32.655	32.670	32.686	32.702	1790
1800	32.702	32.718	32.734	32.750	32.765	32.781	32.797	32.813	32.829	32.844	32.860	1800
1810	32.860	32.876	32.892	32.907	32.923	32.939	32.955	32.970	32.986	33.002	33.017	1810
1820	33.017	33.033	33.049	33.064	33.080	33.096	33.111	33.127	33.143	33.158	33.174	1820
1830	33.174	33.189	33.205	33.221	33.236	33.252	33.267	33.283	33.299	33.314	33.330	1830
1840	33.330	33.345	33.361	33.376	33.392	33.407	33.423	33.438	33.454	33.469	33.485	1840
1850	33.485	33.500	33.515	33.531	33.546	33.562	33.577	33.593	33.608	33.623	33.639	1850
1860	33.639	33.654	33.670	33.685	33.700	33.716	33.731	33.746	33.762	33.777	33.792	1860
1870	33.792	33.807	33.823	33.838	33.853	33.869	33.884	33.899	33.914	33.930	33.945	1870
1880	33.945	33.960	33.975	33.990	34.006	34.021	34.036	34.051	34.066	34.081	34.097	1880
1890	34.097	34.112	34.127	34.142	34.157	34.172	34.187	34.202	34.217	34.232	34.248	1890
1900	34.248	34.263	34.278	34.293	34.308	34.323	34.338	34.353	34.368	34.383	34.398	1900
1910	34.398	34.413	34.428	34.442	34.457	34.472	34.487	34.502	34.517	34.532	34.547	1910
1920	34.547	34.562	34.577	34.591	34.606	34.621	34.636	34.651	34.666	34.680	34.695	1920
1930	34.695	34.710	34.725	34.739	34.754	34.769	34.784	34.798	34.813	34.828	34.843	1930
1940	34.843	34.857	34.872	34.887	34.901	34.916	34.931	34.945	34.960	34.975	34.989	1940
1950	34.989	35.004	35.018	35.033	35.047	35.062	35.077	35.091	35.106	35.120	35.135	1950
1960	35.135	35.149	35.164	35.178	35.193	35.207	35.222	35.236	35.251	35.265	35.279	1960
1970	35.279	35.294	35.308	35.323	35.337	35.351	35.366	35.380	35.394	35.409	35.423	1970
1980	35.423	35.437	35.452	35.466	35.480	35.494	35.509	35.523	35.537	35.551	35.566	1980
1990	35.566	35.580	35.594	35.608	35.622	35.637	35.651	35.665	35.679	35.693	35.707	1990

W5 Re/W26 Re Seebeck Voltages.

EMF in Millivolts

Thermoelectric Voltage in Millivolts

DEG C	0	1	2	3	4	5	6	7	8	9	10	DEG C
0	0.000	0.013	0.027	0.040	0.054	0.067	0.081	0.094	0.108	0.121	0.135	0
10	0.135	0.149	0.162	0.176	0.190	0.204	0.217	0.231	0.245	0.259	0.273	10
20	0.273	0.286	0.300	0.314	0.328	0.342	0.356	0.370	0.384	0.398	0.412	20
30	0.412	0.426	0.441	0.455	0.469	0.483	0.497	0.512	0.526	0.540	0.554	30
40	0.554	0.569	0.583	0.598	0.612	0.626	0.641	0.655	0.670	0.684	0.699	40
50	0.699	0.713	0.728	0.742	0.757	0.772	0.786	0.801	0.816	0.830	0.845	50
60	0.845	0.860	0.875	0.889	0.904	0.919	0.934	0.949	0.964	0.979	0.994	60
70	0.994	1.009	1.024	1.039	1.054	1.069	1.084	1.099	1.114	1.129	1.144	70
80	1.144	1.159	1.175	1.190	1.205	1.220	1.235	1.251	1.266	1.281	1.297	80
90	1.297	1.312	1.327	1.343	1.358	1.374	1.389	1.405	1.420	1.436	1.451	90
100	1.451	1.467	1.482	1.498	1.513	1.529	1.545	1.560	1.576	1.592	1.607	100
110	1.607	1.623	1.639	1.655	1.670	1.686	1.702	1.718	1.734	1.750	1.766	110
120	1.766	1.781	1.797	1.813	1.829	1.845	1.861	1.877	1.893	1.909	1.925	120
130	1.925	1.942	1.958	1.974	1.990	2.006	2.022	2.038	2.055	2.074	2.087	130
140	2.087	2.103	2.120	2.136	2.152	2.168	2.185	2.201	2.217	2.234	2.250	140
150	2.250	2.267	2.283	2.300	2.316	2.332	2.349	2.365	2.382	2.398	2.415	150
160	2.415	2.432	2.448	2.465	2.481	2.498	2.515	2.531	2.548	2.565	2.581	160
170	2.581	2.598	2.615	2.632	2.648	2.665	2.682	2.699	2.716	2.732	2.749	170
180	2.749	2.766	2.783	2.800	2.817	2.834	2.851	2.868	2.885	2.902	2.919	180
190	2.919	2.936	2.953	2.970	2.987	3.004	3.021	3.038	3.055	3.072	3.089	190
200	3.089	3.106	3.124	3.141	3.158	3.175	3.192	3.210	3.227	3.244	3.261	200
210	3.261	3.279	3.296	3.313	3.331	3.348	3.365	3.383	3.400	3.417	3.435	210
220	3.435	3.452	3.470	3.487	3.505	3.522	3.539	3.557	3.574	3.592	3.609	220
230	3.609	3.627	3.645	3.662	3.680	3.697	3.715	3.732	3.750	3.768	3.785	230
240	3.785	3.803	3.821	3.838	3.856	3.874	3.891	3.909	3.927	3.945	3.962	240
250	3.962	3.980	3.998	4.016	4.034	4.051	4.069	4.087	4.105	4.123	4.141	250
260	4.141	4.158	4.176	4.194	4.212	4.230	4.248	4.266	4.284	4.302	4.320	260
270	4.320	4.338	4.356	4.374	4.392	4.410	4.428	4.446	4.464	4.482	4.500	270
280	4.500	4.518	4.536	4.554	4.573	4.591	4.609	4.627	4.645	4.663	4.682	280
290	4.682	4.700	4.718	4.736	4.754	4.773	4.791	4.809	4.827	4.846	4.864	290

W5 Re/W26 Re Seebeck Voltages.

Thermoelectric Voltage in Millivolts

DEG C	0	1	2	3	4	5	6	7	8	9	10	DEG C
300	4.864	4.882	4.900	4.919	4.937	4.955	4.974	4.992	5.010	5.029	5.047	300
310	5.047	5.065	5.084	5.102	5.121	5.139	5.157	5.176	5.194	5.213	5.231	310
320	5.231	5.250	5.268	5.287	5.305	5.323	5.342	5.361	5.379	5.398	5.416	320
330	5.416	5.435	5.453	5.472	5.490	5.509	5.527	5.546	5.565	5.583	5.602	330
340	5.602	5.620	5.639	5.658	5.676	5.695	5.714	5.732	5.751	5.770	5.788	340
350	5.788	5.807	5.826	5.844	5.863	5.882	5.901	5.919	5.938	5.957	5.976	350
360	5.976	5.994	6.013	6.032	6.051	6.070	6.088	6.107	6.126	6.145	6.164	360
370	6.164	6.182	6.201	6.220	6.239	6.258	6.277	6.296	6.314	6.333	6.352	370
380	6.352	6.371	6.390	6.409	6.428	6.447	6.466	6.485	6.504	6.523	6.541	380
390	6.541	6.560	6.579	6.598	6.617	6.636	6.655	6.674	6.693	6.712	6.731	390
400	6.731	6.750	6.769	6.788	6.807	6.826	6.845	6.865	6.884	6.903	6.922	400
410	6.922	6.941	6.960	6.979	6.998	7.017	7.036	7.055	7.074	7.094	7.113	410
420	7.113	7.132	7.151	7.170	7.189	7.208	7.227	7.247	7.266	7.285	7.304	420
430	7.304	7.323	7.342	7.362	7.381	7.400	7.419	7.438	7.458	7.477	7.496	430
440	7.496	7.515	7.534	7.554	7.573	7.592	7.611	7.631	7.650	7.669	7.688	440
450	7.688	7.708	7.727	7.746	7.765	7.785	7.804	7.823	7.842	7.862	7.881	450
460	7.881	7.900	7.920	7.939	7.958	7.978	7.997	8.016	8.036	8.055	8.074	460
470	8.074	8.094	8.113	8.132	8.152	8.171	8.190	8.210	8.229	8.248	8.268	470
480	8.268	8.287	8.306	8.326	8.345	8.364	8.384	8.403	8.423	8.442	8.461	480
490	8.461	8.481	8.500	8.520	8.539	8.558	8.578	8.597	8.617	8.636	8.655	490
500	8.655	8.675	8.694	8.714	8.733	8.753	8.772	8.791	8.811	8.830	8.850	500
510	8.850	8.869	8.889	8.908	8.928	8.947	8.966	8.986	9.005	9.025	9.044	510
520	9.044	9.064	9.083	9.103	9.122	9.142	9.161	9.181	9.200	9.220	9.239	520
530	9.239	9.259	9.278	9.298	9.317	9.337	9.356	9.376	9.395	9.415	9.434	530
540	9.434	9.454	9.473	9.493	9.512	9.532	9.551	9.571	9.590	9.610	9.629	540
550	9.629	9.649	9.668	9.688	9.707	9.727	9.746	9.766	9.785	9.805	9.824	550
560	9.824	9.844	9.863	9.883	9.902	9.922	9.942	9.961	9.981	10.000	10.020	560
570	10.020	10.039	10.059	10.078	10.098	10.117	10.137	10.156	10.176	10.196	10.215	570
580	10.215	10.235	10.254	10.274	10.293	10.313	10.332	10.352	10.371	10.391	10.411	580
590	10.411	10.430	10.450	10.469	10.489	10.508	10.528	10.547	10.567	10.587	10.606	590
600	10.606	10.626	10.645	10.665	10.684	10.704	10.723	10.743	10.763	10.782	10.802	600
610	10.802	10.821	10.841	10.860	10.880	10.899	10.919	10.939	10.958	10.978	10.997	610
620	10.997	11.017	11.036	11.056	11.075	11.095	11.114	11.134	11.154	11.173	11.193	620
630	11.193	11.212	11.232	11.251	11.271	11.290	11.310	11.330	11.349	11.369	11.388	630
640	11.388	11.408	11.427	11.447	11.466	11.486	11.505	11.525	11.544	11.564	11.584	640

W5 Re/W26 Re Seebeck Voltages.

Thermoelectric Voltage in Millivolts

DEG C	0	1	2	3	4	5	6	7	8	9	10	DEG C
650	11.584	11.603	11.623	11.642	11.662	11.681	11.701	11.720	11.740	11.759	11.779	650
660	11.779	11.798	11.818	11.837	11.857	11.876	11.896	11.915	11.935	11.955	11.974	660
670	11.974	11.994	12.013	12.033	12.052	12.072	12.091	12.111	12.130	12.150	12.169	670
680	12.169	12.189	12.208	12.228	12.247	12.267	12.286	12.306	12.325	12.345	12.364	680
690	12.364	12.384	12.403	12.422	12.442	12.461	12.481	12.500	12.520	12.539	12.559	690
700	12.559	12.578	12.598	12.617	12.637	12.656	12.676	12.695	12.715	12.734	12.753	700
710	12.753	12.773	12.792	12.812	12.831	12.851	12.870	12.890	12.909	12.928	12.948	710
720	12.948	12.967	12.987	13.006	13.026	13.045	13.064	13.084	13.103	13.123	13.142	720
730	13.142	13.161	13.181	13.200	13.220	13.239	13.258	13.278	13.297	13.317	13.336	730
740	13.336	13.355	13.375	13.394	13.413	13.433	13.452	13.472	13.491	13.510	13.530	740
750	13.530	13.549	13.568	13.588	13.607	13.626	13.646	13.665	13.685	13.704	13.723	750
760	13.723	13.743	13.762	13.781	13.800	13.820	13.839	13.858	13.878	13.897	13.916	760
770	13.916	13.936	13.955	13.974	13.994	14.013	14.032	14.051	14.071	14.090	14.109	770
780	14.109	14.129	14.148	14.167	14.186	14.206	14.225	14.244	14.263	14.283	14.302	780
790	14.302	14.321	14.340	14.360	14.379	14.398	14.417	14.437	14.456	14.475	14.494	790
800	14.494	14.513	14.533	14.552	14.571	14.590	14.609	14.629	14.648	14.667	14.686	800
810	14.686	14.705	14.725	14.744	14.763	14.782	14.801	14.820	14.840	14.859	14.878	810
820	14.878	14.897	14.916	14.935	14.954	14.974	14.993	15.012	15.031	15.050	15.069	820
830	15.069	15.088	15.107	15.126	15.146	15.165	15.184	15.203	15.222	15.241	15.260	830
840	15.260	15.279	15.298	15.317	15.336	15.355	15.374	15.393	15.413	15.432	15.451	840
850	15.451	15.470	15.489	15.508	15.527	15.546	15.565	15.584	15.603	15.622	15.641	850
860	15.641	15.660	15.679	15.698	15.717	15.736	15.755	15.774	15.793	15.812	15.831	860
870	15.831	15.849	15.868	15.887	15.906	15.925	15.944	15.963	15.982	16.001	16.020	870
880	16.020	16.039	16.058	16.077	16.096	16.114	16.133	16.152	16.171	16.190	16.209	880
890	16.209	16.228	16.247	16.265	16.284	16.303	16.322	16.341	16.360	16.379	16.397	890
900	16.397	16.416	16.435	16.454	16.473	16.491	16.510	16.529	16.548	16.567	16.585	900
910	16.585	16.604	16.623	16.642	16.661	16.679	16.698	16.717	16.736	16.754	16.773	910
920	16.773	16.792	16.811	16.829	16.848	16.867	16.886	16.904	16.923	16.942	16.960	920
930	16.960	16.979	16.998	17.016	17.035	17.054	17.072	17.091	17.110	17.128	17.147	930
940	17.147	17.166	17.184	17.203	17.222	17.240	17.259	17.278	17.296	17.315	17.333	940

W5 Re/W26 Re Seebeck Voltages.

Thermoelectric Voltage in Millivolts

DEG C	0	1	2	3	4	5	6	7	8	9	10	DEG C
950	17.333	17.352	17.371	17.389	17.408	17.426	17.445	17.463	17.482	17.501	17.519	950
960	17.519	17.538	17.556	17.575	17.593	17.612	17.630	17.649	17.667	17.686	17.704	960
970	17.704	17.723	17.741	17.760	17.778	17.797	17.815	17.834	17.852	17.871	17.889	970
980	17.889	17.908	17.926	17.945	17.963	17.981	18.000	18.018	18.037	18.055	18.074	980
990	18.074	18.092	18.110	18.129	18.147	18.166	18.184	18.202	18.221	18.239	18.257	990
1000	18.257	18.276	18.294	18.312	18.331	18.349	18.367	18.386	18.404	18.422	18.441	1000
1010	18.441	18.459	18.477	18.496	18.514	18.532	18.550	18.569	18.587	18.605	18.623	1010
1020	18.623	18.642	18.660	18.678	18.696	18.715	18.733	18.751	18.769	18.788	18.806	1020
1030	18.806	18.824	18.842	18.860	18.878	18.897	18.915	18.933	18.951	18.969	18.987	1030
1040	18.987	19.006	19.024	19.042	19.060	19.078	19.096	19.114	19.132	19.151	19.169	1040
1050	19.169	19.187	19.205	19.223	19.241	19.259	19.277	19.295	19.313	19.331	19.349	1050
1060	19.349	19.367	19.385	19.403	19.421	19.439	19.457	19.475	19.493	19.511	19.529	1060
1070	19.529	19.547	19.565	19.583	19.601	19.619	19.637	19.655	19.673	19.691	19.709	1070
1080	19.709	19.727	19.745	19.763	19.781	19.799	19.816	19.834	19.852	19.870	19.888	1080
1090	19.888	19.906	19.924	19.942	19.959	19.977	19.995	20.013	20.031	20.049	20.066	1090
1100	20.066	20.084	20.102	20.120	20.138	20.155	20.173	20.191	20.209	20.227	20.244	1100
1110	20.244	20.262	20.280	20.298	20.315	20.333	20.351	20.369	20.386	20.404	20.422	1110
1120	20.422	20.439	20.457	20.475	20.492	20.510	20.528	20.546	20.563	20.581	20.598	1120
1130	20.598	20.616	20.634	20.651	20.669	20.687	20.704	20.722	20.739	20.757	20.775	1130
1140	20.775	20.792	20.810	20.827	20.845	20.863	20.880	20.898	20.915	20.933	20.950	1140
1150	20.950	20.968	20.985	21.003	21.020	21.038	21.055	21.073	21.090	21.108	21.125	1150
1160	21.125	21.143	21.160	21.178	21.195	21.213	21.230	21.248	21.265	21.282	21.300	1160
1170	21.300	21.317	21.335	21.352	21.369	21.387	21.404	21.422	21.439	21.456	21.474	1170
1180	21.474	21.491	21.508	21.526	21.543	21.560	21.578	21.595	21.612	21.630	21.647	1180
1190	21.647	21.664	21.682	21.699	21.716	21.733	21.751	21.768	21.785	21.802	21.820	1190
1200	21.820	21.837	21.854	21.871	21.889	21.906	21.923	21.940	21.957	21.975	21.992	1200
1210	21.992	22.009	22.026	22.043	22.061	22.078	22.095	22.112	22.129	22.146	22.163	1210
1220	22.163	22.180	22.198	22.215	22.232	22.249	22.266	22.283	22.300	22.317	22.334	1220
1230	22.334	22.351	22.368	22.385	22.403	22.420	22.437	22.454	22.471	22.488	22.505	1230
1240	22.505	22.522	22.539	22.556	22.573	22.590	22.607	22.624	22.641	22.657	22.674	1240
1250	22.674	22.691	22.708	22.725	22.742	22.759	22.776	22.793	22.810	22.827	22.844	1250
1260	22.844	22.860	22.877	22.894	22.911	22.928	22.945	22.962	22.978	22.995	23.012	1260
1270	23.012	23.029	23.046	23.063	23.079	23.096	23.113	23.130	23.147	23.163	23.180	1270
1280	23.180	23.197	23.214	23.230	23.247	23.264	23.281	23.297	23.314	23.331	23.347	1280
1290	23.347	23.364	23.381	23.398	23.414	23.431	23.448	23.464	23.481	23.498	23.514	1290

W5 Re/W26 Re Seebeck Voltages.

Thermoelectric Voltage in Millivolts

DEG C	0	1	2	3	4	5	6	7	8	9	10	DEG C
1300	23.514	23.531	23.548	23.564	23.581	23.597	23.614	23.631	23.647	23.664	23.680	1300
1310	23.680	23.697	23.714	23.730	23.747	23.763	23.780	23.796	23.813	23.829	23.846	1310
1320	23.846	23.862	23.879	23.895	23.912	23.928	23.945	23.961	23.978	23.994	24.011	1320
1330	24.011	24.027	24.044	24.060	24.077	24.093	24.110	24.126	24.142	24.159	24.175	1330
1340	24.175	24.192	24.208	24.224	24.241	24.257	24.274	24.290	24.306	24.323	24.339	1340
1350	24.339	24.355	24.372	24.388	24.404	24.421	24.437	24.453	24.470	24.486	24.502	1350
1360	24.502	24.518	24.535	24.551	24.567	24.583	24.600	24.616	24.632	24.648	24.665	1360
1370	24.665	24.681	24.697	24.713	24.730	24.746	24.762	24.778	24.794	24.810	24.827	1370
1380	24.827	24.843	24.859	24.875	24.891	24.907	24.923	24.940	24.956	24.972	24.988	1380
1390	24.988	25.004	25.020	25.036	25.052	25.068	25.084	25.100	25.117	25.133	25.149	1390
1400	25.149	25.165	25.181	25.197	25.213	25.229	25.245	25.261	25.277	25.293	25.309	1400
1410	25.309	25.325	25.341	25.357	25.373	25.389	25.405	25.420	25.436	25.452	25.468	1410
1420	25.468	25.484	25.500	25.516	25.532	25.548	25.564	25.580	25.595	25.611	25.627	1420
1430	25.627	25.643	25.659	25.675	25.691	25.706	25.722	25.738	25.754	25.770	25.785	1430
1440	25.785	25.801	25.817	25.833	25.849	25.864	25.880	25.896	25.912	25.927	25.943	1440
1450	25.943	25.959	25.975	25.990	26.006	26.022	26.038	26.053	26.069	26.085	26.100	1450
1460	26.100	26.116	26.132	26.147	26.163	26.179	26.194	26.210	26.226	26.241	26.257	1460
1470	26.257	26.272	26.288	26.304	26.319	26.335	26.350	26.366	26.382	26.397	26.413	1470
1480	26.413	26.428	26.444	26.459	26.475	26.490	26.506	26.521	26.537	26.552	26.568	1480
1490	26.568	26.583	26.599	26.614	26.630	26.645	26.661	26.676	26.692	26.707	26.723	1490
1500	26.723	26.738	26.753	26.769	26.784	26.800	26.815	26.830	26.846	26.861	26.877	1500
1510	26.877	26.892	26.907	26.923	26.938	26.953	26.969	26.984	26.999	27.015	27.030	1510
1520	27.030	27.045	27.061	27.076	27.091	27.107	27.122	27.137	27.152	27.168	27.183	1520
1530	27.183	27.198	27.213	27.229	27.244	27.259	27.274	27.290	27.305	27.320	27.335	1530
1540	27.335	27.350	27.366	27.381	27.396	27.411	27.426	27.441	27.457	27.472	27.487	1540
1550	27.487	27.502	27.517	27.532	27.547	27.562	27.578	27.593	27.608	27.623	27.638	1550
1560	27.638	27.653	27.668	27.683	27.698	27.713	27.728	27.743	27.758	27.773	27.788	1560
1570	27.788	27.803	27.818	27.833	27.848	27.863	27.878	27.893	27.908	27.923	27.938	1570
1580	27.938	27.953	27.968	27.983	27.998	28.013	28.028	28.043	28.058	28.072	28.087	1580
1590	28.087	28.102	28.117	28.132	28.147	28.162	28.177	28.191	28.206	28.221	28.236	1590
1600	28.236	28.251	28.266	28.280	28.295	28.310	28.325	28.340	28.354	28.369	28.384	1600
1610	28.384	28.399	28.413	28.428	28.443	28.458	28.472	28.487	28.502	28.517	28.531	1610
1620	28.531	28.546	28.561	28.575	28.590	28.605	28.619	28.634	28.649	28.663	28.678	1620
1630	28.678	28.693	28.707	28.722	28.737	28.751	28.766	28.780	28.795	28.810	28.824	1630
1640	28.824	28.839	28.853	28.868	28.883	28.897	28.912	28.926	28.941	28.955	28.970	1640

W5 Re/W26 Re Seebeck Voltages.

Thermoelectric Voltage in Millivolts

DEG C	0	1	2	3	4	5	6	7	8	9	10	DEG C
1650	28.970	28.984	28.999	29.013	29.028	29.042	29.057	29.071	29.086	29.100	29.115	1650
1660	29.115	29.129	29.144	29.158	29.173	29.187	29.201	29.216	29.230	29.245	29.259	1660
1670	29.259	29.274	29.288	29.302	29.317	29.331	29.345	29.360	29.374	29.388	29.403	1670
1680	29.403	29.417	29.431	29.446	29.460	29.474	29.489	29.503	29.517	29.532	29.546	1680
1690	29.546	29.560	29.574	29.589	29.603	29.617	29.631	29.646	29.660	29.674	29.688	1690
1700	29.688	29.703	29.717	29.731	29.745	29.759	29.774	29.788	29.802	29.816	29.830	1700
1710	29.830	29.844	29.859	29.873	29.887	29.901	29.915	29.929	29.943	29.957	29.971	1710
1720	29.971	29.986	30.000	30.014	30.028	30.042	30.056	30.070	30.084	30.098	30.112	1720
1730	30.112	30.126	30.140	30.154	30.168	30.182	30.196	30.210	30.224	30.238	30.252	1730
1740	30.252	30.266	30.280	30.294	30.308	30.322	30.336	30.350	30.364	30.378	30.391	1740
1750	30.391	30.405	30.419	30.433	30.447	30.461	30.475	30.489	30.502	30.516	30.530	1750
1760	30.530	30.544	30.558	30.572	30.585	30.599	30.613	30.627	30.641	30.654	30.668	1760
1770	30.668	30.682	30.696	30.710	30.723	30.737	30.751	30.765	30.778	30.792	30.806	1770
1780	30.806	30.819	30.833	30.847	30.861	30.874	30.888	30.902	30.915	30.929	30.943	1780
1790	30.943	30.956	30.970	30.983	30.997	31.011	31.024	31.038	31.052	31.065	31.079	1790
1800	31.079	31.092	31.106	31.119	31.133	31.147	31.160	31.174	31.187	31.201	31.214	1800
1810	31.214	31.228	31.241	31.255	31.268	31.282	31.295	31.309	31.322	31.336	31.349	1810
1820	31.349	31.363	31.376	31.389	31.403	31.416	31.430	31.443	31.457	31.470	31.483	1820
1830	31.483	31.497	31.510	31.524	31.537	31.550	31.564	31.577	31.590	31.604	31.617	1830
1840	31.617	31.630	31.644	31.657	31.670	31.683	31.697	31.710	31.723	31.737	31.750	1840
1850	31.750	31.763	31.776	31.790	31.803	31.816	31.829	31.842	31.856	31.869	31.882	1850
1860	31.882	31.895	31.908	31.922	31.935	31.948	31.961	31.974	31.987	32.001	32.014	1860
1870	32.014	32.027	32.040	32.053	32.066	32.079	32.092	32.105	32.118	32.132	32.145	1870
1880	32.145	32.158	32.171	32.184	32.197	32.210	32.223	32.236	32.249	32.262	32.275	1880
1890	32.275	32.288	32.301	32.314	32.327	32.340	32.353	32.366	32.378	32.391	32.404	1890
1900	32.404	32.417	32.430	32.443	32.456	32.469	32.482	32.495	32.507	32.520	32.533	1900
1910	32.533	32.546	32.559	32.572	32.584	32.597	32.610	32.623	32.636	32.649	32.661	1910
1920	32.661	32.674	32.687	32.700	32.712	32.725	32.738	32.751	32.763	32.776	32.789	1920
1930	32.789	32.801	32.814	32.827	32.840	32.852	32.865	32.878	32.890	32.903	32.915	1930
1940	32.915	32.928	32.941	32.953	32.966	32.979	32.991	33.004	33.016	33.029	33.041	1940
1950	33.041	33.054	33.067	33.079	33.092	33.104	33.117	33.129	33.142	33.154	33.167	1950
1960	33.167	33.179	33.192	33.204	33.217	33.229	33.242	33.254	33.266	33.279	33.291	1960
1970	33.291	33.304	33.316	33.329	33.341	33.353	33.366	33.378	33.390	33.403	33.415	1970
1980	33.415	33.427	33.440	33.452	33.464	33.477	33.489	33.501	33.514	33.526	33.538	1980
1990	33.538	33.550	33.563	33.575	33.587	33.599	33.612	33.624	33.636	33.648	33.660	1990

AUTHOR AND COMPANY INDEX

Page numbers in *italics* indicate pages on which complete references appear.

SUBJECT INDEX